THE UNITED STATES ARMY MARKSMANSHIP UNIT

THE MARKSMANSHIP INSTRUCTORS' AND COACHES' GUIDE

UNITED STATES ARMY

Fredonia Books
Amsterdam, The Netherlands

The Marksmanship Instructors' and Coaches' Guide

by
The United States Army Marksmanship Unit

ISBN: 1-58963-987-1

Copyright © 2002 by Fredonia Books

Reprinted from the original edition

Fredonia Books
Amsterdam, The Netherlands
http://www.fredoniabooks.com

All rights reserved, including the right to reproduce this book, or portions thereof, in any form.

FOREWORD

Acquiring the ability to shoot accurately is no simple matter. One should not assume that the art of advanced marksmanship is fully realized upon reading a training manual. It cannot be completely understood even after hours of advice and instruction from a qualified marksmanship coach or expert shooter.

To become a qualified shooter and be able to produce consistently high scores, one must learn to perform all the fundamentals of shooting, acquire certain definite habits, flawless coordination - and above all - have a capacity for the intense concentration essential to exercising a high degree of mental control. For this, one must <u>train</u>.

It is difficult, if not impossible, to establish a universal system of training in marksmanship which will cover all cases, that is, one that can be adapted to every shooter's technique or special need. Training shooters requires an individual approach. For this reason, knowlegeable coaches take the pecularities of an individual - such as experience, degree of preparation and physical fitness, and other factors into account. Relying on their own experience, they devise a training program which allows special considerations for each shooter's capabilities. In spite of the difference in details, technique or method, there is much that is common to the training of advanced marksmen. This manual endeavors to present, in a detailed, comprehensive manner, these universal applications.

Advanced marksmanship training must of necessity, avoid the involved and exceedingly complex because it is an activity whose participants form a great cross-section of our national life, and the average citizen is its greatest asset.

We are grateful to the United States Army Infantry School for their counsel, advice, and cooperation in the preparation of those portions of this volume which deal with techniques of instruction. Many of their proven instructional techniques have been adapted to marksmanship instruction.

Your constructive comments are invited. They should be addressed to the:

Commander
The United States Army Marksmanship Unit
Fort Benning, Georgia 31905

SIDNEY R. HINDS, JR.
Colonel, Infantry
Commanding

THE USAMU MARKSMANSHIP INSTRUCTORS' AND COACHES' GUIDE

FOREWORD   **PAGE**

SECTION ONE	TECHNIQUE OF MARKSMANSHIP INSTRUCTION	1
Chapter I.	The Learning Process	3
II.	Methods of Instruction	11
III.	Lesson Planning	30
IV.	Effective Speaking	46
V.	Control of Interest	56
VI.	Management of Instruction	63
VII.	Training Aids	72
VIII.	Review, Critique, Examination and Panel Discussion	81
IX.	Marksmanship Instructor Training Course Lesson Outlines	85
GLOSSARY OF INSTRUCTIONAL TERMS		117
SECTION TWO	MARKSMANSHIP COACHING	121
Chapter X.	Service Pistol Coaching I (Attributes, Aids & Duties)	122
XI.	Service Pistol Coaching II (Technique)	133
XII.	Service Rifle Coaching	154
XIII.	International Rifle Coaching	197
SECTION THREE	US ARMY MARKSMANSHIP TRAINING PROGRAM	203
Chapter XIV.	Army-Wide Small Arms Competitive Marksmanship Training	204
XV.	Programs of Instruction	234
	A. Service Pistol Marksmanship	235
	B. Service Rifle Marksmanship	244
	C. International Rifle Marksmanship	249
	D. International Skeet and Trap Marksmanship	251
	E. International Running Target Marksmanship	252
Chapter XVI.	Training Standards and Scheduling Courses of Instruction	253
XVII.	Lesson Outlines for Service and International Pistol Marksmanship	262
	A. Fundamentals I (Attain A Minimum Arc of Movement)	262
	B. Fundamentals II (Sight Alignment)	269
	C. Fundamentals III (Trigger Control)	275
	D. Technique of Fire I (Establish a System)	281
	E. Technique of Fire II (Slow Fire)	288
	F. Technique of Fire III (Sustained Fire)	296
	G. Mental Discipline	302
	H. Technique of Fire IV (Int'l Slow Fire) (Air Pistol-Annex I)	316
	I. Technique of Fire V (International Center Fire)	322
	J. Technique of Fire VI (International Rapid Fire)	329
	K. Technique of Fire VII (International Standard Pistol)	335
	L. Technique of Fire VIII (Combat Pistol Match)	342
	M. Coaching I (Attributes, Aids and Duties)	349
	N. Coaching II (Technique)	363

PAGE

	O.	Physical Fitness I (Conditioning).	372
	P.	Physical Fitness II (Diet)	377
	Q.	Physical Fitness III (Effects of Alcohol, Coffee, Tobacco & Drugs)	381
	R.	Competitive Regulations I (NRA Pistol Match Rules)	385
	S.	Competitive Regulations II (Range Procedure and Safety).	399
	T.	Competitive Regulations III (National Trophy Pistol Match Rules-Earning Distinguished Badge).	404
	U.	Competitive Regulations V* (International Shooting Union Rules for Pistol).	
	V.	Review of Pistol Fundamentals.	417
	W.	Panel Discussion	426
	X.	Pistol Marksmanship Examination.	428
	Y.	Graduation	433

Chapter XVIII. Lesson Outlines for Service Rifle Marksmanship . . 435
  A. Orientation to Marksmanship Instructors' and Coaches' Guide 435
  B. Squad Selection, Organization and Training . . 438
  C. Selection, Care and Cleaning of a Rifle. . . . 446
  D. Effects of the Weather, Use of Scorebook and Telescope. 453
  E. Mental and Physical Conditioning 460
  F. Rifle Range Safety 463
  G. Combat Rifle Match Principles. 466
  H. Precision Combat Rifle Match 476
  I. Shooting Techniques. 479
  J. National Trophy Team Coaching Techniques . . . 495
  K. National Trophy Team Coaching Techniques (Practical Exercise) 508
  L. National Trohpy Team Coaching Techniques (Examination). 510
  M. Panel Discussion 512
  N. Graduation 514

Chapter XIX. Lesson Outlines for International Rifle Marksmanship. 516
  A. Equipment, Procedures and Techniques of Training 516
  B. International Prone, Kneeling and Standing Positions. 522
  C. International Rifle Match Program. 528

Chapter XX. Lesson Outlines for International Skeet and Trap Marksmanship 533
  A. Introduction to International Skeet and Trap Marksmanship 533
  B. International Clay Pigeon (Trap) Marksmanship 540
  C. International Skeet Marksmanship 551

Chapter XXI. Lesson Outlines for International Running Target Marksmanship 559
  A. Running Target Marksmanship Program (Competition, Range Procedures, Equipment and Techniques). . 559

*To be published

iii

PAGE

SECTION FOUR   TEAM ADMINISTRATION

    Chapter XXII. Administration I (Squad Administration).   565
              XXIII. Administration II (Security of Weapons)
                      Equipment and Ammunition)   583

SECTION FIVE   COMPETITIVE REGULATIONS

    Chapter XXIV. Competitive Regulations II (NRA Pistol Range
                      Procedure and Safety)   587

    Chapter XXV. Competitive Regulations III (National Trophy
                      Pistol Match Rules)   591

    Chapter XXVI. Competitive Regulations IV (Requirements for
                      Earning a Distinguished Pistol, Rifle or
                      International Shooter Badge)   602

SECTION ONE.   TECHNIQUE OF MARKSMANSHIP INSTRUCTION

THE CHALLENGE OF EFFECTIVE MARKSMANSHIP INSTRUCTION

Marksmanship instructors become outstanding for three reasons: rich, natural endowment; great opportunity; and a consuming desire to serve. But there is a price that must be paid to achieve success. Here are the four most important demands:

1. <u>Diligent preparation:</u> Many of us aspire to become outstanding in our chosen field but few are willing to make the necessary sacrifices. We will not fully prepare outselves; we won't demand of ourselves high qualification; we fail to become completely involved. A successful teacher is a leader. A leader is a director; a coordinator; an imaginative thinker and an organizer. These accomplishments admit of painstaking preparation. Consider the criteria of one hour of preparation for each minute of speaking. Franklin D. Roosevelt made mine drafts of his famous speeches. The first was rough; the second was improved; the third showed greater improvement; by the eighth draft, only a word or two had to be changed before the final draft was ready. Profound and moving words flow from the prepared mind.

2. <u>Influence the efforts of others:</u> The wise instructor realizes early that what he can accomplish by himself is negligible when compared to what he can do by guiding and influencing the efforts of others. The mushrooming vitality of understanding in the stimulated human mind is a miraculous occurrence. The successful instructor's worth is predicated by what he is able to get other people to do.

A leader or an enterprising marksmanship instructor knows people. He comes to know intimately their motives and ambitions. He helps them to fight their weaknesses. He shows them how to apply their stronger attributes. He must encourage and direct the champion in each step up the ladder of success. It isn't enough to uncover talent; he must help this talent to grow. The hallmark of the potentially great instructor is that his rewarding moments are realized from the successful efforts of those shooters that he has taught the arts and skills of championship shooting.

3. <u>Be persistent:</u> Too often, when the first plateau of success is reached, an effective instructor may start to rest on his laurels. The initial accomplishments are significant, effecting changes and improvements. Then he may level off as if expecting the momentum of past efforts to continue the stimulation of productive ideas and action. He becomes comfortable. As this point, the three tyrants of complacency begin to appear; indifference, inaptitude and indecision. A fur lined rut! Self-satisfaction comes as a guest, lingers under the mantle of self-admiration and stays to enslave us in a comfortable caress of self-esteem.

We must always assume that tomorrow the competition will be stiffer. We must continue to improve, to perfect, to better qualify ourselves. We can never let up.

4. <u>Accept responsibility.</u> If the marksmanship instructor expects to succeed in leading the shooter through what appears to be a trackless maze of techniques and fundamentals toward a true understanding of the demands placed on skilled marksmanship, he must be responsible for his own actions and thoughts. Pursue and be invigorated by new knowledge and exercise this power in impartial improvement of each shooter's skill and judgment. Encourage in others the assumption of responsibility in the marksmanship program. Create a disdain for the proposition of less responsibility and more recreation time. No man has ever become great on an eight hour day.

If you aspire to leadership in the shaping of minds that are destined to work for victory in competition, you will have to plunge into the job and attack vigorously with all the resources at your command.

CHAPTER I

THE LEARNING PROCESS

A. GENERAL

How do students learn? The answer to this question has baffled philosophers for centuries. Psychologists and educators have researched the problem. Yet there is no universally accepted solution. Basically, the learner is exposed to a new situation (or problem) and he reacts or responds to it. When he responds in a way that is satisfactory to him personally or to society generally, we say that he responds correctly. Thus, learning can be reduced to reacting to situations. As an instructor, you are interested in what you can do to assist the student to learn correctly, quickly, and accurately, and to learn thoroughly so that he retains what he learns. Therefore, the emphasis in this chapter will be on controlled learning situations - those which occur within a classroom or training area.

B. STEPS IN LEARNING

1. <u>Learning Steps</u>: We speak of learning, not learn. By using the "ing" form of the word, we imply an action that takes place over a period of time, not something that occurs once and is complete and finished. In itself, learning is a continuing process of improvement and increasing proficiency. Visualize learning as follows:

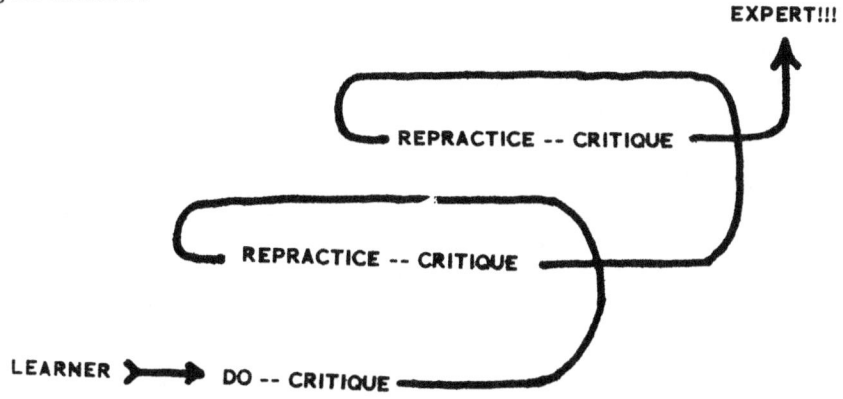

Figure 1-1. Spirals of Learning.

Learning is thus a process of continuing improvement with ascending spirals of intake of information, practice, critique of practice, repractice, critique of repractice. With each new spiral there is an increase in understanding or skill as the learner strives to progress from novice to expert; basically there are four steps in the learning process:

INTAKE	REACTION	CRITIQUE	REPRACTICE
(Observe, Listen)	(Respond by doing; practice)	(Evaluation of performance)	(Improve over first practice)

Step 1: The intake step - whenever the learner is exposed to the situation or problem. He uses his senses to observe, listen, and gather information.

Step 2: The reaction step - whenever the learner reacts or responds to the situation by doing something. Frequently, this doing is practice of the material to be learned - whether this be a manual skill or a mental or problem-solving skill.

Step 3: The critique or evaluation step. Such a critique may be personal evaluation of performance or it may be instructor evaluation of student performance.

Step 4: The repractice step in the light of improvements suggested by the critique step.

NOTE: These steps are repeated as the student gradually improves or progresses toward qualified or expert proficiency.

C. TIPS TO ASSIST LEARNING

1. Suggestions for Step 1, "The Intake Step."

a. During the intake step, the learner must understand what he is to learn and how he is to perform. As instructor, you must supply the purpose or motive for learning the material. You must show that your lesson will satisfy the student's need. Be sure that you understand why this lesson is necessary and that you have answered the students challenge of "why should I learn this?" Learning begins with a clear-cut purpose and a driving need.

b. Before the intake step can be effective, you students must be carefully prepared. They must be set or ready to learn. You do this by gaining and focusing attention in the lesson introduction. Next, you control interest during the intake step so students can get the most out of their observation and listening. With student interest high, their enthusiasm to learn will be high. Therefore, they will put forth more effort and drive to concentrate. Learning comes easier with concentration.

c. Use the "multiple-sense approach." Remember, students learn with their whole being. Insure that in your presentation you appeal to as many senses as possible. Stimulate sight, sound, physical doing, and mental activity by demonstrations, skits, training aids, questions, discussion, practice and problem-solving exercises. The more senses you stimulate, the better the chance for learning. The more vivid impression you make, the greater the learning imprint.

d. Use a variety of techniques to appeal to different groups in the class. One group will learn as a result of your initial statement and supporting material. Some will learn as a result of your rephrasing of the teaching point or revealing additional supporting material. Others will learn by a student's answer to your question. Still other students will learn after you have shown a chart, given an example, or a personal experience. Remember, among your students there are some who learn best by discussing.

e. Supply the big picture or overview to help students acquire insight into the problem to be learned. Use the whole-part-whole technique.

(1) In lesson planning you do this by giving the big picture in the scope and motivation of the introduction, then you develop the "parts" as you present the different teaching points. Once again you present the "whole" as you wrap up the lesson with a comprehensive summary.

(2) Show them the perfection they are striving to reach as you demonstrate expert rifle disassembly, skilled marksmanship or excellent examples of good teaching or public speaking. In this way the student sees what he is striving to achieve and he better understands how smaller learning units fit into the big picture of developing the finished learning product.

2. Suggestions For Step 2, The Reaction and Doing Step.

a. As the instructor you know that learning is not complete until the student practices what he has learned. He learns by doing. Your job is to supply meaningful practice - training the student under simulated conditions with practice or problem exercises designed to teach manual or mental skills clearly and effectively.

b. Divide the practice materials into small logical learning units or exercises. One of the few confirmed findings in educational psychology is this: FREQUENT SHORT PRACTICES ARE MOST EFFECTIVE FOR RAPID LEARNING AND LONGER RETENTION.

c. Don't introduce similar materials simultaneously that might confuse students. Students frequently learn and remember through association. They learn new things by relating them to things previously learned. Build up from known to the unknown (from that already learned to that which is to be learned). To study the nomenclature of two similar weapons simultaneously will confuse the average student.

d. Watch for learning plateaus. These are leveling-off places in the student's learning curve. At that time students are making no apparent improvement. This is normal for all students when learning a new skill. Explain this temporary stop-over to the student so he does not get discouraged. It is a good time to schedule review, examination, or critique sessions. Here are examples of how students normally learn easy and difficult materials:

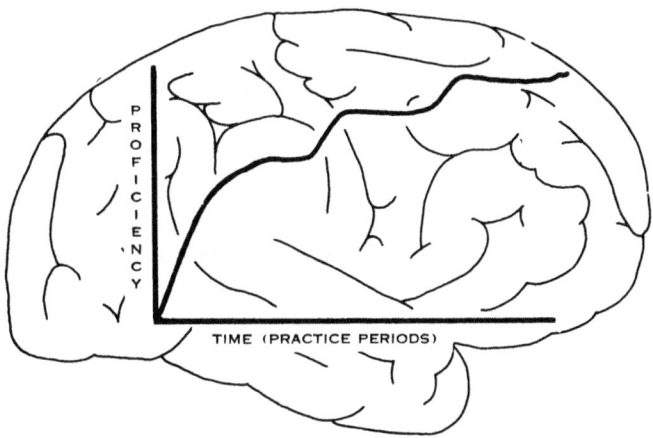

Figure 1-2. Normal Learning Curve - Easy Material.

NOTE: With easy materials there is a sharp initial spurt followed by learning plateaus and gradual slow improvement in proficiency. With learning most motor skills such as marksmanship golfing, bowling, ans swimming, good progress is made initially then it becomes increasingly difficult to reach the expert level.

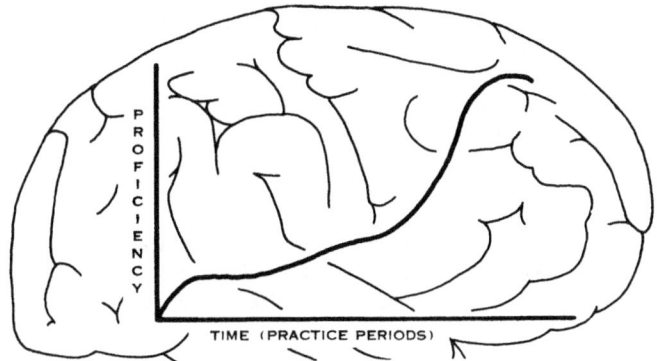

Figure 1-3. Normal Learning Curve - Difficult Material.

3. Suggestions During Step 3, The Critique Step.

a. The critique or evaluation step may be self-evaluation of performance or instructor critique of student performance. Here are suggestions for the latter:

b. Praise correct student performance as soon as possible. In this way the student will know what he is doing correctly. Encouragement is necessary early in student performance for every individual is apprehensive when he first practices. After all, he has seen expert demonstrations to set standards of performances and despairs of achieving such excellence.

c. Point out only the gross errors at first. You must do more than this however, as time passes. Show students exactly what they did wrong and try to explain in detail why they performed incorrectly. Emphasize that such errors are normal with beginners.

d. If the same student errors are repeated frequently, if they are dangerous errors, or when they are caused by inattention, take time to emphasize to the student that he must eliminate such errors because a continuation will not be tolerated.

e. Test student performance frequently. Do this both for your benefit as the instructor and for the student's benefit. Both members of the learning team must know the status of student progress. As the student progresses, make your critiques more detailed and severe as you strive to smooth out the rough edges of performance.

4. <u>Suggestions During Step 4, Repractice Step.</u>

a. Topflight shooters agree that the three cardinal rules for expert proficiency are: practice, repractice and more practice. However, this repractice should be supervised, not blind repractice. With blind repractice, the student may only ingrain the same wrong habit. Supervised repractice following a careful critique helps to set the correct way of performing and gradually eliminates errors.

b. At some stage in practice and repractice, under the eye of professional, a student finally acquires "insight" into the correct way of performing. In popular language we say he "gets the hang of it." Once the student realizes how to perform correctly and the common pitfalls to avoid, he can practice with only occasional supervision.

D. CONTROL OF LEARNING

1. <u>Learning "Control Measures."</u>

a. A good instructor has various control measures to insure the success of his teaching mission. Many of these control measures for learning have been discussed in detail in other portions of this handbook; they are mentioned briefly here to show how they affect learning. Each or all of them have a significant impact on learning. Used correctly they can "make" the learning situation successful, neglected or misused they can "break" the learning situation.

b. Motivation: The student must need or have a motive for learning material; otherwise, he quickly forgets. For example, students who learned meaningless or nonsense words or phrases forgot them as quickly as they had learned them because they had no reason to retain them. When the student knows the purpose for learning material he learns it in order to apply such learning. This conditions the way in which he learns the material. Competition is one form of motivation. Individuals like to excel; Americans enjoy competition. Competition adds interest and encourages whole hearted participation. It can also provide valuable training in cooperation as long as you don't let competition get out of hand.

c. Interest: Motivation and interest are closely related. Since learning requires student action and concentration, you, as the instructor, must anticipate lack of interest, motivation and periodically remotivate, create the best teaching conditions and establish a favorable atmosphere for learning.

d. Action: Remember, learning is an active process, it doesn't occur by passive sitting and absorption. Students must perform, question, restate, analyze, argue, discuss, explain, and elaborate. Provide them with opportunities for such student action in the lessons you prepare. Get students to participate and react to the learning materials.

e. Logical Order: The order in which materials are presented will frequently affect learning success. As the instructor you <u>guide</u> the learning process. Arrange teaching materials in the simplest and most logical sequence. Make proper transitions from Step 1 to Step 2 so that the student understands how these steps relate or tie together. Some common sequences are;

(1) Chronological or orderly in time. With learning which involves procedures or steps, be sure that you arrange them in the time order in which they occur.

(2) Whole-part-whole: This is the developmental order that psychologists claim is best to learn new materials. Give the big picture, then the details and then restate or summarize with the overview or big picture.

(3) Simple to complex. Go from the less difficult materials and build up to the complex. When students grasp fundamental ideas they can more readily apply them in advanced or complex situations.

(4) From known to the unknown. By this order you begin with concepts or principles already learned by the student. By comparison or analogy he learns how the new (Unknown) material is similar to something he already knows. Tie in the new concept with past or everyday experiences.

(5) Orderly progression. Develop your explanations of objects so that you proceed from the left to the right, from top to bottom or from outside to inside. This will make student learning easier.

f. Repetition: Have the student repeat the learned action again and again. This helps to set a mental and muscular pattern. Tell him to make a conscious effort to remember as he performs. Use the principle of distributed practice - that is to space student application into frequent short practices so that fatigue does not waste training time. Don't interrupt. Be sure the student completes the action even though he is making mistakes and then critique. If you interrupt as you see errors he never has a chance to practice the action as a complete unit.

g. Critique or Evaluation: Evaluate progress periodically. This is important for the learner puts forth more effort if he knows he will be tested or checked regularly. Avoid general comments, be specific in your critique.

h. Emotion: Watch for strong feelings or emotions during performance and critique. Fear, anger, resentment, discouragement, and worry impede the learning process.

i. Individual Differences: Students are individuals. The nature and degree of individual differences in learning ability is important to the instructor. Class instruction is based on the assumption that people are enough alike that learning groups can be organized. Within these groups, however, you will find large differences in individual rate of learning. The fact that all individuals go through the same steps in learning does not mean that they all go through at the same speed, for the same reasons, with the same emotional reaction, or with the same net results.

Students differ in:

(1) Intelligence. Differences in individual ability are recognized in the use of aptitude area scores for classification and assignment in the Army. Wide differences in capacity for learning various types of performance will be found even in small groups.

(2) Level of Aspiration: Every student has his own level of aspiration. Some students set this so high that they are disappointed with their successes. They feel that they should be in the top ten of the class and if they place lower they are extremely disappointed with their achievement. Many students, on the other hand, set their goals so low that they are not motivated to do a creditable job. The goal-setting behaviour of your students is influenced by their past experiences of success and failure. All students, to be learners, need success experiences commensurate with their abilities.

(3) Anxiety: Some students, the highly anxious group, are the worriers. They constantly worry about how they are making out in their studies. Experiments have demonstrated that this group often does more poorly on complex learning tasks than the less anxious group. It seems that the pressure on them to do better actually impedes their performance. When some pressure is put on the less anxious group, those who don't worry excessively about their school work, it spurs them to improve. Even after a task is learned, failure or the threat of failure to excel may produce frustration and ruin a performance. Look at the shooters who "blow up" during stiff competition. Stress can produce varying results on students. Some do better under stress because they are more highly motivated. Others fold up under stress. Anxiety does affect learning.

(4) Past Experience: Relearning is easier than mastering new material. Sometimes a student will learn a skill in a tenth of the time it takes another student of equal ability to learn the same skill. The first student is probably drawing on past learning while the second is encountering new material. This can occur even though the student who is learning it for the first time has more responses, and has a better expectation of the learning goal.

E. PRODUCTS OF LEARNING

1. <u>Training goals.</u> There are five common learning goals. The products of learning are skills, facts, concepts, preferences, and critical thinking ability.

a. Skills: Skills are acquired by drill. In firing a weapon, fingers and hands learn to work together by actually performing the act over and over again until it becomes established as a pattern. Most skills require coordination, which means that many muscles must work together to produce the desired action. Repeated practice is necessary to achieve coorperation from several muscles - some contracting and some relaxing - to produce the skill pattern.

b. Facts: Facts are names, relationships, dates, and laws. They are often based on mental associations. These associations are fixed by meaningful repetition, the product is memory. Examples of such facts are: the elevation and windage rule, the cycle of operation of the weapon bei fired.

c. Concepts or Principles: Concepts differ from facts primarily in depth of meaning tha a student attaches to a word or other symbol. Concepts are usually the result of combining many basic facts together.

d. Preferences or Tastes: Preferences may be the result of both emotional and logical thinking. Usually they are the by-products of experience from other learning situations. A certain preferred way of acting brings satisfaction. Graduates of a technical or scientific school have preferences which differ markedly from those of the graduates of a liberal arts school. The high power shooters' preferences and tastes vary from those of the small bore rifle or a pistol shooters'.

e. Critical Thinking Ability: Critical thinking involves manipulation of concepts and principles to arrive at logical conclusions. It has two phases: analysis, to dissect the problem into its elements and synthesis, to build up satisfactory solutions. It includes identifying basic issues, classifying evidence, and drawing sound deductions from data. Analytical thinking, or problem-solvi is learned by witnessing demonstrations of problem-solving techniques and by supervised practice in these techniques. Critical thinking in a specific area is learned by practice in evaluating concepts ar weighing values in that area. Learning is fixed by experience in analyzing data, drawing colcusions, and determining appropriate courses of action.

F. HOW YOU REMEMBER

1. <u>Types of Remembering:</u> Psychologists who study learning recognize three types of remembering: recollection, recall and recognition.

a. Recollection: When you recollect, you reestablish an earlier personal experience on basis of a partial clue. For example, the odor of a certain flower may bring back the memory of you first high school date who wore that flower to the school dance. Sometimes you can recollect the orchestra, decorations, names of school chums, where you went after the dance, etc. The details may or may not be complete. Recollections are distinguished from other kinds of remembering by reconstruction of a past occasion.

b. Recall: Recall is the human ability to perform some activity in the present which is based on past learning. You jump into water and swim although you may not have done so since child hood. You still remember the basic skills necessary. You can recite a poem you learned in grade school. Recall differs from recollection in that you cannot even remember the circumstances under which you learned to do this thing - all you can do is perform.

c. Recognition: In recognition, you recognize someone or something as familiar. It ma be the way a person walks, the pitch of his voice, his accent, or it may be an odor, seeing a street o a river. You have had experiences similar to this: you are walking down a street of a strange town and say to yourself, "I have been here before. There will be a drugstore on the corner and the theate will be down there." In this case, our memory is generalizing from similar street scenes and signalling faulty recognition. The present scene, though actually strange seems familiar to us.

2. <u>The Curve of Remembering:</u> The curve of remembering usually follows this pattern (See Figure 1-4). There is a rapid forgetting at first, followed by a leveling off or slowly decreasing loss Frequently there is a slight gain after a short period before we begin to forget. We may lose as high 50% of the material in this initial early period.

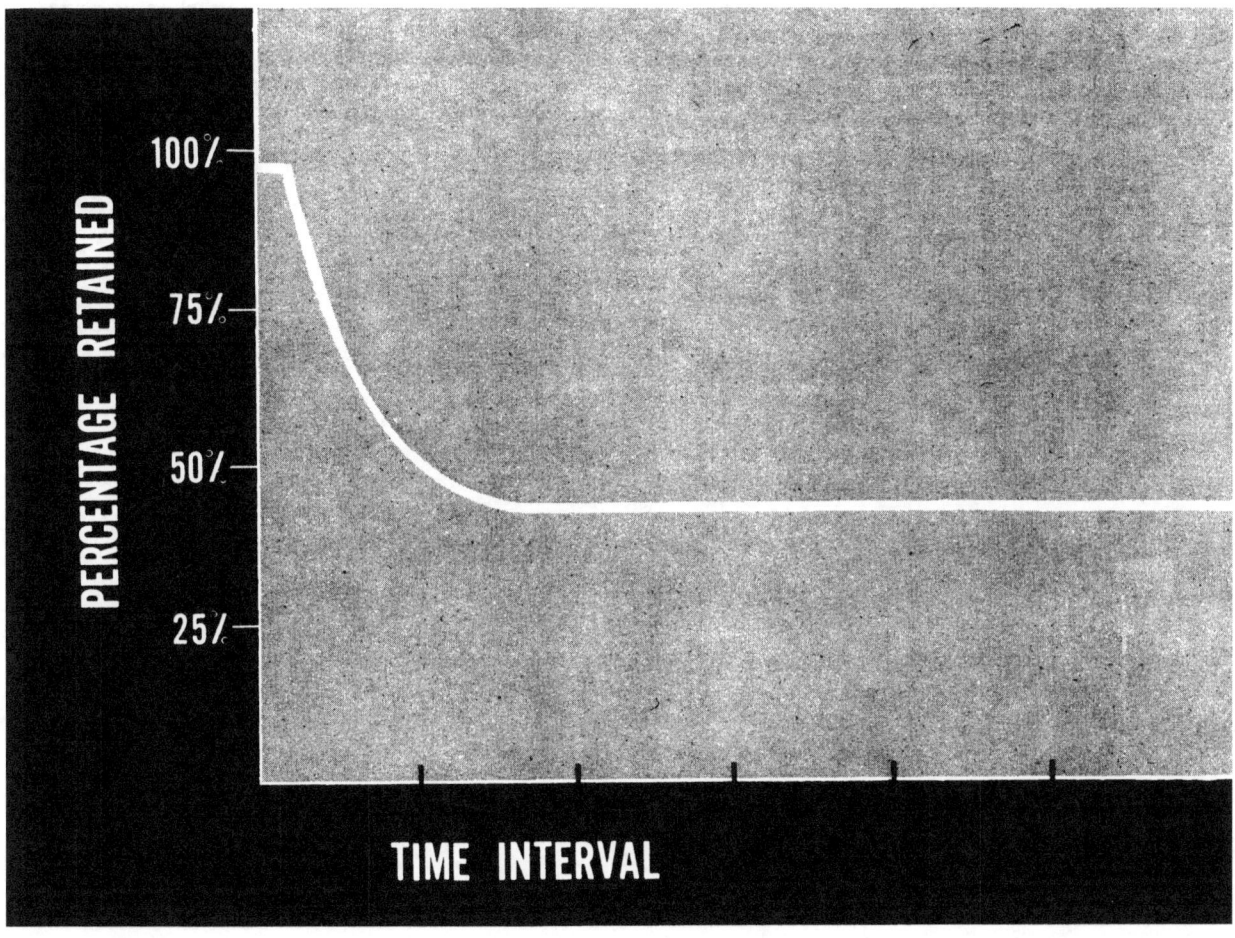

Figure 1-4. How You Forget.

G. WHY YOUR FORGET

    1. <u>Forgetting</u>: You forget because of:

        a. Disuse: The first cause for forgetting is the decay of the skill or memory pattern through lack of use. Although nothing is lost to the mind the student cannot call it back.

        b. Active Desire to Forget: Many things you forget because your mind wants you to. Emotions enter into the picture to help you repress unplesant or embarrassing experiences. Experiments have shown that we remember pleasant experiences of childhood more than unpleasant ones. This is why it is essential that you do everything possible to make student learning of your material as pleasant as possible. You remember some unpleasant experiences because you enjoy telling them to your friends, not because they were unpleasant.

        c. Interference Learning:

            (1) Psychologists call this retroactive inhibition. For example, you learn to drive a car with a regular transmission. Later, you learn to drive a car with an automatic transmission. When you return to driving the car with conventional transmission, the learning received while driving a car with automatic transmission has interfered with the old skill you had and you must relearn to drive with a manual shifting transmission if you expect to perform smoothly without clashing the gears

(2) Interference learning has implications in studying for an examination. Don't study different subject material between the time you do your studying and the time you take the test. Psychologists have found that you forget less when you sleep through eight hours than when you are awake for the same period, since new impressions gathered while awake are interfering with the older material learned. Study for your exam before you go to bed and you will be best prepared to take it first thing the next morning. If you have an afternoon exam, study or review during the noon hour.

CHAPTER II

METHODS OF INSTRUCTION

A. GENERAL

1. **Purpose.** This chapter is designed to give the marksmanship instructor a working knowledge of the methods of instruction used, the advantages and limitations of these methods, how to conduct effective instruction using the several methods, and the wide scope of the planning necessary to use the various methods of instruction.

2. **Principal Methods of Instruction.**

   a. **Lecture.** The lecture is a method of instruction in which the instructor orally develops his subject without student discussion. Because of the lack of student participation, the lecture should be used for an entire period of instruction only when no other method or combination of methods is appropriate. Normally, instructors lecture during most periods of instruction to give direction, present lesson introductions and conclusions, and to guide or summarize student discussion.

   b. **Demonstration.** The demonstration is a method of instruction which assists student learning by showing correct procedures and expected standards.

   c. **Conference.** The instructional conference is a method of instruction by which the teaching points are developed primarily through student discussion. The instructor initiates student discussion by asking thought-provoking questions. He guides discussion through timely use of follow-up questions, subsummaries and transitions. The emphasis during the conference is to achieve learning through student participation. The conference uses a combination of lecture, demonstration, and performance where appropriate to assist student understanding.

   d. **Performance.** The performance method emphasizes student practice and application of principles and procedures. Students learn much by doing. Learning is frequently not complete without practice. Student performance consists of problem exercises to solve or routine practice to correct errors and acquire skill.

3. **Selection of Method.**

   a. What is the best way to communicate the necessary information to the student? Is any one method best or will it require a combination of methods?

   b. The relative degrees of instructor-student activity vary according to the method used. Can your ideas best be communicated to and retained by the student from action limited to the instructor only, as in a lecture? Must the student participate as in a conference or performance type instruction where most of the activity is on the part of the student? Figure 2-1 illustrates this relationship.

   (1) The lecture is nearly all instructor activity (telling) while the student listen.

   (2) The demonstration requires slightly diminished instructor activity (showing) while the student sees the action as well as hearing the instructor.

   (3) The conference sharply reduces instructor activity (guide discussion) as he takes part in the discussion of the subject by the students.

   (4) Performance instruction is minor activity by the instructor (observing), while the student is (doing) something under the instructor's supervision.

   c. Since all marksmanship instruction is aimed toward effective student performance, emphasis will be placed upon practical work rather than upon theoretical instruction.

   (1) Lessons should be reviewed periodically to insure that they include a maximum of student performance activities.

(2) Conference techniques should be used to the maximum for all oral presentations.

(3) Instructors should not lecture throughout the whole instructional period when any other method of instruction is appropriate.

d. In all instruction, the accent is on student learning, and the students learn best by participating actively in instruction and accepting their share of responsibility for the teaching-learning success. Because of the diversity of training, recognize that no one method will suffice to accomplish all types of training. The method must fit the student learning outcomes desires.

(1) If the objective is a skill, then you must include demonstration and ample student performance.

(2) If the objective is a general knowledge or understanding, then you could use a combination of lecture and conference.

(3) There is no sharp line of separation among methods; usually the instructor uses a combination of methods to achieve the lesson objective.

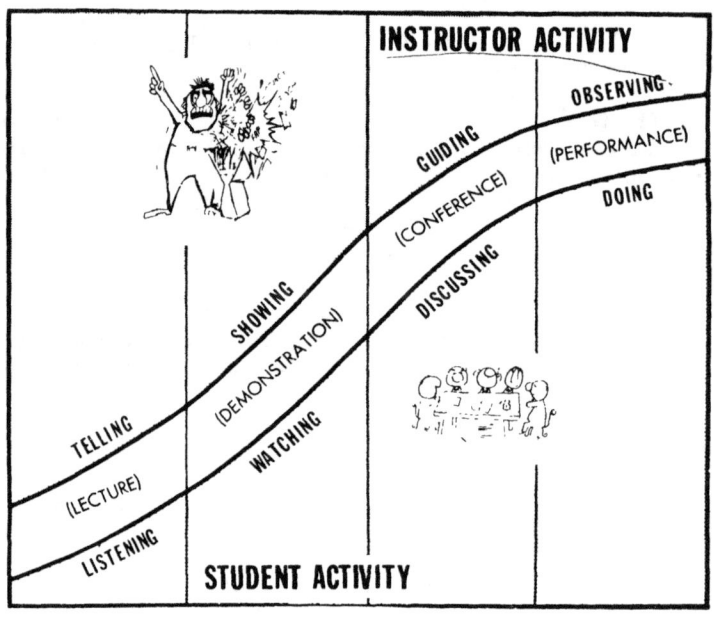

Figure 2-1. A Guide to Selection of Method.

B. LECTURE

1. <u>The "Telling" Method</u>. In the lecture method, you tell facts, principles, theories, or relationships which you want your student to learn and understand. When the word lecture is mentioned, students and instructors immediately think of a very formal discourse on a subject read by an expert from a prepared manuscript. Such a formal lecture is only one type of telling. They overlook that a lecture can be delivered by memorizing the key ideas with only occasional glancing at notes. It could even be extemporaneous or impromptu telling. The lecturer may have check-up questions planned and asked at appropriate intervals. He may encourage student questions during his telling. Essentially, the lecture is one-way oral presentation of ideas. Accordingly, the lecture uses an instructor-centered approach since the teller plans no (or limited) student participation. Some instructors regard any use of questions by the principal instructor as evidence of an instructional conference, but this is incorrect since the primary emphasis is still on telling.

2. <u>Advantages of the Lecture</u>. Many persons believe that lecture is an unproductive method of instruction. This is true only if the presentation is so dull that student attention wanders. As an instructor, you will lecture to introduce, subsummarize, and conclude your presentation, to direct and

Figure 2-2. The Lecture - One-Way Presentation of Ideas.

critique student performance, to narrate demonstrations, to explain and illustrate principles and procedures, to give examples, testimony, definitions, and statistics in support of teaching points. The lecture in combination with other methods of instruction is an indispensable ingredient of effective instruction.

    a. With the lecture you have an effective method of presenting many facts and ideas in a short period of training time. You are in control of the time and after rehearsals you can make optimum use of this time. CAUTION: You must be careful not to "run ideas by" so fast that you confuse the student. You can minimize this danger by checking on student learning frequently, by providing opportunities for students to question points they do not understand, and by soliciting such questions when you have completed a complex teaching point. If student answers reveal a general lack of comprehension, you will have to rephrase your ideas, add substantiating material or further examples, and then recheck comprehension.

b. Place emphasis specifically where you desire it. Since you control the learning situation closely, develop only those points you need to develop. Through your experience and research into the instructional problem, you select the important facts to emphasize and subordinate less important facts and ideas.

c. The lecture is very effective in presenting new or complex information to provide students with the background of knowledge they need to participate in subsequent discussions or in performance exercises.

d. The size of the listening audience does not restrict the lecture. You can deliver the lecture to twenty, two hundred, or two thousand students.

e. The lecture is an effective method for guest speakers or those who present constantly changing material. It is most effective when presented by a real authority or an expert on the subject matter. When the speaker is well informed and uses good presentation techniques, he can hold the

THE LECTURE IS EFFECTIVE IN....

1. PRESENTING MANY FACTS AND IDEAS
2. PLACING EMPHASIS WHERE THE INSTRUCTOR DESIRES
3. PRESENTING BACKGROUND INFORMATION
4. SPEAKING TO LARGE AUDIENCES
5. PRESENTATIONS BY GUEST SPEAKERS
6. PRESENTING CLASSIFIED MATERIAL
7. ADDING VARIETY TO OTHER METHODS

THE LECTURE IS LIMITED IN....

1. PROVIDING STUDENT PARTICIPATION
2. DETERMINING WHAT THE STUDENT IS LEARNING
3. GEARING INSTRUCTION TO THE NEEDS OF THE STUDENT
4. MAINTAINING STUDENT INTEREST

Figure 2-3. Advantages and Limitations of the Lecture.

interest and stimulate the imagination of his listeners. The speaker who projects an interesting personality into his presentation, encourages student questions, periodically checks student interest, uses training aids properly, and arranges his material in an understandable way, will convey his message effectively.

   f. The lecture adds variety to other methods of instruction. A dynamic instructor can lecture certain details or teaching points in an extremely interesting way. This provides a welcome change of pace when interspersed among other methods. It will provide a "time cushion" since you can streamline your explanation or amplify it as training time permits.

  3. <u>Limitations of the Lecture.</u>

   a. The major limitations of the lecture are the lack of active student participation, lack of complete free two-way communication between instructor and student, and the problem of maintaining high student interest level. Lectures foster overdependence upon the teacher as the expert and encourage student passivity and uncritical acceptance of everything the speaker says.

   b. It is relatively difficulty to determine what the student is learning. In a straight lecture, you are not sure whether the student grasps the material or not. By using check-up questions, you only sample class comprehension. Those you called upon may be the outstanding students or the poorest students in the class; this is a matter of chance.

   c. You have difficulty in gearing your instruction to the needs or progress of the student. Usually you plan your presentation one way and present it that way. You have little or no flexibility to adapt to the learning level of the students as you proceed.

   d. Student interest is difficult to maintain. It requires a skillful, dynamic speaker, aware of the need for control of class interest, with a fine change of pace to maintain class interest for extended periods of time. As one critic stated it: "The mind can absorb only as much as the seat can endure."

  4. <u>Planning the Lecture.</u>

   a. Organize your ideas in a logical sequence.

   b. You must be specific in explaining and illustrating each supporting idea or difficult point.

   c. You must limit the number of teaching points to the time allowed for complete coverage of the subject.

   d. You must amply support each of these teaching points with sufficient facts to establish understanding.

   e. You must devise illustrations to supplement your descriptions - either verbal or pictorial.

   f. After you have specifically explained each supporting idea, sub-summarize before you move on to the next idea.

   g. Usually, you can improve your lecture by asking two or three thought-provoking questions; these enable you to introduce some discussion into the presentation.

   h. In going to the next teaching point, make a good transition. A transition not only alerts the student to the fact that you are leaving one point and beginning another, but it also shows the student the relationship between ideas.

   i. Since the lecture does not provide for active student participation, you must <u>stimulate</u> the students sight as well as his hearing. Use training aids which reinforce learning.

j. Many marksmanship subjects lend themselves to an interesting demonstration or problem-solving exercise. <u>Combine</u> the lecture with other methods to increase student participation and overcome this major disadvantage of the lecture method.

k. List the personnel and equipment needs that will amply support the presentation of the subject, i.e., sound-equipment, chalk-boards, projection equipment, assistant instructors, etc.

l. The instructor must provide for student comfort, i.e., ample seating, ventilation, lighting, etc.

5. <u>Conducting the Lecture</u>. Student interest is difficult to maintain in a lecture. For this reason, keep your lecture time brief by varying lecture with conference, demonstration, or student performance. You have no doubt listened to lecturers who were able to hold student interest for one or even two hours. Perhaps you attribute this ability to the interesting personality, vast knowledge, or effective delivery of the instructor. These qualities are of course important, but if you listen carefully, you will usually discover that this interesting lecturer plans for and liberally uses the interest factors discussed in Chapter 5. He tells stories, he may use humor, and his presentation is always sprinkled with "for example............"

a. In the lecture, students learn by what they hear. Therefore, it is extremely important that you present your ideas in an interesting, enthusiastic manner.

b. At the beginning of the lesson, you state the purpose of the lesson and emphasize why it is important to the student to learn this material in order to do his job effectively. The student then is motivated--he realizes the need to learn. He gives his full attention to what the instructor is saying and doing.

c. Outline the key points. You give direction to student's attention by orienting him on the specific things to be learned.

d. By various methods which involve hearing and seeing, you present and develop the teaching points while maintaining attention and interest.

e. It is the instructor's responsibility to provide for student participation. If the student loses interest, he no longer gives his full attention or concentration to the lesson and learning breaks down. Questions and discussion stimulate interest.

f. Humorous stories and illustrations relax students and helps them to remember the points emphasized by humor.

g. Use demonstrations to illustrate various points in the subject matter. Showing how to accomplish a specific action in marksmanship is immeasurably more effective than the mere telling or verbal description.

h. Usually you can improve your lecture by asking two or three thought-provoking questions; these enable you to introduce some discussion into the presentation.

i. In going to the next teaching point, make a good transition. A transition not only alerts the student to the fact that you are leaving one point and beginning another, but it also shows the student the relationship between ideas.

j. Since the lecture does not provide for active student participation, you must <u>stimulate</u> the student's sight as well as his hearing. Use training aids which reinforce learning.

C. THE DEMONSTRATION - "THE SHOWING METHOD"

1. <u>General</u>. Teaching is a process of telling, showing, and doing. Demonstrating, or showing, is used in combination with other methods, and is usually preceded and accompanied by lecture and conference and followed by student performance, critique, and conference. The student must know what to observe or look for in the demonstration. When your lesson objective involves the development of skills or the practice of procedures, your demonstration should be followed by student performance.

2. **Purposes.** While the demonstration is usually regarded as a fore-runner of student performance, it can also be used very effectively in supporting teaching points. Sometimes a demonstration is used purely to illustrate or support a teaching point. They are not designed to be followed by a student performance or application.

Uses of the demonstration are as follows:

a. To teach manipulative operations (how to do it).

EXAMPLE: Use a demonstration to give students a visual impression of how to disassemble the .45 cal pistol. Demonstrate step-by-step. The combination of seeing the demonstration and hearing your comments reinforces student learning through the use of two senses.

b. To teach principles and theories (why it works).

EXAMPLE: Fire a M14 rifle suspended on cords to demonstrate the recoil principle.

c. To teach operation and functioning (how it works).

EXAMPLE: Use a cutaway M14 rifle to demonstrate the phases in the cycle of operation.

d. To teach team organization during team matches (why men work together).

EXAMPLE: Have demonstration shooters arranged in team firing position to illustrate the important features that aid in team control.

e. To teach team procedures (how men work together).

EXAMPLE: Use a skit to portray range safety and procedures.

f. To teach appreciations (how skilled must a man be to use the weapon effectively).

EXAMPLE: Have demonstrators fire weapon for accuracy to develop an appreciation for the skills involved.

3. **Advantages of the Demonstration.**

a. Demonstrations save time. A brief demonstration of the proper method of applying trigger pressure is more effective than a lengthy discourse. Showing is usually simpler than telling. To illustrate this, try to explain briefly, with words alone, how to tie a shoelace or how to make a Windsor knot in your tie. Notice how much clearer and simpler it is to demonstrate. Demonstration will make your explanation more concrete in the minds of your students.

b. Demonstrations insure thorough understanding through their appeal to sight as well as to hearing. They clear up student confusion through their appeal to several senses.

c. Demonstrations stimulate interest and therefore learning through their realism, dramatic appeal, and the variety which they add to other methods of instruction.

d. Demonstrations set the stage for student performance by illustrating correct methods, by setting standards of performance, and by giving the student confidence that attainment of the skill is possible.

4. **Limitations of the Demonstration.**

a. During the presentation period, demonstrations do not normally provide for active student participation. You can overcome this limitation by asking questions of students and by encouraging student questions between steps in a demonstration, and by providing for student performance following the demonstration.

b. The more elaborate demonstrations require additional personnel and equipment and time for rehearsals. The increase in student learning, however, will more than compensate for the expense involved since demonstrations can be presented to large groups. Some demonstrations are presented to a thousand students simultaneously.

c. Range demonstrations are frequently affected by weather conditions. Indoor substitutes are usually inferior or less effective.

5. <u>Forms of the Demonstration</u>. There are five general forms of the demonstration:

a. The Procedural Demonstration. This is the form of demonstration used to show and explain the operation and functioning of weapons. This type of demonstration may be conducted indoors or outdoors and is used widely throughout basic and advanced training.

b. Displays. Displays must be planned so that students can view them quickly. This requires arranging the displayed materials so that each item can be seen by all students at the same time. For large classes use duplicate displays or divide the class into sections, with the sections rotating from one exhibit to another. This is known as the "Country Fair" system.

c. Range Demonstrations. Complicated demonstrations can be shown one part at a time; later, the complete performance can be shown. One phase must be properly assimilated before the next phase claims the student's attention.

d. Motion Pictures. Training films provide ready made demonstrations by experts. Here the student has the opportunity to see internal operation of weapons, or matches being conducted - things he would otherwise have to imagine.

e. Skits. Instructors or assistants may act out operations or procedures. This form of demonstration has proved an effective means of portraying range safety. Skits guide student appreciations and attitudes. Skits may be designed to show the wrong way; however, you must insure that the right way is obvious, or show the correct way later. Skits must be carefully planned and smoothly presented; this requires repeated rehearsals.

6. <u>Planning a Demonstration</u>. The success of a good demonstration depends ninety percent upon planning and rehearsal, and ten percent upon execution. (See Figure 2-4.) Here is a checklist of things to consider:

a. Based upon the lesson objective and the specific teaching point, decide what to demonstrate. Limit the scope so that the demonstration has a specific purpose. A lengthy, involved demonstration will confuse your students. Remember, the essence of effective demonstration is precision, timing and conciseness.

b. If your demonstration involves several operations, list these operations and demonstrate them one at a time. If you wish students to learn more than one way of performing an operation, plan a separate and distinct demonstration for each method.

c. Prepare a scenario. Include an introduction, an explanation and a summary, and incorporate training aids (charts and mock-ups) to aid in explaining the steps being demonstrated. Attach the scenario as an inclosure to the problem lesson plan.

d. Make a list of personnel and equipment needed for the demonstration and arrange for their availability for rehearsals and performance.

e. Arrange for and check the site of the demonstration. Insure that students can see and hear, and arrange for sound equipment if necessary.

f. Rehearse the demonstration at the actual site to be used. Utilize initial rehearsals to check equipment and timing and to develop the proficiency of participants. Request the presence of other members of your team at your dress rehearsal to obtain constructive criticism and to insure high standards of performance.

IN PLANNING A DEMONSTRATION....

    LIMIT THE SCOPE

    PLAN TO DEMONSTRATE STEP-BY-STEP

    PREPARE A SCENARIO

    LIST PERSONNEL AND EQUIPMENT

    CHECK FACILITIES

    REHEARSE

    MAKE FINAL CHECK

Figure 2-4. Planning a Demonstration.

    g. Prior to the actual performance, check to insure that personnel and equipment are present and that classroom arrangements are complete.

7. <u>Conducting a Demonstration.</u>

    a. During a demonstration, you obviously want the student to watch the demonstration, yet you do not entirely forfeit your job of oral communication. If you are demonstrating an item of equipment, speak to the students, not the equipment. If, on the other hand, you are explaining while an assistant instructor demonstrates the equipment, direct the student's attention to the demonstrator. It is not uncommon for students to watch an instructor and fail to realize that an assistant instructor is demonstrating the steps of the operation.

    b. Insure that all students can see the demonstration. You may find it necessary to repeat a demonstration in a different position or at a different angle. Be careful that neither you nor your assistants block the student's view.

    c. Introduce the overall demonstration carefully to insure that the students understands the purpose and what to look for. Knowing what they should get out of the demonstration affects the way students observe it and assists learning.

    d. Explain the demonstration one step at a time. Introduce each step by telling the students what you will do next. During the step, explain what is being done and why it is done that way. Summarize after a complex step or after several steps during a lengthy demonstration. Ask questions to check student understanding. Encourage students to ask questions between steps, but do not allow students to interrupt the demonstration of a step.

    e. Where sequence is important or where an operation involves several difficult steps, o line each step on the blackboard or with a chart before you explain and demonstrate the step.

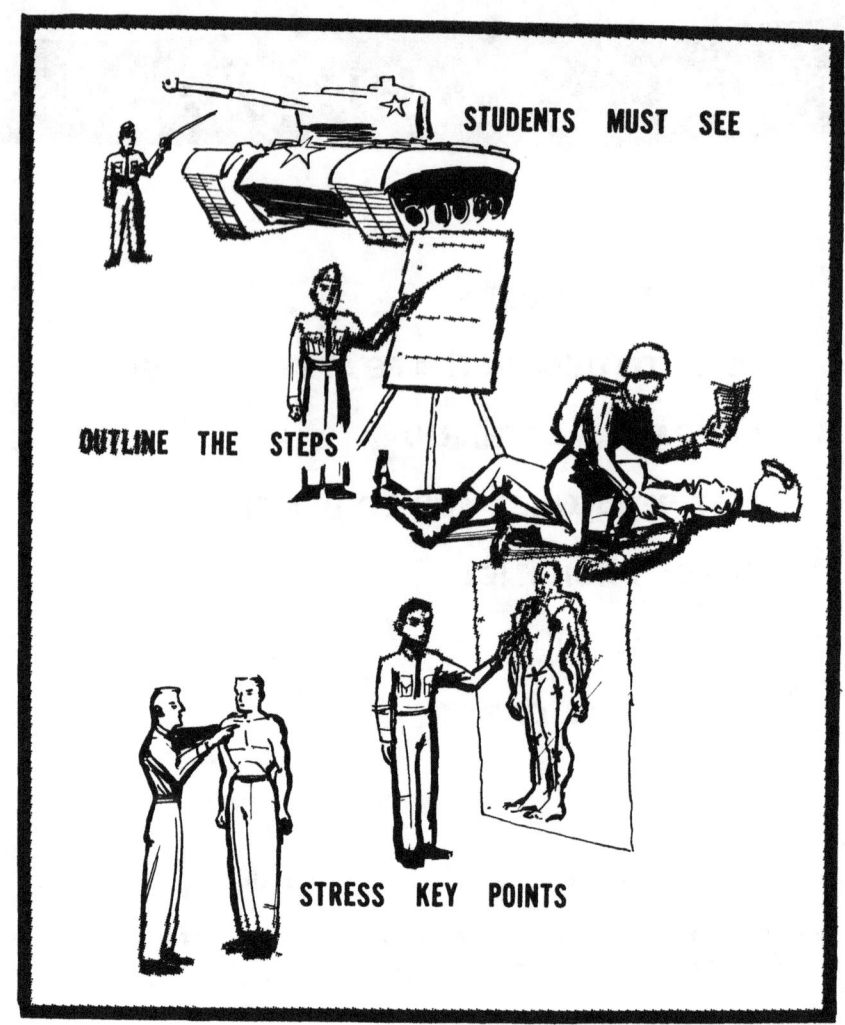

Figure 2-5. Conducting the Demonstration

D. CONFERENCE

1. The "Discussing" Method.

    a. As education has placed increased emphasis on student activity for optimum student learning, the instructional conference has proved effective in securing good student participation. As a leader of an instructional conference, you will discover that the conference is not an easy way out. You must research every facet of your subject because new questions will arise from students in each succeeding discussion. Your teaching task will require that you exercise a high degree of intelligence, tact, alertness, and ability to think on your feet. A poor lecture is at worst boring; but a poorly led conference can be devastating when measured in terms of mislearning, student antagonism or frustration.

    b. The word "conference" has many meanings. In industry or in the military, a pure conference is "an assembly of a group of individuals in which the group members contribute information and ideas toward accomplishing a common purpose." The goal or common purpose of the conference may be a problem to solve, a decision to make, or learning or information to be shared. Regardless of the goal, the outcome of this type of conference is not preconceived or preplanned by the conference leader. Each member contributes in theory to the group discussion, solutions, or conclusions. In short, the conference leader does not attempt through skillful manipulation or guidance of the discussion to lead the group towards preconceived solutions or learning.

c. In marksmanship training, we must standardize doctrine and administrative procedures. Since this is the case, how can we possibly use the conference method to teach students doctrine and procedures? In marksmanship instruction, not only must the objective of each lesson be predetermined but also each of the teaching points. The key to the answer is in the instructional conference. Here the teaching points are developed, not initiated, by student discussion. The instructor initiates, stimulates, and guides this student discussion. When sufficient pertinent discussion occurs, certain conclusions can be drawn from this discussion. These conclusions are the student's understanding of the supporting material that makes the teaching points valid and sound. The method is actually inductive: from a series of student opinions, solutions, or experiences, the class builds up the principles to be learned. In the lecture method, the instructor frequently uses the deductive process; he states a principle or teaching point and then proceeds to prove it by a series of examples, illustrations, statistics, or other supporting material. The technique in the instructional conference leadership is to stimulate the free flow of ideas between instructor and students. This guided discussion promotes quicker student understanding.

THE INSTRUCTIONAL CONFERENCE:

MAINTAINS ACTIVE STUDENT PARTICIPATION

DRAWS UPON STUDENT EXPERIENCE

STIMULATES CRITICAL THINKING

PROVIDES CHECK OF STUDENT UNDERSTANDING

DEVELOPS STUDENT SENSE OF PERSONAL RESPONSIBILITY

TRAINS STUDENTS TO COOPERATE

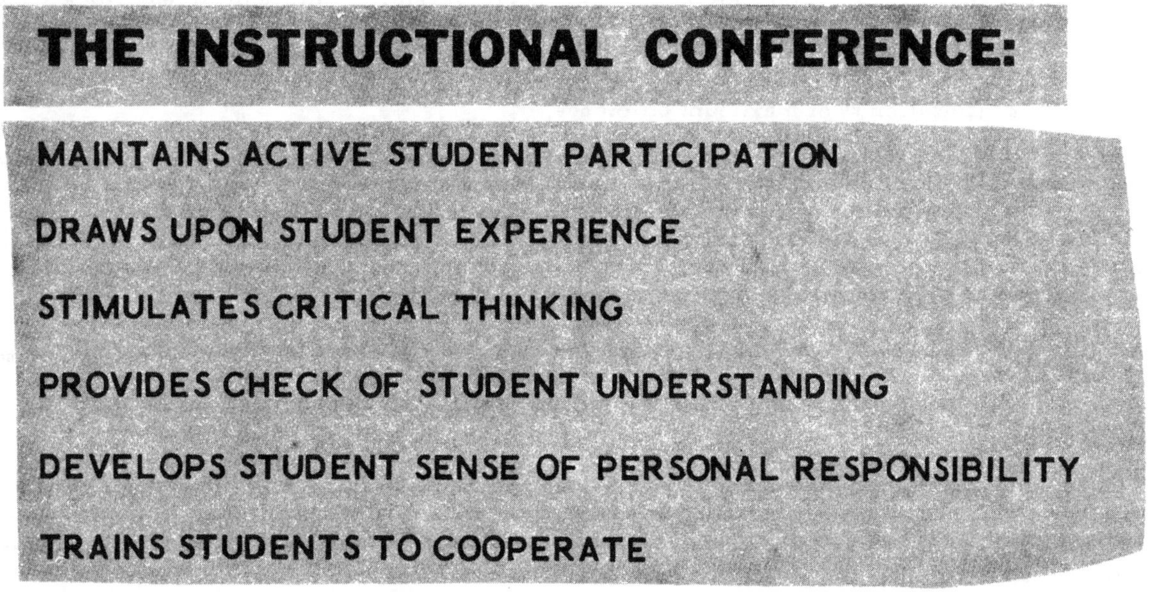

Figure 2-6. Advantages of the Instructional Conference.

2. Advantages of the Conference.

a. The instructional conference or guided discussion method encourages active student participation and maintains interest. When students discuss, probe, disagree, or answer provocative questions they are concentrating, thinking, and learning actively. Such active learning makes a greater mental impression and is remembered longer.

b. During the guided discussion students often contribute new ideas and new applications from their background of experience. No instructor no matter how well informed can match the cumulative wealth of experience of the class. These student-originated ideas not only make instruction more meaningful, but they result in course improvements.

c. Discussion stimulates reflective thinking and reasoning. Students become accustomed to thinking critically, to making comparisons, and to relating ideas and doctrine to their experiences and previous learning. The result of a good conference is that students consider all aspects of the problem. Many points which might otherwise still be doubtful are resolved by this questioning and discussion.

d. The conference method provides you with frequent opportunities to check student comprehensions of the subject. This enables you to gear instruction to the proper learning rate for that group. You can get an immediate reaction to how well students are absorbing the materials to be learned.

e. Since active student participation is essential for an instructional conference students must assume responsibility to assist in their own learning. The burden of learning responsibility shifts from the instructor's back and becomes a shared cooperative task. Both the instructor and the students must insure that learning takes place.

f. A bonus effect in the group participation is that the conference trains students in the skills of cooperative effort, group thinking on a common problem, self-expression, and tolerance of the opinions of others.

3. <u>Limitations of the Instructional Conference.</u>

a. The instructional conference method consumes more training time than does the lecture to cover a specified amount of instructional material. Ideas must be discussed, analyzed, accepted, modified or rejected. CAUTION: Make very sure that you have selected for conference only those portions of your lesson that are extremely significant, controversial, or difficult to learn or accept. Don't waste time on a conference on the obvious or on material which does not permit several points of view. For example, beginning instructors sometimes try to conduct a conference on the organization of a team. This is a fact and no discussion is possible. However, a discussion might be conducted on whether this organization is of maximum efficiency.

b. The conference method is employed for classes containing up to one hundred students with the belief that it is better to have some discussion rather than none. It is felt that even limited student participation will bring out the major differences of thought or usual difficulties. It is most effective in small groups of fifteen or twenty students where all students voice their opinions. When you use the conference method with one hundred students you only sample student opinion and only the more aggressive students will volunteer to express their opinions.

c. To conduct a guided discussion you must have students with some knowledge of the subject to be discussed. Your advance study assignment should help to furnish their basic background knowledge. Advance sheets must provide the review of principles, help to get the student "read into" the problem.

d. Many gaps in learning or even some incorrect learning may occur if the instructor is inexperienced in planning and conducting a good conference. Compared with the lecturer, the conference instructor must possess more comprehensive knowledge of the subject, more tact, greater versatility and flexibility, and an ability to think on his feet to seize the key points of student discussion and to summarize skillfully.

4. <u>Planning a Conference.</u>

a. Planning the introduction and conclusion for an instructional conference is similar to planning of any lesson and is discussed in Chapter 3. The primary concern in this paragraph is planning for discussion in the body of the lesson.

b. Since a discussion presupposes an adequate student background knowledge of the subject matter, you must analyze student background in terms of previous instruction and previous experience. For example, you know that few novice shooters have previous experience. You can, however, take advantage of the instruction that these students have received to date, as well as their backgrounds. You can supplement the student's knowledge by issuing advance study assignments, by lecturing initially, or by using both techniques.

c. In preparing your lesson plan, consider the teaching points as the framework of the conference, with each teaching point constituting a potential discussion area. Prepare questions to provoke thought, to stimulate discussion, and, above all, to insure thorough understanding of the teaching points. Avoid the tendency to ask only "what" questions. Have the students consider also "why", "how", "when", or "where". For example consider the discussion potentials of these questions: "What is the primary function of the gas system?"

NOTE: For a treatment of types of question and their characteristics, see Chapter 6.

IN PLANNING A CONFERENCE....

**DETERMINE STUDENT BACKGROUND AND
SUPPLEMENT WHEN NECESSARY**

**PREPARE QUESTIONS TO INSURE ACCOMPLISHMENT
OF THE STUDENT PERFORMANCE OBJECTIVES**

**CONSIDER LIKELY STUDENT RESPONSES
TO QUESTIONS**

**PLAN INTEREST FACTORS, TRAINING AIDS,
SUMMARIES AND TRANSITIONS**

Figure 2-7. Planning a Conference.

d. Taking each of your prepared questions in turn, consider likely student responses to these questions. While student responses vary with the individual, they will usually fall into general response patterns. By considering likely student responses, you are better able to ask good followup questions and to lead the students to a thorough understanding of the teaching points. It is impossible for you to anticipate every student response; therefore, you must display versatility and ability to think on your feet when conducting a conference. You must evaluate student responses as they occur and use them to maximum advantage in advancing the conference or insuring student learning.

e. As with other instructional methods, a good conference requires a skillful use of interest factors, training aids, summaries, and transitions. Although questions are good interest factors in themselves, you will frequently find a need for examples, illustrations, and training aids to stimulate discussion and to support teaching points. Consider whether a teaching vehicle might assist student understanding. (See Chapter 5.) Frequent subsummaries are especially necessary in a conference. In the planning stage, you will not be able to determine the frequency with which subsummaries will be needed nor the comments that will occur during discussion; however, you can plan for summaries of teaching points and important supporting ideas, for you know if the discussion has been long and animated there must be a subsummary before moving to a consideration of the next teaching point. Similarly, you can plan transition between important ideas, and deliver other transitions extemporaneously.

f. Plan advance sheets which contain thought-provoking questions or the initial discussion of the general situation so that valuable class time is not spent in preparatory reading.

5. <u>Conducting a Conference</u>.

a. Student Participation. As a student you have probably attended an instructional period listed as a "conference" wherein the instructor could not get a discussion started. At no time did a student volunteer an answer, nor did students appear interested. The instructor was not able to secure student participation for maximum learning. This instructor probably attributed the lack of student participation to a "bad day" or to a "slow class." It is more likely that the instructor either failed to understand or lacked the ability to apply effective conference techniques. Proper use of the techniques which follow will help you to arouse student interest and to obtain optimum student participation and learning.

    b. Permissive Atmosphere.

        (1) A conference is based upon student participation, and is most effective when all students are motivated to think reflectively and to enter into the discussion. Students will think more freely and will enter into a discussion more readily when you establish a permissive atmosphere, wherein you encourage a free flow of ideas. We do not have a single word in our language which aptly describes this relationship between instructor and student, so we go to French and we refer to instructor-student rapport: a feeling of mutual cooperation and understanding, of harmony and congeniality. You establish rapport by your sincerity and enthusiasm in your introduction, in your conduct of the discussion, and even in your administrative directions. You maintain rapport by the manner in which you ask questions, call upon students for contributions, and give credit for student ideas.

        (2) For example, if you ask a question, pause, and then bark out a student's name in a drill sergeant's command voice, the student is likely to jump up, give the verbatim answer that he thinks you are looking for, and sit down. He has merely attempted to recall an answer from memory and has not attempted to think reflectively and come to a logical conclusion. This has the earmark of a grade-school type of recitation, not of a discussion among mature individuals.

        (3) You can often stifle a good discussion by careless unthinking remarks. For example, you ask a question and call upon a student, who gives you a good answer, but one that you had not planned for. If, as happens all too often, you reply, "That is true, but it isn't what I have in mind," you are stifling student thought. You are, in effect, telling the student not to think, but to give you the correct response, to guess the "right" answer, to read your mind. In evaluating student responses keep in mind your objective - student learning - and make every attempt to promote such learning by stimulating student thinking and free expression of ideas.

    c. Stimulating Discussion. By planning good thought-provoking questions and by establishing rapport with students, you have laid the foundation for discussion. You will find, however, that discussion will falter unless you take definite steps to stimulate further discussion. By making a startling or controversial statement or quotation, followed by a good question, you can often obtain enthusiastic participation once the students are aroused by these devices. Student stimulation by good questions is a technique used so successfully by the ancient Greek philosopher Socrates in his teaching.

    d. Guiding a Conference.

        (1) Based upon an analysis of the lesson objective and the teaching points, you have planned thought-provoking questions, questions to check student learning, subsummaries, and transitions. Within this framework, you must guide the discussion so as to insure student understanding of each teaching point and thus ultimately attain the lesson objective. How you guide and control the discussion is largely extemporaneous. Based upon student answers or contributions, you will frequently ask questions requiring further explanation or elaboration to insure student understanding. Use this technique when a student is obviously bluffing or when a student uses terms which you realize are unfamiliar to some of the other students. Don't generally permit a student to make a bold statement without drawing from him his reasons for believing that way. In this way both you and the class can evaluate the soundness of the belief.

        (2) After you are successful in stimulating a good discussion, your next role is that of moderator. You must keep the discussion on the teaching point. When the discussion is obviously beginning to go "off on a tangent," summarize the comments and, by asking a question, bring the discussion back to the subject. Insure that participation is distributed amont the students, that a few of the more enthusiastic students do not dominate the discussion. By checking off each student contributing on a class roster, you can single out nonparticipants and bring them into the discussion. When two students threaten to monopolize a discussion by getting into an argument, stop the argument by summarizing the opposing points of view, and bring a third student into the discussion for his opinion.

Remember that at the end of each discussion your students expect you to voice the instructors position on this matter or summarize what they as students should take away as a result of the discussion.

(3) Your contribution to student understanding depends largely upon the effectiveness of your subsummaries. A good subsummary of the student's discussion will briefly point out the ideas expressed, resolve conflicting points of view, relate ideas to the teaching point, point out application of principles, and state the instructor position, as appropriate. While subsummaries are planned for, their frequency and their content will depend upon the length and scope of the discussion. Subsummaries give an instructor an excellent opportunity to make transitions into his next teaching point or supporting idea. (For further discussion of subsummaries and transitions, see Chapter 6.)

(4) Your role as discussion moderator is not an easy one. You must be firm in keeping the discussion on the subject, yet tactful in your control. A mark of a good conference leader is his ability to guide discussion toward desired learning outcomes without dominating the discussion. He realizes when there is need for a definition of terms, clarification, arbitration or summary and sees that these needs are filled.

NOTE: See Chapter 6 for specific techniques used in asking questions, handling student response, accepting voluntary contributions, and handling student questions.

E. PERFORMANCE

1. *Student Participation*. The performance method emphasizes student doing in order to learn and improve physical and mental skills. Sometimes it is routine practice of physical skills such as those used in weapons firing. Sometimes it involves problem solving skills in applying principles, techniques, or procedures to a realistic training situation. The performance method takes any one of two basic forms: the practice exercise and the problem exercise. You can introduce variety by combining these basic forms.

a. Practice Exercise. An applicatory exercise designed to teach procedures and to develop basic skills. The practice exercise is used widely to teach subjects such as maintenance and operation of weapons, and marksmanship. Variations of the practice exercise include:

(1) Controlled Practice. Practice by group in which group members practice the same thing at the same rate and time to gain the correct concept and to develop accuracy. The PI explains each step as the students perform. AI's control the practice and give aid. Controlled practice is especially suited for the initial step in learning a skill.

(2) Individual Practice. Routine practice by individual students until a skill becomes automatic. This practice is supervised, but not closely controlled since each student works at his own speed.

(3) Coach and Pupil. Used for teaching students who have learned the fundamentals of a skill or procedure. Both coach and pupil are students whose roles are periodically reversed. The coach and pupil technique is especially useful in marksmanship.

(4) Team Practice. An exercise in which students serve as members of a team. Team practice is normally conducted in two phases: first, a demonstration by the members in which techniques are emphasized; second, a phase allowing the application of techniques.

b. Problem Exercise. Characteristics: The problem exercise employs the problem solving process and is used when solutions are based solely upon the application of principles, techniques, or procedures to the problem situation. It may be a simple situation presented orally and followed by questions, or it may be a written exercise.

2. *Advantages of the Performance Method:*

a. Allows full student participation in the learning.

b. Repractice and repetition results in improvement of skill.

  c. Allows individual practice as in the actual situation.

  d. Allows team practice for improvement of unity of action by a group.

  e. Applies mental skill to problem solving in realistic training situations.

  f. Promotes enthusiasm and competition among students.

 3. <u>Disadvantages of Performance Type Instruction</u>: Performance instruction has some disadvantages but they do not limit the effectiveness as much as in the other methods.

  a. May become sterotyped or unchanging.

  b. Complete evaluation of instruction is difficult except for the training of small groups.

  c. It is difficult for one instructor to control the complete operation. The larger the training group, the more assistance needed to maintain standards. Each student group must remain small for effective supervision.

  d. Supervision of performance requires extensive training or ample previous experience on the part of the assistant instructors.

 4. <u>Planning Performance Type Instruction</u>: Most of the deficiencies can be offset in the planning stage.

  a. Check for required equipment.

  b. Establish realistic standards for students.

  c. Set a time limit on completing a performance task or action.

  d. Give step-by-step specific directions to students before attempting action.

  e. Demonstrate the action prior to student performance.

  f. Plan for close supervision of each individual student and observe safety controls.

  g. Train assistants by step-by-step instruction in the procedures to be followed.

 5. <u>Conducting Student Performance.</u> Careful planning is critical to the successful conduct of performance type instruction.

  a. Motivate. Because a student is performing an act does not mean that he will learn well. Tell students specifically what they are to learn and why. Men will meet a challenge, so let them know what standards you expect of them, and tell them how well they are progressing. Maintain interest through competition. Overcome monotony by varying your procedures and by gradually increasing performance standards. Make performance realistic.

  b. Explain and Demonstrate. In explaining a procedure, use conference techniques and encourage student questions, not only to insure understanding but to make the student interested in learning the procedure. A good demonstration will make the procedure clear and will set standards for the student to attain.

  c. Control and Supervise. Make your directions specific, and encourage students to ask questions if they do not understand. Brief your assistants thoroughly so that they help you control student activity. Supervise constantly to insure that students understand how to perform correctly and are making satisfactory progress. When students learn to perform the operation correctly, insure that they are kept busy learning to perform it better and faster. Prevent students from forming faulty habits; but, at the same time, allow students to use initiative and resourcefulness. Your students will learn better if you and your assistants demonstrate that you are there to help them to learn not to harass them. For example, most men like to fire weapons, but too many are harassed into hating range work by oversupervision and over control. Keep your students wanting to learn and keep them challenged to do better.

    d. **Critique.** It takes common sense and tact to critique student performance properly. Frequently, both the student and the instructor are tired and exasperated by poor performance. You must encourage the student and help him to maintain self-confidence under the trying conditions of learning a new skill. At the same time you realize that a certain minimum standard must be attained to insure his safety. Skill in human relations is essential in this highly personal learning situation. Help, not harassment, is what the student needs. Praise the correct portions of performance. Select the major weaknesses to be improved first and offer definite suggestions as to how they can be improved. Let the student know periodically how he is progressing. As you detect definite steady improvement, bear down more heavily to achieve desired standards.

IN CONDUCTING STUDENT PERFORMANCE....

MOTIVATE

- STATE WHAT AND WHY
- SET STANDARDS
- EVALUATE PROGRESS
- PROVIDE COMPETITION
- VARY PROCEDURES
- INCREASE STANDARDS
- PROVIDE REALISM

EXPLAIN AND DEMONSTRATE

CONTROL AND SUPERVISE

- GIVE SPECIFIC DIRECTIONS
- HAVE ADEQUATE ASSISTANTS
- SUPERVISE CONSTANTLY
- KEEP STUDENTS CHALLENGED
- PREVENT FAULTY HABITS
- HELP – DO NOT HARASS

CRITIQUE AND SUGGEST

Figure 2-8. Conducting Student Performance.

 LECTURE, CONFERENCE

 SLIDES & FILMS, ACTUAL OBJECTS
DEMONSTRATION

 PERFORMANCE, PROBLEM METHOD

Figure 2-9. A Good Lesson Combines Several Methods.

CHAPTER III

LESSON PLANNING

A. GENERAL. An outstanding and proper application of all material contained in this handbook is necessary to accomplish a high quality of instruction. Underlying the discussion of each instructional method, aid and technique is one basic premise--that the instructor has a complete, usable lesson plan. The heart of the usable lesson plan is the <u>lesson outline</u>. It is from this outline that the instructor prepares his presentation or lectern notes. The complete lesson plan includes the lesson outline, research materials, instructional handouts, copies or descriptions of training aids used, and the preparation data section listing physical facilities, personnel, sound, transportation, ammunition, and other requirements to present this lesson successfully. Lesson planning will occupy more time than any other single step in the instructional staircase (see Figure 3-1). Thorough understanding of lesson preparation is essential for all instructors.

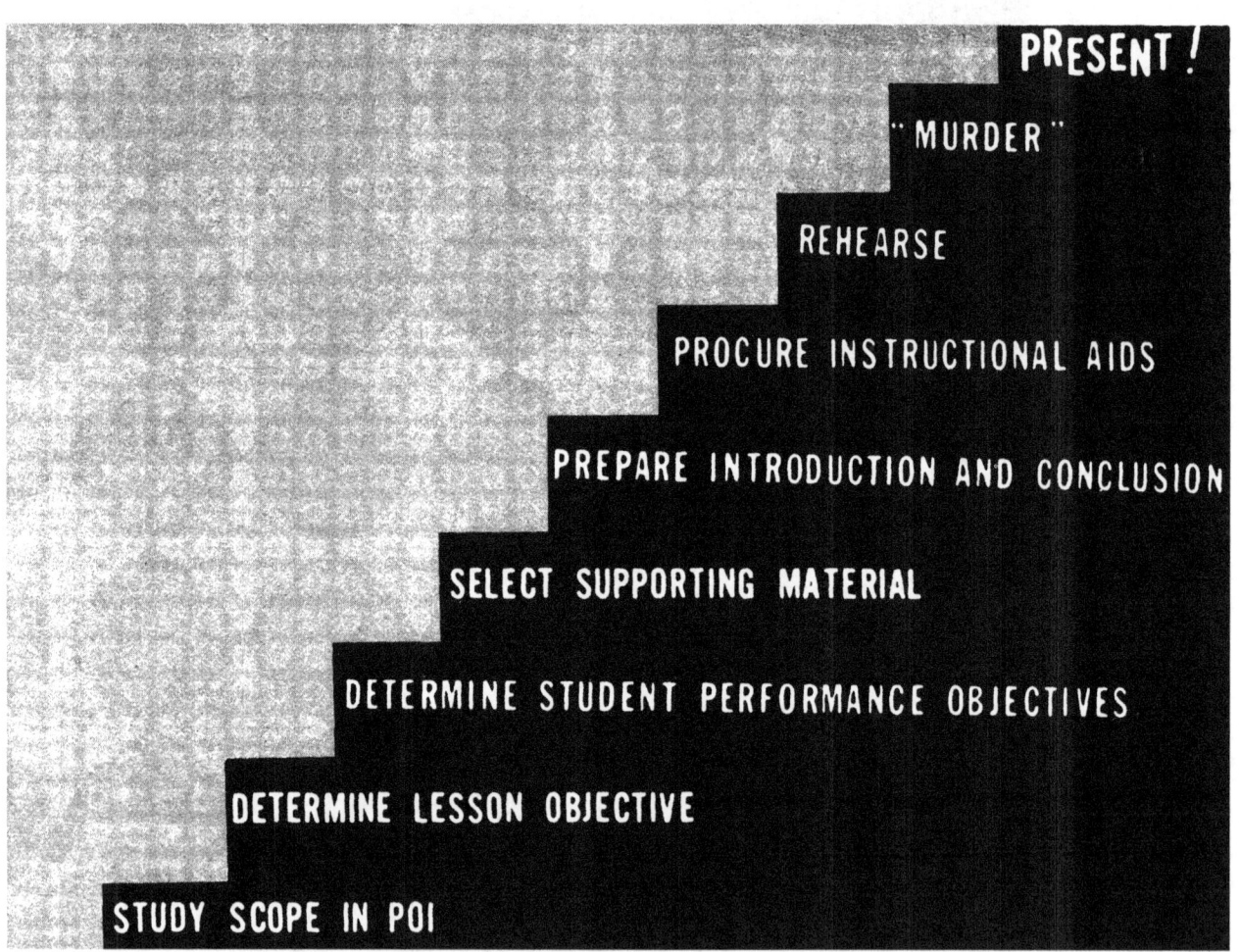

Figure 3-1. Instructional Staircase to Success.

1. <u>Estimate of the Teaching Situation</u>:

    a. General. In planning the lesson there are certain considerations that help you decide the best way to achieve the lesson objective. These considerations will largely determine the degree of student learning which will result from your instruction. Some of them will be your decisions, some will be determined by official directive. In making your estimate you will consider these major factors.

    b. Subject Text. This provides you with the essential information to be taught. Plan preparation, rehearsal, and presentation and consider the relationship of this lesson with other lessons in the same POI. The subject text serves as the foundation for all reference material to be included in the period of instruction.

c. Program of Instruction (POI). This document is the source directive for all instruction. The program of instruction establishes the subject and overall scope of the lesson. It should be the first reference.

d. Students. Another consideration is the types of students to be taught. Novice shooters will not receive the same level of instruction as that given to advanced students. The methods of presentation, the speed with which material is presented, and the complexity or depth of subject coverage will vary considerably. Consider the experience level and how much previous instruction this class has received prior to your instruction. You do this so that you can properly relate your instruction to previous instruction and avoid needless repetition.

e. Time Available to Prepare and Present. How much time will the instructor have to prepare his period of instruction? This time will be divided into intermediate target dates. This will allow time for last minute revisions of training aids, handout materials, and final rehearsals to polish up rough spots. Allocate time for your detailed coordination, and for preparing the completed vault file. Another important consideration is the number of hours allocated to present the instruction. This will influence the amount of information you can present, how deeply you will go into the subject and the degree of student proficiency you can reasonably expect. In a compressed marksmanship training schedule your task will be to insure student understanding within the time available.

f. Equipment, Facilities, and Personnel. In planning the lesson you must consider the type of instructional area required and make a reconnaissance of this instructional area prior to completing your plans. By doing this you can insure that the instructional area will support the type of lesson and methods of instruction you plan to use. The availability of training aids and assistant instructors will also be important considerations.

2. Lesson Objective:

a. The lesson objective is a brief statement of the student learning to be achieved as a result of this period of instruction.

b. Necessary Elements of the Lesson Objective. To be a workable guide for the instructor, the lesson objective must contain these elements:

(1) The type of students who will receive the instruction; designate these students by identifying them by the specific course they are attending.

(2) Learning to be achieved or level of student proficiency or action expected as a result of this instruction. Specify the knowledge, appreciation, skills that the student should have learned and be able to perform as a result of this instruction. These learning outcomes are covered by further amplification or explanation in the student performance objective.

(3) Subject. Here it is necessary to be specific. Pinpoint the portions of the subject you will teach.

c. Levels of Student Proficiency. There are three levels of student proficiency. They are:

(1) General Knowledge. That level which makes the student aware of a subject and its general application to a major field. It provides the student with knowledge of the existence of certain fundamental facts and principles in a degree sufficient to enable him to recognize their implications and where to locate further information when the need arises.

(2) Working Knowledge. That level which makes the student sufficiently familiar with the primary purposes, major functions, or principles of employment of a subject to permit routine practical applications. It provides the student with sufficient knowledge and skill for him to apply without further training or experience. In addition, the student acquires limited proficiency to supervise or train others in that skill. It must be augmented by on-the-job training or further schooling for full qualification.

(3) Qualified. That level which provides the student with a comprehensive knowledge of the subject that permits skilled performance and authoritative interpretation of the principles involved.

  d. Selecting a Desired Level of Student Proficiency. You should consider these factors:

   (1) The nature of the subject. Is it simple or complex? Is this the first formal instruction the student receives in this subject area or does he have previous instruction or on-the-job training? What is the training purpose as stated in the program of instruction? Will this instruction be an orientation or are the students expected to become technically proficient?

   (2) Time available to present the subject. How many hours have been allocated in the POI to present the subject?

   (3) Student backgrounds. What is the educational background and experience of the students? The instruction must be so presented to enable the average student in the class to comprehend, not to bore the more experienced, and to satisfy the needs of the less experienced. (See Chapter IX for sample lesson objective.)

3. <u>Student Performance Objectives:</u>

  a. Explanation. Once the lesson objective is formulated, you are ready to consider what must be taught in order to accomplish this objective. In other words, certain key facts or elements of knowledge must be understood. Military and civilian schools call them specific objectives, main thoughts, ideas or points, student learning outcomes, or primary areas to be taught or discussed. Once the student understands and can use these ideas the lesson objective has been accomplished. We shall call these elements of knowledge student performance objectives. Each must be a necessary and critical part of the overall lesson objective. The sum of the student performance objectives constitute the lesson objective.

  b. Definition. A Student Performance Objective is a statement in complete sentence form of a specified and significant principle, item of doctrine technique, skill, or element of knowledge that students must understand and be able to apply as a result of a period of instruction.

  c. How Many Student Performance Objectives. The instructor preparing a lesson is always concerned with how many student performance objectives there should be in any given lesson. The decision should never be based solely on number. Rather, select them by considering how many important parts or elements will be needed to accomplish the objective. There could be as many as ten sub-objectives in a two-hour period of instruction or only one in an eight-hour series of classes. In addition to consideration of the elements of knowledge needed to teach the objective, there are other aspects of the problem to consider in selecting student objectives. Certainly the degree of learning dictated by the lesson objective will have a bearing on the number as well as the complexity of the performance objectives. Whether the lesson is complete in itself or one of a series of related lessons will be considered. Since student performance objectives are the heart of the lesson, give them much thought before selecting each one. Check carefully to insure that each is clear, concise statement of fact. Of course, the proof of each performance objective will be developed in the actual presentation of your supporting material.

  d. How to State Performance Objectives. State them clearly and briefly. Make sure they are complete sentences. Much care and work will be necessary to make them simple and brief, yet meaningful. A performance objective doesn't have to carry the whole burden of significance, however, sometimes the supporting material will provide the critical backup. Be careful that you do not elevate supporting material to the level of an objective. Frequently a new instructor prepares a lesson with eight or ten sub-objectives. After he studies it carefully, he discovers that several of these so-called points are really supporting material for one overall teaching objective. By further thinking through his problem, he discovers that he really has only three or four main points. Remember, a student performance objective is a significant and critical element of knowledge - refined and distilled. It is the product of prolonged careful thinking and considered judgement. (See Chapter IX for sample student performance objectives.)

4. <u>Supporting Material</u>. After your performance objectives have been established and listed in the most meaningful sequence you must then decide how you can best achieve understanding of these points. You do this by the use of good supporting material. We might compare student performance objectives and supporting material to the presentation of a law case by a lawyer. The lawyer states in his introductory remarks certain elements or facts which he will prove as the trial develops. These elements or facts correspond to our subobjectives. Throughout the presentation of his case he proves or creates an understanding of each fact previously presented. This would be supporting material. So you too, in deciding upon how to prove or teach your points, must consider what will be the most efficient manner of presenting your case. The common ways are: by explanation, demonstration, discussion, skits, use of historical examples, quotations from famous people, practice exercises, and the use of visual aids (see Figure 3-2). As a well prepared instructor, you have a number of varied illustrations, examples, and restatements in reserve. Which of these you will use in supporting each performance objective will depend upon the degree of understanding desired, the amount of time you have to create this understanding, and the complexity of the point you are supporting. Don't over-teach or over-explain. You should insure that you have enough supporting material to explain the spe fic objective completely and adequately. How much supporting material you use will be determined by how soon most of your class grasps the projected point. This you can evaluate by using checkup questions. When satisfied that the class understands the first objective, proceed to the next. If the students do not grasp the point, restate, and supply additional example or illustrations until learning occurs.

Figure 3-2. Ways to Support Student Performance Objectives.

B. ORGANIZATION OF THE LESSON. Up to this point in lesson planning, you have studied the problem directive, examined the program of instruction, determined the lesson objective, and carefully selected the student performance objectives and appropriate supporting material.

    1. <u>Determine how to organize this material</u> to insure effective student learning. Check these points:

        a. Will the sub-objectives selected provide the student with the knowledge essential to accomplish the lesson objective?

        b. Can these points be further consolidated or clarified?

        c. Are the performance objectives arranged in a logical, easy-to-understand order that will enhance student learning? Are they arranged from the simple to the complex, known to unknown, easy to difficult?

        d. Is the supporting material sufficient to prove the point? Have I used various types of supporting material to maintain student interest?

        e. What method(s) of instruction will I use to present the student performance objectives? (See Chapter 2)

    2. <u>Lesson Outline</u>:

        a. The lesson outline indicates what is taught, in what order or sequence it is taught, and what teaching methods are used. The outline insures that teaching is properly planned and not haphazard or impromptu. Proper organization helps to ensure effective presentation.

        b. You prepare a full sentence outline for each problem you present. This lesson outline will follow the prescribed format as illustrated in the Sample Lesson Outline (See Annex I) and will be filed in the Lesson Outline Section of the Vault File for that problem.

        c. You do not teach directly from the lesson outline. After completing the outline you may use it for the first or second rehearsal. Then you prepare a set of teaching notes to keep on the lectern during the presentation. These notes are prepared in the way most convienent for your use. They may contain phrases, abbreviations, color coded key ideas, or whatever symbols you wish to assist you during your presentation.

        d. This outline should be complete enough so that another instructor could prepare his own teaching notes or teach the subject if the emergency arose which precluded his doing the necessary original research prior to presenting.

        e. The heart of the lesson outline consists of the introduction, the body, and the conclusion. In preparing the lesson outline you will normally prepare the body first. This is the principal part of the lesson because your first decision was what must be taught to accomplish the lesson objective. When you have decided this, you are ready to prepare a proper introduction. As a final step you write the conclusion.

    3. <u>The Body of the Lesson Outline</u>:

        a. The body is that portion of the lesson outline where you develop carefully and completely the material to be taught. The body contains the student performance objectives together with the supporting material, leadoff questions with anticipated student responses, transitional sentences or paragraphs between major divisions, notes indicating where and when training aids are used, and planned sub-summaries.

        b. The first step in preparing the body is to place the sub-objectives and the material supporting each point in a logical order. Frequently, the sequence of presentation is obvious; however, it may be necessary sometimes to experiment with the sequence before you actually decide the order in which you will present the various points. The two principal orders of presentation are the inductive or deductive order. In the inductive order you start with the individual specific examples or cases and build up to a conclusion or principle based on these examples. In the deductive order you state the basic principle or rule and proceed to prove or illustrate it by a series of examples. Remember this: if the order of presentation is not too clear to you it will completely confuse your students!

   4. The Lesson Introduction:

      a. The lesson introduction consists of four mandatory elements: gain attention step, lesson tie-in, motivation, and scope. Two other elements may be included when they are appropriate; student application of this instruction and the methods of instruction which will be used.

      b. Gain Attention Step:

         (1) Gaining attention may suggest to some persons the use of a gimmick, a trick, or the use of a startling device. If so, the real need for and purpose of an attention-gaining step is misunderstood. Before you can expect to have students learning, or even to stimulate student desire for learning, you must have the attention of the student. You do this by gaining attention-focusing of student attention upon the subject to be learned, and not distracting students by a novel gimmick.

         (2) Some ways to gain attention are to tell a story (humorous or historical), ask a rhetorical question, present a skit or conduct a demonstration, or merely walk to the center of the platform, pause until things are quiet, and commence your instruction. If the latter is a method of gaining attention, then we might ask ourselves, "Why is there a need for the other methods mentioned?"

         (3) Sometimes you will be faced with the problem of presenting introductory periods or perhaps you are scheduled for the first period after a holiday or the first period in the afternoon of a very hot day. These instances are not extreme. You must be sure that the attention of the student is on you and on the subject which you are to present and not on some personal problem or activity. The method which you choose to gain attention will be selected only after a careful consideration of the class, the subject and the time allowed.

      c. Lesson Tie-In. In the lesson tie-in you show the relationship of the problem you are presenting to other instructional problems which the students have received or will receive. It is easier for students to understand material if they can see the relationship of the present lesson to the overall block of instruction which they will receive. The fact that a problem is an introductory problem does not dismiss the need for a lesson tie-in. An accepted procedure for an introductory problem is to show the students the relationship of problems which they will receive in this instructional block to the accomplishment of the overall course objectives.

      d. Motivation:

         (1) Another mandatory element in the introduction is student motivation. This all-important step cannot be slighted. In stimulating your students, it is not necessary to "fire" them up as some people might think, but it is absolutely necessary to insure that the students know why they are receiving this instruction. They must appreciate the importance of the material being presented. You will never be assigned to teach a class that is not important; unimportant materials are not presented. The importance of this subject then, must be communicated to the class.

         (2) One of the ways in which you can arouse motivation is to relate this lesson to the skill which the student will eventually perform. In doing this, however, you must be careful not to cite absurd situations.

         (3) Another method of motivating is to appeal to the pride of the class or make reference to the competitive marksmanship spirit that this class should possess. You can also arouse motivation by relating this instruction to goals which appeal to the personal and professional ambitions which virtually all student-shooters possess.

         (4) A type of fear motivation (which may not stimulate long-term retention) is the statement that the students will be held responsible for knowing enough of this material to pass an examination.

e. Scope:

(1) A mandatory element of the introduction is the statement of the scope. Here you tell the students specifically what you and they are going to accomplish during this period. You inform them of the learning which they should acquire during the period. This statement of the scope will be derived from the lesson objective which you selected when you initiated your lesson planning, but it will be more specific and developed in more detail than was that lesson objective.

(2) After you have stated and discussed the scope, the student should have no questions as to what will be accomplished during the ensuing period. This discussion of the lesson objective(s) is a very important part of the general student orientation for the period and must not be hurried. Here you explain how the objective will be accomplished by giving the students an overview of what points will be covered in this lesson. The students can adjust mentally to how the lesson will be developed and thus be able to follow more easily.

(3) Some educators call this "the whole-part-whole method." In the introduction the student sees the big picture, then you develop the parts which make up that big picture in the body. Once again you wrap up these parts with another (but now a fuller) understanding of the "whole" as the lesson conclusion.

f. Sequence of Presentation:

(1) Any or all of the elements in the introduction may be combined. You are not restricted to presenting these parts in the order in which we have discussed them. Of course, you must gain attention before you present the other elements. Some instructors prefer to develop the motivation step immediately after gaining attention. They argue that it is good to motivate while interest is high. Present them as you will all of your instruction--in the most logical sequence, the way that seems best to you.

(2) Whichever sequence you employ, you must insure that before you go into the body of your presentation, the students understand what instruction is to be presented, how it ties into their instructional program, and why they personally need to know this. These are the three mandatory parts of the orientation.

g. Optional Portion of the Introduction. An additional element of the introduction is explanation of methods of instruction. Orient students on the methods of instruction only when multiple methods will be used either in this lesson or later in the block of instruction. For example, if you plan to use a training conference during a portion of the course and then use a demonstration during another portion and finally to require the students to perform, this should be announced to the students. This helps to get your students mentally set for what will follow. They will listen more carefully if they realize that they will be performing later in the problem.

5. Preparing the Conclusion:

a. The final part of the instructional planning is the lesson conclusion. This part of the period is unfortunately the part which is most frequently slighted. It is slighted because some instructors do not understand the need for an effective conclusion. They do not understand the real purpose of the conclusion. The conclusion, when properly presented, will point out the importance of the material which the students have learned and re-emphasize the main points. It will leave the students with the big picture of how these points all fit together to fulfill the lesson objective and what main ideas they should take away with them.

b. There are four parts to an effective conclusion: Retain student attention, summarize, point out application, and make strong closing statement. You do not need to present them in the order in which we will discuss them; you may combine and present them in the most logical or meaningful sequence.

(1) Retain Student Attention. When moving into your conclusion, be sure that you have the full attention of the students. This is the first step of the conclusion. A common place for interest

to sag is near the end of the period. Students know when a period or problem is about to be concluded and unless you insure that you have their attention, the effectiveness of your conclusion may be lost. One of the ways to maintain attention is by the effective use of a transition which will be discussed in detail in Chapter 6. Other ways of maintaining or raising student attention have been discussed in the paragraph on the introduction.

(2) <u>Summarize</u>. When you are sure that you have retained or stimulated the attention of the class, you then re-emphasize or summarize the main points of your lesson. Your statements or restatement of performance objectives, together with carefully selected key supporting material, constitute the lesson summary. Point out each main idea in the summary. Remarks such as: "in conclusion," "during this period we have covered," "let me see what we have discussed," "what have we learned today," are statements which through excessive use generally cause a loss of student interest. In approaching your summary, treat it as the third essential part of a good lesson rather than as a windup or just a way of dismissing your class. A good summary should challenge student imagination to use the main ideas presented in the problem; it is the climax of the learning situation. Effective learning must influence the students' future thinking on this subject.

(3) Point Out Application: In this portion of the conclusion show students when, where, and how they will use the materials learned. If appropriate, state the immediate student application of these principles during their course and also the future application. The application step in the conclusion closely parallels the motivation step in the lesson introduction. Then, you mentioned <u>why</u> it was important to learn this material; now, in the conclusion you show <u>how</u> students will apply what they have learned.

(4) Strong Closing Statement. Finally, prepare a strong closing statement. Carefully plan this statement to stress again the importance of the material which the students have learned. Rehearse this closing remark until you can deliver it with maximum effectiveness. Remember, the final impression you make upon your students is as critical as the initial impression. It is your final chance to drive home the importance of this learning. Make the most of it. Many good instructors make the final statement "the action step," what students should do about the principles or techniques they have just learned. Some conclude the lesson with an impressive demonstration. Here is a golden opportunity to impress the student with the importance of what you have taught him during the period. While you have his interest, drive home your message!

6. <u>Instructor's Teaching Notes</u>:

a. You never teach directly from a lesson outline. In the first place it is too detailed to permit you to orient yourself rapidly if you lose your train of thought. Secondly, as a written document it contains more formal diction that you would use in speaking. The essence of good teaching is to speak in a way natural to you. Therefore, you use the lesson outline as a source for preparing a personal series of teaching notes. These will be on the lectern for ready reference.

b. Most instructors find that good teaching notes are an efficient aid to presenting instruction. Simply knowing that they are available builds confidence in the beginning instructor. Do not use them unless you have to, for you may develop the bad habit of using them as a crutch and never rehearsing your presentation sufficiently. These notes should be organized in such a way that you can easily present your instruction without awkward stops to relocate yourself in the lesson. Make them large enough to be read at a glance.

c. Other instructors use key words or phrases as notes to cue them into the main portions of their lesson. Place the notes in a spot where they are readily visible for your use, but do not distract students. Effective instructors color code the significant parts of their notes to indicate student objectives, training aids, handouts, or other presentation cues.

7. <u>Prepare Instructional Aids and Handouts</u>. After the lesson has been planned you must decide the number of assistant instructors needed and then orient and rehearse them. If you are going to need advance sheets and/or new training aids, prepare for these materials early. You accomplish all this after your lesson has been planned. It is wise to prepare a checklist for your problem when you begin your lesson planning and as you accomplish each step in the overall planning of your problem, check off this item as completed.

8. __Rehearsing__:

a. Once you have planned and written out your lesson outline and established all of the problem requirements, you are ready to rehearse. The type of problem is a prime consideration in establishing your rehearsal schedule. It is possible to rehearse too much, just as it is possible to rehearse too little. The number of rehearsals which you conduct before presenting a problem will depend upon how well you have researched your problem and how familiar you are with the technique which you will employ.

b. As a minimum, three major rehearsals are required for each lesson. You should have a learning rehearsal which you conduct by yourself and for yourself. You should have an evaluation or murder rehearsal during which all of your training aids, handouts, and assistant instructors are used. The evaluation rehearsal should be attended by your supervisor and selected team members. As a result of this rehearsal, you will receive comments and constructive criticisms from experienced personnel and make revisions in your lesson plan for class improvement. Finally, you should have a full dress rehearsal using every device, method, and technique just as you will use them during the actual conduct of the class. You may have to conduct more than three rehearsals but the decision depends on you. You alone know when you are ready to go. Knowing your material thoroughly will reduce apprehension and stage fright on the day of actual presentation. Rehearsals are the only way to check your knowledge of the subject and build up and retain self-confidence.

9. __Common Errors in Lesson Outlining__:

a. Instructors involved in lesson outlining should be aware of the following common errors. Use this as a checklist to review and revise your draft outline.

b. General:

(1) Instructors fail to use the approved lesson outline format properly.

(2) They omit the necessary indentations.

(3) They fail to insert the necessary note on when the handout is issued or when a specific training aid is used. Sometimes they insert the note but do it incorrectly.

(4) They fail to write the planned questions at the appropriate places in the outline or fail to write out the expected student responses as supporting material.

c. Lesson Objective:

(1) Frequently, the lesson objective is vague or too broad. The "what to teach" is so big or general that it cannot be taught in the time allotted. Actually the instructor plans to teach only a portion of the objective in that period. For example, he states his objective as "working knowledge of the pistol" when he means, " a working knowledge of the cycle of operation of the pistol cal . 45 (service).

(2) Instructors frequently omit one of the three necessary elements of the lesson objective: type of students; level of proficiency expected or action to be taken; or pin pointing the portions of the total subject to be taught in this lesson.

(3) Frequently, the expected level of proficiency is too high for the time allowed for the training period. For example, it may be "to qualify" and the instructor is allotted only a fifty-minute period to this qualification training.

d. Lesson Introduction:

(1) The introduction may be too vague, too brief.

(2) Frequently the attention gaining portion is not written out. Sometimes there is only a brief reference. Remember, the outline should be complete enough so that another instructor could intelligently use it if such an emergency arose.

(3) Sometimes <u>motivation</u> is incomplete or not slanted specifically to the needs of this class.

(4) The <u>scope</u> may not be specific enough to orient students so they can get mentally set for the instruction to come. New instructors frequently think that the scope is simply a statement of the student performance objectives.

(5) The <u>transition into the body</u> is usually too abrupt or sketchy. This transition is very important since it should smoothly set the stage for absorption of the first student performance objective.

e. <u>Student Performance Objectives</u>:

(1) Frequently the performance objectives selected are not the major items needed to accomplish the planned lesson objective.

(2) They are incompletely stated.

(3) They are vaguely stated.

(4) They are not in complete sentence form - merely phrases.

(5) They do not pertain to the objective as it is stated.

(6) They are not repeated in the lesson body at the appropriate place as the outline.

(7) They are actually supporting material for another objective and should be so subordinated.

f. <u>Supporting Material</u>:

(1) It may be too brief to substantiate the sub-objective or insure student understanding of the point advanced.

(2) It may be incomplete with key ideas assumed or inferred. Another instructor would not understand the development. Remember, you should have much more supporting material in the outline than you use during the presentation. These are the examples in reserve in case this class does not grasp the point as presented initially.

(3) The supporting material may be unrelated to the objective to be learned. Perhaps the relationship was not shown by a good transition.

g. <u>Subsummaries</u>:

(1) Frequently, several extensive teaching points are developed without any planned subsummaries.

(2) These subsummaries are not adequately written out.

h. <u>Conclusion</u>:

(1) The instructor sometimes neglects to plan for a good <u>attention-retaining step</u>. He doesn't use the same degree of imagination as he did to gain attention initially. He fails to spend as much thought and effort in planning this step.

(2) The <u>application step</u> may not be specifically slanted to this class. Sometimes the instructor simply says, "Remember this and apply it when you get on the range."

(3) The <u>summary</u> may be too general. The instructor fails to reteach briefly the individual performance objectives and key supporting material. It may be too brief because he fails to allow sufficient time. He may over emphasize the later points and fail to mention the early ones.

(4) The closing statement may be good but the instructor overelaborates on it, explaining it in more detail and thus weakening its effects. To be good, a closing statement should be short. Brevity is the soul of effec veness.

ANNEX I TO CHAPTER III - LESSON PLANNING

LESSON OUTLINE FORMAT   Class Number (SAFS____)
                              Class Length (Min)
SUBJECT TITLE   Date (Month & Year)

LESSON OUTLINE

I. LESSON OBJECTIVE: To enable _____ student to _____
                              (Course)   (What
_____ _____
    Action)   (Subject Matter)

II. STUDENT PERFORMANCE OBJECTIVES: As a result of this instruction, students must be able to accomplish the following student performance objectives:

    A. Write out complete SPO (state a condition, action expected and standards to be met).

    B. Write out complete SPO.

III. ADVANCE ASSIGNMENT: (Study requirements.)

    A. <u>Gain Attention</u>: (Attract student interest.)

    B. <u>Orient Students</u>:

        1. <u>Lesson Tie-In</u>: (With previous or future lessons.)

        2. <u>Motivation</u>: (Why the student should learn this.)

        3. <u>Scope</u>: (What will be taught or done during this period of instruction.)

V. BODY:

    A. <u>First Student Performance Objective</u>: (Write in complete SPO A from Section II.)

        1. Material supporting performance objective (list all items of supporting material under the performance objective).

QUESTION: 2. What are the . . . . ?

ANSWER: . . . . Answer(s) to question.

        3. Supporting material.

        4. Supporting material.

NOTE: SHOW SLIDE #1 (TITLE OF SLIDE)

        5. Additional material supporting the performance objectives.

TRANSITION: Written before each SPO.

    B. <u>Second Student Performance Objective</u>: (Write in complete SPO B from Section II.)

VI. CONCLUSION:

    A. <u>Retain Attention</u>: (Revive student interest.)

    B. <u>Summary</u>: (Brief review of main points taught during this period.)

C. <u>Application</u>: (How can this instruction be used?)

D. <u>Closing Statement</u>: (A forceful, significant statement that will be remembered.)

ANNEX II TO CHAPTER III - LESSON PLANNING

ACTION VERBS ASSOCIATED WITH JOB-ORIENTED PERFORMANCE

ACCEPT	BALANCE	CORRECT
ACCOMPLISH	BRACKET	CRITICIZE
ACCOUNT	BRIEF	DEADLINE
ACHIEVE	BUDGET	DEBRIEF
ACCOUNT FOR	BUILD	DECIDE
ACT	CALCULATE	DECIPHER
ADAPT	CALIBRATE	DECLASSIFY
ADMINISTER	CARRY ON	DECODE
ADOPT	CARRY OUT	DEFEND
ADJUST	CARRY THROUGH	DELETE
ADJUDGE	CATALOG	DELINEATE
ADVANCE	CERTIFY	DELIVER
ADVISE	CHALLENGE	DEMONSTRATE
AID	CHARACTERIZE	DEPLOY
AIM	CHARGE	DERIVE
ALERT	CHOOSE	DESIGN
ALINE	CLASSIFY	DESIGNATE
ALLOCATE	CLOSE	DETECT
ALLOT	COMBINE	DETERMINE
ALTER	COMMIT	DETONATE
ANALYZE	COMMAND	DEVELOP
ANNOTATE	COMMUNICATE	DEVISE
ANSWER	COMPARE	DIAGRAM
APPEAL	COMPILE	DIAGNOSE
APPLY	COMPLETE	DICTATE
APPOINT	COMPLY	DIFFERENTIATE
APPROVE	COMPOSE	DIRECT
APPROXIMATE	COMPUTE	DISARM
ARRANGE	CONCLUDE	DISCLOSE
ARM	CONDUCT	DISCHARGE
ASSAULT	CONFIRM	DISCRIMINATE
ASSEMBLE	CONSTRUCT	DISENGAGE
ASSESS	CONSUME	DISPATCH
ASSIGN	CONTRIBUTE	DISPENSE
ASSIST	CONTROL	DISPERSE
ASSOCIATE	CONTRAST	DISPLACE
ATTACK	CONVERT	DISPLAY
AUTHENTICATE	COORDINATE	DISTINGUISH
AUTHORIZE	COPY	DIVIDE

DOCUMENT	INVENTORY	PREPARE
DRAFT	INVESTIGATE	PRESCRIBE
DRILL	ISOLATE	PRESENT
EDIT	ITEMIZE	PRESIDE
EMPLACE	JUDGE	PREVENT
EMPLOY	LEAD	PROBE
ENCODE	LOCATE	PROCESS
ENCIPHER	MAINTAIN	PROCURE
ENFORCE	MAKE	PRODUCE
ENGAGE	MAKE USE OF	PROGRAM
ENTER	MANAGE	PROJECT
ENVELOP	MANIPULATE	PROPOSE
ERECT	MAP	PROSECUTE
ESTIMATE	PARK	PROVE
ESTABLISH	MAXIMIZE	PUBLISH
EVACUATE	MEASURE	PUT
EVALUATE	MOBILIZE	QUALIFY
EXECUTE	MODIFY	RAISE
EXERCISE	MONITOR	REBUILD
EXHIBIT	MOUNT	RECOMMEND
EXPLOIT	MEND	RECONNOITER
EXPOSE	MANEUVER	RECONDITION
EXPRESS	NAVIGATE	RECORD
FILE	NOTIFY	RECOUNT
FIND	NUMBER	RECOVER
FIRE	OCCUPY	REFER
FORECAST	OPERATE	REFUTE
GAUGE	ORDER	REJECT
GOVERN	ORGANIZE	REGISTER
GUIDE	ORIENT	REGULATE
IDENTIFY	OVERHAUL	REINFORCE
IMPOSE	PARAPHRASE	RELATE
INDEX	PATROL	RELAY
INDICATE	PERFORM	RENDER
INFORM	PICK	REORGANIZE
ISSUE	PLAN	REPAIR
INSTALL	PLOT	REPLACE
INSTITUTE	POINT OUT	REPLY
INSTRUCT	POSITION	REPORT
INTERPRET	POST	REPRESENT
INTERROGATE	PREDICT	REQUEST

REQUISITION	SKETCH	TOTAL
RESERVE	SLATE	TRACK
RESOLVE	SOLVE	TRAIN
RESPOND	SORT	TRANSACT
RESTORE	SPECIFY	TRANSCRIBE
ROUTE	SPOT	TRANSLATE
SALVAGE	STABILIZE	TRANSPORT
SCAN	STAGE	TRANSPOSE
SCHEDULE	STANDARDIZE	TRAVERSE
SCRUTINIZE	SUBSTANTIATE	TROUBLESHOOT
SEARCH	SUPERVISE	TYPE
SELECT	SUPPLY	USE
SEPARATE	SURVEY	UTILIZE
SERVICE	SET	WARN
SELECT	TABULATE	WIELD
SET FORTH	TAKE (ACTION)	WORK
SIGHT	TAKE (STEPS)	WRITE
SIGN	TALLY	
SIGNAL	TEST	

CHAPTER IV

EFFECTIVE SPEAKING

A. GENERAL. Clear Communication is Essential: Why should you as a shooter be concerned with your ability to speak? You've been speaking for years and have had no difficulty in making yourself heard. But being heard is not enough. No matter what your purpose in speaking, you, as a contributing member of society, want to make certain your thoughts and ideas are understood, not just heard. You want people to understand your words as the first step toward a richer goal: that your thoughts and ideas be believed, felt, learned, and remembered. All this is true of the person who realizes his social responsibility. But as a marksmanship instructor, it is not only desirable that you make people believe, feel, learn, remember and act, it is essential that you influence student shooters to be champion shooters. You must make ideas vivid in your student's mind --- so vivid that a chain reaction of new understanding is jarred loose, and the student takes your lead and thinks for himself. You must make him feel the power that new knowledge gives him: the power of confidence and assured success. You must teach him what you know! And how will you do this? There is only one answer--<u>by learning to</u> communicate effectively! You must speak with your entire being, your voice, your body and your mind if you are to speak effectively.

B. THE VOICE - THE MASTER CONTROL OF SPEECH. Teachers of speech have compared the voice to an exceptional type of musical instrument. Like other instruments, it has pitch, range and volume; it can communicate feeling as well as sound. However, with the instrument of your voice you can convey emotions such as rage, reverence, amusement, love and seriousness. When voice techniques are appropriate for the gravity or impact of the words to be spoken, the combination produces a potent communications tool.

1. <u>Articulation</u>. Good articulation is the combining of distinct sounds, that is, the syllables, to form distinct words.

    a. There are two basic types of speech sounds: vowels (a, e, i, o, u, and sometimes y) and the remainder of the alphabet, the consonants. Articulation, for our purposes, is defined as the production and combination of separate sounds to produce intelligible speech. It involves the articulators, the tongue, teeth, lower jaw, and lips acting on the breath stream to form specific sounds. You achieve loudness of speech through the vowel sounds; you achieve intelligibility mainly through the consonant sounds. Hence, the maxim: "The vowels give beauty, the consonants give clarity." Say the vowels to yourself and notice that the breath stream is shaped to obtain the difference in sound, without interruption. Now say the consonants and notice that each one involves constriction of the breath stream sometimes to the point of interruption by the tongue, lips, or throat muscles.

    b. To be a good speaker you must have as your goal distinctness in articulation. You must avoid slurring and mumbling of syllables. The indistinct syllable creates the indistinguishable word. With the exception of mental speech difficulties such as stuttering, the most common cause of poor articulation is laziness!

    c. Practice saying words as you know they should be said. "Explode" the p in pull, the <u>t</u> in talk, and carry this through to your everday speech. Relax your throat, tongue, jaw and lips and use them to clip off crisp sounds and blend sounds into clean, bright, clear words; don't be lip lazy!

    d. Articulation can be improved by being conscious of how you say words. Clear speech will establish efficient communication with your audience.

2. <u>Pronunciation</u>. Good articulation can produce a word made up of distinct syllables but the combination of sounds may not result in a properly pronounced word. Correct pronunciation is a necessity if you desire to be understood.

    a. Pronunciation is the sounding, or articulating, of a word with the accent on the correct syllable in accordance with good English usage. The principal difficulties associated with pronunciation arise from sheer ignorance and the influence of regional accents or colloquialisms. Ignorance of correct pronunciation can be overcome by acquiring the habit of listening carefully to cultivated speakers and by using the dictionary when in doubt.

If you say jist, git, gonna, whot, whatcha, hafta, or any of the many other commonly heard "easy" ways of saying words, your speech is sloppy and you are faced with a cleanup job. Just pay attention to what you're saying, as you expect other people to do; remember, they won't if you don't. You must convey the intended meaning of all the words you speak.

    b. If you have a regional accent, such as a Southern drawl or a New England twang, don't try to eliminate it - make the most of it! It's part of your personality. Just don't let it get out of hand to the extent that people from other parts of the country can't understand you. A small degree of accent is pleasant to the ear and adds interest and personality to your speech.

    c. Practice in placing the accent on the proper syllable will help you form the habit of pronouncing words correctly.

  3. <u>Grammar</u>. The incorrect choice of words and faulty sentence construction can reduce a listener's understanding of what you say.

    a. Grammar is concerned with the correct usage of the spoken or written word to express an idea. A sense of rightness or wrongness of grammar comes more from familiarity with good literature and association with well-spoken people rather than from memorization of academic rules and principles of the English language. Above all, you must want to speak correctly. You must be mentally alert to what you are saying and make conscious and continual effort to maintain an acceptable level of grammar. Association with persons using poor grammar insidiously affects your speech unless you guard constantly against the tendency to slip into careless grammar.

    b. Glaring errors, such as "him and me is going", "I seen", "he give" and " it run" should be attacked immediately and corrected. But good spoken grammar must be based primarily on a long-range program. You must acquire a continuous consciousness of right and wrong usage, not for a period of weeks or months, but for a lifetime. Read as much as you can of the world's good writing, listen closely to people whose speech is cultivated and make use of books on grammar and correct usage to answer any special questions which arise. Then practice good grammar. Your speech, and your personality, will show a definite change for the better.

  4. <u>Rate.</u> Rate of speech is the speaker's speed of delivery. It helps the voice to convey the speaker's emotions. The average rate of speech is 125 words a minute. However, each person has his own "normal" rate in conversational speech. This is, to a large extent, a result of nationality and geographic background - the environment in which you grew up, the section of the country where you spent the most time. There is no standard, proper rate of speech. You can't afford to speak at a slow, plodding rate that puts your listeners to sleep; neither can you rattle off words so rapidly that they run together and cannot be understood. As a general rule, speak fast enough to be interesting, slow enough to be understood. Just as a good baseball pitcher keeps the batter on his toes with a slow ball or a fast ball, take advantage of a vocal "change of pace" to hold the interest of your audience. Let your rate of speech be governed by the complexity of the thought, idea, or emotion you are trying to communicate. You can give motion to word-pictures by a rate of speech appropriate to the picture described. For example, use a fast rate for joy, excitement, vigorous action. Depending on your personality and the idea you wish to give to your listeners, add emphasis by either slowing or speeding your rate. Interest is added by a variation of rate. You can place emphasis on the important points you want to bring out.

  5. <u>Common Rate Difficulties and Suggestions for Improvement:</u>

    a. Slow, Ponderous Rate: Force yourself to think faster so that you may also force yourself to speak faster. Plan to express ideas rapidly or slowly, as suits the mood of the idea. Read aloud and interpret the meaning of the words by the rate at which you speak them. Use a recorder to listen to your rate and then do it over to cut down the total time.

    b. Fast, "Machinegun" Delivery. Curb your impatience to blurt out your ideas. Take time to be clear. Force yourself to slow down. Recognize the listener's need to absorb your ideas, and give him time to do so by saying words clearly and pausing longer between ideas to let the ideas "sink in." Read aloud and observe the marks of punctuation. Express the meaning of the words carefully at the rate which fits your interpretation. Taking care to enunciate more precisely will generally slow your rate.

c. Halting, "Choppy" Rate. Concentrate on speaking complete ideas or sentences. Take a deep breath before you begin a sentence and do your breathing between, not in the middle of ideas or phrases. Sometimes a choppy rate is the result of tenseness or nervousness. Work off excessive energy through physical activity just prior to talking or by exchanging pleasantries or comments with persons around you before you begin your presentation.

6. <u>Pauses.</u> Words which express a single idea are grouped together into phrases.

a. In writing, we use periods, commas, question marks, and other punctuation marks to separate thoughts and ideas and to give the desired meaning and emphasis to our words. In speaking, the same functions are accomplished, to a large degree, by pauses. This pause is the punctuation mark of speech. Pauses are also used to gain certain effects: humorous, dramatic, thought-provoking. Pauses are also a means of giving subtle emphasis to the meaning of an idea.

b. The proper use of pauses accomplishes four things. Listeners are able to absorb ideas more easily; you get a chance to concentrate on your next point; you give emphasis, meaning, and interpretation to your ideas; and you get a chance to breathe!

7. <u>Common Pausing Difficulties and Suggestions for Improvement:</u>

a. Not Enough Pauses. Begin by reading aloud something you like and force yourself to pause between ideas; at periods, commas, and other punctuation marks. Try to adapt the attitude of the artist who makes a few brush strokes then steps back to evaluate the results.

b. Too Many Pauses. This difficulty is caused usually by a lack of knowledge of the subject, failure to organize material thoroughly, or inadequate rehearsals. The speaker may not be certain of the sequence of his ideas and so must take the time to organize them while on the platform. Study your material, organize it on paper, then rehearse until your thoughts and words flow smoothly. Thorough familiarity with the subject matter increases personal fluency.

c. Over-Use of "uh". Don't be afraid to pause! Silence, properly placed in the flow of speech, is more effective than any number of words; it certainly is more effective than meaningless, gutteral sounds. Use the same techniques suggested for b. (Too Many Pauses) and leave out the "uh". Snip off these "speech whiskers" to improve your effectiveness.

d. Pausing in the Middle of Ideas. Here, your problem is to coordinate two necessary functions: thinking and breathing. Think of your listener as you speak; concentrate on making each idea clear by pausing only between ideas. Then, coordinate your breathing with the phrasing of your words - short pauses between phrases, deep breaths or long pauses between ideas.

8. <u>Inflection:</u>

a. Just as musical notes become melody when they are arranged in different relative positions on the musical scale, your voice becomes more interesting and your words more meaningful when you make use of changes in pitch; this is <u>inflection,</u> or "vocal variety."

b. Here is an example to illustrate how inflection on different words changes the meaning of a question. Say the question to yourself, raising your pitch on the underlined words, as indicated.

Figure 4-1. Inflection - Key to Expression.

c. Inflection is the master key to expression of all kinds - emotional, persuasive, convincing. With it, you can move an audience to tears or laughter, imprint your ideas indelibly on their minds; without it, you will put them to sleep.

d. Like the pause, inflection is a way of punctuating speech; it puts the question mark at the end of a question, makes a statement of fact more positive, and helps to put an exclamation mark at the end of a strong statement. Inflection is the principal difference between just saying words and speaking ideas with meaning.

9. <u>Common Inflection Difficulties and Suggestions for Improvement</u>:

a. Monotone (No Inflection). We have said that inflection conveys feeling and meaning. But this is a situation similar to the old problem of which came first, the chicken or the egg, because feeling produces inflection. You must be willing to show your feeling about what you say. This is what is meant by personalizing your speech. Read aloud to a listener and practice using inflection to show him the meaning of the words. First, analyze "What is the emotion or feeling of this selection," then, convince yourself of the necessity for you to communicate this emotion through your inflection. A tape recorder is an excellent device to improve your inflection because here you must communicate emotion entirely through the voice; there are no gestures or visible expressions to help.

b. Misplaced Inflections. Generally speaking, use downward inflection at the end of sentences to express positiveness and conviction. Be careful of downward inflection within the sentence itself, for each downward inflection gives a sense of finality to the thought phrase, and creates a mental break in the listener's thoughts. Use slight upward inflection within the sentence to indicate the thought is not yet complete; this serves to bind ideas together and to give unity to the thought. Give an upward inflection at the end of sentences only when you are implying question or uncertainty.

10. <u>Force</u>:

a. Forceful speech combines the volume or carrying power of the voice with the demonstrated vitality and strength of conviction of the speaker. It includes the proper placement of stress or emphasis on key words and phrases. Like rate, pauses, and inflection, force is a way of conveying feeling, giving meaning, or adding emphasis. Yet, unlike the three factors previously considered, force cannot be set apart distinctly. It involves rate, pauses, and inflection <u>plus</u> carrying power, fullness of tone (or body), and proper regulation of loudness.

b. Listeners will not respond properly to constant banging of their ears by a speaker who shouts and is insensitive to their feelings. Neither will they be convinced by the cool, detached manner conveyed by a speaker who is consistently calm, quiet, conversational, or patronizing. In order to communicate, we must awaken ractions and feelings in our listeners.

c. Knowledge of subject and a firm grasp of the sequence in which you plan to present ideas will enable you to project yourself into good mental contact with your audience, leading their thoughts now by calmness, then driving home a point with power, then letting pure silence underline the significance of your words.

11. <u>Common Difficulties with Force and Suggestions for Improvement</u>:

a. Lack of Volume. Concentrate on making the person farthest away hear you. Practice in a classroom and convey your meaning to an imaginary someone seated in the last row of seats. Most people can obtain this carrying power by raising the pitch of their voice and adding more than their normal amount of nasal quality. Above all, do not shout! Do not force your voice. Keep your chest and throat relaxed because if your throat muscles tighten too much, your pitch will rise too high. Take a deep breath and sound off. Volume comes from the diaphragm not the throat.

b. Dropping Volume at End of Words or Sentences. This is usually the result of incorrectly associating a drop in volume with downward inflection. Develop the habit of paying attention to the sound of your own voice and you will be able to judge whether you are being heard. Practice lowering the pitch of your voice without dropping the volume. Record your voice so that you can sit back and analyze how you sound to others. Read aloud, and concentrate on projecting every word in a thought or idea to an imaginary listener seated in the rear of the room.

c. *Failure to Give Emphasis to Main Points and/or Key Words.* Know your subject! Identify your main points; then practice putting them across with a spurt of energy in your voice. You can't expect your listeners to do the work of sorting out the main ideas from the subordinate ideas. You must interpret for them by stressing key words and phrases with volume, pitch, rate, and pauses.

12. *General Recommendation for Speech Improvement:* Although various specific means of correcting common difficulties have been indicated in this section, there are four ways in which you can improve all aspects of your speaking voice:

a. First, use a tape recorder periodically to hear how you sound to others. This will enable you to make on-the-spot corrections of errors which have gone unnoticed.

b. Secondly, read aloud to one or more persons. If you have children, or a patient wife, or both, they will make a good audience. Choose selections you particularly enjoy, interpret the meaning and communicate this meaning by the way you use your voice.

c. Thirdly, listen closely to polished speakers on radio and television (Eric Sevareid, Douglas Edwards, Howard K. Smith, Walter Cronkite, Chet Huntley). Do not try to make your voice or manner of speaking exactly like that of one of these men, but notice how they use their voices to give meaning to their words and emphasis to their ideas.

d. Finally, listen to yourself! Acquire the habit of evaluating constantly how you use the speech factors listed in this section and you cannot help but improve your effectiveness.

C. THE BODY: THE BODY SPEAKS. With improved speech comes the realization that the voice alone is not enough. If the voice communicated perfectly, television never would have become popular. Seeing a person as he speaks adds immeasurably to the listener's appreciation of and receptiveness to ideas, if the speaker uses his body to reinforce, emphasize, and clarify his ideas as he speaks them. Skilled pantomine actors are able to communicate every human emotion through the clever use of body, gestures and facial expressions alone.

Figure 4-2. The Body Speaks.

1. *Posture.* Posture is the speaker's stance. Your position should be comfortable without being slouchy, erect without being stiff. Let the weight of your body rest slightly more on the balls of the feet rather than on the heels. Stand erect with the assurance of command! The way you stand is an outward manifestation of an inner attitude. It will definitely affect the way your students receive your instruction.

2. *Movement:*

a. Movement is the motion of the whole body as it travels about the presentation area. One effect of movement is to attract the attention of the listener; the eye instinctively follows moving objects and focuses on them. Movement can greatly assist in conveying the thought of the speaker. Transitions from one point in the speech to another often can be indicated and made emphatic merely by shifting the weight from one foot to the other or by lateral movement of a step or two. Such a movement is an informal signal that "I am finished with that point; now let's turn our attention to another." Always start lateral movements with the foot on the side toward which you are going; this avoids awkward crossing of the feet. Then walk a step or two, naturally, in that direction. Forward and backward movements imply the degree of importance which you don't want your audience to miss; this emphasizes the point. A backward step suggests that you are willing for your listeners to relax and let the last idea take root before your go on to the next one.

      b. The basic rule is moderation: Don't remain glued to one spot, and don't keep on the move all the time. When you do move, move briskly and with purpose. As your skill and experience increases, you will find your movement becoming less obvious and more meaningful, and you will learn to modify the degree of movement to make if natural and meaningful.

3. Gesture:

      a. Definition. A gesture is a natural movement of any part of the body to convey a thought or emotion, or to reinforce oral expression. Your arms, hands, and body are your principal tools of gesture. Practice gestures as natural parts of your speaking manner, but never rehearse specific gestures to use at definite points in your presentation; they should arise spontaneously from enthusiasm conviction and emotion. There are two basic types of gesture: conventional and descriptive.

      b. Conventional Gestures:

      (1) Pointing. The index finger has been used universally to indicate direction and to call attention to objects at which it is pointed. Use it to reinforce an accusation or challenge by pointing directly at the audience or at an imaginary person.

      (2) Giving or Receiving. If you were to hand someone a sheet of paper or to hold out your hand to accept one offered to you, the palm of your hand would be facing upward. This same gesture is used to suggest the giving of an ideas to the listeners or to request that they give you their support. Sometimes it is combined with the pointing gesture - the idea is held out in one hand while the other hand is used to point to it.

      (3) Rejecting. A forward movement of the arm with hands raised and palms turned toward the audience can be used to reinforce such statements as: "This cannot be tolerated," "Hold everything," or "Just a minute."

      (4) Clenched Fist. This gesture is generally used with expressions of strong feeling such as anger, power, or determination. The clenched fist may be used to emphasize such statements as: "We must put every ounce of our energy behind this plan." Sometimes the clenched fist is pounded into the other hand or upon the lectern or table.

      (5) Dividing. Move the hand from side to side with the palm vertical to indicate the separation of facts or ideas or to divide the audience into imaginary opposing factions.

      (6) Restraining. Extend your hand at shoulder height, palm facing outward and downward. This is the gesture to use when saying, for example: "Now, take it easy," "Just a minute," or "We're coming to that."

      c. Descriptive Gestures:

      (1) Descriptive gestures portray an object or illustrate an action. The speaker describes the size, shape, or movement of an object by imitation. A vigorous punch is shown by striking with the fist; height, by holding the hand at the desired level; speed, by a quick sweep of the arm; complicated or humorous movement, by pantomine as you describe it. The curves of a beautiful girl are described with both hands outlining the lateral figure. Churchill's "V" formed with two fingers was symbolic of victory. Even teenagers graphically describe a "personality" by forming a square with their two hands.

      (2) The gestures described above will give you an idea of what might be done with gestures. Your gestures will depend to a large extent on whether your personality is vigorous and dynamic or calm and easygoing. But, regardless of your personality, gestures will add to the effec-

tiveness of your speech if you <u>relax</u> your shoulders, arms, and hands and concentrate on communicating to your audience the meaning and importance of your ideas. When the gesture is natural, it is effective. If the gesture is artifical, posed or strained it detracts rather than reinforces.

    4. <u>Facial Expression.</u> A facial expression is a type of gesture,. but it is considered separately here because of its importance. If you are to sway people's minds, inspire them, or even interest them your face must show what you are feeling and thinking. The most common fault in facial expression is the "dead-pan" or total lack of expression. You can overcome dead-pan by looking over your audience until you find someone who is smiling or has a naturally pleasant face. Smile back at him and you will find **yourself** warming up to all your listeners. Unconsciously, your face will take on a more meaningful expression. Another common expression difficulty is that of the constantly intense expression, usually manifested by a frown. Overcome this by relaxing all over, then use your intensity only on key ideas. Finally, remember that you are neither a wooden Indian nor a clown - you are a human being; and the more natural and human you appear and act, the more you will influence your listeners. The presentation area is no place for a poker face.

DISTRACTION CAN BE CAUSED BY UNNATURAL OR EXAGGERATED PLATFORM BEHAVIOR SUCH AS ....

Figure 4-3. Improper Platform Behavior

5. Avoid Exaggerated Platform Behavior. Don't ham it up! Skilled pantomine actors are able to communicate every human emotion through the clever use of body movement and facial expression. Don't put on an act, just accent your words with the appropriate body movement.

Figure 4-4. Improper Platform Behavier. (Continued)

D. THE MIND. THE PROPER SPEAKING ATTITUDE

1. General:

    a. We cannot think or conjure up images in the mind except in the form of words or objects that have already been classified and typed by a word. We think in terms of words. Every word in the English language has greater or lesser emotional overtones to each listener, depending upon his background. Words like "mother, home, school, hometown" evoke strong emotional reactions in your listeners. Semanticists know the emotional impact of words. The dynamic speaker is aware of word power and a master at using the forceful word at the right time.

    b. Speech is a means of communicating ideas and thoughts. Sometimes these ideas and thoughts originate in the mind; sometimes they start with outside stimuli which are evaluated and interpreted by the mind and then communicated. In speech you do more than report facts; you also interpret facts and express opinions. Every word has two effects: what it _means_ or denotes and what it _implies_ or connotes. If speaking were merely the production of objectively precise language with no emotional implications nor possible misinterpretations, listening would be an unpleasant and boring task indeed! No speaker would communicate anything of his personality. He would not reveal his feelings, his laughter, his tears. His ability to stimulate, convince, persuade, or interest his audience stems primarily from the emotional effect he produces; not necessarily emotion in the sense of joy, sorrow, love and hate, but, more often, emotion in the sense of strong belief, conviction, earnestness, quiet sincerity, or dynamic enthusiasm. Your speech carries with it emotional overtones, how you feel about what you say as well as the emotional impact of your speech upon others. Thus, emotion is the indicator of how a person feels about all that surrounds him - or more simply, it is the indicator of his _attitude_. His attitude affects the words he chooses to use. There are four specific indicators of a proper speaking attitude: sincerity, confidence, enthusiasm, and humor.

2. Sincerity:

    a. Sincerity, from the speaker's point of view, is the apparent earnest desire to convince your audience of the truth and value of your ideas. There are two sources of sincerity. (1) Your personal, intense belief in your subject; (2) Your belief in the value of your subject to your listeners. The first of these sources is ideal since all personal belief which is intense enables the sincerity to flow forth naturally from every word or gesture of the person. The second source is more rational than emotional. You are convinced that this teaching material is extremely valuable and you present it in an honest and forthright manner. You do not rely on gimmicks or questionable reasoning to make your presentation look good. The word sincerity means "without wax" (sine/cera) and is derived from the Roman sculptor's trick of covering up chiseling errors by filling them with wax. You won't pad your presentation if you develop belief in the value of the material to the students. You'll present it simply and honestly.

    b. Assuming that your sincerity originated from one of the sources described, you must show that sincerity! If you appear to believe in what you say, you have taken a major step in convincing your students of the importance of the subject. Sincerity shows in a number of ways: directness of manner, facial expressions, clarity of explanation, the proper combination of humility and authority, effective use of the voice and body to reinforce and emphasize ideas. Regardless of the source of sincerity influencing you, no matter how your personality dictates that you show your sincerity, the important thing is that your students see, hear, and feel your belief in what you say.

3. Confidence:

    a. Confidence is a personal attitude of a feeling of assurance, a belief in your ability to perform a task well. In order to be confident, you must have three basic prerequisites: knowledge of your subject, belief in your ability to speak, and the power to control nervousness or "stage fright." Knowledge of subject you obtain through research and study. Belief in your ability comes from rehearsal and experience. Both of these requirements are entirely up to the individual to accomplish in his own way. But the third factor requires further explanation.

b. Control of obvious stage fright increases with proper practice. Steady, regular breathing has a calming effect on the whole body. Relaxing the muscles of the body as much as possible will help to maintain poise and more complete control. Some speakers find that purposeful walking or moving about gets rid of stored-up nervous energy and brings back normal balance and ease. Others conduct a short rehearsal or "warm-up" just before their presentation by running through their introductory remarks. Remember that the problem is control of a natural reaction. If you feel no stage fright or tingling anticipation, you are not taking your job seriously enough to accomplish it effectively.

c. If you stand erect, move purposefully, look your listeners in the eye, and let your ideas flow freely and clearly without stumbling and awkward hesitation, you will appear confident to your audience - and, best of all, you actually will be! Rehearse your introduction thoroughly, for stage fright is usually strongest in the first few minutes of a presentation. If you get off to a good beginning, tension will soon disappear.

4. Enthusiasm:

a. The word enthusiasm was mentioned previously in describing other aspects of effective speaking. Enthusiasm is the outward manifestation of sincerity and confidence. From the speaker's standpoint, enthusiasm is defined as strong personal excitement or feeling about a cause or a subject.

b. Enthusiasm is not shouting; it is not affected, over-dramatic speech; it is not waving of the arms and leaping about on the platform! Rather, it is the way you show your belief in your subject. You can, indeed you must, be enthusiastic - but remember - each in his own way! Vigorous, dynamic persons will show enthusiasm by brisk, energetic movement, sweeping gestures, a rapid rate of speech, widely-varying inflection, and plenty of vocal force. Instructors of a more subdued nature will move and gesture with less energy, speak in more measured tones, use force only on the key words and ideas, make more use of the pause for effect, and maintain a calm, pleasant, but confident and authoritative manner. Others, probably a majority, will be enthusiastic by combining various characteristics from these two extremes. Don't hide behind a pretense of dignity that takes the form of stiff, deadly monotonous mouthing of words. Loosen up - mentally, emotionally, physically! Stop at nothing, within the bounds of common sense and decency, to persuade, convince and teach! How you show enthusiasm will by governed by your personal characteristics or traits, - but do it you must!

5. Humor:

a. It is entirely possible that an instructor who has the necessary attributes of sincerity, confidence, and enthusiasm may still be lacking in proper attitude to gain student attention, interest, and understanding. A person who lacks a sense of humor will give listeners the impression that he is unreal, inhuman, or very conceited. Humor is the quality that shows you are, after all, just another human being, that you have a warm, lively interest in all that goes on around you. Having a sense of humor does not necessarily imply the ability to tell funny jokes, although there is certainly a place in good instruction for humorous stories of the appropriate type. The thigh-slapping, belly laugh type of humor has its place, but not as a steady diet.

b. A more effective type of humor is spontaneous classroom humor. Take advantage of unexpected humorous classroom situations which sometimes arise by making a brief comment, a well-placed pause, or by a single open smile. Humor directed at yourself is one of the most effective types. Remember that one of the basic things that make people laugh is the sight of something or someone regarded as important or pompous appearing ridiculous.

c. In addition to decency, the only rule to follow in using humor is that of good judgment. Take care in directing humor at individual students for he or some of his classmates may resent this. Be sure your humor is good natured and lightly done. Clean humor is as American as the hot dog and will frequently assist student learning.

E. YOU HAVE THE TOOLS. Your goal of becoming an outstanding instructor and making the most of what you have can be accomplished by: (1) Understanding how to speak effectively; (2) Planning improvement; and (3) Practice - all day, every day, wherever you are and to whomever you speak. You have in your hands the means by which you can attain that goal. And while you are practicing, keep this always in mind: It is more difficult to be plain in speech than to be fancy. So be simple rather than artificial, be what you are rather than pretend to be what you are not.

CHAPTER V

CONTROL OF INTEREST

A. GAIN, MAINTAIN, AND RETAIN STUDENT INTEREST

1. There is probably none among us who has not had the following experience: You come into a classroom and take your seat. You are one of several students. As you look about you, you notice that the curtains are neatly arranged, the lighting is good, the instructional area is orderly, and there are no distracting influences. The classroom looks right for learning. An instructor begins. He is a young, energetic fellow with a wide variety of experience. He has a pleasing manner and an excellent speaking voice. The instructor begins the lesson with an introduction which stimulates your interest and suggests that something worthwhile is to follow. You recognize that the subject is one that is vital to you. You lean back in your chair, cross your arms and get set to learn. The instructor explains that there are five main points which you must learn during this period in order to be able to use the subject. He explains the first of these five points and discusses several considerations that influence an understanding of this first point. You follow the instructor through point #1 and are with him when he starts to explain point #2. Gradually it happens: The chart appears foggy. The instructor's voice fades away. You are no longer in the classroom. You are busy replaying the third shot on the seventh hole yesterday, and wondering whether the trouble was in your grip or in your back swing. Or perhaps you are buckling on your skis for a long run down a snow-covered slope. You are no longer learning. Periodically you snap out of this reverie and try hard to follow the presentation. But again and again you drift down through channels which offer far less resistance than the subject to be learned.

2. This situation is one which could happen to you as an instructor. A good lesson plan is not worth the effort if learning does not take place - and learning will not take place unless you can keep students interested in your presentation. "Interest" in instruction doesn't just happen; it must be carefully gained, maintained and retained in lesson presentation. To help you understand how you plan to control interest, consider the following:

    a. How does interest affect learning?

    b. What are interest factors?

    c. Is there a span of interest?

    d. How is control of interest applied to teaching?

B. HOW INTEREST AFFECTS LEARNING

1. You can make a shooter do many things but you can't force him to learn: A marksmanship student will concentrate, apply, and learn only when he is interested. Some student interest will be self-generated: he may enter your class with a desire to learn more about shooting, a personal interest in marksmanship, and an understanding of why the subject is important to him. To other students you must explain the importance of this subject. Even after student motivation, attention, and concentration has been accomplished, you must give direction, provide usable information and guide the learning in an interesting way. An instructional period is interesting because of what you as instructor say and do and because of what the student does mentally and physically.

2. At the beginning of the lesson you state the purpose of the lesson and emphasize why it is important for the student to learn this material in order to do his job effectively. The student is motivated--he realizes the need to learn, so he gives his full attention to what the instructor is saying and doing. You have given direction to the students attention by orienting him on the specific things to be learned.

3. By various methods which involve hearing, seeing, and doing, you present and develop the teaching points which the student must understand and apply to accomplish the learning.

4. Interest is the continued focusing of student attention upon a specific felt need until that need is satisfied. It is the instructor's responsibility to control student interest throughout the lesson. If the student loses interest, he no longer gives his full attention or concentration to the lesson and learning breaks down.

5. **How do you maintain student concentration and control interest?** By using presentation techniques which appeal to the student and lead to rapid and thorough learning. You must hold student interest by planning for interesting student learning activities--interesting hearing, seeing, and doing. These interesting learning activities are called Interest Factors.

B. INTEREST FACTORS

1. Interest Factors are not Gimmicks. Interest factors are the presentation methods and techniques used to gain, maintain or increase student interest in instruction. These are not gimmicks thrown into instruction merely to startle students or gain momentary attention. Interest factors are ways by which the instructor communicates his teaching points and supporting ideas - they must assist the presentation. To be effective, interest factors must help to maintain attention and concentration on the material being presented. Analyze different interest factors according to the physical processes by which they lead to learning: hearing, seeing and doing.

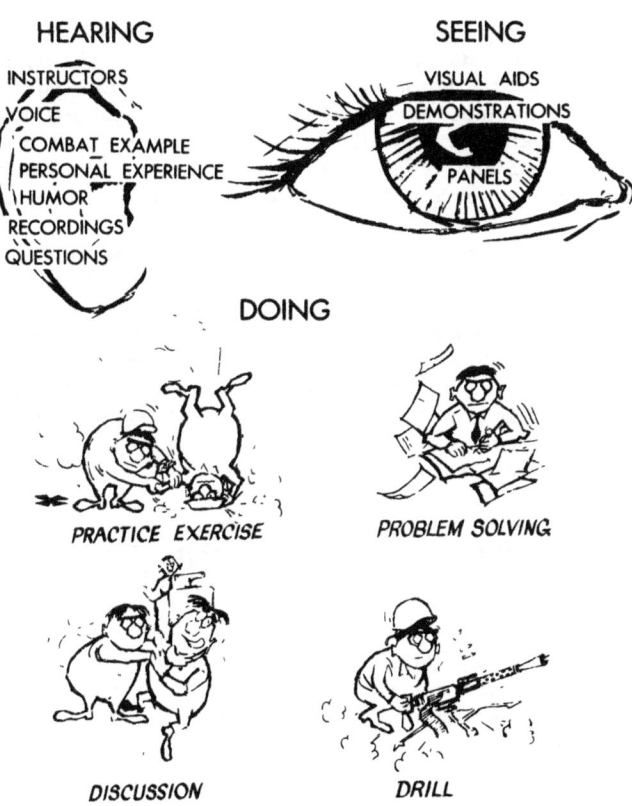

Figure 5-1. Control Class Interest.

2. <u>Hearing</u>. The Introduction to this chapter described how the instructor talked for the entire period. He failed to control interest because continuous verbal explanation is not interesting. Explanations can be and must be brought to life and made interesting by combat (or historical) examples, personal experience, humor, recordings, and questions.

a. Instructor's Voice. Your voice can be an excellent interest factor or a boring feature of your instruction. Radio and television announcers, commentators and speakers make their living through skillful voice techniques. They captivate listener interest by varied inflection, significant pauses, clear articulation and pronunciation, and skillful variations in rate and volume. A good portion of your instructional success will depend on your voice delivery. You may possess all the knowledge in the world but unless you communicate this knowledge in an interesting, dynamic manner you are an ineffective instructor. The mere sound of your voice may gain interest initially but how long will your voice hold interest? If you speak monotonously or mumble or swallow your words, you will lose your student audience within a few minutes. No student will struggle to understand you no matter how important your lesson may be.

b. Personal Experiences. Personal experiences lend the weight of authority to the points you are making. They may be taken from your own experience or may be quotations from other sources. To be interesting and effective for learning, personal experiences must be based on valid authority. Personal experience must be based on impressive experience or observation over a long period. You must exercise care in giving personal opinion to be sure that your experience in the field which you are discussing is sufficient to make such opinion valid. This limitation does not apply to factual experience, particularly that encountered in combat. Factual personal experiences are just that: facts that you have experienced.

c. Humor.

(1) Humorous stories and illustrations relax students and help them to remember the points emphasized by humor. Marksmanship instruction deals with deadly serious ideas. Unrelieved tension leads to the mental ward. Use humor to relax student tension momentarily. It will assist students to maintain interest in the subject. Don't restrict humor to joke telling alone, include skits, demonstrations, cartoon slides, and training films. Capitalize on spontaneous humor when it arises in a teaching situation, as from a funny, unexpected reply during a discussion. The instructor who displays a keen sense of humor has won most of the battle for control of student interest.

(2) Humor must be clean and must be related to the instruction. The group that gathers behind the club house to exchange dirty stories is a volunteer group. Members may leave at any time when they think they've had enough. Students in class are a captive audience; they do not have the option of leaving when foul or obscene jokes are told. For this reason, if for no other, the instructor should respect the rights of all his students and should use only humor which is clean. He should not ridicule any race, religion or nationality. To do so alienates those students concerned and others sympathetic to those offended. From another standpoint, the instructor must set a high standard in his humor to retain the respect of students. Telling dirty stories is a cheap appeal for popularity with the coarsest members of the class.

(3) Humor will not be used solely for entertainment. Humor must teach or it may distract, destroying the student's concentration on the subject. Tie in humor very closely with a specific idea in your lesson so that when the student recalls the humor, he also remembers the point you made.

d. Recordings. Recordings add realism and variety to your presentations; they bring into the classroom voices and sounds not otherwise available. Use recordings alone; or in conjunction with skits, demonstrations, or projections. Bring into the training area the voices of such leaders as MacArthur, Eisenhower and Churchill. Record a conversation and thus avoid the need for many assistants to be physically present to act out such a conversation. Recordings are easy to prepare and use.

e. Questions. Thought-provoking questions force students to concentrate on the instruction to prepare an appropriate answer if called upon. There are several types of questions you can use, depending on your purpose. These are discussed in Chapter 6. Generally speaking, the best questions are those which stimulate student analysis of a situation and require him to reason through to a second solution. The effect of a question is to alert the student to the possibility that he may be called upon. He concentrates on the material being presented so as not to be caught unaware and perhaps embarrassed by not knowing the answer to a question. Therefore, a good question should be asked in such a way as to give most students ample time to formulate a reply. If you begin your questions with "What," "Why," "How", etc., you gain the students' attention and they are alert to the critical part of your question. Don't put these pronouns at the end of the question.

3. _Seeing._ "A picture is worth a thousand words." This is an old but true saying. Some students are able to grasp ideas that are developed by verbal explanation and discussion alone. But for most students, real understanding comes only when they can "see" the ideas being discussed. In some learning situations, it is fairly obvious to the instructor that he must "show" something for full student learning. In many cases, however, this need to "show" is not immediately apparent. Use imagination and ingenuity to provide ways for students to learn by seeing as well as by hearing. Otherwise, students will lose track of the idea being developed. If the idea is explained in conjunction with a visual demonstration of the point, student concentration will be reinforced by the "seeing" activity and better learning will result because students are able to follow the idea visually. Ways of holding interest by "seeing" are visual aids, demonstrations, skits and panels.

    a. Visual Aids. Visual aids may be interest factors, but not all visual aids are necessarily interest factors. A well selected aid, properly used, will be interesting because it assists student understanding. The types of aid and their uses are covered in detail in Chapter 7, "Training Aids." As far as interest is concerned, visual aids must focus student concentration on the points to be learned. A simple, clear aid will focus student thought on that idea. A complicated aid will only confuse the student with too many ideas and distract his attention from the point you want to emphasize. Color in an aid may be used to direct student attention to a particular point. An aid must be large enough to be seen by all; if it is too small to be seen clearly, student interest is lost. The aid must be introduced at precisely the correct, logical point in the learning situation, and must accurately portray the idea being developed by explanation.

    b. Demonstrations, Skits, Panels. These are all forms of the same thing: a dramatic representation of an idea. Demonstrations show how something is done. Skits may portray a procedure or operations, usually with a humorous slant; they are particularly appropriate to demonstrate personal interrelationships in team coaching. A panel is a group discussion, usually among various instructors, although sometimes it includes students. Panels will maintain student interest if the discussion is closely controlled so that it does not wander from the subject. All three of these forms of presentation must be well planned and rehearsed. They must be designed primarily to teach, not merely to entertain. Use humor to increase interest in demonstrations, skits, and panels but don't overuse it.

4. _Doing._ Discussion and Practical Work.

    a. Students Learn by Doing. Interest is high when students are doing. Doing activities include discussion and all types of practical exercise. Doing may be mental or physical.

    b. Discussion. Student discussion is a form of mental doing. Discussion, when animated by thought-provoking questions, is an excellent interest factor. When students become involved in instruction to the point of freely expressing their ideas in your discussion, class interest is high. To learn, the student must become involved in the subject intellectually and emotionally. However, you must control and direct the discussion so that the major points of the lesson are developed. A wandering discussion may be interesting, but students will learn little unless you channel the discussion and make periodic summaries. This is your job as monitor of student discussion.

    c. Practical Exercises.

    (1) Most marksmanship instruction teaches practical knowledge and skills - the ability to perform specific tasks. Practical work provides the best means to teach these skills and knowledge. Too often, the instructor spends too much time on lengthy explanation, and allocates too little time to student doing. Some orientation and direction must preface learning to do things. But the sooner the student begins doing, the higher his interest will be. You may use practical exercises with any subject.

    (2) To be interesting, practical work must give all students a specific task, and the time allotted for the task should be just enough to complete the job. With too much time, students lose interest; with too little time, students become frustrated.

5. _Controlling Interest vs. Entertaining._

    a. Interest factors are interesting ways or techniques of presenting the ideas of a lesson. This chapter discusses only a few; there are many other interesting methods, limited only by imagination. Interest factors used with imagination together with good instructional techniques help students learn. Some years ago, an instructor of Army marksmanship was widely acclaimed as the most interesting instructor most shooters had ever heard. But few students could tell you what he taught. He had entertained extremely well but students did not learn much about his subject. He overdid interest to the detriment of learning. Don't overuse interest factors to the point that students remember the entertainment and overlook the subject that is taught. They are tired of too many jokes, extended periods of practical work, too much discussion, too many training aids. Once a point is learned, students want to move on to another. If a point has been well learned, through explanation, examples and discussion, further explanation of the same idea will bore the student.

      b. Overuse of one type of interest factor will lessen its appeal as the lesson progresses. We know that one interest factor holds high student interest only for a short time - then interest decreases unless it is skillfully revived by the instructor. It will decline again and again as the explanation of each supporting factor exceeds the interest span of the student.

C. INTEREST SPAN

Interest span is the length of time that a student can give undivided attention (interest) to instruction.

    1. *Length of Interest Span.* The interest span depends upon many factors. One very important factor is the instructor's presentation techniques. Some subjects interest students more than other subjects. Personal physical comfort of students affects their ability to concentrate; weather conditions influence the span. Students concentrate longer and better in the early hours of the day when their minds and bodies are fresh and rested. Other factors being equal, the interest span shortens when the instructor is talking, and the span lengthens when students "do." The instructor with a knowledge of human reactions quickly senses a drop in class interest and promptly modifies his actions to regain student interest.

    2. *Bridging the Interest Span.* Instruction cannot consist solely of student performance. Before the student can perform effectively, he must learn what and how to perform. This type of instruction involves instructor presentation by lecture, conference, and demonstration. In this instruction, too often the instructor lectures at great length without concern for student interest. As a result, students lose interest after a certain amount of lecturing - they reach the end of their interest span for that particular part of the instruction. Carefully analyze the materials for probable dull spots in your lesson and plan to maintain interest by logical integration of interest factors to bridge the interest lapses. For instance, you may begin with an attention gaining introduction, then find that the next fifteen minutes are devoted to lengthy explanations. You know that, at best, you can hold interest for about five to ten minutes while lecturing. So you must plan to use an example, an aid, humor, or other interest factors early in this period to keep student interest high and to maintain it throughout this fifteen minutes. Throughout the lesson, interest tends to drop after students learn a point. The instructor must re-motivate students periodically with additional interest factors to reduce or eliminate places in his instruction where interest is low. In the introduction to this chapter, we described a situation where the instructor has a good outline of ideas but failed to control student interest. A graph of what happened appears in Figure 5-2. If he used appropriate interest factors, a graph of interest control during his lesson appears as shown in Figure 5-3.

      a. In Figure 5-3, he used an example in his introduction. Then he began to explain the first point. While he was talking, interest went down. So he used a humorous skit to demonstrate his point. Interest rose, and he capitalized on it by asking questions and stimulating discussion. Interest dropped slowly as the discussion progressed. At this point, he involved the students in a practical exercise. Interest was high again, but fell off as students solved the requirement. He again picked up interest with student discussion of solutions. He concluded the discussion with a tape recording of a sound solution to the exercise.

      b. He issued another practical work exercise requirement followed by a short discussion of solutions. At this point, the student interest span was shorter, since the class was almost over. So he used a humorous story to raise interest high, and concluded with a skit which re-emphasized the main points of the lesson.

c. As you can see in the graph, interest remained above the "good" level throughout. He had learned that students can concentrate on instruction only when he assisted them by making his presentation interesting. By picking up student interest at critical points in his lesson, he assured continued student concentration and its logical consequence - more effective student learning.

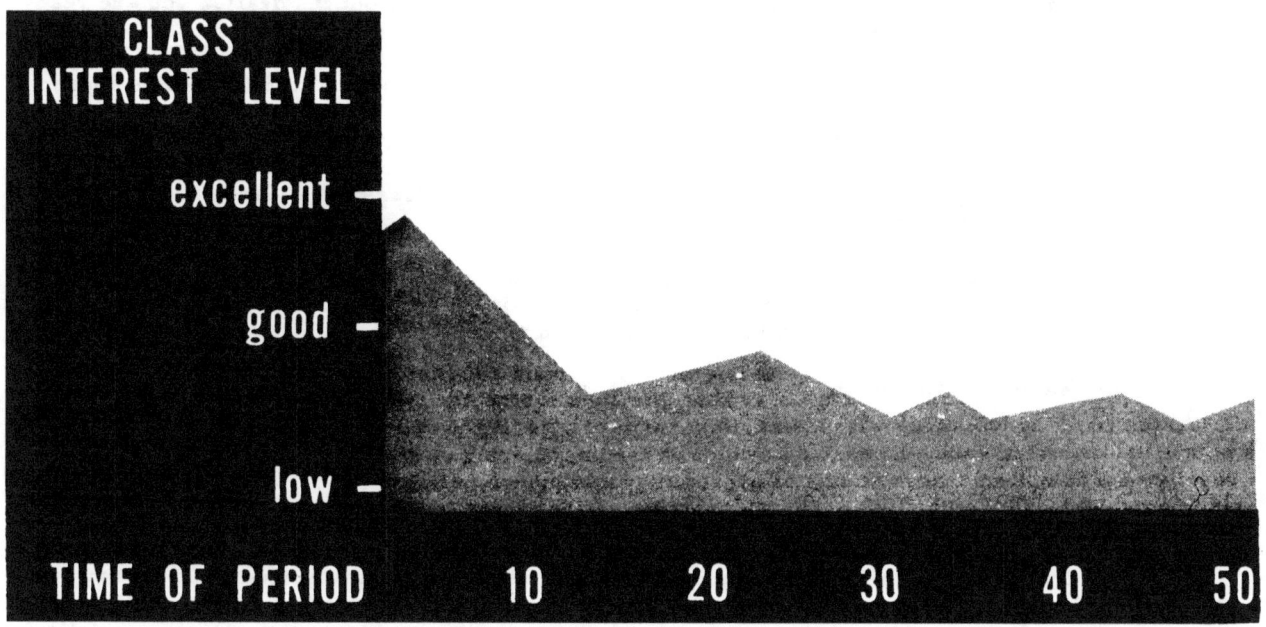

Figure 5-2. Graph of Lesson with Poor Control of Interest

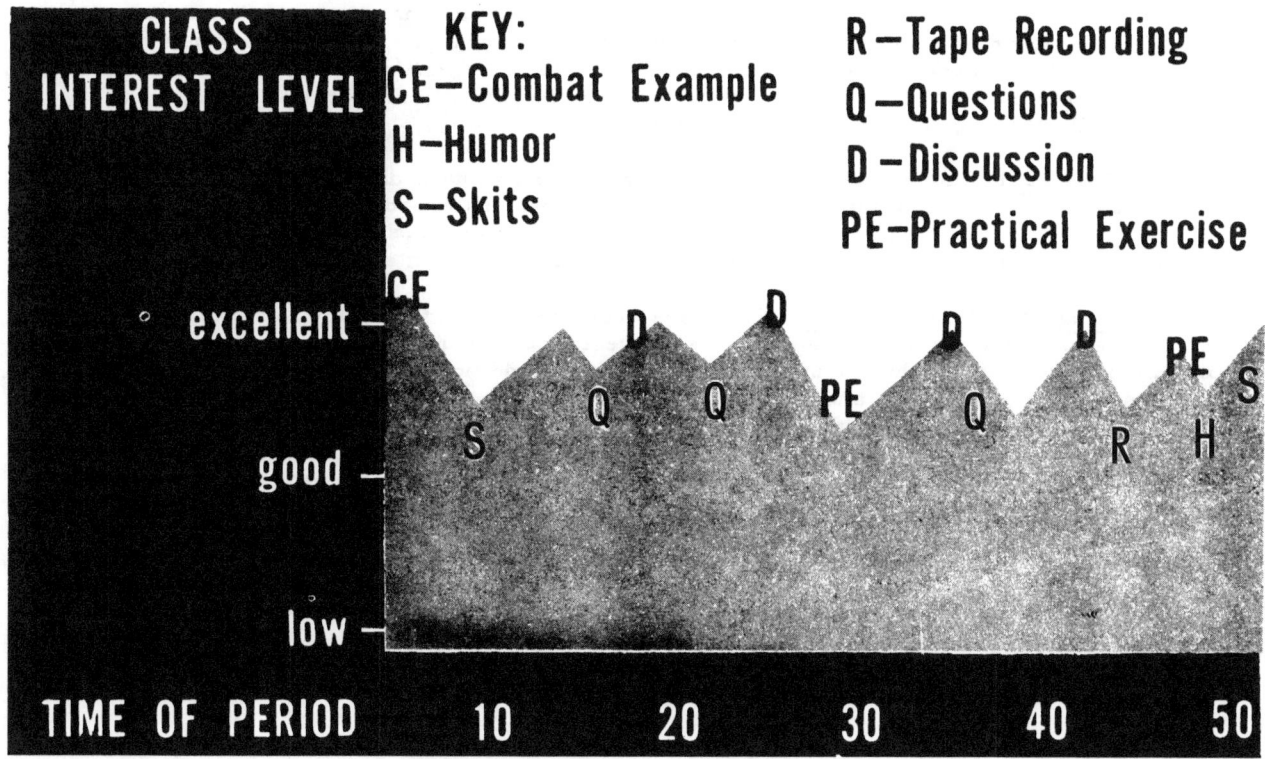

Figure 5-3. Graph of Lesson with Good Control of Interest

D. APPLYING CONTROL OF INTEREST

    1. <u>Control of Interest Follows Lesson Outlining.</u> We have discussed interest, interest factors interest span, and teaching vehicle separately. How do you apply them to instruction? You don't just sit down and write an interesting lesson. You plan for interest after drafting a good sentence outline with your ideas completely expressed and logically organized. With this outline drafted you are ready to apply techniques to control interest.

    2. <u>Select a Teaching Vehicle.</u> First, consider the use of an overall teaching vehicle appropriate for the whole lesson. This will require imagination and ingenuity. The situation may be true or imaginary, but it should be realistic, one which students may later face. Teaching vehicles lend themselves well to the use of humor, skits, and demonstrations which relate to the basic story. This vehicle will serve as a logical framework upon which to develop the ideas of the lesson and supply a reference point for smooth transitions.

        a. Actual Equipment. Students follow easily the use of actual equipment, where the natural relationship of various parts and their operation are physically apparent. The discussion of points should follow some logical pattern in the construction and operation of the equipment.

        b. One interesting method is to use the background and history of the equipment to show its development from its earliest form. For example, one instructor taught the steering mechanism of a 2 1/2 ton truck by beginning with the earliest form of steering used in horse-drawn wagons. From this point, he discussed each successive advance in the engineering of steering mechanism until he had developed a complete understanding of present day equipment. He used historical development as the teaching vehicle.

        c. Continuing Situations. Continuing situations help to teach any subject. They may take the form of handouts, skits, or just explanation. Using a continuous situation to present problems for students to solve and discuss not only creates interest, but makes the subject matter more realistic and learning more effective.

        d. Fictional Characters and Situations. An instructor with imagination may create a fictional situation and character to fit his problem. Place your fictional character in continuing situations to develop teaching points in a realistic and interesting manner. Few people would read an involved analysis of the psychological impact of the Civil War. Yet millions of people paid to read the story of Scarlett O'Hara and Rhett Butler, in "Gone with the Wind." Margaret Mitchell, by creating a fictional family of this period, became world famous.

    3. <u>Analyze the Interest Spans.</u> Next, analyze your outline to determine places where it is known that the interest span is short, where interest will obviously tend to decrease. These portions of the outline will require the use of other interest factors. Interest is usually lowest when for example, you are lecturing for extended periods of time with no other interest factors being used.

    4. <u>Integrate Interest Factors.</u> For these areas where interest is low, plan to integrate interest factors to bridge the interest span. The interest factors must relate to the material, and should, if possible, tie into the teaching vehicle. By using additional interest factors, you will raise student interest at these points, and thus enable you to emphasize your main ideas while interest is high. You try to keep student interest at a high minimum level throughout the lesson, but you must insure it is highest at those times when you stress the main points of your lesson. It is not expected and it is not considered possible to maintain student interest at a consistent peak.

CHAPTER VI

MANAGEMENT OF INSTRUCTION

A. GENERAL

1. <u>The Need for Classroom and Instructional Management.</u>

   a. Classroom and instructional management are necessary factors in good instruction. The instructor must be concerned with the comfort of his student, the effective and economical use of materials, the timely use of equipment, and the full utilization of all available training facilities. Furthermore, the principles of instructional management and administration are of primary importance in preventing difficulties and in solving problems that develop in a learning situation. As an instructor, you will be called upon to exercise judgment and work out solutions which do justice to everyone concerned. The principles and procedures you adopt in handling this part of your job will go a long way toward making you a better instructor.

   b. Instructional management and classroom administration may seem to be only indirectly related to instructional effectiveness; never the less, these administrative aspects of your work are of great importance. They set the tone for the entire instructional job and can determine, to a large extent, whether your students learn. Attention to the conditions of instructional environment, to the needs of students, and to the proper supervision of their activities will lubricate the machinery of instruction to reduce classroom friction or confusion. The better you systematize the routine tasks associated with teaching and the more you eliminate physical, mental, and personal distractions, the more time you will have to concentrate on the most enjoyable part of your work - teaching.

   c. Although most of your instructional time will be spent in lesson preparation, the best prepared lesson will fail if you do not provide proper classroom and instructional management. During the actual presentation, many situations will occur for which you can not possibly plan. It is essential, then, that you be aware of the common types of difficult situations and be flexible enough to cope with them. Section B will discuss the problem of questions used during a conference type presentation. Sections C, D, and E will be concerned with the instructor's control of the class through subsummaries presentation of solutions, transitional techniques, and handling of difficult conference situations. Section F will consider the need for good student/instructor relations in teaching, and finally, Section G will outline some suggestions for better classroom administration.

B. QUESTIONS

1. <u>Good Questions Are the Heart of the Conference.</u>

   a. Good questions are the heart of the conference since they are used to stimulate student thinking, to check understanding and retention, and to hold attention. The primary purpose of questions during a conference is to guide thinking or lead to understanding through discussion; secondarily, they create interest and hold attention.

   b. Questions should have a specific purpose, be clearly worded, and thought provoking. The specific purpose of most conference questions is to guide student thinking along pre-determined lines which lead to sound reasoning and understanding. Phrase the questions in simple form so students will have no difficulty in recognizing what is required. Simple and well worded questions will not just happen the first time you write them out (and you <u>must</u> write them out.) Phrase them and rephrase them until the questions ask what you want. Like <u>good</u> test questions, simple, clear conference questions are the result of much instructor effort. They should stimulate thoughtful reasoning by the student - emphasizing the "how" and "why" - rather than asking for rote memory or recall.

   c. Try out the questions on your colleagues. See how they answer them. If your fellow instructors are confused and request further clarification your students will undoubtedly be confused too.

   d. There are four types of planned questions which you will be using during a conference: lead-off questions, follow-up, check-up, and rhetorical.

Figure 6-1. Types of Conference Question.

2. **Lead-Off Question.**

   a. The first of these conference questions is the lead-off question. This is the question which initiates discussion on a selected teaching point. The lead-off question is general in nature to guide thinking along broad lines which may encompass one or more of your planned teaching points, or the whole subject. It is used to focus attention on "the big picture" as a preliminary to more specific discussion of particular points. Therfore it is a good technique to address the lead-off question to several students so as to generate broad introductory discussion of the subject. Your lead-off question must be carefully planned so that student interest is immediately aroused and a variety of responses is anticipated.

   b. The lead-off question should be provocative and should be so planned and stated that "yes" or "no" answers or short obvious answers are not possible. Normally, questions which begin with "why", "how", and "under what circumstances" are questions which will generate thinking and discussion. Should it be necessary in planning a lead-off question to have one which will require a "yes" or "no" answer, follow up immediately with a "why" or "why not." Lead-off questions, like teaching points, will be preplanned and inserted in the lesson outline in the place where they normally will be used. A good question is like a good transition. Once it has been proved effective, it seldom needs to be changed or revised. Sometimes conferences prove ineffective solely because the led-off questions were not thoroughly planned.

   c. An example of an acceptable lead-off question might be: "Why is it necessary to make sight changes?" This type of lead-off question is throught-provoking and would normally generate discussion.

3. **Follow-Up Question.**

   a. Another type of question used in conducting a conference is the follow-up question. Follow-up questions are questions asked as a result of student responses to lead-off questions. The instructor uses follow-up questions to guide the discussion toward an understanding of the teaching point by a consideration of the more detailed or specific aspects of that particular point.

   b. Certainly, you cannot know in advance the actual responses which you will receive from students. Thus you cannot plan the exact wording of your follow-up questions in advance. However, you should consider possible student answers to lead-off questions so that you can have some advance idea of the type of follow-up question which will be required.

   c. Flexibility is required in order to ask effective follow-up questions. Thorough research is needed in order to become aware of the usual student answers. Follow-up questions will not be an integral part of the lesson outline but reference to possible student answers and suggested follow-up questions will be placed in the outline in the form of notes.

   d. An example of a follow-up questions follows: If a student answered a lead-off question with a short answer which did not indicate the process which he considered in stating his answer you might follow-up with, "Explain your reasoning in arriving at this conclusion." Or you might ask, "Under what circumstances would your answer not be appropriate?" Another type of follow-up question which generates discussion is one which asks another student to comment upon a student answer.

   e. In conducting a conference you must insure that by student discussion or by your explanation, you can accomplish this understanding.

4. **Check-Up Question.**

   a. A third type of question which is asked during a conference is the check-up question. Check-up questions are factual type questions which check understanding, retention, and recall of points previously studied or developed. They are summary-type questions and are equally useful during a lecture or conference.

   b. Use check-up questions to evaluate how thoroughly students understand a teaching area which has been developed. Check-up questions assist you in determining how much more time should be spent in the development of the teaching point and how much reteaching may be necessary later on in the conference. Some check-up questions should be planned in advance and others should be asked whenever the instructor senses a need to check student understanding. If you suspect that the class may be confused or if student answers are incorrect, ask a check-up question to evaluate the present state of class learning. Planned check-up questions will be inserted in the outline at the place where they normally may be asked. Whether you ask them depends on how the class is progressing.

5. Rhetorical Question.

a. The fourth type of question which is asked during the instructional conference can be used with virtually any method of instruction. This is the rhetorical questions, a question which is asked and answered by the instructor. One effective use of a rhetorical question is to assist in gaining the student's attention during the lesson introduction. Rhetorical questions are used frequently as a transition or to introduce a new area of discussion.

b. Rhetorical questions are planned and placed in the lesson outline. The technique of asking a rhetorical question is similar to that used in asking any type of question: ask the question, pause, but then answer it yourself. Don't overuse rhetorical questions or you will lessen their effectiveness.

6. Technique of Asking A Question.

a. The A. P. C. Technique of questioning is the approved technique. "Ask the question, Pause, and then Call on the student to answer." This procedure is the only acceptable procedure to use in asking lead-off questions. The pause is necessary in order to provide time for the students to formulate their response.

b. This questioning technique is not an inflexible procedure. Once a discussion has been generated, interest in the discussion may lessen if the instructor pauses between each follow-up question. At such a time you should vary the technique, insuring only that the question is understood and that the person who is designated to answer the question knows that he is required to answer it.

c. Initially you should call on students to answer questions after consulting the class roster A technique for insuring that your pause is timely enough is to ask the question, then go to the roster, select the man, check off his name, and call upon the student. In using a class roster it is wise to skip from page to page and not go through the roster systematically one page at a time. After a class has been with you for a short time, each student learns the page of the roster which contains his name and thus he is aware when he may or may not be called upon. This may detract from the effectiveness of the discussion. When asking follow-up questions, it is perfectly permissible to accept volunteer answers. Care should be taken to insure that a small group of volunteers do not monopolize the student discussion. You should use the class roster to insure that this does not occur.

d. As a general rule, don't repeat questions for the benefit of inattentive students, as this will lessen the overall attention holding the effect of the questioning. If students know you will repeat everything, they pay less heed to what you are asking initially.

e. When asking questions, demonstrate a sincere and enthusiastic manner. Indicate a real desire to hear the student's viewpoint. Your attitude will affect the answer you will receive and influence the amount of student participation. Interest and enthusiasm are contagious.

7. Handling of Student Questions.

a. Discussion thus far has concerned the questions which the instructor will ask during a period of instruction. There has been no discussion of the type of question that a student may ask. If you have properly motivated the class, interest in the subject will be high and you certainly can expect student questions from time to time as the discussion develops. Your response to student questions will play an important part in determining the overall effectiveness of your presentation.

b. When a student asks a question that seems unrelated or if he seems to challenge or question something you have said, your first reaction may be "Oh, a sharpshooter - well I'll cut him down to size." Don't give in to the temptation to be sarcastic. The chances are that since you are somewhat tense while presenting instruction, you may misinterpret an innocent question as an attempt to "shoot you down." Listen to the student carefully and give him the benefit of the doubt. By his second or third question you can decide whether he meant to be antagonistic. Even if you discover that the student is baiting you, don't lose your temper, for then everthing is lost. Tell him to see you at the break and you will discuss it further. Your duty to the class is to avoid personal bickering.

        c. If a discussion is interesting and student interest is obviously high, it is a good technique to ask another member of the class to answer a question posed by the student. This is especially true if students raise hands immediately following a student question. They probably are eager to respond to the student's question and this is one way to generate lively discussion. If the question is one that you feel the average student may not be able to answer adequately, answer the question yourself. Your guide should be to adopt the course of action which maintains high student interest in the discussion.

        d. Even though you have properly prepared yourself and have carefully researched each area which will be developed, from time to time students will ask questions which you cannot answer without further research. When this happens, the accepted course of action is to inform the students that you will find the answer and let them know. After you have committed yourself to do this, you must insure that the students get the information.

    8. <u>Reaction to Student Answers</u>.

        a. Remember that when a student is called upon, he should rise, face the majority of the class, state his name and speak loud enough for all to hear. Use tact in this regard. If the summer fans are on, it may be worthwhile to use your mike power to summarize or repeat his words. Indicate when you are through with a student so he may be seated; don't let him stand while you discourse.

        b. Select from the varied student responses those ideas or facts which you can draw on for further questions and discussion. Recognize a good answer by complimenting the student; this will encourage further participation. Remember that students are human beings who like and need praise. Don't ridicule a poor answer; try to pick out some good point in the reply which you may use to further the class discussion and understanding.

        c. While the student is answering, don't turn your back or look at notes. Show an alert positive interest in what he is saying because you must evaluate this response as he is talking and plan your next move as soon as he is finished. Watch that the response and follow-up questions do not revert into a private conversation between you and one student.

C. SUB-SUMMARIES

    1. <u>Sub-Summaries Aid Student Understanding</u>.

        a. Sub-summaries are extremely important when using the conference method of teaching. After active student participation and discussion, the student frequently needs someone to wrap up what has been learned as a result of this discussion and relate it to the present teaching point or to the overall lesson objective. In the student discussions, some comments and opinions will contain key ideas and others will be interesting but not too important or relevant. This is a good time to step in and present a summary of this discussion, indicating which ideas are worthy of retention and how they supported the teaching point discussed or the lesson objective. Since this is not the final summary (which is part of the lesson conclusion,) it is called a sub-summary.

        b. A sub-summary is one way that the instructor controls discussion and prevents the class from wandering afield or overdiscussing a ceratin teaching point. As you sub-summarize you have class control and can use the sub-summary to transition into your next teaching point. After a discussion wherein varying or opposed viewpoints have been expressed, students look to you for comment and, if appropriate, to present the approved position.

D. TRANSITIONS

    1. <u>Class Control by Transition.</u> In each problem you develop the teaching material in the way that makes sense to you and, you hope, makes sense to your students. You arrange certain teaching points or supporting materials before others because the lesson is more understandable that way. However, you cannot assume that every class member will see this logical pattern instinctively. This is where well planned transitions are valuable. They are inserted to help bridge over the train of thought from idea #1 to idea #2. They are most essential and must be skillfully used when teaching by the

conference method. By good transitions you control or guide the student's thinking along the proper channels and assist him in understanding the logical relationship between key ideas. An understanding of this relationship is essential for a student learning instructional material. Smooth transitions are a sign of professional skill. New instructors frequently fail to make smooth transitions either because they haven't planned them carefully or because they themselves aren't too aware of the relationship of the ideas.

    2. Transition by Sub-Summary. Section C explained how to use a sub-summary to pin down the key ideas which arose from the student discussion and then move on to the next point. A sub-summary is probably the most common way of transistioning into the next point.

    3. Transition by Teaching Vehicle. In Chapter 5 the teaching vehicle was defined and discussed. By referring to the teaching vehicle you lend unity to the lesson and indicate how the next teaching point should follow logically at this time. You show the relative position of this teaching point or student performance objective within the context of the whole lesson.

    4. Transition by a Slide. Some instructors place the teaching points in a logical sequence on a slide. They use the slide in the introduction to show the scope of the period, then refer to it to transition to each new teaching point. They use the slide again during the conclusion to summarize what has been discussed or learned during the period.

    5. Transition By Rhetorical Question. Another method of transition is to use a rhetorical question. Instructors frequently transition in this way: "Now that we have discussed A and B, what is the next logical point to consider?" (pause) "C". Then they explain why C is logically next.

    6. Transition By Natural Pattern or Sequence. Certain instructional material lends itself to a definite natural pattern of development. For example, in discussing a motor vehicle it would be natural to consider various systems in order, such as ignition, fuel, brakes. In a problem situation you would consider events in a chronological pattern of what happens first, then what happens. These subjects contain natural transitions. To aid understanding all you have to do is indicate that you are finished with that phase and are now going into a new phase. Again, here the chart or training aid may help as a transitional device. However, a word of caution, don't simply display a new chart without making some introduction or transition into it. Students should be adequately introduced to new material whether it is visual or verbal material.

E. RESOLVING DIFFICULT CONFERENCE SITUATIONS

    1. Students won't Discuss.

        a. Occasionally you will encounter a class wherein the students simply reply "I don't know." Try to diagnose the situation on the spot. This is where your human relations skill as an instructor is important. Perhaps you might relax them with a humorous anecdote. Adopt a friendly sympathetic attitude. Rephrase the questions so as to give students additional time to think. Ask them what is their opinion on the subject being discussed or what action they would take rather than what action should be taken. Anyone can express a personal opinion on the subject but not everyone is confident that he has the correct answer. Then get another student to comment or tell what action he would take.

        b. Perhaps you have not yet sufficiently clarified the issue or point. Amplify your explanation and then try the lead-off question again.

        c. Give an illustration which is absolutely contrary to doctrine. This should arouse class discussion because some students should see the error of such reasoning. Then you can lead the class around to the correct procedure.

    2. The Student Answers with the Complete Solution. Sometimes a student will respond with a complete solution and all the supporting material. He apparently has stolen all your thunder. In this case, thank him, and praise the answer. Then you may decide to resummarize what he has stated, adding some additional illustrations or examples. Your summary will be your words but will reinforce the same key ideas. Again you may decide to jot down his statements on the balckboard and discuss each in more detail. Stay flexible and keep your sense of humor. Don't get panicky. A statement of the teaching points by a student does not mean that they have been taught to the class. Another possible tack is to propose alternate actions contrary to the solution and ask students why they would not follow these alternates.

3. <u>The Student Talks too Much.</u> Sometimes a student is enamoured with his own voice and rambles along endlessly. You cannot encourage or permit lengthy solo dissertations since the conference is designed to get opinions and discussion from many class members. Hear him out for a while while and if it becomes apparent that he doesn't intend to conclude, interrupt him and ask him to condense his comments or summarize. If he still continues, thank him in a polite, definite and forceful way and call upon someone else to comment.

4. <u>The Discussion Turns Into Argument.</u> Sometimes a lively discussion turns into an argument either between students or with the instructor. Some emotional comments may be worthwhile in a sincere, lively discussion. However, don't let comments degenerate into personal attacks. Step in and rechannel the discussion. Either call upon another student for his opinion, settle the issue by a comment reflecting the approved position, or ease the situation with a light remark or a summary.

5. <u>Students Who Oppose Everything.</u> A student may try to object to everything you say. Rather than argue with him directly, check to see if any hands go up. The chances are that they are from students who wish to object or answer the obnoxious student. Whenever you get students to answer and reflect the approved policy it eases the burden of explaining everything yourself or of putting yourself on the defensive. If the student still persists, remember that (as principal instructor) you are in complete charge of that class period. Ask him to see you after class for further discussion. Remind him that he is taking up class time. If this extremity is necessary, ask him to leave.

6. <u>The Discussion Generates Too Late in the Period.</u>

    a. A class may get into a lively discussion just before the period is scheduled to end. If you have this same class for the next consecutive period there is no problem. You can let the discussion continue for five or ten minutes more. Then you can give them their break a bit later but you will have to make up time lost in your next period. However, do not shorten the class break time.

    b. If the discussion starts late in the period and you have only enough time for a rapid condensed conclusion, you break in and begin your conclusion. That is where a flexible conclusion is useful. Perhaps you planned five or ten minutes to use a check-up question type of summary. As the discussion was lively, switch to a summary slide and hit only the high points. As a good instructor you must be ready for either eventuality.

F. HUMAN RELATIONS IN TEACHING

  1. <u>The Instructor - Student Relationship.</u>

    a. An instructor must establish a receptive, cooperative, working relationship if his instruction is to be effective. If learning is to be student-centered, the student must be treated as an adult and progressively given more and more responsibility; otherwise, he will gain little or no confidence in performing the task.

    b. In addition to assisting the student in gaining a sense of responsibility and confidence, the instructor must develop the sense of belongingness so important for effective student learning and participation. You must take a personal interest in the student and continually strive to raise his personal standard as well as his performance standard.

    c. Students respond to and tend to reflect your attitude. If you are natural, helpful, and enthusiastic, your students will be friendly, eager, and ready to learn. Learn their names, talk with them personally whenever you have the chance. Show that you are trying to help them. You are working with men, not machines. Remember, you desire students to be good shooters through the exercise of their own initiative rather than through a conformity which is forced upon them by a superior. If a student is to respect rather than to fear or resent authority, the instructor must be fair, firm, and friendly. Some suggestions which will aid you in acquiring these very important human relations qualities follow.

  2. <u>Some Suggestions for Maintaining Good Relationships:</u>

    a. Show no partiality or favoritism. Nothing destroys student readiness or receptiveness as partiality. Never reprimand a class for the wrong doing of a few.

b. Never try to fluff. Students soon learn when you are trying to fool them. Acknowledge a mistake. The simple admission, "I do not know, but will find out," or "You were right and I was wrong" can do much to develop and maintain a good working relationship.

c. Be loyal. To your class, your superiors, and to existing policies. Correct at once any errors of administration affecting your students, but carry out the full extent of the teaching directive from your superiors.

d. Act decisively. In making a decision, give full consideration to all aspects of the problem and then act with conviction. State the approved position firmly and positively.

e. Abide by decisions. A fair decision, once made, should be carried through, and a proper order, once given should be executed. Students will respect you for your decisiveness.

f. Keep the student headed toward his objective. See that all activities of each lesson are toward achieving the established goal.

g. Respect the rights of your students. Always place your class ahead of your personal convenience and consider their rights as students. Be sensitive to how you emotionally affect others. Emotion is present in any job of teaching situation and may work for you or against you.

h. Be courteous. Correct student mistakes as they are made, but do so in a straightforward, impersonal manner. Never be sarcastic or make your criticisms personal.

i. Be cheerful and enthusiastic. The enthusiasm you have for your teaching job is directly reflected in an increase in student interest in learning and in a decrease in disciplinary problems. If you are cheerful, students will reflect your attitude. Know when to use humor in the classroom; good clean humor will create good will and help you fulfill your instructional mission.

j. Be business-like. Your job is important and there is no time for foolishness. Encourage initiative and self-reliance. A student who has learned to think for himself will be able to handle problems for which he has been taught no prescribed procedure. Welcome the assistance of students who have exceptional backgrounds, and use them as an aid in arousing the interest of less experienced members of the class. Each instructional problem is a cooperative learning experience.

3. <u>Physical Comfort</u>.

a. No matter how interesting your presentation, or how good an instructor you are, students will have difficulty paying attention if they are uncomfortable. Physical conditions constantly influence every minute of instruction. You must foresee possible unfavorable conditions and take every action to make students comfortable when ever and wherever possible.

b. Here is a physical environment checklist:

(1) Room temperature satisfactory?

(2) Room well ventilated?

(3) Chart or chalkboard glare eliminated or reduced?

(4) Lighting where it is needed?

(5) Distracting materials removed from walls?

(6) Distracting outside noise eliminated or reduced?

4. <u>Student Safety</u>. The importance of safety practices cannot be overemphasized. Whenever you demonstrate a new skill to be learned, show students the reasons for safety practices. Safety consciousness is an attitude which must be developed in students. Insist that your students observe common sense precautions. Be sure that you have established all controls in your power over the human element that causes accidents.

G. CLASSROOM ADMINISTRATION

   1. Good Classroom Management.

   a. The principles and procedures you adopt in handling necessary administrative details will contribute to your success or failure as an instructor. Start well before class time, check the learning area, classroom, shop, and problem area to be sure that all necessary equipment, supplies and training aids are in proper condition and ready to be used.

   b. Some suggestions that will assist you in handling administrative details:

   (1) Begin and end class promptly.

   (2) Check attendance.

   (3) Make students responsible for routine details such as clean up of an area.

   (4) Check appearance and conduct of assistant instructors.

   (5) Report all inadequacies of equipment: i.e., sound, classrooms, training aids, or heating.

   (6) Do not permit students to "close shop" during the final summary.

CHAPTER VII

TRAINING AIDS

A. GENERAL

1. Training Aids Should be Used to Aid Understanding - to assist the student in learning. When you present an idea to students by means of words alone, students must picture in their minds what you are trying to convey through these words. Psychological research has demonstrated that most students learn more easily through the sense of sight than through any other sense. Realizing this, capitalize on this principle by using visual aids whenever they will help to present the subject in a more understandable manner. Training aids will make instruction more meaningful by reinforcing the student's sense of hearing with an appeal to another sense. Training aids will not take the place of verbal explanation but they will help to make such explanation clearer to your students.

2. The Object of this Chapter is to explain the purpose, types, and value of training aids. The techniques of proper use and future trends in training aids will also be discussed.

B. PURPOSES

1. The major purposes of training aids are to reinforce explanation, to direct student thinking to a specific item, to aid retention, and to save training time.

2. Training aids reinforce explanation by appealing to many of a student's senses simultaneously. Besides hearing your explanation, students can see what you mean. Through use of both senses clearer understanding will result. For effective instruction all students must have a comprehensive mental picture of the subject being discussed. One way you can make sure that all students receive a similar mental picture is by using a training aid which helps to standardize the visual picture.

3. Training aids focus student thinking to a specific item. You know that when a visual training aid is being discussed, explained, or demonstrated, students are concentrating on the aid because objects which appeal to the sense of sight have great power in holding student attention. When you are lecturing, students may appear to be listening but they are possibly thinking of many other things.

4. A good training aid assist student learning by increasing retention. Since impressions secured through visual means are more lasting and vivid than those acquired through hearing, visual aids play a dominant role in the learning process by insuring greater retention. The greater the impact of learning on our senses, the deeper the impression with increased chances of retention.

5. Training aids frequently save time and replace long involved explanations by the instructor. It is virtually impossible to explain some technical ideas or processes by means of words alone. A good chart, an operating model or mock-up will put across the ideas more effectively and more quickly. When training operates within a compressed time schedule, any device which speeds up average learning time is extremely valuable.

C. SELECTION OF AIDS

Let's assume at this point that you have done the research and organizing necessary to complete your lesson outline. Your outline is apparently all set, but there are several teaching points or instructional areas which are complex and difficult to explain by words alone. You remember that the function of a training aid is to assist in increasing student learning and retention of material.

1. The First Step in Selecting Training Aids is to sit down and carefully examine the instructional objective and teaching points because a training aid must definitely aid the students in understanding this objective. If it fails in this respect, it ceases to become an aid and becomes a distraction.

2. There is a Limit to the amount of material that a student can understand at any one time even with the help of accurate training aids. Too many ideas presented at one time or too many details on any one training aid will create confusion and hinder student learning. It is far better to present a few simple ideas on a chart and to use several charts than to attempt to crowd everything into one aid. In planning training aids simplicity is an important word to remember. A chart is not supposed to be

a condensed version of your teaching notes. New instructors frequently err in trying to place too many notes on a chart or projectual. The aid should contain only the key noun, verb, or phrase. It is not designed to stand alone without explanation. However, words or phrases will never have the impact upon student senses, the gateways to the mind, that an appropriate picture of visualization would. Strive to visualize the concept in your training aids. Too many charts and projectuals are merely lists of words that could just as easily be written on the chalkboard by the instructor during the class; this would be more interesting and dramatic. Don't make a slide to use as a memory crutch.

3. If Your Instructional Objective is to Familiarize your students with the operation of the M14 rifle gas system, one of the best aids would be a large schematic drawing of the M14 rifle but you would not want to use this as the sole means of getting across your objective. You could project slides to show parts of the system or even bring in a training film that explains how it works. Perhaps a scale model would help to explain some difficult point. Keep in mind that you can use various types of aids to explain the same idea. By using a variety of aids during the period instead of all charts, all slides, or all film, you increase student interest. Of course, there is a point of diminishing returns here as well, since you do not want to make your instructional hour like a circus - flipping from one type of aid to another indiscriminately.

4. Careful Evaluation of your Instructional Problem will reveal where a chart, model or other training aid will promote better student understanding. Remember that there are many types of aids and each has certain advantages and limitations. Frequently it will be necessary to use more than one type of aid to present a lesson effectively. It is of utmost importance to weave the aid into your lesson pattern carefully. The actual use of the aid must not interrupt the flow of ideas during lesson presentation. To be effective, training aids must be a natural, intrinsic portion of the sequence of instruction.

5. A Good Rule of Thumb to follow when determining whether to use a training aid or not is to ask yourself, "Could I achieve the lesson objective as well without the aid?" If the answer is yes, then either you have selected a poor aid for your subject or no aid is necessary at that particular time. Remember, training aids are not "crutches" for you to lean on; they are incorporated in the lesson to assist student learning. They supplement instruction, they do not replace it. The criterion, "Does it help student understanding?', is the main standard by which the training aid should be measured. The other advantages such as "to appeal to more than one sense", "to focus student attention", to increase student interest", apply to a good training aid but should never be the main consideration.

D. TYPES OF TRAINING AIDS

Training aids in use include actual equipment, models, charts, chalkboard illustrations, training films, other types of projectuals, and recordings.

1. Actual Equipment.

a. The actual equipment is the most realistic training device. It always creates a lasting impression to have the actual object present during instruction. However, sometimes the actual object might be unavailable or inappropriate for class use. It may be too large or too small for formal classroom instruction.

b. At times it may be desirable to have both the actual object present in combination with other training aids to illustrate the object. Certain features can be shown on a model or chart that cannot be shown with the actual equipment since they may be internal portions or features of the object. Study your lesson to see if it is possible to use the actual object in your period of instruction. Remember that if the weapon is before them, students will show greater interest and they will certainly gain a better understanding of what you are teaching.

2. <u>Models (Training devices.)</u>

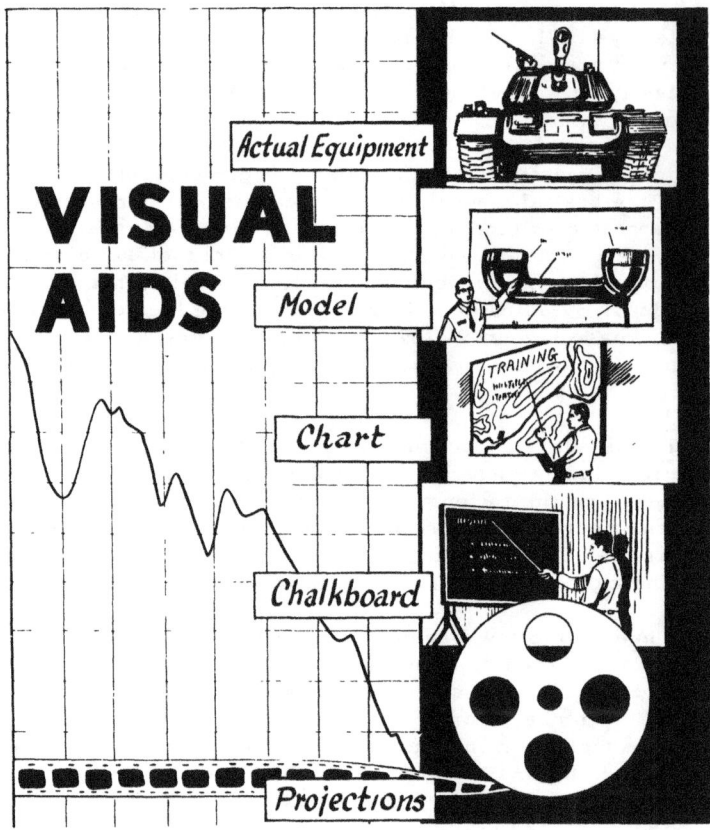

Figure 7-1. Visual Techniques.

  a. Models are of little practical value unless they are made to scale and give an accurate representation of the actual object. They must be of suitable size for the teaching purpose and they must be convenient to handle. Above all, they must be appropriate and necessary to the fulfillment of the teaching objective. A model suitable for a ten-man class may not be effective for a 50-man class.

  b. There are many different sizes and types of models. Analyze your training situation carefully before you decide the type of model to suit your purpose. Here are some of the more common types:

   (1) Actual size model - a model of actual equipment made to exact scale. One reason to use actual size models in lieu of actual equipment is that the model may be made of lighter material or not necessarily operable, hence it is much cheaper for use in instruction. You don't have to tie up the actual equipment when the model would serve the purpose. It can be field stripped exposing parts not otherwise accessible.

   (2) Enlarged scale model - one made to a scale larger than the actual equipment. For example, an enlarged model of the sights to instruct classes in the correct methods of obtaining a correct sight alignment.

   (3) Reduced scale model - here the model is made to scale but smaller than actual equipment. Usually this type is used for small group work and is more portable and storable than the actual equipment. Frequently, for range procedure classes, a reduced scale model of a range may be used.

   (4) Working model - a model where the parts move to simulate the actual operation.

(5) Cut-away model - a sectionalized or cut-away view of a piece of equipment to display the internal mechanism in such a manner that students can see the parts and operation.

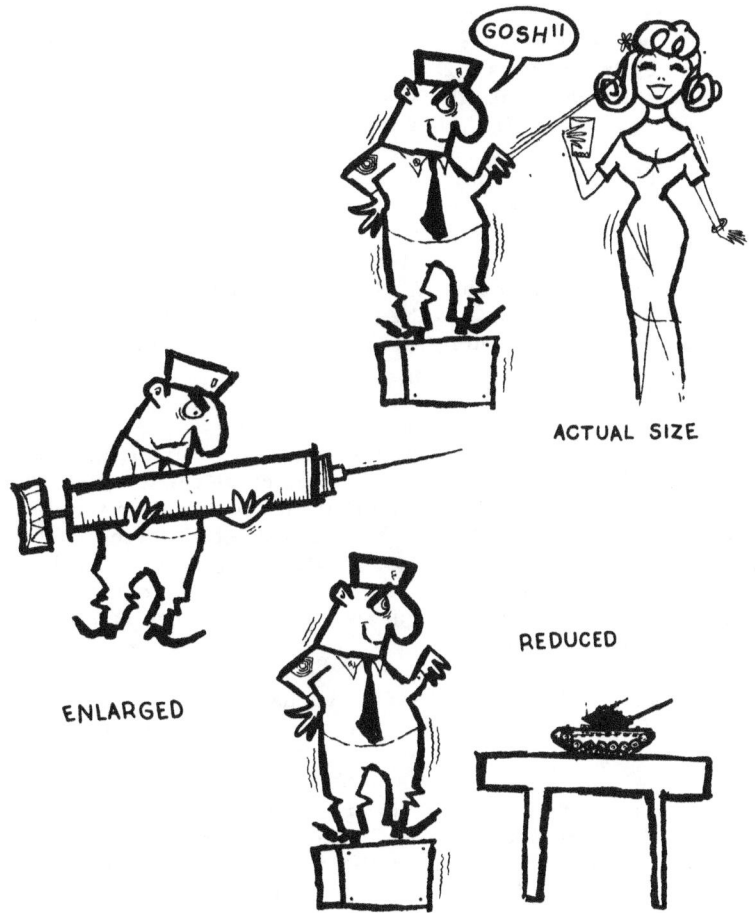

Figure 7-2. Models Are Excellent Training Devices.

3. <u>Training Films</u>.

a. Training films have won a well earned place in instruction; when properly designed and used, they are extremely interesting and vivid training aids. By using the training film an instructor can bring into the classroom all of the impressions of outdoors. With training films, specific situations can be employed which would otherwise be difficult or impossible to present. Films can bring the experience of experts into the classroom to help explain difficult portions of your problem. A good film will assist student learning and save training time when carefully planned for and used. Films are excellent for both standardizing instruction and accelerating learning, but there never was a training film that could replace a well-prepared instructor. Don't use a film as the sole means of instructing but as an aid to your teaching. Before you use a training film carefully follow these steps:

(1) Preview the film several times to be thoroughly familiar with its content. Select the portions of the film that you wish to show and do not show irrelevant portions, no matter how interesting

(2) Prepare notes and pertinent questions about the film to use in class orientation and post presentation discussion.

(3) Introduce the film to students carefully, directing attention to the purpose of the film and the specific items to look for.

(4) Show the portions of the film as planned. Remain in the classroom while the film is being shown.

(5) After showing the film, review the main points which it brought out. This summary will reinforce the teaching ideas and increase the instructional value of the film. Answer student questions and ask questions to check student learning.

(6) If time permits, reshow portions of the film. This is an extremely valuable step which is all too often overlooked by instructors. Experimental studies have demonstrated that students get much more out of the second showing of the film than the first showing. It is amazing how many important details students overlook during the first showing.

b. Training films have certain disadvantages which you, as an instructor, should be aware of and guard against. When they are projected front view, you must darken the room. This creates an uncomfortable, hot situation in non-airconditioned classrooms. Since there was a time when training films were overused in mobilization training, many students were unfavorably conditioned against films and they look upon film time as a good time to sleep. Never show a film simply to entertain students - never show portions of a film that are irrelevant to the teaching points being considered. When used sparingly, they are an excellent training aid; when overused they are a drug.

4. Film Strips.

a. A film strip consists of a length of standard motion picture film containing still pictures of a specific subject. A film strip is most effective when a series of related pictures must be presented in a definite order or sequence. A good example of a film strip use would be the disassembly or assembly of a mechanical unit such as a M14 or 45 cal pistol. An advantage in using a film strip is that you operate the strip projector yourself which permits a personalized approach. Again, you control the rate of presentation and can project the various pictures to dovetail with your words. If there are any student questions, the frame can be reshown and discussed. The sequence can be stopped at any time to discuss and to clear up points not understood.

b. The main disadvantage of using film strips is that it is generally necessary to have the room blacked out and the blinds closed. This reduces personal contact between instructor and the class and may encourage sleeping.

5. 35MM Slide. The 35mm slide or film strip projector is becoming more and more popular as an instructional aid. 35mm slides are not so expensive to make or reproduce as large wall charts. Yet they serve the same purpose. Using a 1000 watt bulb you can show a number of pictures, graphs, maps or diagrams projected up to 18' x 26' in size at 100 feet distance and yet you carry the slides to class in your pocket. A disadvantage of front view projection of the 35mm slide projector is that the room has to be darkened. This hampers students from making notes readily and you have difficulty in seeing group and individual reactions so you miss that on-the-spot evaluation of how your instruction is progressing.

6. Overhead Projector.

a. The overhead projector accommodates clear 10" x 10" acetate slides. The overhead projector is a fast, convenient and flexible means of projecting instructional material. As such, it has earned its rightful place as one of the most useful training devices. With this projector you can use as many charts or drawn projections as you wish. You can display material quickly and easily and the expense and effort necessary to prepare slides is not great. Color can be used on transparencies to add interest. A single mounted transparency can be reproduced in several layers with each layer containing several colors.

b. Another variation of the overhead projected transparency is the "flip over" transparency. Here is a number of transparencies which are hinged to the side of a single mount and each side is flipped on as you call it. With the 1000 watt projector you have as many as six or eight flip overs without materially affecting the quality of the picture. A special situation can be built up very effectively by means of such flip overs. It assists sequential or step-by-step presentation of learning materials.

c. The major advantage of the overhead projector over other types of projector is that it can be used in lighted classrooms or normal daylight conditions. Students can make notes and you can observe their reactions. The projector is excellent for formal or informal instruction. Its few movable parts make it extremely serviceable and easy to maintain.

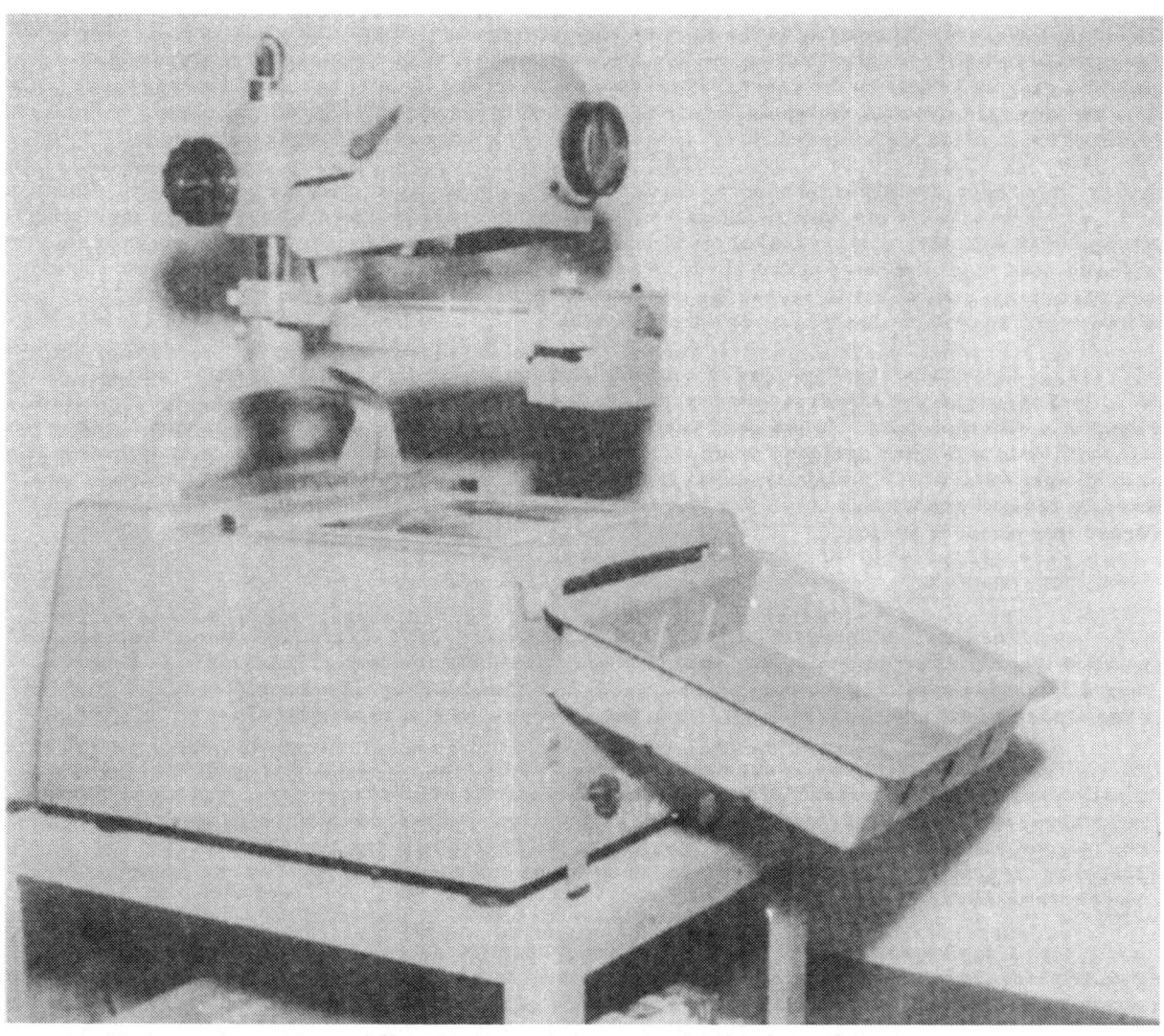

Figure 7-3. Overhead Projector.

7. <u>Wall Charts</u>.

    a. The wall chart is a drawing or sketch on medium heavy drawing paper which generally represents a schematic or topical outline of ideas. A well designed and skillfully used chart may frequently make the difference between a highly effective and interesting lesson and one that is dull, uninspired or confusing. Sometimes a wall chart is preferable to a projected aid especially if extended discussion is to follow each point, because the use of a wall chart eliminates the problem of a partially darkened room with attendant lack of ventilation and student attention.

    b. Wall charts help to build ideas or concepts cumulatively and assist the logical thinking of students. They add visual reinforcement since they appeal to another student sense. One good technique is to develop the fundamental ideas by means of the chart, then put the chart away as the conference or discussion develops and lastly, to reuse it in the summary to reemphasize the fundamental ideas of the lesson.

    c. Each classroom should be equipped with chartboard panels. In placing your charts on these panels be sure that the lectern or curtains are not blocking the view of the charts. Pay particular attention to the centering of the chart on the panel. If the chart paper has been cut squarely, you correctly center it by keeping the distance from the edge of the chart to the edge of the sliding panel uniform. Every chart should have a simple, yet comprehensive title describing the subject material portrayed on the chart. Simplicity is the keynote for good charts. Work deliberately to reduce your teaching points to brief, simple language; don't clutter the chart with unnecessary words or details. Put only the essential facts on the chart. Phrases on a chart are usually condensed to keywords. The chart is not expected to teach the whole lesson for you and is not expected to be completely understandable without explanation by the instructor. It is an aid, not a copy of the instructor's notes.

d. There are different ways to focus student attention to an item on your charts. One way is to use a pointer to point out specific ideas being discussed. Another interesting way is the "strip tease" method. With the "strip tease" method you conceal portions of the material by cover strips until you actually need them. When you are ready to proceed to the new point in your lesson, you remove the next strip from your chart to reveal the next point. Still another variety is the venetian blind technique where you open each slat as you develop that idea.

e. Key words sometime assist student learning and make charts more interesting. In this key word technique you use the first letter of the first word of each idea. Sometimes this first letter is printed in a different color. When read vertically the letters should make some simple English word. If the word ties in with your subject matter so much the better. It will reinforce student memory. Use a key word only when it will definitely assist learning. Use key words only where rote memory of a sequence is desired since many times key words tend to confuse students and many instructors tend to overuse this memory device.

8. <u>Chalkboards.</u>

a. The most valuable training aid is the chalkboard. All classrooms should be equipped with at least two chalkboard panels and portable boards available for outdoor instruction. Hence a chalkboard is always available for instructor use. It is a flexible, inexpensive method of presenting ideas and students will learn more readily when they are able to see as well as hear.

b. Another advantage of the chalkboard is that it is an excellent way to promote class participation by helping the instructor and the class develop the subject together. It is an ideal aid for the conference method of instruction. As the class gives various comments, suggestions, or answers to your questions, you place their ideas upon the chalkboard and students feel that they have contributed to the class. Then as you summarize at the end of the period you can use the ideas they gave you to write on the board.

c. A chalkboard enables you to make any picture or sketch you wish or you can hand-make your own training chart on the spot by preparing the board in advance. A useful technique that many instructors use to reproduce material on the chalkboard is to project the material from an overhead projector directly on the board, then copy in pencil the important lines. These pencil lines will not be seen by your class but you can copy over them in chalk when the time comes for developing them.

d. Here are some suggestions regarding the proper use of a chalkboard:

(1) Plan your chalkboard work carefully in advance. Take into consideration such things as: the amount of space available, proper layout of your material, and adequate visibility by the most distant students.

(2) Pencil-in in advance any required sketches on the chalkboard. Here you may use a straight edge and any other aid in a calm and precise manner so that when you copy one in chalk the sketches will be of fair quality despite the fact that you will be chalking-in rather hastily during the actual class.

(3) When you put material on the board, print in large, legible block letters. Use only one type letter throughout and make your printing neat and legible.

(4) Put your material on the board in a simple brief manner but be sure that it is complete enough to be understood by the average or weak students. Abbreviations may be used as long as they are the usual standard abbreviations. Don't use too many abbreviated words one after the other or students may lose the sense of the idea.

(5) Some instructors use the wet board technique very effectively. Here you get the board very wet and then write in soft chalk. If you time your presentation accurately as the board begins to dry, the key word or ideas appear as if by magic on the board. Timing of this method is effective. A variation of this is the "strip tease" method discussed under wall charts; paper strips can be used on chalkboards very readily.

(6) Don't erase until all students have had time to see the material and copy if they so desire.

e. There is no more dramatic way of presenting material visually than by use of the chalkboard. Persons love to see things or ideas developed before their eyes since it stimulates their creative instincts. Remember how often as a child you watched a cartoonist at work in such movie programs as "Out of the Ink Well" or how you watched spellbound as sand or snow sculptors developed their material. Writing teaching points on the chalkboard as you secure them from student discussion is doubtless the most effective way of conducting an instructional conference for the student feels that this lesson was a result of a stimulating interchange of opinions between instructor and student and among various students.

9. Recordings.

a. Any instructor may make tape recordings of part of his class-instruction. These recordings may be played directly through the sound system installed in the classroom or through portable sound systems for outdoor problems. Such tape recordings will add drama, interest and variety to your instruction and increase student learning.

b. By using recordings you can introduce other voices into the classroom to add interest and authority to the presentation. There is a dramatic appeal to the recorded word since most of the members of your classes have been conditioned to radio and television listening. Recording the skit will save repeated use of AI's and other instructional personnel as your present the problem time after time.

c. Recordings force the student listener to recreate the situation and to use his imagination. This frequently appeals to many students who have a sense of the dramatic. Be careful not to overuse recordings since what we learned through our ear is not so readily understood and retained as what we learn through seeing. When you use imagination in your teaching through the use of bold and dramatic training aids, student learning is stronger since the initial impression is deeper.

E. PROPER USE OF TRAINING AIDS

1. General. After selecting the desired aids, you must know how to use them properly. Nothing destroys class morale more than watching an instructor fumble with a chart not properly displayed, struggle with equipment not ready for use, or discover at the last minute that all of his training aids are not at the class site. Follow these general rules in using any type of training aid:

a. Preparatory Preview. In this step make sure that all of your aids are clean and in good repair. Remember, a shoddy training aid indicates a careless instructor. When you have training aids made, check the content of charts and transparencies carefully for possible mistakes in spelling and grammar. Because you submitted a perfect copy does not mean you don't have to recheck. The time to correct errors is before the class begins and not while you are presenting the problem. Explain carefully to your assistant instructor what you want him to do while you are using this training aid; where you want it displayed; when you intend to use it; and how you want him to help. Rehearse this with him to insure smoothness.

b. Introduce the Aid Skillfully. Since a training aid is used to assist learning, some discussion should precede the showing of the aid to give a logical reason why a student should look at or listen to this aid. Too many instructors simply flash a training aid before a startled class and begin talking about it without properly introducing it and conditioning the students to the features of the aid. The aid should appear at the strategic time because student need for the training aid has been built up by proper introduction to it.

c. Show the Aid. Display your training aid smartly and skillfully without undue distraction. Rehearse the use of aids before your presentation. Show the aid so that the entire class can see it. Position yourself to avoid obstructing student view. Use a pointer, a spotter, or the strip method to focus attention on the specific portion of the aid under discussion. While showing the training aid talk to your class and not to the aid. Generally, it is not good to talk when revealing or putting away an aid.

d. Review the Aid. Frequently the question is asked, "How many times should I show a particular training aid during a one or two hour problem?" Show it as many times as it is necessary to assure yourself that the class understands and that you have reached your instructional objective. Many instructors give a quick summary by reshowing their training aids near the end of each problem. Some problems may require that the training aid continue to be exposed during the instructional conference but it is generally a good practice to put away the aid or cover it as soon as you are done with it.

CHAPTER VIII

CRITIQUE, REVIEW, EXAMINATION AND PANEL DISCUSSION

A. CRITIQUE AND REVIEW - A STAGE OF INSTRUCTION

1. The critique is so important that it may be considered a stage of instruction in itself. However, it is most valuable when it becomes, in effect, a part of another stage. Every period of presentation, application, and examination should include a well-integrated critique. The effectiveness of this stage depends upon the flexibility with which the instructor employs it.

2. The Critique is the Final Stage of Instruction. It is designed to review the lesson and reemphasize the teaching points. Practical exercise should always be followed by a critique, otherwise students may not have a clear, orderly idea of what was done right and what was done wrong. Good instruction includes intelligent, tactful, and constructive criticism. This criticism can be given most effectively in a group discussion held after a practical exercise or classroom instruction. The critique can be used to:

   a. Sum up and clarify a situation developed in the lesson or test and point out correct or incorrect methods of execution.

   b. Provide an overall view of the entire applicatory operation.

   c. Indicate the strong and the weak points of a performance and methods or procedures to be used in correcting errors or mistakes.

   d. Reemphasize the fundamental points of the lesson.

   e. Develop among personnel a spirit of unity and an appreciation of the cooperation and teamwork necessary in team activities.

3. Human Relations Are Important. In conducting a critique, the instructor must not be sarcastic; he must make criticisms or comments in a straightforward, impersonal manner. If deemed necessary, he should criticise individuals in private, praise them in public. Students should leave the critique with a favorable attitude toward the training and have a desire to improve.

4. The Critique Should Relate the Instruction to the Subject or Course. It should emphasize the continuous nature of training by calling attention to what has been done earlier and to the relationship of the instruction just completed to the subject or course of which it is a part.

5. Specific Points Should Be Covered. Procedures used, examples of personal initiative, type of errors and ways for correcting them, and fundamental teaching points should be covered specifically.

6. Fundamentals Should Be Emphasized. The critique should indicate the various acceptable solutions; it must not give the impression that there is but one correct method of solving the problem. Such a misconception leads to the adoption of stereotyped solutions and to attempts to guess the approved solution, resulting in loss of initiative and independent thought. The critique should emphasize the fundamental principles and should criticize and evaluate the different solutions on the basis of their completeness, effectiveness, and observance of the fundamentals.

7. Participation Should Be Encouraged. In almost every class there will be individuals who can contribute experience that will emphasize and illustrate key points. Too, a well-controlled class discussion makes the students feel that the critique is a period for learning rather than a time set aside for criticism of their performance.

8. Instruction Should Be Foremost. The critique must be conducted as a stage of instruction and part of the course. A good critique might be said to "nail the lid" on the store of knowledge the student has gained during the course of instruction.

9. **Steps in the Conduct of the Critique or Review.** The critique cannot be planned as thoroughly as other stages of instruction because the points to be covered are influenced directly by the members of the class. Advance planning can include the general outline; during other stages of instruction the instructor can take notes to guide his critique; but detailed planning is not practical. However, the instructor can insure complete coverage of the essential elements by following this general procedure.

Step 1: Restate the Objectives. This will enable the class to start its consideration of the instruction on a common ground. This step is necessary because some students may have become concerned only with a particular aspect of the subject and may have forgotten the overall objective.

Step 2: Review the Teaching Points. In this step, briefly summarize the methods used, or the teaching points brought out, and answer the student questions: "What was this all about?" "What did we do?" "How was it done?"

Step 3: Evaluate Performance. This is the most important part of the critique. Using notes taken during the exercise, the instructor points out and discusses the strong points. Then he brings out the weaker points and makes specific suggestions for improvement. He must be careful not to "talk down" to the group and must not expect a standard of performance beyond the capabilities of the students, considering their state of training. All remarks must be specific; students will not profit from generalities.

Step 4: Control the Group in Discussion. The instructor should encourage the class to discuss the points he has mentioned and to suggest other points for discussion. All the techniques of conducting a directed discussion apply in this step to insure that criticism is constructive and that discussion is to the point.

Step 5: Summarize. The critique should be concluded with a brief but comprehensive summation of the points brought out. The instructor can reemphasize teaching points and suggest specific practices to overcome certain deficiencies. The critique should be businesslike; it must not degenerate into a harangue.

B. WRITTEN EXAMINATION

The use of tests or examinations to evaluate student performance is a necessary step in the teaching process. Instructors must use tests to determine overall training progress and they must also use tests to check on the effectiveness of instruction. It cannot be assumed that men have learned until the examination stage of instruction has revealed a desirable standard of achievement.

1. **Purposes of Tests or Examination.**

   a. Tests aid in improving instruction by:

   (1) Discovering gaps in learning. Properly constructed tests reveal gaps and misunderstandings in student learning. If frequent tests are given, such weaknesses can be discovered and instructors can correct them by reteaching their material.

   (2) Emphasizing main points. A test is actually a valuable teaching device in that students tend to remember longer and more vividly those points which are covered in an examination. Tests encourage students, as well as instructors, to review the materials that have been presented and to organize various phases of instruction into a meaningful set of skills, techniques, and knowledge.

   (3) Evaluating instructional methods. Tests measure not only student performance but also instructor performance. By studying the results of tests, instructors can determine the relative effectiveness of their various methods and techniques.

   b. Tests provide an incentive for learning. Students learn more rapidly when made to feel responsible for learning. For example, they are more likely to pay close attention to a training film if they know a test will be given when the picture is over. Generally, instructors who more frequently give tests will find that their students will be more alert and learn more. There is a danger, however, in overemphasizing tests and test scores as a basic motivation for learning. Student interest in test scores is superficial and one which can easily lead to efforts "hit the test" rather than learn the subject matter for its value in the future. Students who study primarily to pass tests may forget what they learn much faster than those who are interested in learning because of the real value to be derived. The instructor should give rigid tests and give them frequently, but they should be designed to require the student to make application of what he has been taught.

    c. Tests provide a basis for evaluation. Another purpose of testing is to determine which students have attained the minimum standard of performance and which have not. In many cases it is desirable to indicate the extent to which students exceed or fall below the standards required. Students learn different amounts; the grade recorded for each student should be an accurate index of what he has learned. Unless a sound testing program is employed, it is impossible to determine the relative achievement of students.

    d. Tests furnish a basis for selection for further advanced training. The results of training tests furnish valuable supplementary information if tests are actually a measure of student performance. The test results become a valuable basis for determining whether a student should be placed in another activity or whether he should receive advanced training.

2. **Characteristics of an Examination.** There are six important factors which affect the quality of an examination. These factors, while not considered to be separate and distinct, are defined and discussed separately in order to develop a clear understanding of the characteristics of an examination.

    a. The test must be valid.

        (1) The test must measure what it is supposed to measure; this is its most important characteristic. A test designed to measure what students have learned in a specific training program should measure achievement in that training program and nothing else.

        (2) The instructor should, whenever possible, invite the opinion of other competent persons as to the validity of his tests. The test results obtained should be compared with other measures of student achievement. A variety of tests and other evaluating devices must be used in obtaining a valid measure of achievement.

    b. The test must be reliable.

        (1) A test is said to be reliable when it measures accurately and consistently. If the test measures in exactly the same manner each time it is administered, and if the factors that affect the test scores affect them to the same extent every time the test is given, the test is said to be highly reliable. This characteristic of a test is especially important when tests are used to compare the proficiency of several classes.

        (2) There are several factors which affect the reliability of a test. In general, the reliability of a test can be raised by increasing its length. The more responses required of students, the more reliable is the measurement of their achievement. Test items should be designed to make it difficult to guess the correct answer. Also, the way in which a test is administered, and the conditions under which it is given should be consistent.

        (3) Other characteristics of the test, such as validity and objectivity, also contribute to its reliability.

    c. The test must be objective. A test is objective when instructor opinion, bias, or individual judgment is not a major factor in scoring it. Objectivity is a relative term. Some tests, such as written examinations which are machine graded, are highly objective; others, such as essay examinations, written exercises, and observation techniques, are less objective. Sometimes observation is the only effective way of determining proficiency. In such cases the instructor must strive to make his observations as objective as possible.

    d. The test should discriminate. The test should be constructed in such a manner that it will detect or measure small differences in achievement or attainment. This is essential if the test is to be used for ranking students on the basis of individual achievement or for assigning marks. It is not an important consideration if the test is used to measure the level of the entire class or as an instructional quiz where the primary purpose is instruction rather than measurement. As is true with validity, reliability, and objectivity, the discriminating power of a test is increased by concentrating on and improving each individual test item. After the test has been administered, an item analysis can be made which will show the relative difficulty of each item and, of greater importance, the extent to which each discriminates between good and poor students. Often, as with reliability, it is necessary to increase the length of the test to get clear-cut discrimination. Three things will be true of a test that has discrimination:

(1) There will be a wide range of scores when the test is administered to the students who have actually achieved amounts that are significantly different.

(2) The test will include items at all levels of difficulty. Some items will be relatively difficult and will be answered correctly only by the best students; others will be relatively easy and will be answered correctly by most students.

(3) Each item contained in the test will possess discrimination. If all students answer an item correctly, it is probably lacking in this respect.

e. The test must be comprehensive. It must sample liberally all phases of instruction which are covered by the test. It is neither necessary nor practical to test every point that is taught in a course; but a sufficient number of points should be included to provide a valid measure of student achievement in the complete course.

f. The test must be readily administered and scored. It must be so devised that a minimum amount of student time will be consumed in answering each item. The test items must also be constructed so that they can be scored quickly and efficiently.

3. Types of Tests:

a. Written examination requiring written answers.

b. Oral examination in the form of a quiz, and questions or directions given verbally in conducting practical exercises.

c. Practical exercise or performance type testing where questions or situations are written and action is taken by the student for evaluation by the instructor.

NOTE: An example of a marksmanship examination is in Chapter XII - Evaluation of a Team Shooter.

C. PANEL DISCUSSION

A panel discussion may be used in conjunction with the critique to provide greater latitude in subject coverage. The panel will be composed of the instructor-coaches and shooters who presented the course of instruction and various members of prominent pistol and rifle teams. By their experience and knowledge of advanced marksmanship, these individuals can contribute materially to the progressive understanding of all phases of the techniques and fundamentals of pistol and rifle marksmanship.

1. Introduce guest speaker(s).

2. Introduce pistol or rifle instructors and name the classes presented by them.

3. Questions asked should be directed to the principal instructor who presented the subject. You may ask for additional comments from other panel members.

4. Please ask questions only about subjects taught during the course of instruction.

5. In large class groups when you have a question, please have the student raise his hand so that an assistant instructor can reach him with a microphone. The question must be heard by everyone in the class.

CHAPTER IX

MARKSMANSHIP INSTRUCTOR TRAINING COURSE LESSON PLANS

A. INSTRUCTOR TRAINING I
(Methods of Instruction)

AAMI & CC705
50 Minutes
Mar 1975

LESSON OUTLINE

I. LESSON OBJECTIVE: To enable the marksmanship student to define each of the four (4) methods of instruction; lecture, demonstration, conference, and performance; explain the advantages and limitations of each method and disclose the procedures used in planning and conducting each type of instruction.

II. STUDENT PERFORMANCE OBJECTIVE: As a result of this instruction, the student must be able to accomplish the following student performance objectives:

    A. Given the definition, advantages, limitation, the procedures for planning and conducting the lecture method of instruction; DEFINE the lecture, EXPLAIN the advantages and disadvantages and DISCLOSE the planning and conducting procedures as shown in Chapter II, USAMU Marksmanship Instructors' and Coaches' Guide.

    B. Given the definition, advantages, disadvantages, the procedures of planning and conducting a demonstration; DEFINE the demonstration as a method of instruction, EXPLAIN the advantages and limitations and DISCLOSE the procedures for planning and conducting a demonstration as shown in Chapter II, USAMU Marksmanship Instructor' and Coaches Guide.

    C. Given the definition, advantages, limitations, the procedures for planning and conducting an instructional conference; DEFINE the conference as a method of instruction, EXPLAIN the advantages and disadvantages and DISCLOSE the procedures for planning and conducting an instructional conference as shown in Chapter II, USAMU Marksmanship Instructors' and Coaches' Guide.

    D. Given the definition, advantages and disadvantages, the procedures for planning and conducting performance type instruction; DEFINE the performance method of instruction, EXPLAIN the advantages and limitations, and DISCLOSE the procedures for planning and conducting a period of performance type instruction as prescribed in Chapter II, USAMU Marksmanship Instructors' and Coaches' Guide.

III. ADVANCE ASSIGNMENT: Chapter II, USAMU Marksmanship Instructors' and Coaches' Guide.

NOTE: SHOW INTRODUCTION SLIDE, "THE MARKSMANSHIP INSTRUCTOR" AS STUDENTS ARE BEING SEATED.

IV. INTRODUCTION:

    A. Gain Attention: Competitive pistol and rifle shooting must retain its dynamic quality as a sport. It must continue to be a factor in promoting the general readiness of the citizen to meet the defense needs of the community and nation. This objective can be accomplished by having well trained pistol and rifle marksmanship instructors available in each of the many shooting clubs, Police departments and military installations scattered throughout the United States.

    B. Orient Students:

        1. Lesson Tie-In: Your presence at the Coach/Instructor clinics attests to the high degree of skill necessary to successfully compete in marksmanship competition. If it is your desire to assist other people to become expert shooters as you are, you must also be an expert instructor. If the instructor fails to teach, the student fails to learn.

        2. Motivation: The great majority of shooters are part time enthusiasts. Shooting and training time is limited. When marksmanship instruction is given, it must be an interesting and efficient presentation. The shooting club members that absorb the greatest proportion of the essential

information offered in the classroom and on the practice range, will be the group in strongest contention for winning honors.

NOTE: SHOW SLIDE #1 (METHODS OF INSTRUCTION)

    3. <u>Scope</u>: You can learn to use the four (4) principal methods of instruction effectively by being able to:

        a. Define the lecture, demonstration, conference and performance.

        b. Explain the advantages and limitations of each method.

        c. Disclose the procedures for planning and conducting each of the methods.

<u>TRANSITION</u>: The characteristics of the principal methods of instruction furnish an indication of the best methods that should be used to meet a given training situation.

NOTE: SHOW SLIDE #2 (INSTRUCTOR-STUDENT PARTICIPATION)

    V. BODY:

      A. <u>First Student Performance Objective</u>: Given the definition, advantages, limitations, the procedures for planning and conducting the lecture method of instruction; DEFINE the lecture, EXPLAIN the advantages and disadvantages and limitations and DISCLOSE the procedures for planning and conducting a lecture as shown in Chapter II, USAMU Marksmanship Instructors' and Coaches' Guide.

NOTE: SHOW SLIDE #3 (LECTURE - THE "TELLING" METHOD)

        1. <u>The "Telling" Method</u>. In the lecture method, you tell facts, principles, theories or relationships which the student must learn and understand. Essentially the lecture is a one-way oral presentation of ideas. The instructor developes the subject without any or with only negligible student participation. For this reason it is the least effective method for teaching marksmanship.

NOTE: SHOW SLIDE #4 (ADVANTAGES OF THE LECTURE)

        2. <u>Advantages of the lecture</u>: Many persons believe that the lecture is an unproductive method of instruction. This is true only if the presentation is so dull that student attention wanders. As an instructor, you will lecture to introduce, subsummarize, conclude your presentation, direct and critique student performance, or to narrate demonstrations. The lecture in combination with other methods of instruction is an indispensable ingredient of effective instruction.

          a. With the lecture, you will have an effective method of presenting many facts and ideas in a short period of training time.

          b. Place emphasis specifically where you desire it. Since you control the learning situation closely, develop only those points you need to develop.

          c. The lecture is very effective in presenting new or background information to provide students with the knowledge they will need to participate in discussions or in performance exercises.

          d. The size of the listening audience does not restrict the lecture. You can deliver the lecture to twenty, two hundred, or two thousand or more students.

          e. The lecture is an effective method for guest speakers. It is more effective when presented by a real authority on the subject. When the speaker is well informed and uses good presentation techniques, he can hold the interest and stimulate the imagination of his listeners.

TRANSITION: The lecture adds variety to other methods of instruction and provides a welcome change of pace when interspersed among other methods. It will provide a "time cushion" since you can shorten your explanation or amplify it as training time permits.

NOTE: SHOW SLIDE #5 (DISADVANTAGES OF THE LECTURE)

3. <u>Limitations of the lecture</u>:

    a. The major limitation of the lecture is the lack of active student participation. There is no two - way communication between instructor and student. Lectures foster overdependence upon the teacher as the expert and encourages uncritical acceptance of everything the speaker says.

    b. It is relatively difficult to determine what the student is learning. In a straight lecture, you are not sure whether the student is absorbing the material or not. By using check-up questions, you only sample class comprehension.

    c. You have difficulty in gearing your instruction to the needs or progress of the student. Usually you plan your lesson in a certain way and present it that way. You have little or no flexibility to adapt to the students as you proceed.

    d. Student interest is difficult to maintain. It requires a skillful, dynamic speaker to maintain class interest for extended periods of time. As one critic stated it: "The mind can absorb only as much as the seat can endure."

TRANSITION: The handicaps of the lecture can be minimumized by proper planning.

NOTE: SHOW SLIDE #5 a (PLANNING THE LECTURE)

4. <u>Planning the Lecture</u>:

    a. Organize your ideas in a logical sequence.

    b. You must be specific in explaining and illustrating each supporting ideas or difficult point.

    c. You must limit the number of teaching points to the time allowed for complete coverage of the subject.

    d. You must amply support each of these teaching points with sufficient facts to establish understanding.

    e. You must devise illustrations to supplement your descriptions-either verbal or pictorial.

    f. After you have specifically explained each supporting idea, sub-summarize before you move on the next idea.

    g. Usually, you can improve your lecture by asking two or three thought-provoking questions; these enable you to introduce some discussion into the presentation.

    h. In going to the next teaching point, make a good transition. A transition not only alerts the student to the fact that you are leaving one point and beginning another, but it also shows the student the relationship between ideas.

    i. Since the lecture does not provide for active student participation, you must <u>stimulate</u> the students sight as well as his hearing. Use training aids which reinforce learning.

* *

j. Many marksmanship subjects lend themselves to an interesting demonstration or problem-solving exercise. <u>Combine</u> the lecture with other methods to increase student participation and overcome this major disadvantage of the lecture method.

k. List the personnel and equipment needs that will amply support the presentation of the subject, i.e., sound-equipment, chalk-boards, projection equipment, assistant instructors, etc.

l. The instructor must provide for student comfort, i.e., ample seating, ventilation, lighting, etc.

<u>TRANSITION:</u> When all the advantages and limitations are taken into consideration and the plans are made, the lecture must then be conducted effectively.

<u>NOTE:</u> SHOW SLIDE #5 b (CONDUCTING THE LECTURE)

5. <u>Conducting the Lecture:</u>

In the lecture, students learn primarily by what they hear. Therefore, it is extremely important that you present your ideas in an interesting, enthusiastic manner.

a. At the beginning of the lesson, you state the purpose of the lesson and emphasize why it is important to the student to learn this material in order to do his job effectively. The student then is motivated---he realizes the need to learn. He gives his full attention to what the instructor is saying and doing.

b. Outline the key points. You give direction to student's attention by orienting him on the specific things to be learned.

c. By various methods which involve hearing and seeing, you present and develop the teaching points while maintaining attention and interest.

d. It is the instructor's responsibility to provide for student participation. If the student loses interest, he no longer gives his full attention or concentration to the lesson and learning breaks down. Questions and discussion stimulate interest.

e. Humorous stories and illustrations relax students and helps them to remember the points emphasized by humor.

f. Use demonstrations to illustrate various points in the subject matter. Showing how to accomplish a specific action in marksmanship is immeasurably more effective than the mere telling or verbal description.

* *

g. Usually you can improve your lecture by asking two or three thought-provoking questions; these enable you to introduce some discussion into the presentation.

h. In going to the next teaching point, make a good transition. A transition not only alerts the student to the fact that you are leaving one point and beginning another, but it also shows the student the relationship between ideas.

i. Since the lecture does not provide for active student participation, you must <u>stimulate</u> the student's sight as well as his hearing. Use training aids which reinforce learning.

<u>TRANSITION:</u> When the lecture is supplemented by other methods of instruction, it adds interest and variety.

B. <u>Second Student Performance Objectives</u>: Given the definition, advantages, disadvantages, the procedures for planning and conducting a demonstration; DEFINE the demonstration as a method of instruction, EXPLAIN the advantages and limitations and DISCLOSE the procedures for planning and conducting a demonstration as shown in Chapter II, USMAU Marksmanship Instructors' and Coaches' Guide.

NOTE: SHOW SLIDE #6 (DEMONSTRATION - "THE SHOWING METHOD")

1. <u>Demonstration - The "Showing Method"</u>: This method assists student learning by showing correct procedure and expected standards. Demonstrating of "showing" is used in combination with other methods, and is usually preceded by lecture and conference and followed with student performance.

<u>TRANSITION</u>: The demonstration has significant advantages.

<u>NOTE</u>: SHOW SLIDE #7 (ADVANTAGES OF THE DEMONSTRATION)

2. <u>Advantages of the Demonstration</u>:

a. Demonstrations save time. A brief demonstration of a proper method is more effective than a lengthy discourse. Showing is usually simpler than telling. Demonstration will make your explanation more concrete in the minds of your students.

b. Demonstrations insure through understanding through their appeal to sight as well as to hearing.

c. Demonstrations stimulate interest.

d. Demonstrations stimulate learning through their realism and dramatic appeal.

e. Demonstrations add variety to other methods of instruction.

f. Demonstrations set the stage for student performance by illustrating correct methods and by setting standards of performance.

g. Demonstrations give the student confidence that attainment of the skill is possible.

<u>NOTE</u>: FIRE A 45 CAL. PISTOL AT AN ENLARGED 50 YD. PISTOL TARGET. (Use blanks)

<u>NOTE</u>: SHOW A PISTOL OR RIFLE TARGET WITH A GROUP OF FIVE SHOTS, IN AND NEAR THE CENTER.

<u>NOTE</u>: SHOW SLIDE #8 (DISADVANTAGES OF THE DEMONSTRATION)

3. <u>Limitations of the Demonstration</u>:

a. During the presentation period, demonstrations do not normally provide for active student participation. You can overcome this limitation by asking questions of students and by encouraging the student to ask questions between steps in a demonstration.

b. Demonstration type instruction poses a difficulty in determining the degree of student learning as a result of observing a demonstration.

c. The scope of coverage in a given demonstration is restricted because of the ease with which a prolonged or involved demonstration can confuse students.

<u>TRANSITION</u>: Most of the limitations can be compensated for by careful planning.

<u>NOTE</u>: SHOW SLIDE #10 (PLANNING A DEMONSTRATION)

4. <u>Planning a Demonstration</u>: The success of a good demonstration depends ninety percent upon planning and rehearsal, and ten percent upon execution. Here is the checklist of things to consider.

        a. Limit the scope so that the demonstration has a specific purpose. Remember, the essence of effective demonstration is brevity and precise timing.

        b. If your demonstration involves several operations, list these operations and demonstrate them one at a time. If you wish your students to learn more than one way of performing an operation, plan a separate and distinct demonstration for each method.

        c. Prepare a scenario. Include an introduction, an explanation and a summary, and incorporate training aids in explaining the steps being demonstrated.

        d. Make a list of the personnel and equipment needed for the demonstration and arrange for their availability for rehearsals and for the actual presentation.

        e. Arrange for and check the facilities at the site of the demonstration to insure that the students can see and hear. Arrange for sound equipment if necessary.

        f. Rehearse the demonstration at the actual site to be used. Utilize initial rehearsals to check equipment and timing and to develop the proficiency of assistants. Request the presence of other members of your team at your dress rehearsals to obtain constructive criticism and to insure high standards of performance.

        g. Immediately prior to class time, make a final check to insure that personnel and equipment are present and that classroom arrangements are complete.

TRANSITION: Planning should be continued as rehearsals reveal short comings in initial planning.

NOTE: SHOW SLIDE #11 (CONDUCTING A DEMONSTRATION)

    5. <u>Conducting a Demonstration:</u>

        a. Introduce purpose of demonstration and outline important points.

        b. During a demonstration, you obviously want the student to watch the demonstration, yet you <u>do not entirely forfeit your job of oral communication.</u> If you are demonstrating an item of equipment, speak to the students, not the equipment.

        c. If, on the other hand, you are explaining while as assistant instructor demonstrates the equipment, direct the students' attention to the demonstrator. It is not uncommon for students to watch an instructor and fail to see an assistant instructor demonstrating the steps of the operation.

        d. Insure that all students can see the demonstration. You may find it necessary to repeat a demonstration at a different angle. Be careful that neither you nor your assistants block the students' view.

        e. Explain the overall demonstration step-by-step to insure that the students understand the purpose and know what to look for. Knowing what they should be getting out of the demonstration affects the way students observe it. Stress the key points.

        f. Ask questions to check student understanding. Encourage students to ask questions between steps, but do not allow students to interrupt the demonstration.

        g. Summarize after completing a step or after several steps or following a series of demonstrations.

TRANSITION: Lecture and demonstration do not lend themselves to ample student participation. If interest, motivation and learning are to be maintained, the student must become closely involved in the teaching process.

    C. <u>Third Student Performance Objective:</u> Given the definition, advantages, limitations, the procedures for planning and conducting an instructional conference; DEFINE the conference as a method of instruction, EXPLAIN the advantages and disadvantages and DISCLOSE the procedures for planning and conducting an instructional conference as shown in Chapter II, USAMU Marksmanship Instructors' and Coaches' Guide.

NOTE: SHOW SLIDE #12 (CONFERENCE - THE "DISCUSSING" METHOD)

    1. <u>Conference</u>: The "discussing" method in which the teaching points are developed with student participation.

As education has placed increased emphasis on student activity for optimum learning, the instructional conference has proved effective in securing good student participation. As a leader of an instructional conference, you will discover that the conference is not an easy way out. You must research every facet of your subject because new questions will arise from students in each succeeding discussion. Your teaching task will require that you exercise a high degree of intelligence, tact, alertness, and ability to think on your feet.

NOTE: SHOW SLIDE #13 (ADVANTAGES OF THE INSTRUCTIONAL CONFERENCE)

    2. Advantages of the conference:

        a. The conference or guided discussion method encourages active student participation and maintains interest. When students discuss, probe, disagree, or answer provocative questions, they are concentrating, thinking, and learning actively. Such active learning makes a greater mental impression and is remembered longer.

        b. During the guided discussion, students often contribute new ideas and new applications from their background of experience. It is seldom that any instructor, no matter how well informed, can match the cumulative experience of the class. These student-originated ideas not only make instruction more meaningful, but they result in course improvements.

        c. Discussion stimulates creative thinking and reasoning. Students become accustomed to thinking critically, to making comparisons, and to relating ideas and doctrine to their experience and previous learning.

        d. The conference method provides you with frequent opportunities to check student comprehension of the subject. This enables you to gear instruction to the proper learning rate for that group. You get an immediate appraisal of how well students are absorbing the materials to be learned.

        e. Since active student participation is essential for an instructional conference, students assume responsibility to assist in their own learning. The responsibility for learning shifts to a degree from the instructor and become a shared, cooperative task. Both the students and the instructor must insure that learning takes place.

TRANSITION: A bonus effect in the group participation is that the conference trains students in the skills of cooperative effort, group thinking on a common problem, self-expression, and tolerance of the opinions of others.

NOTE: SHOW SLIDE #14 (LIMITATIONS OF THE INSTRUCTIONAL CONFERENCE)

    3. <u>Limitations of the Instructional Conference</u>:

        a. The instructional conference method consumes more training time than the lecture to cover a specified amount of instructional material. Ideas must be discussed, analyzed, accepted, modified, or rejected.

        b. It is more effective in small groups of fifteen or twenty students where all students voice their opinions. When you use the conference method with one hundred or more students you only sample student opinion and only the more aggressive students will volunteer to express their opinions.

        c. To conduct a guided discussion, you must have students with some knowledge of the subject.

        d. Many gaps in learning or even some incorrect learning may occur if the instructor is inexperienced and there are deficiencies in planning and conducting the conference.

    e. Compared with the lecturer, the conference instructor must possess a more comprehensive knowledge of the subject, more tact, greater versatillity and flexibility.

    f. Control is rather difficult to maintain if there are disagreements, wandering from subject or a student or group with definite convictions wishes to dominate the discussion.

<u>TRANSITION</u>: Most of these limitations can be offset by good planning.

<u>NOTE</u>: SHOW SLIDE #15 (PLANNING A CONFERENCE)

  4. <u>Planning a Conference</u>:

    Planning the introduction and conclusion for an instructional conference is similar to planning of any lesson. The primary concern is planning for the discussions that takes place in the body of the lesson.

    a. Since an informative discussion requires an adequate background knowledge of the subject matter, you must investigate student background and take advantage of the instruction that these students have received to date. You can supplement the students knowledge by issuing advance study assignments, by lecturing initially, or by using both techniques.

    b. Prepare questions to provoke thought, to stimulate discussion, and, above all, to insure thorough understanding of the student performance objectives.

    c. Taking each of your prepared questions in turn, consider likely student responses to these questions. While student responses vary with the individual, they will usually fall into general response patterns. You then are better able to ask good follow-up questions and to lead the students to a thorough understanding of the teaching points.

    d. As with other instructional methods, a good conference requires a skillful use of interest factors, training aids, summaries, and transitions. Plan for frequent subsummaries in conducting a conference.

<u>NOTE</u>: SHOW SLIDE #17 (CONDUCTING A CONFERENCE)

  5. <u>Conducting a Conference</u>:

    Proper use of the techniques which follow will help you to arouse student interest and to obtain optimum student participation and learning.

    a. A conference is based upon student participation, and is most effective when all students are motivated to think constructively and to enter into the discussion.

    b. Students will think more freely and will enter into a discussion more readily when you establish a permissive atmosphere, wherein you encourage a free flow of ideas. Promote a feeling of mutual cooperation, harmony and congeniality by your sincerity and enthusiasm. You maintain this atmosphere by the manner in which you ask questions, call upon students for contributions, and give credit for student ideas.

    c. Thought provoking questions stimulate discussion: By planning good thought-provoking questions and by establishing rapport with students, you have laid the foundation for discussion.

    d. You can often stifle a good discussion by careless, unthinking remarks. For example, you ask a question and call upon a student, who gives you a good answer, but one you had not planned for. If, you reply, "That is true, but it isn't what I have in mind," you are stifling student thought.

    e. You will find that discussion will also falter unless the instructor takes frequent, definite steps to stimulate discussion. By making a startling or controversial statement, you can often obtain enthusiastic participation once the students are aroused by these remarks.

    f. After you are successful in stimulating a good discussion, your next role is that of moderator. You must keep the discussion on the teaching point. When the discussion is obviously beginning to go "off on a tangent," summarize the comments and, by making a discussion stimulating remark, bring the discussion back to the subject.

    g. Your contribution to student understanding depends largely upon the effectiveness of your subsummaries. A good subsummary of the students' discussion will briefly point out the ideas expressed, resolve conflicting points of view, relate ideas to the teaching point. The mark of a good conference leader is his ability to guide discussion toward learning outcomes without dominating the discussion.

TRANSITION: The reaction to the intake of information is not complete until the student does something besides talk about it. Physically "doing" something increases the manual skill and a mental task improves problem solving ability.

  D. Fourth Student Performance Objective: Given the definition, advantages and disadvantages, the procedures for planning and conducting performance type instruction; DEFINE the performance method of instruction, EXPLAIN the advantages and limitations, and DISCLOSE the procedures for planning and conducting a period of performance type instruction as prescribed in Chapter II, USAMU Marksmanship Instructors and Coaches' Guide.

NOTE: SHOW SLIDE #18 (PERFORMANCE OR THE "DOING" METHOD)

   1. Performance - The "Doing" Method: The performance method emphasizes student participation in order to learn and improve physical and mental skills. Sometimes, it involves problem solving skills in applying principles, techniques, or procedures to a realistic training situation.

NOTE: SHOW SLIDE #19a (ADVANTAGES OF PERFORMANCE TYPE INSTRUCTION)

   2. Advantages of the Performance Method:

    a. Allows full student participation in the learning.

    b. Repractice and repetition results in improvement of skill.

    c. Allows individual practice as in the actual situation.

    d. Allows team practice for improvement of unity of action by a group.

    e. Applies mental skill to problem solving in realistic training situations.

    f. Promotes enthusiasm and competition among students.

TRANSITION: Performance instruction has some disadvantages but they do not limit the effectiveness as much as in the other methods.

NOTE: SHOW SLIDE #19b (DISADVANTAGES OF PERFORMANCE TYPE INSTRUCTION)

   3. Disadvantages of Performance Type Instruction:

    a. May become sterotyped or unchanging.

    b. Complete evaluation of instruction is difficult except for the training of small groups.

    c. It is difficult for one instructor to control the complete operation. The larger the training group, the more assistance needed to maintain standards. Each student group must remain small for effective supervision.

    d. Supervision of performance requires extensive training or ample previous experience on the part of the assistant instructors.

NOTE: SHOW SLIDE #19c (PLANNING PERFORMANCE TYPE INSTRUCTION)

    4. **Planning Performance Type Instruction:**

        a. Check for required equipment.

        b. Establish realistic standards for students.

        c. Set a time limit on completing a performance task or action.

        d. Give step-by-step specific directions to students before attempting action.

        e. Demonstrate the action prior to student performance.

        f. Plan for close supervision of each individual student and observe safety controls.

        g. Train assistants by step-by-step instruction in the procedures to be followed.

TRANSITION: Careful planning is critical to the successful conduct of performance type instruction.

NOTE: SHOW SLIDE #20 (CONDUCTING STUDENT PERFORMANCE)

    5. **Conducting Student Performance:**

        a. *The student must be motivated.*

            (1) Tell students specifically what they are to learn and why.

            (2) Men will meet a challenge, so let them know what standards you expect of them.

            (3) Tell them how well they are progressing.

            (4) Maintain interest through competition.

            (5) Overcome monotony by varying your procedures.

            (6) Gradually increase performance standards.

            (7) Make performance realistic.

        b. *Explain and Demonstrate:* In explaining a procedure, use conference techniques and encourage student questions, not only to insure understanding, but to make sure the student is learning the procedure. A good demonstration will make the procedure clear and set standards for the student to attain.

        c. *Control and Supervise:* Performance instruction must be closely controlled and supervised at all times.

* *

            (1) Make your directions specific, and encourage students to ask questions if they do not understand.

            (2) Brief you assistants thoroughly so that they help you control student activity.

            (3) Supervise constantly to insure that students understand how to perform correctly and are making satisfactory progress.

            (4) When students learn to perform the operation correctly, insure that they are kept busy learning to perform it better and faster.

(5) Prevent students from forming faulty habits; but, at the same time, allow students to use initiative and resourcefulness. It takes common sense and tact to critique student performance properly.

NOTE: SHOW TOP LINES ONLY OF SLIDE #21 (COMBINING SEVERAL METHODS)

VI. CONCLUSION:

A. <u>Retain Attention</u>: Combining methods of instruction yields great and satisfactory results When the instructor appelas to the several senses of the student, he provides for participation and student initiative. Learning is enhanced.

B. <u>Application</u>: Marksmanship instruction for the shooter must be efficient, interesting and informative if the instructor hopes to materially affect the shooters skill with the information he has to offer. The shooter's time is valuable and he cannot devote endless hours of plodding, dull instruction.

NOTE: SHOW SLIDE #21 (RIGHT SIDE ONLY-LECTURE, DEMONSTRATION, CONFERENCE AND PERFORMANCE)

C. <u>Summary</u>: During this period of instruction, we have defined the principal methods of instruction, explained the advantages and limitations of each method of instruction and disclosed the manner in which a lecture, demonstration, conference and performance type instruction is planned and conducted.

NOTE: SHOW ALL OF SLIDE #21 - 1. "TELL" BY LECTURE, 2. "SHOW" BY DEMONSTRATION, 3. "DISCUSS" BY CONFERENCE, 4. "DO" BY PERFORMANCE.

D. <u>Closing Statement</u>: Don't waste the shooter's valuable time by trying to teach him to improve his shooting skill without proper organization and preparation of instruction.

B. INSTRUCTOR TRAINING II AAMI& CC706
(Control of Interest and Effective Speaking) 50 Min
Mar 1975

LESSON OUTLINE

I. LESSON OBJECTIVE: To enable the marksmanship student to employ interest factors and effective speaking techniques when presenting courses of instruction.

II. STUDENT PERFORMANCE OBJECTIVES: As a result of this instruction, the student must be able to accomplish the following student performance objectives:

    A. Given the factors and technique of maintaining student interest, EMPLOY these methods of interest control as sown in Chapter V, USAMU Marksmanship Instructors' and Coaches' Guide.

    B. Given the factors which make up effective speech, EXPLAIN how the voice, body and mind of the speaker effect human communication as outlined in Chapter IV, USAMU Marksmanship Instructors' and Coaches' Guide.

III. ADVANCE ASSIGNMENT: Chapter IV and V, USAMU Marksmanship Instructors' and Coaches' Guide.

IV. INTRODUCTION:

    A. Gain Attention: What qualities make an interesting, effective speaker? These men know.

NOTE: PLAY THE CHURCHILL-KENNEDY TAPE

NOTE: SHOW CHURCHILL SLIDE #1A., DURING SOUND OF HIS VOICE, SWITCH TO KENNEDY SLIDE #1B., WHEN HIS VOICE COMES ON, TURN OFF PROJECTOR AND TAPE RECORDER WHEN DIALOGUE IS COMPLETED.

"These two men were able to influence the destiny of nations with the power of the spoken work. Their listeners were interested in every syllable they uttered during a speech."

    B. Orient Students:

        1. Lesson Tie-In: In this second hour of instructor training course, we will explore the techniques of effective speaking and holding the interest of the student.

        2. Motivation: The student learns only when he is interested in what the speaker is saying You, as an instructor can hold the students attention and communicate your ideas to him if your speech has a magnetic, vivid quality that is projected effectively into his mind. A mental chain reaction must be initiated in the student's mind to jar loose new ideas as well as bring understanding. He must feel the power of new knowledge. He must be stimulated to think.

NOTE: SHOW SLIDE #1C (INSTRUCTION TECHNIQUE) SHOW EACH ITEM AS IT IS MENTIONED

        3. Scope: The vital ways of infusing life, meaning, and brilliance into your presentation of marksmanship instruction are:

            a. Employ the methods of interest control.

            b. Vary the methods of interest control.

            c. Reflect your personality, but avoid exaggerated platform behavior.

            d. In speaking, use the voice, body and mind to communicate effectively.

NOTE: TURN OFF PROJECTOR

TRANSITION: Gaining attention is relatively easy compared to maintaining interest.

V. BODY:

   A. First Student Performance Objective: Given the factors and techniques of maintaining student interest, EMPLOY these methods of interest control as shown in Chapter V, USAMU Marksmanship Instructors' and Coaches' Guide.

QUESTION: Define an interest factor.

ANSWER: An interest factor is a technique used to gain, maintain or increase student interest by appealing to one or more of the senses; hearing, seeing, or doing (touching).

QUESTION: Name one important factor in hearing interest and state why you think it is important.

ANSWER: "Voice technique is an important interest factor if the voice being heard captivates the listener by skillful use of inflection, pauses, clear articulation and pronunciation, and variations in rate and volume."

NOTE: SHOW SLIDE #2 (USE INTEREST FACTORS)

NOTE: PLAY FREDRIC MARCH TAPE

NOTE: SHOW SLIDE #3 (THE HEARING INTEREST FACTORS)

   1. Hearing the instructors' voice is the primary means of communication. Speech is the factor that ties all the elements of good instruction together and gives them meaning and purpose.

      a. Combat examples provide the instructor with a realistic and dramatic method of presenting or emphasizing a point.

      b. Personal experiences lend the weight of authority to the points you are making.

      c. Humorous stories and illustrations relax the student and help him to remember the points emphasized by humor.

         (1) Humor must be clean.

         (2) Humor must be applicable to the technique point.

      d. Thought provoking questions force students to concentrate on the instruction.

         (1) A question stimulates a student to analyze a situation and arrive at a solution.

         (2) A question alerts the student that he is responsible for an answer.

         (3) A question should be asked so as to give the student time to formulate a reply - ask, pause, call.

         (4) Questions can lead to discussion by students.

         (5) Different voices in classroom appeal to the sense of hearing.

      e. Recordings add realism and variety.

         (1) And bring voices and sounds into classroom not otherwise available.

NOTE: PLAY BACKGROUND MUSIC AND SOUNDS TAPE

         (2) Relatively easy to produce.

         (3) Can be amplified over regular public address system.

NOTE: SHOW SLIDE #2A (USE INTEREST FACTORS)

2. Seeing: "A picture is worth a thousand words."

    a. Real understanding comes to most students when they can "see" an idea illustrated or demonstrated.

    b. The instructor must use imagination and ingenuity to promote student learning by means of sense of sight as well as by hearing.

    c. The student may lose track of the idea being developed if the steps of explanation are not visually depicted by:

        (1) Visual training aids.

        (2) Demonstrations and skits.

3. Doing: Students learn best by doing.

    a. Interest level is high when students are doing something.

    b. Doing includes discussion and all types of practical exercises.

    c. Perform mental tasks as well as physical action.

    d. Drill or extensive repetition developes skill.

TRANSITION: Do not overuse any one interest factor. Avoid exceeding the interest span. The interest span is the time a student can give undivided attention to the instruction. Change techniques periodically.

NOTE: SHOW SLIDE #4 (LIMITED VARIATION OF METHOD AND INTEREST FACTORS - MONOTONOUS, WITH POOR CONTROL OF INTEREST) REMOVE SLIDE #4, AFTER EXPLANATION: SHOW SLIDE #5 (WIDE VARIATION OF METHOD AND INTEREST FACTORS - GOOD CONTROL OF INTEREST)

4. The instructor must follow lesson outline in applying control of interest. List each of the interest factors as used.

    a. Student Performance Objectives - SPO

    b. Combat Example - CE

    c. Humor - H

    d. Skits - SK

    e. Sub-Summary - SUB

    f. Tape Recording - R

    g. Questions - Q

    h. Discussion - D

    i. Practical Exercise - PE

    j. Summary - SU

NOTE: CUT OFF PROJECTOR

TRANSITION: The transmission of the instructor's thoughts, ideas, and emotions cannot be fully accomplished by interest factors alone. The speakers voice must be able to sway the audience.

B. **Second Student Performance Objective:** Given the factors which make up effective speech, EXPLAIN how the voice, body, and mind of the speaker effect human communications as outlined in Chapter IV, USAMU Marksmanship Instructors' and Coaches' Guide.

NOTE: SHOW SLIDE #6 (EFFECTIVE SPEAKING) (VOICE, BODY, MIND)

    1. The entire being, to include the voice, the body and the mind, is needed if we are to speak effectively.

    2. Voice sound is produced when the chest muscles and the diaphragm force air from the lungs up the windpipe and through the voice box. The vocal cords produce sound when they are vibrated by the air passing through them. The voice factors are: articulation, pronunciation, grammar, tempo, pauses, inflection, and force.

NOTE: SHOW SLIDE #7 (ARTICULATION)

    a. Articulation is the combining of sounds, i.e., syllables, to form distinct words. The articulators are used to do this.

    (1) In distinct speech, the sound is molded and formed into syllables by the action of the jaw, tongue, teeth, and lips. Vowel sounds are for loudness - consonants for intelligibility or clarity. Each syllable then becomes part of a clearly spoken work.

    (2) Poor articulation causes slurring of the syllables which in turn results in muffled, indistinct, mumbled words.

    (3) Articulation can be improved by the speaker being conscious of how he says words. Speaking clearly will establish efficient communication with your listeners.

TRANSITION: Good articulation can produce a word made up of distinct syllables but the combination of sounds may not result in a properly pronounced word.

NOTE: SHOW SLIDE #8 (PRONUNCIATION)

    b. Proper pronunciation of a word requires the placing of stress or accent on the proper syllable or syllables in accordance with established standards, usually a dictionary. The speaker is then able to convey the intended meaning as the word is spoken.

    (1) Correct pronunciation is an absolute necessity if you desire to be clearly understood.

    (2) Correct pronunciation may be learned from listening to people whose speech is cultivated, by consulting the dictionary and conforming to sectional manners of speech (Colloquialisms).

    (3) Practice in placing accent on proper syllables will form the habit of pronouncing words correctly.

TRANSITION: The incorrect choice of words and faulty sentence construction can reduce understanding.

NOTE: SHOW SLIDE #9 (GRAMMAR)

    c. Good grammar is the correct usage of the spoken or written word to express an idea.

    (1) A basic knowledge of grammar is obtained from academic training in the principles of the English language.

    (2) The familiarity with good literature and association with well spoken people are aids to grammatical excellence.

(3) The speaker must have the desire to use good grammar in his speaking.

TRANSITION: The effective speaker not only wants to be heard and understood, he wants his ideas and thoughts to be believed, felt, learned from and remembered. These last four (4) factors help the voice to convey the speaker's emotions. They are rate, pauses, inflection and force.

NOTE: SHOW SLIDE #10 (RATE OF SPEECH)

    d. Tempo or rate means speak fast enough to be interesting, yet slow enough to be understood.

QUESTION: What is the average rate of speech?

ANSWER:

(1) The average rate of speech is 125 words a minute and varies greatly among individuals.

(2) Interest is added by variation of rate. Place emphasis on important points.

(3) Tempo of the word flow reinforces the meaning of certain statements by helping to create word pictures.

(4) Improve the effectiveness of the rate factor by:

    (a) Listening to yourself on a tape recorder.

    (b) Practice thinking before you speak. Say exactly what you mean to say.

    (c) Gear your tempo of speaking to fit the mood and complexity of the thoughts and ideas you are expressing.

        1. Thoughts in a lighter vein are expressed with a gaiety that fits into a comparatively fast rate.

        2. More serious ideas are expressed in a rather profound atmosphere and need a slower pace of delivery.

        3. Complicated explanations require a slow, mechanical rate.

TRANSITION: The words which express a single idea are grouped together into phrases.

NOTE: SHOW SLIDE #11 (PAUSES)

    e. Pauses accomplish this grouping of ideas for the speaker. The pause is the punctuation mark of speech.

(1) Pauses accomplish four things:

    (a) The listener has a chance to absorb what has been said.

    (b) The speaker has a chance to think of his next point.

    (c) The speaker can take a breath.

    (d) Subtle emphasis is given to the meaning of ideas. The speaker gains the dramatic effect.

NOTE: PLAY THE ROOSEVELT TAPE. SHOW ROOSEVELT PICTURE SLIDE #11A DURING DELIVERY OF VOICE.

(2) Improve the effectiveness of the pause by:

(a) Reading aloud the series of ideas that express a concept. Carefully use pauses to help divide one idea from another.

(b) Do not be afraid to use pauses. Do not inject meaningless sounds such as "uh" to fill the pause.

(c) Coordinating the functions of thinking and breathing prevents untimely pauses from interrupting the expression of an idea.

(d) Avoiding the use of too many pauses because they result in a choppy delivery. The student has difficulty in following the development of an idea.

TRANSITION: There is another punctuation mark of speech, the change in the pitch of the voice.

NOTE: SHOW SLIDE #12 (INFLECTION)

f. Inflection is the change in the normal pitch or tone of the voice. The master key of expression. You can move an audience to tears or laughter.

(1) Vocal variety makes your words more meaningful.

(2) Skillful inflection of your voice imprints your ideas indelibly on the students mind. Without it you put them to sleep.

(a) Where pauses punctuate major segments of an idea, inflection punctuates the lesser division of thoughts used in developing each idea.

(b) Inflection injects the question mark at the end of a sentence.

(c) Change in tone makes a statement of fact more positive.

(d) Variation in pitch puts the exclamation mark at the end of a strong or an emotional statement.

TRANSITION: Each of the preceding factors is a necessary part of clear, interesting speech. None of them are more important than just being heard.

NOTE: SHOW SLIDE #13 (FORCE)

g. Forceful speech combines the volume or carrying power of the voice, the quality of voice tone and the vitality and strength of conviction of the speaker.

(1) There must be regulation of loudness. Students will not respond readily to a constant, high level of sound volume. The speaker should not shout and bang the lectern constantly. Neither will the audience be convinced by a cool, detached manner of speaking that is consistently calm, quiet, conversational or patronizing.

(2) Combine the degrees of volume with the rate of speaking, pauses, and the skillful employment of inflection to convey conviction, meaning or added emphasis.

(3) Knowledge of the subject will enable the speaker to project himself into good mental contact with the audience. He can show that he knows what he wants to say.

(4) A firm grasp of the sequence in which the main points are to be made will allow the speaker to alternately:

(a) Lead the students thought by calmness.

(b) Drive home a point with power.

(c) Let pure silence underline the significance of your words.

(5) Make all statements with firmness and certainty. Display the force of your convictions by a confident manner and purposeful movement.

NOTE: SHOW SLIDE #14 (SPEECH IMPROVEMENT)

(6) Improve your speech by using tape recorder to listen to your voice, read aloud to yourself, and listen to polished speakers.

TRANSITION: The voice factors of effective speech are articulation, pronunciation, grammar, tempo or rate, apuse, inflection and force. The voice alone, however, is not sufficient to communicate perfectly, otherwise radio would have never been replaced by television.

NOTE: SHOW SLIDE #15 (BODY MOVEMENT AND GESTURES)

3. **The body speaks:** The speaker must use his body to reinforce, emphasize and clarify his thoughts as he speaks them. Body movement and gestures add immeasurable to the listener's appreciation and receptiveness of the ideas being expressed.

    a. Posture is the speakers stance. Your stance can be comfortable without a slouch; and erect, without being stiff.

(1) Shift your weight forward toward the balls of the feet. Maintain a nimble, flexible balance.

(2) Stand erect with the body form of assurance and a commanding presence. Poise is the descriptive word for this.

(3) The way you stand is an indication of your inner attitude. Your image affects the manner in which you will be received by the student.

TRANSITION: Standing in one place during all or most of the presentation, even through you speak well, will become tiring to your audience.

    b. Movement is the motion of the instructors body as he moves about the speaker's platform or presentation area.

(1) The speaker's movement is to attract the listeners attention. The eyes of the student instinctively follow a movement by the instructor.

(2) Movement greatly assists in conveying the thoughts of the speaker. Move briskly and with purpose.

(a) Transitions between teaching points can be indicated and emphasized by a lateral shifting of the body weight from one foot to the other.

(b) Lateral movement of a step or two is an informal signal of the closing out of a point and proceeding to the next. Always start lateral movement with the foot toward the direction you intend to move. This avoids awkward crossing of the feet.

(c) A step forward implies the importance of the point being emphasized.

(d) A backward step suggests that the speaker is terminating a point. The listener can relax a bit and consider the point before the speaker goes to the next one.

(e) All movement must be natural and spontaneous. Contrived movement is obviously false and results in the speaker appearing stiff and unnatural.

    c. A gesture is a natural movement of any part of the body to convey a thought, an emotion or to reinforce oral expression.

NOTE: TURN OFF PROJECTOR

       (1) Conventional gestures.

          (a) Pointing is used to reinforce a challenge or an accusation. Point finger at persons and use a pointer to point out objects.

          (b) Giving or receiving. Palm of the hand is facing upward.

          (c) Rejection. Palm of the hand is vertical and toward the audience.

          (d) Anger, power or determination. The clenched fist toward the audience, pounded into palm of other hand or upon lectern.

          (e) Division of points or audience. Palm is vertical with finger edge toward audience.

          (f) Restraint. Extend both arms at shoulder height, palm outward with fingers up.

       (2) Descriptive gestures portray an object or illustrate an action. The speaker describes by imitation:

          (a) Size

          (b) Shape

          (c) Movement

       (3) Avoid exaggerated platform behavior. "Don't ham it up!"

NOTE: SHOW SLIDE #16 (EXAGGERATED BEHAVIOR)

TRANSITION: Skilled pantomime actors are able to communicate every human emotion through the clever use of body movement and facial expression.

NOTE: SHOW SLIDE #17 (FACIAL EXPRESSION AND ATTITUDE)

    4. The Mind: Some ideas and thoughts originate in the mind of the speaker and other thoughts are initiated externaly. All, however, are evaluated and interpreted mentally and then communicated. The speaker then, does more than report the facts. Every word spoken has two effects: What it means and what is implies. The attitudes present in the mind of the speaker are expressed in this manner:

       a. Facial Expression: To sway people, inspire them, or even interest them, your face must show what you are feeling or thinking. Facial expression stimulates listener reactions and helps to maintain a proper relationship with the audience.

          (1) Dead pan or Poker Face. Scan the audience for a smile or a plesant expression. You will unconsciously warm to your class and your expression will become pleasant and meaningful.

          (2) The constantly intense expression or frown can be overcome by removing the anxiety caused by lack of full preparation. Relax as you cover the material and save the intensity for emphasizing the key ideas.

          (3) The more natural and human you appear and act, the more you will influence your students. Your face must reflect your interest in them.

TRANSITION: Emotion is the indicator of how a person feels about all that surrounds him. More simply it indicates his attitude.

b. Sincerity is a priceless attitude for a teacher.

    (1) There are two sources of sincerity:

        (a) Personal, intense belief in your subject.

        (b) Belief in the value of the subject to the listeners.

    (2) The word sincerity is derived from the Latin, sine cera, or without wax. Wax was used by ancient Roman sculptors to cover chiseling errors.

    (3) You can show your sincerity by:

        (a) Directness of manner.

        (b) Facial Expression.

        (c) Clarity of explanation.

        (d) Have a proper balance of humility and authority.

        (e) Effectively use the voice and body.

    (4) The student must see, hear and feel your belief in what you say.

c. Confidence is an attitude of personal assurance. It is your belief in your ability to do a job well.

    (1) There are three sources of confidence:

        (a) Knowledge of subject through research and study.

        (b) Belief in your ability comes from rehearsal and experience.

        (c) Control of stage fright comes from proper practice of calming factors.

            1. Deep breathing

            2. Relaxation of muscles

            3. Get rid of excess energy by exertion.

            4. Short, partial warm-up rehearsal immediately before class convenes.

    (2) Show you are confident by:

        (a) Standing erect

        (b) Moving purposefully

        (c) Looking your listeners in the eye

        (d) Letting ideas flow freely without hesitation

        (e) Rehearsing your introduction throughly so as to carry you through the first minutes of greatest tension.

    (3) Appear confident to your audience and you actually will be.

        d. Enthusiasm is the outward manifestation of a sincere and confident attitude caused by a strong personal excitement and feeling about the subject.

          (1) Show your belief in the subject your way:

            (a) Vigorous, dynamic people show it by brisk, energetic movement, sweeping gestures, rapid rate of speech, widely variable inflection and high vocal force.

            (b) People of a more subdued nature show it by less energetic movement, speaking in measured tones, force is used only on key words and ideas. More use is made of the pause and a calmer businesslike manner.

        e. Humor is the quality that makes something seem funny, amusing or comical. A person whose disposition or temperament includes the ability to recognize this quality in events, conversations and acts is said to have a sense of humor.

          (1) A sense of humor shows that you have very human traits, a warm outgoing personality, and a lively interest in your environment.

          (2) Ability to tell funny jokes and stories relevant to the instruction.

            (a) Humor must be clean; vulgarity and obscenity have no place in the classroom.

            (b) Good judgment is that the comedy is not directed at any one person or group

            (c) Good taste - The belly laugh is not a good steady diet for any audience.

            (d) Be good natured - Keep the class well disposed toward you.

          (3) Have the ability to appreciate and respond to spontaneous humor.

          (4) A humorous attitude helps to gain student interest, attention and understanding.

NOTE: TURN OFF PROJECTOR

    VI. CONCLUSION:

      A. <u>Retain Attention:</u> The student is going to listen attentively only when he is interested in what the speaker is saying. Effective speaking is effectively communicating with the student.

      B. <u>Summary:</u>

NOTE: SHOW SLIDE #1 (INSTRUCTION TECHNIQUES)

        1. An interest factor is a technique used to gain, maintain or increase student interest by appealing to one or more of the senses:

          a. Hearing

          b. Seeing

          c. Doing (touching)

        2. Vary the use of interest factors.

        3. Reflect your personality and avoid exaggerated platform behavior.

NOTE: SHOW SLIDE #6 (EFFECTIVE SPEAKING)

4. Effective speaking is accomplished by the use of the whole being; voice, body, and mind.

    a. The voice factors are articulation, pronunciation, grammar, tempo or rate, pauses, inflection and force.

    b. Body movement and gestures help to emphasize the instructor's ideas.

    c. The mind of the instructor is revealed by his attitude. The attitude is demonstrated by facial expressions, sincerity, confidence, enthusiasm and a sense of humor.

NOTE: TURN OFF PROJECTOR

C. <u>Application</u>: Your Rifle and Pistol Club, Police Department or military organization will benefit greatly if your instruction is well planned and delivered. Training time is limited and every word and minute count.

D. <u>Closing Statement</u>: The key to effective instruction is to know what to do and then do it enthusiastically.

C. INSTRUCTOR TRAINING III AAMI&CC707
(Lesson Planning and Training Aids) 50 Minutes
March 1975

LESSON OUTLINE

I. LESSON OBJECTIVE: To enable the Marksmanship student to organize a lesson plan, select and use appropriate training aids.

II. STUDENT PERFORMANCE OBJECTIVES: As a result of this instruction, the student must be able to accomplish the following student performance objectives:

A. Given the requirements of a lesson plan, ORGANIZE a period of instruction as outlined in Chapter III, USAMU Marksmanship Instructors' and Coaches' Guide.

B. Given the qualities of a good training aid, EXPLAIN how to use training aids to assist student understanding as defined in Chapter VII, USAMU Marksmanship Instructors' and Coaches' Guide.

C. Given a description of several training aid field expedients, DEMONSTRATE the use of instructor-made training aids that can be used in place of prepared training aids as shown in Chapter VII, USAMU Marksmanship Instructors' and Coaches' Guide.

III. ADVANCE ASSIGNMENT: Chapters III and VII, USAMU Marksmanship Instructors' and Coaches' Guide.

IV. INTRODUCTION:

A. <u>Gain Attention</u>: Lesson planning insures proper organization of instruction material and proper use of adequate training aids will reinforce the student's understanding.

B. <u>Orient Students</u>:

1. <u>Lesson Tie-In</u>: When a method or methods of instruction is chosen that will best meet the training mission, student understanding is further aided by complete planning of the lesson and effective use of training aids.

2. <u>Motivation</u>: The student will learn more about marksmanship during the time available for training if the instructor carefully plans and illustrates the lesson.

3. <u>Scope</u>: This period of instruction will be devoted to:

a. Explaining the requirements of organizing a lesson plan.

b. Explaining how properly selected training aids assist student understanding.

c. Explaining and demonstrating field expedient training aids.

V. BODY:

A. <u>First Student Performance Objective</u>: Given the requirements of a lesson plan, ORGANIZE a period of instruction as outlined in Chapter III, USAMU, Marksmanship Instructors' and Coaches' Guide.

QUESTION: What is the purpose of a lesson plan?

NOTE: SHOW SLIDE #1 (PURPOSE OF A LESSON PLAN)

<u>ANSWER</u>: 1. The purpose of a lesson plan is that:

a. Lesson planning insures complete coverage of the subject.

      b. Careful planning affords a wiser selection of supporting material.

      c. Detailed planning helps to develop a logical sequence of teaching points.

      d. Organization of lesson will permit control of time involved in teaching class.

      e. Analysis of student objectives assists in selection and use of proper training aids.

      f. Research and study of subject matter refreshes instructor's memory.

      g. The lesson plan furnishes a record of subject coverage for developing an examination.

NOTE: SHOW SLIDE #2 (PLAN THE LESSON)

QUESTION: How is a lesson planned?

ANSWER: 2. Steps in planning the lesson:

      a. Make an estimate of the teaching situation (determine the best way to achieve the lesson objective.)

          (1) Determine the scope of instruction.

          (2) Establish the relationship of the lesson to other instruction on the same subject.

          (3) Resolve the desired degree of proficiency.

              (a) Levels of proficiency

                  1. General Knowledge

                  2. Working Knowledge

                  3. Qualified

              (b) Factors to consider in selecting a level of proficiency.

                  1. Is subject simple of complex?

                  2. Is this the initial instruction on subject?

                  3. What is the established training purpose?

                  4. Is this an orientation or must students become technically proficient?

                  5. How much time is allowed to present subject?

          (4) Determine the type of student.

              (a) What is the Educational and Military background?

                  1. Advanced students will receive a different level of instruction than basic students.

                  2. Method of presentation will vary according to students knowledge and experience with subject matter.

                  3. Complexity of coverage should reflect student experience on subject.

                  4. Speed of coverage will vary in proportion to amount of previous instruction.

(b) Is the subject of primary interest to student?

(5) Consider the type and availability of facilities and support. Will class site and proficiency of support personnel be suitable for method of instruction and training aids to be used?

TRANSITION: The instructor must now determine what the student must learn in order to achieve the required level of proficiency.

b. The lesson objective is a brief statement indicating the action expected of the student as a result of the instruction. It must contain these elements:

(1) The type of students who will receive the instruction.

(2) The specific action expected of the student.

(3) The subject matter that the student will be expected to learn.

NOTE: READ THE LESSON OBJECTIVE OF THIS LESSON OUTLINE AS AN EXAMPLE.

c. The instructor must decide which is the best method or methods of instruction that will accomplish the training mission. This decision is made by analysis of instructor-student participation.

NOTE: SHOW SLIDE #3 (INSTRUCTOR - STUDENT PARTICIPATION)

(1) The lecture is nearly all instructor activity (telling) while the student listens.

(2) The demonstration requires slightly diminished instructor activity (showing) while the student sees the action as well as hearing the instructor.

(3) The conference sharply reduces instructor activity (guide discussion) as he takes part in the discussion of the subject by the students.

(4) Performance instruction is minor activity by the instructor (observing), while the student is (doing) something under the instructors' supervision.

TRANSITION: Will telling or showing accomplish your objective or will it be necessary for the student to talk about it and/or do it to meet the training mission?

NOTE: SHOW SLIDE #2 (PLAN THE LESSON)

d. The instructor must determine the teaching points or student performance objectives that will fulfill the lesson objective.

(1) Analyze the lesson objective.

(2) List all necessary student actions.

(3) Combine actions that accomplish a major step in achieving the lesson objective.

(4) Arrange steps in logical sequence.

(5) Convert major action steps into teaching points or student performance objectives. Each SPO must contain these elements:

(a) Give a condition.

(b) Describe the action or behavior expected of student.

(c) Establish a standard to be attained.

EXAMPLE: Under conditions of total darkness, students must be able to fieldstrip and reassemble an M-14 Rifle within three minutes. Procedures must conform with those stated in FM 23-8.

    e. The instructor must select the supporting material necessary to explain the teaching points or student performance objectives.

NOTE: SHOW SLIDE #4 (SUPPORTING MATERIAL)

    (1) Define the terms used in SPO.

    (2) Restate the other terms.

    (3) Cite examples of the action.

    (4) Make comparisons with known facts.

    (5) Use testimony of experts.

    (6) Quote statistics to enforce statements.

    (7) Illustrate points with visual aids.

    f. The instructor must organize the lesson outline.

NOTE: SHOW SLIDE #5 (THE LESSON OUTLINE)

    (1) The major parts of the lesson outline are:

        (a) The introduction

        (b) The body

        (c) The conclusion

    (2) The format for a complete lesson outline. An example of the lesson outline format will be available as a handout at the end of class.

NOTE: SHOW SLIDE #6 A, B, C, & D IN ORDER (LESSON OUTLINE FORMAT)

        (a) I Lesson Objective

          II Teaching points or Student Performance Objective

NOTE: READ THE FIRST STUDENT PERFORMANCE OBJECTIVE FROM THIS LESSON

        (c) III Advance Assignments

        (d) IV Introduction

            1. Gain Attention

            2. Orient Students with:

                a. Lesson Tie-In (with previous or future lessons)

                b. Motivation (why student should learn this)

                c. Scope (what will be discussed or done during this period of instruction)

    (e) V Body

      1. First Student Performance Objective (Supporting material and transition)

      2. Second Student Performance Objective (Supporting material and transition)

      3. Include questions and notes, etc.

    (f) VI Conclusion

      1. Retain Attention

      2. Summary

      3. Application (immediate or future use)

      4. Closing Statement

TRANSITION: The lesson outline is not an adequate podium reference for the instructor.

NOTE: SHOW SLIDE #2A (PLAN THE LESSON)

  g. The instructor must prepare teaching notes.

    (1) The lesson outline is too detailed and hard to follow.

    (2) Ready reference builds confidence.

    (3) Use sparingly. Do not use as a crutch.

    (4) List in same sequence as lesson outline.

    (5) Write large enough to be read at a glance.

    (6) Use cue words.

    (7) Set up a color code for various steps.

      (a) Black-dialogue

      (b) Red-transition

      (c) Blue-questions

      (d) Green-notes

NOTES: DISPLAY MANILLA FOLDER WITH LARGE GREASE PENCIL NOTES

  h. Instructional handouts must be prepared.

    (1) Reference handouts not to be used in class.

    (2) Advance study assignments handouts should be issued to students the day prior to instruction.

NOTE: SHOW ITEM #9, SLIDE 2A, PROCURE INSTRUCTIONAL AIDS.

TRANSITION: The teaching points or student performance objectives are better understood if adequately illustrated and demonstrated.

B. **Second Student Performance Objective:** Given the qualities of a good training aid, EXPLAIN how to use training aids to assist student understanding as defined in Chapter VI, USAMU Marksmanship Instructors' and Coaches' Guide.

QUESTION: What are the qualities of a good training aid?

NOTE: SHOW SLIDE #7 (TRAINING AIDS)

ANSWER: 1. The qualities of a good training aid are:

    a. Provide a supplement to the instruction, reinforce the explanation.

    b. Assist student understanding and retention.

    c. Appeal to more than one sense.

    d. Focus attention on a specific point.

    e. Increase student interest.

    f. Save training time.

TRANSITION: Adequate training aids cover a wide range of materials. (Actual equipment, models, charts, chalkboard, projector, recording)

NOTE: TURN OFF PROJECTOR AS INSTRUCTOR MOVES TO TABLES

NOTE: HAVE EXAMPLES OF THE FOLLOWING TRAINING AIDS ON STAGE AND UNCOVER EACH AS IT IS MENTIONED.

    2. Types of training or visual aids.

    a. Actual equipment (pistol).

    b. Models: actual size M-16 Rifle, larger or smaller size claymore mine, working model of M-14 rifle, cut-a-way or mock-up of M-14 rear sight.

    c. Charts - Prepared

    d. Chalkboard (Flurescent chalk, illustrations, colored chalk) (Have list on chalkboard of factors of minimum arc of movement: stance, position, grip and in pencil only, breath control)

    e. Visual projections (films, slides, vugraphs) (Turn on 35mm slide projector)

    f. Recordings (Tape, records or dubbed sound track)

NOTE: (MAKE ADMINISTRATIVE ANNOUNCEMENT) "I HAVE AN ADMINISTRATIVE ANNOUNCEMENT TO MAKE CONCERNING THE ADDITION OF CLASSROOM SPACE FOR GROUP WORK IN LESSON PLANS. TO REACH THE CLASSROOM AREA, START FROM THE FRONT DOOR OF THIS BUILDING. WALK OUT TO LAWRENCE ROAD AND TURN LEFT AND PROCEED WEST PAST THE ARCADE, THE CHAPEL, AND THE STATISTICAL OFFICE. WHEN YOU REACH THE INTERSECTION OF DAVY RD AND LAWRENCE RD, TURN LEFT AND MOVE SOUTH PAST SMALL ARMS FIRING SCHOOL AREA #6. UPON REACHING THE INTERSECTION OF DAVY RD AND SOMMERS RD, TURN LEFT OR EAST AND PROCEED TO THE SAFS HEADQUARTERS PARKING LOT, TURN SOUTH AGAIN AND THE CLASSROOMS ARE IN THE FIRST BUILDING TO YOUR RIGHT. ARE THERE ANY QUESTIONS?"

NOTE: INSTRUCTOR VIEWS STUDENT DISTRESSED FACES. "IT APPEARS I HAVE LOST A FEW OF YOU ENROUTE. IF MY DIRECTIONS ARE GOING TO HELP YOU, I WILL HAVE TO REPEAT THEM. AND THIS TIME I WILL USE A TRAINING AID IN THE FORM OF A PROJECTED MAP."

NOTE: SHOW SLIDE #8 (STRIP MAP) (REPEAT THE INSTRUCTIONS)

NOTE: TURN OFF PROJECTOR.

TRANSITION: "I am sure my second explanation was far superior to my first one. There are two reasons for this, I spoke too fast and did not visually assist you to follow my directions. You can disregard the announcement of additional classrooms. No building other than this building will be used for instructor training classes."

      3. Proper use of training aids.

NOTE: SHOW SLIDE #9 (PROPER USE OF TRAINING AIDS)

        a. Rehearse with training aids.

        b. Keep aids clean and in good repair.

        c. Check for mistakes in spelling and composition.

        d. Introduce the training aid.

        f. Review or summarize course with training aids.

NOTE: TURN OFF PROJECTOR

TRANSITION: All training aids need not be fancy or perfect examples of workmanship.

    C. Third Student Performance Objective: Given a description of several training aid field expedients, DEMONSTRATE the use of instructor-made training aids that can be used in place of prepared training aids as shown in Chapter VII, USAMU Marksmanship Instructors' and Coaches' Guide.

QUESTION: What is a field expedient training aid?

ANSWER: An instructor-made training aid that can be used in place of a prepared training aid.

NOTE: TAPE WRAPPING PAPER TO CHALKBOARD

      1. Use wrapping paper taped to a wall or chalkboard and write in or draw illustrations with a magic marker or grease pencil.

        a. Use for small class groups.

        b. Can be disposed of when class is completed.

        c. Can be prepared in advance.

NOTE: TURN ON 35 MM PROJECTOR AND SHOW PREVIOUSLY INSERTED SLIDE.

      2. Use 35mm slide projector with instructor-made slides.

        a. Show pictures of personal experience.

        b. Show slides of hand made illustrations or drawings.

NOTE: LIST FACTORS OF MINIMUM ARC OF MOVEMENT ON CHALKBOARD

      3. Draw illustrations or list teaching points on chalkboard.

        a. Prepare before class or during class.

        b. Erase information and reuse indefinitely.

NOTE: TURN ON VU-GRAPH PROJECTOR AND LIST SHOOTING EQUIPMENT ON BLANK SLIDE. SLIDE #9A

    4. Use a blank Vu-graph slide to illustrate or list factors during class.

        a. List items as mentioned in discussion.

        b. Erase information and reuse indefinitely.

NOTE: TURN ON DIRECT IMAGE PROJECTOR AND SHOW ILLUSTRATION FROM MARKSMANSHIP MANUAL.

    5. Use an artist's direct image projector.

        a. Low priced.

        b. When placed over a book page or an illustration, it will project a direct image.

        c. Use for small class groups.

VI. CONCLUSION:

    A. <u>Retain Attention:</u> There are a number of points I want to draw your attention to on the instructional staircase to success.

    B. Summary:

NOTE: SHOW SLIDE #10 (INSTRUCTIONAL STAIRCASE TO SUCCESS)

    1. Make an estimate of the teaching situation.

    2. Determine the lesson objective.

    3. Determine the Student Performance Objective.

    4. Select supporting material.

    5. Prepare lesson outline.

        a. Introduction.

        b. Body.

        c. Conclusion.

    6. Procure instructional training aids.

        a. Qualities of a good training aid.

        b. Types of training aids.

        c. Proper use of training aids.

    7. Rehearse the lesson.

    8. Present the lesson.

    C. <u>Application:</u> Lesson planning and proper use of training aids results in efficient use of instruction time.

    D. <u>Closing Statement:</u> A good instructor is good at managing all the factors that make up a successful presentation.

ANNEX I TO INSTRUCTOR TRAINING III
(Lesson Planning and Use of Training Aids)

LESSON OUTLINE FORMAT Class Number (SAFS___)
                       Class Length (Min)
SUBJECT TITLE   Date (Month & Year)

LESSON OUTLINE

I. LESSON OBJECTIVE: To enable _____ student to _____
                                    (Course   (What

_____ _____.
Action)   (Subject Matter)

II. STUDENT PERFORMANCE OBJECTIVES: As a result of this instruction, students must be able to accomplish the following student performance objectives:

    A. Write out complete SPO (State a condition, action expected and standards to be met)

    B. Write out complete SPO

III. ADVANCE ASSIGNMENT: (Study requirements)

IV. INTRODUCTION:

    A. <u>Gain Attention:</u> (attract student interest)

    B. <u>Orient Students:</u>

        1. <u>Lesson Tie-In:</u> (With previous or future lessons)

        2. <u>Motivation:</u> (Why the student should learn this)

        3. <u>Scope:</u> (What will be taught or done during this period of instruction)

V. BODY:

    A. <u>First Student Performance Objective:</u> (Write in complete SPO A from Section II.)

        1. Material supporting performance objective (list all items of supporting material under the performance objective.)

QUESTION: 2. What are the . . . . ?

ANSWER: . . . . Answer(s) to question.

        3. Supporting material.

        4. Supporting material.

<u>NOTE:</u> SHOW SLIDE #1 (TITLE OF SLIDE)

        5. Additional material supporting the performance objectives.

TRANSITION: Written before each SPO.

    B. <u>Second Student Performance Objective:</u> (Write in complete SPO B from Section II.)

VI. CONCLUSION:

    A. <u>Retain Attention:</u> (Revive student interest)

    B. <u>Summary:</u> (Brief review of main points taught during this period)

    C. <u>Application:</u> (How can this instruction be used?)

    D. <u>Closing Statement:</u> (A forceful, significant statement that will be remembered)

ANNEX TO SECTION ONE, "TECHNIQUE OF MARKSMANSHIP INSTRUCTION"

GLOSSARY OF INSTRUCTIONAL TERMS

ADVANCE ASSIGNMENT: Study or other work required to be performed by students prior to class. The advance assignment may be prescribed in the schedule or included in an advance sheet issued prior to the presentation.

ARTICULATION: Production and combination of separate sounds to produce intelligible speech. You violate good articulation when you run words together or fail to sound all parts of the word.

BODY (OF LESSON): The main part of a lesson, the lesson body, contains each student performance objective with appropriate supporting material arranged in outline form.

CONCLUSION (OF LESSON): The final portion of a period of instruction during which you reemphasize the main points, tell students how they will apply this new learning, and close with a strong statement to leave a lasting impression of the importance of the subject learned.

CONFERENCE: The instructional conference is a method of instruction which develops the learning material primarily through student discussion. It includes lecture, demonstration and student performance when applicable.

DEMONSTRATION: The "showing" portion of a period of instruction; demonstrations may be followed by student performance or they may be designed as supporting material to assist student understanding.

ENTHUSIASM: An outward display which conveys the impression of intense, eager and sincere interest in the subject. You frequently reveal enthusiasm thru vigorous, confident, forceful movement, gestures and speech. However, you do not convey enthusiasm by loud, bombastic, boisterous delivery.

EVALUATION: Analysis and interpretation of the results of student proficiency measurements in terms of acceptability or the accomplishment of training objectives.

FLUENCY: A smooth flow of well-chosen words arising from clear logical thinking.

FORCE: Demonstrated vitality and strength of conviction. Aggressive decisiveness displayed thru authoritative confident manner, movement, voice and language.

GENERAL KNOWLEDGE: Level of proficiency which makes the student aware of a subject and provides certain fundamental facts and principles.

GESTURE: A natural movement of any part of the body to help convey a thought or emotion or to reinforce oral expression. A gesture does not draw attention to itself but seems a natural aid to expression.

GRAMMAR: Correct usage of spoken or written words with respect to accepted principles of number, case, and tense.

INTEREST FACTOR: Something used to gain, maintain or increase student interest in the instructional presentation. An interest factor could be a startling statement, stimulating question, skit, teaching vehicle or training aid. It is not an instructional "gimmick" to startle or entertain students but must relate to the subject matter being learned.

INTEREST SPAN: The length of time that most students in a class can give uninterrupted or continuous concentration upon the instructional presentation.

INTRODUCTION: The beginning of a period of instruction which serves to gain attention, to tell the student <u>what</u> he is to learn, to show <u>why</u> he should learn it, and to explain <u>how</u> the subject relates to past, current or future instruction and how it will be presented.

LEARNING: A change in the student's knowledge, understanding, skill or appreciation which causes him to act differently as a result. To be effective, learning must be purposeful activity which results in a change in behavior.

LECTURE: A method of instruction designed to present orally a large amount of material in a relatively short period of time with no or a minimum of student participation. Such participation is generally secured by asking brief factual questions at stated periods.

LESSON OBJECTIVE: A brief statement (in infinitive form) of the mission of the instructional period, the type of class receiving instruction and the student learning proficiency desired (general knowledge, working knowledge, qualified). The objective is the broad purpose of the lesson; the teaching points are the smaller, essential units which insure fulfillment of the objective.

LESSON OUTLINE: The systematic arrangement of the material to teach in full sentence outline form. Annex A is a sample lesson outline with Roman numerals indicating the six main portions: lesson objective, teaching points, advance assignment, introduction, body and conclusion.

LESSON PLAN: The contents of the vault file. It includes the following instructional materials: preparation date section, lesson outline section, training aids used, handouts used, official directives and comments, bibliography of references, research materials and pertinent portions of former lesson plans.

LESSON PLANNING: All the steps you follow in preparing a period of instruction from the time you receive the training directive until you actually present the lesson. These steps are divided: careful analysis of the mission, estimate of the teaching situation, preparation of lesson objective, study and research, selection of teaching points and proper supporting materials, organizing teaching points and supporting materials in a logical sequence in the written outline, deciding appropriate methods of instruction, determining good transitions and training aids, accomplishing necessary administrative details, preliminary rehearsals, revisions and final rehearsals.

MANNERISM: A characteristic movement made by an individual which draws the attention of others to the movement. It normally constitutes a distraction rather than an aid to oral communication.

MEASUREMENT: The act of determining a student's knowledge, understanding, or skill by means of a device such as a performance exercise, paper and pencil test or rating form.

MOTIVATION: The process of stimulating action towards satisfying a need or reaching a goal. The creation or arousing of a desire to learn.

MURDER BOARD: A rehearsal for members of the group, who critique in detail the instructional techniques, materials, content, and estimated effectiveness for learning.

PITCH: The tone level of the voice. Your natural pitch in the midpoint of your range. This may be compared to singers who are tenor, bass, or baritone, yet each has a range of notes at his command.

PRINCIPLE: A settled rule of action or law of conduct or doctrine; a fundamental truth.

PROGRAM OF INSTRUCTION (POI): A training directive which is a combination of master training program and subject schedule. It includes course number and title, purpose, prerequisites, length, location and special annexes or blocks of instruction which in turn list the subject and problem number, length of period and type of instruction (lecture, conference, demonstration, or practical work), scope of instruction and specific study references.

QUALIFIED: Level of proficiency which includes comprehensive knowledge of subject that permits skilled functioning in a specific job and the ability to supervise or train others in that job.

QUESTIONS:

    a. LEAD-OFF QUESTION: A Question (of general nature) used to start an overall discussion concerning the entire problem or broad aspects of a teaching point.

    b. FOLLOW-UP QUESTION: A question designed to narrow the area of discussion to more specific material supporting the teaching point and/or to develop ideas expressed by students responding to lead-off questions.

    c. CHECK-UP QUESTION: A question to review or summarize instruction, and to check student understanding at various points throughout the lesson.

    d. RHETORICAL: A question which is not meant to be answered by students, but by the instructor himself. It is designed to stimulate thought, to capture attention, and to lead into an area of discussion or explanation.

RAPPORT: The mutual feeling of "oneness" wherein the instructor and his students are cooperatively working toward the instructional objective.

RATE: The speaker's speed of delivery.

RESEARCH: The process of recognizing new problem situations and attempting to solve them through critical and intelligent inquiry and collection of pertinent data. This data is analyzed and evaluated in order to reach specific meaningful conclusions.

SCHOOL SOLUTION: A position, technique, or explanation which the instructor presents as the best solution for a given problem or situation based on all information and experience available at the time. The school solution is usually presented as "a school solution," implying and frequently stating that other solutions may be or are acceptable.

SCOPE OF INSTRUCTION: The problem description in the POI annexes which indicates the principal areas of that subject which will be discussed during the instructional period.

SENTENCE FORM OUTLINE: A type of lesson outline in which each main point and all supporting points are written down as complete sentences so that their meaning and relation to other points are made completely clear.

SINCERITY: An observable trait in which the speaker feels that the knowledge he possesses is valid, accurate and of great importance to others. He expresses himself naturally, simply and directly in an earnest desire to convince his listeners of the truth or value of his ideas.

SKILL: The student's ability to use his knowledge effectively. Skills may be manual or mental.

SUB-SUMMARY: A review of all or a portion of the instruction which has been presented to that point. Its purpose is to assist student understanding. Sub-Summaries may be accomplished by brief statements, by asking check-up questions, by a summary chart or projectual, by practical exercises or by having a student restate the student performance objective in his own words.

SUMMARY: That portion of the conclusion wherein the instructor reviews and reemphasizes the performance objectives of the lesson, together with the essential supporting material.

SUPPORTING MATERIAL: Any material or device which clarifies, amplifies, or reemphasizes a main idea and serves to develop maximum student understanding. Supporting material may consist of explanation, analogy, illustration, example, statistics, testimony, quotations, demonstrations, and student practice exercises.

STUDENT PERFORMANCE OBJECTIVE: A specific statement (in full sentence form) of a principle, technique, procedure or an element of essential knowledge, skill, or appreciation which students should learn as a result of a period of instruction. These points should be developed and interrelated throughout the instructional period in a logical and meaningful order.

TEACHING VEHICLE: A device which binds instruction around a central theme or story and helps to give life and movement or logical order to a presentation. Frequently, it serves as an interest factor.

TOPIC FORM OUTLINE: Similar in content to the sentence outline except the main ideas and supporting material are reduced to key words or phrases that serve to recall or suggest the complete idea.

TRANSITION: A means by which the instructor maintains logical continuity or progresses from point to point in a lesson; it is used to show relationships which assist understanding. A good method of transition is to use a teaching vehicle to which the instructor may refer from time to time to show the relative position of teaching material in context to the whole lesson. A rhetorical question or a subsummary is also a good method of transition.

VAULT FILE: The complete official reference folder relating to an instructional presentation. The problem folder includes current and former lesson plans, student handouts, descriptive copies of training aids and all data pertinent to the problem.

VISITORS FOLDER: A folder to familiarize visitors with the problem being observed. It includes problem title, length, method(s) of instruction, type of class(es) receiving instruction, lesson objective(s), the teaching points, time breakdown by teaching point, and instructional materials issued to students.

WORKING KNOWLEDGE: A level of proficiency indicating sufficient familiarity with the primary purposes, major functions, or principles of employment of a subject to permit routine practical applications.

SECTION TWO

MARKSMANSHIP COACHING

CHAPTER X

ATTRIBUTES, AIDS TO COACHING AND DUTIES OF A PISTOL TEAM COACH

A. PERSONAL ATTRIBUTES OF A COACH

A coach's moral character and personal dignity must always serve as a model for those he is training. He must have a profound knowledge of the theory and practice of marksmanship and a serious attitude toward his training responsibilities. A love for the sport, a respect for and the sincere desire to help his shooters will in turn instill in his team members a respect for him.

A coach must be strict in his demands upon his shooters and consistent in what he requires. He must insist always on observing discipline and adhering to the day's program. This respect and discipline together with the knowledge that lapses in effort will not go unnoticed, will spur his team members to conscientious work.

There are no perfect pistol coaches. However, there are those who are outstanding because there is something in their makeup that induces excellence of performance from those who shoot under their guidance. Minor shortcomings do not impair the working loyalty of a progressing shooter who has found a coach with merit. To recognize merit in another person is in itself an essential of character. The coach or shooter who scorns all others for even their minor flaws and thinks no one is worth the effort of leading or following, parades his own inferiority before his team mates and competitors.

There are attributes, moral and mental, that the pistol coach must have that will accelerate the shooter's progress and prevent a lapse into habits that may lead to a decline in performance.

    1. <u>Temperate</u>. A coach must be temperate in all things. He must have the moral fiber to refrain from loose living. Intestinal fortitude and will power will enable him to deny himself things that are damaging to his character and may compromise the example he is to set. Nothing should preclude the attainment of what he knows to be his prime objective - WIN!

    2. <u>Dedicated</u>. Since shooting skill has somewhat less bearing on ability to coach than on being a top shooter, the search for a capable coach should not be directed necessarily toward the top pistol shooter, but toward the dedicated, observant, self-controlled man who possesses a wide knowledge of shooting technique. This does not imply that a Champion Shooter cannot have these desirable qualities. He can, but too often the top shot has time only for improving his own skill rather than in extensively coaching others. The attributes mentioned as desirable in a coach are contagious and will transmit themselves to the shooters.

    3. <u>Self Control and Patience</u>. Most successful coaches feel that self control and an infinite amount of patience are the main attributes of a pistol coach. Many times during team matches, under pressure, things could be done and said which would have an adverse effect on the team. The ability of the coach to control himself and the other members of the team will affect greatly the outcome of the team match. For example, irreparable damage can be inflicted upon team morale if the coach loses his temper with a man on his team and calls him uncomplimentary names. From that time forward, there will be a wall between the two. Patience, especially while coaching new shooters will help the shooter to improve more rapidly. It may be necessary for the coach to stand beside a new shooter or an old one, and watch him jerk the trigger on shot after shot. He may prove to the shooter by use of ball and dummy that he is jerking, and then have to bear up under a continuance of a faulty performance Should he give up? NO! He must keep trying; he must have patience and confidence in the man. Tomorrow may be the day that the developing shooter begins to grasp the meaning of points the coach has continued to stress.

    4. <u>Compatibility</u>. This is simply a big word for getting along with the shooters on the team. If the coach cannot get along with his shooters, he is worse than useless to the team. The coach must be quick to accept blame for a losing performance and equally quick to give the shooters credit for a winning match.

5. _Inspire Confidence._ The successful coach must be able to inspire confidence in his shooters. First he must have confidence in himself and show his team members that he has complete confidence in them and in their ability, both as individuals and as team shooters. Then, and only then, can he expect them to have confidence in their abilities and just as important, confidence in him.

6. _Enthusiasm._ The coach must show enthusiasm, not only when the team is winning, but when the team has lost a match. The team must have an enthusiastic desire to win to be able to come back and shoot better in the next team match. The desire to win and unflagging enthusiasm are the things that keep us from giving up in disgust when everything seems to be going wrong.

7. _Observant._ The pistol coach will be observant of anything in his shooter's performance that can be improved to make his scores better. This applies not only in practice, when the time can be taken to improve, but also should be observed in matches. He must determine if the training program is achieving the required results. During a match he must be aware of any unusual conditions, noise, or movement that could be distracting to his shooters. A close check of all phases of the match such as scoring, alibis, challenges, and block officer decision must be made.

8. _Wide Knowledge._ The pistol coach must have a wide knowledge of employment of the shooting fundamentals and coaching techniques. In addition he must have specific knowledge of match rules and range procedures. The good line coach does not necessarily have to be a top shooter but he should have sufficient match experience to be familiar with the problems encountered in competition and most of the remedies needed by his shooters. The coach must catalogue in his mind the traits of each shooter. He must know how to handle each member in every shooting situation so that successful results can be obtained.

9. _Exacting Standards of Team Performance._ To excel in pistol competition, the pistol coach must set exacting standards for the team to follow. His actions, as a coach, and as a member of the team, must be above reproach at all times. He must require that his team members act in such a way as to reflect credit upon themselves and the US Army. As always, any Army team at a match represents the entire Army. One step out of line will cause some persons to say, "Well, there is the Army making an ass out of itself again."

The coach must require his team to utilize practice sessions to improve their shooting. This is the time to try new ideas, not just as a chance to burn up ammunition. During the shooting, the coach should require a great amount of effort by each man toward correcting his faults. He must not allow the shooter to accept an average performance. Strive for perfection in shooting, but at the same time retain exacting standards of personal and team conduct.

10. _Open, Progressive Mind._ Because a man has been shooting for 30 years does not necessarily mean that he knows more about shooting or has the best coaching techniques of anybody with less longevity. This business of "It was good enough for the last 30 years, and it's good enough now," is not applicable anywhere today, least of all in pistol shooting and coaching. The good coach should accept, with proper frame of mind, constructive advice from experienced marksmen. He should continually strive to improve himself with better techniques. An open, progressive mind applies also to his approach to the difference in personality and shooting ability in his shooters. No two individuals are alike. The coach who is flexible in his dealings with his shooters will in the long run have the best functioning team.

11. _Avoid Being Over-Eager._ An efficient pistol coach recognizes there are times when his effectiveness is enhanced by being silent and observing the shooter's actions. For example, if a shooter has averaged 98 points for each string of slow fire in the past three months, there is little to be gained by pressing hard for a possible score from that shooter. Know when to let your presence on the firing line be quietly sufficient.

B. AIDS TO COACHING

There are a number of aids to coaching that will enable the pistol coach to accomplish his mission of developing a winning pistol team.

1. _The Shooters' Worksheet_ is designed for both slow and sustained fire checkout. As the shooter progresses through the factors that control the delivery of an accurate shot, there must be no omission of essentials. This aid is used mainly in the training phases.

2. __Preparation Check Sheet__. This sheet should be stapled into the gun box lid for ready reference in conducting complete preparation. Items of readiness are included that must be implemented in the assembly area, firing line and after the command "Load."

3. __Sight Adjustment Card__. This aid is used for a threefold purpose. By knowing the size of the scoring rings and how much one click on the sights will move the strike of the bullet, the shooter can make a bold sight change and move his subsequent shots immediately to center of target. Secondly, the shooter can record his sight setting, in elevation and windage, for all caliber weapons in pistol competition; and last, there is reference on all of the popular makes of adjustable sights as to which direction the strike of the bullet moves when the adjusting screw is turned clockwise. This sheet should be removed from the booklet and stapled in the gun box lid.

4. __Adjustable Rear Sight__. Each shooter should know the capability of each of his adjustable sights. How far will one click on the sight in elevation or windage move the strike of the bullet at 50 yards or 25 yards? For instance, he must know which way to turn the windage screw to move the strike of the bullet to the right or left.

5. __The Pistol Score Book__. The pistol scorebook is your personal record of your shooting. Date, time, place, ammunition, weather conditions, sight setting, wind direction, light direction and most important, a shot by shot record, by stages, in practice or registered match of every shot fired in the current season. The worksheet booklet contains sample pages from the USAMTU Pistol Scorebook.

C. THE LINE COACH'S DUTIES

The line coach is responsible for the close supervision and the detailed, exhaustive measures necessary to improve the performance of the four shooting members of his team.

1. __Before the Match:__

a. __Be Available Between Individual Matches__. During the firing of the individual matches, the line coach should check the performance of his team members. If any of them are falling below standard, talk to the shooter concerned, closely observe his performance and try to find out what the trouble is. The coach should try to steer the shooter's thinking back in the right direction. Remember, the individual matches are used as zeroing matches for the team matches. Any individual matches won are incidental to the major task of winning the team matches. The coach must be available to his men to assist them in every way possible.

b. __Supervise Physical Preparation__. After the individual matches and prior to the team matches, the coach should have a team meeting. During this meeting, he should insure that all the team members have the right type and amount of ammunition. Are their weapons clean and in perfect working order? Check everything. (For details of this check-out, refer to the Coach's Worksheet, an Annex to this chapter.)

c. __Assign Relay and Target Numbers__. A point to be covered during the team meeting is the assignment of target and relay numbers. Be sure there is no misunderstanding on these numbers It is demoralizing for a shooter to show up behind target number 40 and find none of his teammates there. By the time the range officer says "load" the shooter discovers that the team is on Target 4. The coach assigns additional duties to members.

The coach assigns additional duties to members of his team such as scoring the adjacent team in the event this requirement has been announced by the range officer. This duty includes the posting of the score board; carrying flags, charts, and scopes up to the firing line; and any other duties the coach feels necessary. These matters should have been established during the previous training program and simply confirmed at the team meeting.

d. __Supervise Mental Preparation.__ From the time the coach begins working with his four shooters, he should be guiding their mental preparation toward winning this team match. Now he must do his best to get their minds off all miscellaneous interests and get them thinking only about this all important team match. (For details of this mental preparation check-out, refer to the Coach's Worksheet, an Annex to this chapter.)

2. During the Match:

   a. Make Decisions. During the actual firing of a team match, decisions regarding the team must be made by the coach. For example, many shooters are prone to accept a slightly misplaced group rather than move their sights. Their attitude seems to be that the setting on their sights is correct (the zero was perfect during the previous matches) and they will not volunteer to change it. Here the coach must give firm, positive directions to the shooter. Another example is the decision as to whether or not to refire a slow fire string when the shooter's target has excessive hits. Naturally the coach would consult the shooter as to his preference, but the final decision lies with the line coach. He has the responsibility for the team's performance and therefore he has the final authority for decisions made on the firing line in the heat of battle.

   b. Stresses Safety. The coach should be constantly aware of any breaches of safety.

   c. Caters. The word may mean different things to different people. What is meant here is that the coach must provide the shooter what he wants and needs, thereby aiding him to achieve his best possible score. As has been mentioned previously the coach should know just how each shooter likes to be coached and what the coach can do for him that is helpful. If this means setting the shooter's gun box on the line and preparing his equipment for shooting, then the coach should be glad to do it. It may also mean for the coach to sit back and watch, and not say a word. The coach must always remember that he is there for the shooter's benefit and not the other way around.

   d. Encourages. The coach must encourage the members of his team at all times. When they are shooting well, he must encourage them to shoot better. When they don't shoot so well, he must assist them to mentally prepare themselves to shoot better in the next match. This assistance is in reality, encouragement. The shooter is being shown that the coach believes the team member can do the job required and that he wants to help him to correct any faults that have lowered his performance. Never say anything to discourage any member of the team. By the same token, he cannot use false praise.

   e. Scoring Record. All scoring of record practice and registered matches must be supervised by the coaches. This doesn't mean that every shot fired must be counted for score. The shooters must be given free practice to allow them to try new procedures such as different position, etc. However, when record scores are fired they must be made a part of the shooter's permanent record. From this record the head coach can ascertain whether the shooter is improving or not, and what weapons or stages he needs more work on. All of the oral and written tests, etc., still won't indicate how well the man can shoot. This can only be determined by the shooter's performance on the firing line.

   f. Irregularities. The coach should be a close observer of events and conditions during a pistol match and intercede by his good offices if an argument, protest or an infraction of the match rules concerns an Army team member.

3. After the Match Critique the Team. There are several things a coach can do after a match. It does no good for him to stand in front of his shooters, rant and rave and tell them (if they didn't win) that they fired like a bunch of ring-tailed baboons. If they did actually shoot that way, the coach does not have to remind them, they already know it. The proper technique at this point should be to:

   a. Discuss and Recommend. For example, the coach and team should go over their performance and suggest ways to improve this shooter's rapid fire or that shooter's slow fire when they return to their practice range. If there was a clash of personalities due to the grueling stress of the competition, the ruffled feathers should be smoothed. The opinions and methods of the individual shooter revealed at this time are the building stones of this manual.

   b. Evaluation and Observation. Direct coaching of the shooter is only half of the coach's job. True, it might seem to be the most rewarding, but the coach must plan what to train him, how to train him, and evaluate his performance to find out if the training is effective. The evaluation of a shooter is achieved primarily from the line coach and from the results of practice and match scores.

The line coach will evaluate the shooter's potential ability, since he will be in the best position to see how the shooter reacts to the problems confronting him when trying to achieve a good score. From his observation he will accurately fill out "The Individual Information and Evaluation Sheet" on each shooter on his team. This evaluation will show the training needed to make him a better shooter. The aggregates recorded therein can serve as a guide when considering the shooter for future placement on a top team.

ANNEX I TO CHAPTER X

A. WORKSHEET FOR PISTOL TEAM COACHES

1. <u>Coach is in charge of team.</u>

2. <u>Each shooter will follow the same complete procedure for each shot or string.</u>

3. <u>Coach has responsibility for assisting shooter</u> between individual matches as well as in team matches.

4. <u>Assemble team members</u> on assembly line a sufficient time to complete preparation before match time or relay is called. Systematic application of the following factors in pistol line coaching will prevent a haphazard approach to the instruction and thereby properly influence the developing shooter to exert progressively better control over his shooting performance.

   a. Physical Preparation:

     (1) Designate relay and target number.

     (2) Check for clean, proper functioning, and lubricated weapons.

     (3) Check for sufficient proper caliber ammunition.

     (4) Check for proper sight setting and weapon zeroed.

     (5) Blacken sights.

     (6) Use ear plugs and shooting glasses.

     (7) Stop watch for time check.

     (8) Score book and pencil.

     (9) Chair.

     (10) Obtain scorecard.

   b. Mental Preparation:

     (1) Review shot sequence (slow or rapid fire worksheet).

     (2) Encourage.

     (3) Think and observe.

     (4) "Maintain confidence that a controlled, uniform and exacting performance will produce good results."

     (5) Talk shooter through relaxation.

     (6) Let coach worry about irregularities.

     (7) Review techniques of shot analysis and positive correction of errors.

     (8) Remind shooter to exercise care and safety.

     (9) "Damage from mistakes is minimized by continuing to work hard."

     (10) "Carefully planning the delivery of each shot will minimize the effect of tension and pressure."

(11) "Concentration on maintaining sight alignment and holding the smallest possible sustained arc of movement while applying positive trigger pressure will result in a surprise shot break that will strike the target within the shooter's ability to hold."

5. <u>Move to firing line</u> with team member when relay is called.

    a. Coaching equipment:

      (1) Ear plugs.

      (2) Pencil and scoresheets.

      (3) Scope and stand (or binoculars).

      (4) If practice, .22, .38, or .45 Cal scoring plugs.

      (5) Stop Watch.

      (6) Shooting glasses.

      (7) If practice, shooter's worksheet.

      (8) Guidon (Team Flag) (Registered team matches only).

      (9) Extra pistols and magazines of all calibers available.

      (10) Chairs or stools.

      (11) Staple gun and staples.

6. <u>During the three (3) minute preparation period</u> have shooter:

    a. Focus scope on proper target. Check target for holes or loose target.

    b. Adjust ear plugs or protectors.

    c. Load ammunition into magazines.

    d. Assume stance.

    e. Practice breath control.

    f. Dry fire for natural position that will enable center hold on aiming area without tendency to settle to either side of bull's-eye.

    g. Dry fire for natural grip that will enable front and rear sights to be in alignment without artificial correction by wrist, arm or head movement.

    h. Mentally rehearse the steps involved in delivering a controlled shot.

7. <u>Plan shot sequence.</u> (Refer to slow fire or rapid fire worksheet.)

    a. The coach should require the shooter to give his shot plan verbally. (Except for 2650 level master shooters.)

    b. Coach and shooter will converse in low tones (coach will refrain from talking during actual shooting as this may disturb shooter's concentration).

8. <u>Relaxation.</u> (No unnecessary muscular tension.)

    a. Relax all major portions of body not needed to maintain stance.

b. The coach will remind the shooter to relax before attempting each shot or string.

9. <u>Deliver shot as planned.</u> (No compromise.)

    a. Follow through (continue to employ all control factors until bullet leaves barrel).

    b. Shot should be fired as a surprise. (No reflex action.)

    c. Do not hold too long. (Bench weapon and start over.)

10. <u>Complete shot analysis</u> after each shot or string.

    a. Call shot (based on sight alignment, <u>not</u> sight picture).

    b. Scope for hit location (compare with shot call).

    c. If shot or call is in error, determine cause.

    d. If shooter is unsure or in error on shot call, have him draw a picture of the front and rear sight relationship as it appeared at the time the shot was fired.

    e. If shot and call are good, review the successful method employed, in an effort to duplicate a good performance.

    f. The coach should require the shooter to plot shot calls and hits on appropriate worksheet and enter shot values in scorebook.

11. <u>Positive corrective measures</u> to be taken on each shot or string:

    a. Coach and shooter should agree on corrective measures to be taken. (Student understanding is enhanced when coach explains reasons for specific corrective measures.)

    b. Include corrective measures in plan for next shot or string (prompt application).

12. <u>Additional Duties:</u>

    a. Time check after each shot.

    b. Watch shooter or his weapon, not the target. (Do not use shooter's scope while he is firing.)

    c. Have team captain or alternate available for scoring between stages of fire.

    d. Have a non-shooting team member of alternate post score on team score board after each stage.

    e. Check and validate team score card with signature if acting as team captain.

    f. Police firing point at completion of firing.

    g. After all relays have completed firing, the coach will conduct a group critique to resolve problems that became apparent during training. Answer questions and explain techniques. The degree of understanding exercised by the student is in direct proportion to the progress he can expect to bring to improving his shooting skill.

B. COACHING OF ZEROING, SLOW FIRE, AND THE TIMED AND RAPID FIRE EXERCISES

<u>NOTE:</u> (REVIEW CHAPTER IV, THE PISTOL MARKSMANSHIP MANUAL, "ESTABLISHING A SYSTEM.")

1. <u>Coach starts supervision of group preparation</u> in the assembly area, before the first relay is called to the firing line.

NOTE: THE COACH WILL INFORM EACH MEMBER OF GROUP TO BE PRESENT ON RANGE THIRTY (30) MINUTES PRIOR TO THE TIME FOR CALLING OF THE FIRST RELAY TO THE FIRING LINE.

    a. The coach will have students check their squadding tickets for relay and target number.

    b. The coach will supervise preparation to fire and have each student check off each completed action during preparation on an appropriate worksheet. Remove the preliminary preparation checklist from the shooter's worksheet booklet and staple it on the inside of the gun box lid for easy reference. All assembly area preparation should be complete before the relay is called to the firing line.

    c. Coach will review zeroing procedure.

        (1) Use the shot group method. Minimum of three (3) shots before original sight adjustment, except in team matches of registered competition. Here the sight adjustment is changed when the shooter and coach deem it immediately necessary.

        (2) Make a bold sight change. Do not creep to center of target a click or two at the time in an effort to center the group. Make a bold sight change and accomplish the necessary sight adjustment immediately. Reference will be made to sight adjustment card to determine the total distance your shot group is located away from the center of the target. Divide the distance the shot group must be moved, either in elevation or windage, by the distance one click on your sight moves the strike of the bullet. This figure represents the number of clicks of sight adjustment necessary to center your group.

        (3) Fire a timed fire string to confirm zero.

        (4) Do not "hold off" or use "Kentucky Windage" to establish zero.

        (5) Review the effect on zero, of wind and other adverse conditions.

        (6) Mark the zero sight setting by:

            (a) A lead or colored pencil, marked temporarily on sight adjustment screws for both elevation and windage.

            (b) Remove sight adjustment card from worksheet booklet and staple it on the inside of the gun box lid. Use this card to make a record of normal sight settings after zeroing is completed.

            (c) Recording the conditions under which this zero is obtained in the spaces provided on the sample score book pages. (Last page in worksheet booklet.)

            (d) After zero is confirmed, mark sights (25 yard setting) permanently with finger nail polish or a quick drying paint such as airplane dope.

            (e) Record the 50 yard sight setting by noting on sight adjustment card the number of clicks necessary to center group at 50 yards range. (Elevation and Windage.)

2. The coach and first relay shooter move to firing line when directed by the range officer. All firing line preparation should be completed in the three (3) minute preparation period allowed.

    a. Set up equipment on assigned firing point but don't handle weapons until directed to do so. When line is clear for handling weapons, the three (3) minute preparation period will start. Sequence of actions taken during three (3) minute preparation period is listed on Slow Fire and Rapid Fire worksheets and on preparation checklist stapled in lid of gun box.

b. Coach will review planning of the delivery of an accurate shot (Sequence of actions in delivering a shot or a string of shots is listed on the Slow Fire or Rapid Fire worksheet).

3. <u>When command "load" is given,</u> the coach and student will strictly adhere to the rules for safe handling of weapons. (The method of loading the weapon will follow the procedure demonstrated during the coaches' briefing class.)

    a. Students will check out the employment of the fundamentals that will enhance attaining a minimum arc of movement: stance, position, grip, and the practice of breath control as previously checked out during the three (3) minute preparation period.

    b. Student will check for proper target number as he extends shooting arm to recheck for natural position which gives him a hold centered in aiming area on the target and effortless sight alignment attained by proper grip and head position.

    c. Relax with pistol at bench rest. Do not change position or allow shift of grip during this period.

    d. Student will continue to mentally review planning of shot. Having the mental processes so engaged will aid in conditioning the mind to control the delivery of the next shot or string. Concentrate on shot sequence. Visualize perfect sight alignment. Mentally reconstruct the smooth trigger control necessary to fire a shot without disturbing sight alignment.

"YOU - ARE - READY!"

4. <u>When commands are given to "Commence Firing"</u> or the targets are faced as a signal to begin firing, the coach and student will be required to utilize all available time allowed in completing the training exercise. After the required number of shots have been fired, the remaining time will be devoted to review and critique of performance. The coach and student will be allowed time, two (2) minutes between five (5) shot strings of timed and rapid fire where scoring is not in progress, to devote to plotting calls and hits, analysis of performance, correction of errors and planning next shot or string. The one (1) extra minute allowed per shot in each slow fire exercise is sufficient to perform the following functions:

    a. The coach will require the student to give his shot plan verbally before shot or string of shots is fired.

    b. The coach will remind the student to relax before attempting to fire each shot or string of shots.

    c. The coach will not talk or use the students' spotting scope during the delivery of a shot or string. (Observe the shooter or his weapon, not the target.)

    d. The coach will require the student to plot shots calls and hits in the square provided on the Slow Fire or Rapid Fire Worksheet. (Draw a small circle in each square to represent the bull's-eye. Plot the shot calls with an "X" and hits with a "Dot.")

    e. The coach will require the student to conduct an analysis of each shot or string of shots fired to determine cause of error, if any.

    f. The coach will require the student to come to an agreement on corrective measures, if necessary, to prevent reoccurrence of error. (Incorporate corrective measures into the plan for the next shot.)

    g. The coach will require student to make a shot value entry on sample scorebook page for each shot fired.

    h. Offer encouragement to student for poor results and compliment him on a good performance in order to stimulate confidence.

NOTE: THE COACH WILL DIRECT THE STUDENT TO FIRE ON A BLANK TARGET (no bull's-eye) IF LESS THAN 50% HITS ARE RECORDED DURING ANY ZEROING STAGE. IF A SCORE OF LESS THAN 75x100 POINTS FOR ANY TEN (10) SHOT STAGE DURING SLOW, TIMED, OR RAPID FIRE EXERCISES IS POSTED BY THE STUDENT, A BLANK TARGET WILL ALSO BE USED FOR THE NEXT TEN (10) SHOT STAGE OF FIRE. SIGHT ALIGNMENT WILL BE STRESSED DURING THESE BLANK TARGET SESSIONS, AS WELL AS REVIEW OF THE OTHER FUNDAMENTALS OF STANCE, POSITION, GRIP, BREATH CONTROL (attaining a minimum arc of movement), AND POSITIVE TRIGGER CONTROL.

CHAPTER XI

TECHNIQUE OF COACHING A PISTOL TEAM

A pistol coach must give direction to the development of a shooter. The rate at which a shooter progresses and the quality of his performance is to a great extent dependent on his determination to become a champion. However, the coach must guide the development. Imparting knowledge, encouraging and motivating the potential champion will insure that those inclined to achieve and maintain championship laurels are not lost due to lack of progress and waning interest.

The shooter's ideas and techniques are greatly influenced by personal coaching. Research, experimentation, analysis and studying the advice of top shooters or his team mates' techniques will aid in avoiding the discouragements that affect many newly developing shooters. Observing, evaluating and correcting his performance will broaden his basis of shooting knowledge and add skill of control in employing shooting techniques. From improved control comes confidence, the most valuable factor in the champion shooter's arsenal. Confidence gives him the power to win in competition.

A. FACTORS IN SYSTEMATIC COACHING TECHNIQUE

    1. Preparation. Preparation for a pistol match actually starts when the decision is made to enter the pistol match. The program is received and at that time preparation begins. Your entry is sent in, arrangements made for absence from work, family affairs taken care of, and travel arrangements are made. Preparation means careful and complete physical and mental preparation. Close attention to detail is required in preparing to shoot so that nothing is left out.

        a. A basic zero must be established for each weapon. A basic zero is one that with perfect sight alignment will place the strike of the bullet in the center of the target on what you consider a normal shooting day.

            (1) When initially zeroing a new weapon start at a short range, twenty-five (25) yards, and fire a three (3) shot group and adjust your sight setting. Continue firing single shots until your zero is established.

            (2) Make bold sight changes. Move the strike of the bullet to center of target with as few sight changes as possible.

            (3) Become familiar with your adjustable sight. Experiment and find out just how far one click moves strike of bullet in both elevation and deflection at fifty (50) and twenty-five (25) yards. Fire ten (10) shots for a group on the target. Move the rear sight adjustment twenty (20) clicks in elevation or windage and fire ten (10) more shots. Measure the distance from shot group center to center and divide by 20.

            (4) Mark sights and record basic setting on the sight adjustment card in gun box and on the proper space in the scorebook sheet.

            (5) Confirm your zero. When shooting on other than your home range, try to get to the range a day before the match starts and confirm your zero. Confirmation may also be necessary under extreme changes of weather and light conditions on the home range.

        b. In the assembly area. Equipment must be made ready which includes weapons and ammunition. Make sure that you arrive at the pistol match with all your equipment. You should have a list of everything that you need during the match. Put a check mark by each item as you load it in the car. On the day of the match, arrive early enough to complete your assembly area checks prior to being called to the firing line.

            (1) Physical action:

                (a) Check squadding for relay and target.

                (b) Have proper gun and ammunition.

(c) Check sight settings.

(d) Carbide lamp w/carbide.

(e) Blacken sights.

(f) Ear plugs inserted.

(g) Screwdriver.

(h) Stop watch.

(i) Scorebook.

(2) Mental action:

(a) Stimulate confidence.

(b) Expect to work hard.

(c) Plan actions on the firing line.

(d) Do not be upset by range irregularities.

(e) Think only of shooting.

(f) Mentally review shot sequence.

(g) Be ready and waiting for your relay to be called to the firing line

c. On the firing line:

(1) Place shooting box on correct firing point.

(2) Scope the correct target.

(3) Inspect target for holes.

(4) Adjust ear protectors.

(5) Load magazines w/proper ammunition.

(6) Check sight blackening.

(7) Locate accessory shooting equipment.

(8) Take a few deep breaths.

(9) Assume stance.

(10) Check position.

(11) Assume grip on pistol.

(12) Trigger pressure straight to rear when "dry firing".

(13) Be ready for the Command "Load" to be given.

d. After the command "Load":

(1) Verify stance.

    (2) Load weapon.

    (3) Safety off.

    (4) Recheck grip.

    (5) Recheck for proper target.

    (6) Relax with pistol at bench rest.

    (7) Resume mental process of planning shot delivery.

2. <u>Planning the Shot Sequence</u>. If there is a magic formula or a secret to accurate shooting, it is in the planning of the shot sequence. Plan carefully and in minute detail the exact actions required to deliver controlled shots on the target. This is a joint effort between the shooter and the coach. The following sequence is the shot plan for slow fire used by The USAMTU Pistol shooters:

    a. Extend arm and breathe deeply.

    b. Settle into minimum arc of movement.

    c. Pick-up sight alignment in aiming area.

    d. Take up trigger slack - apply initial pressure.

    e. Hold breath at normal respiratory pause.

    f. Maintain sight alignment and minimum arc of movement.

    g. Start positive trigger pressure.

    h. Concentrate point focus on front sight.

    i. Follow thru.

3. <u>Relaxation</u>. After planning the shot sequence the shooter must relax. Relaxation delays fatigue and aids muscular control. Relax your shooter's mind and relieve any unnecessary muscular tension. For example, with some shooters, you can tell a joke, with others you can remind them of a past great performance. With a few of the more able shooters, you just stand and listen to them talk between the shots and strings of fire.

    a. Methodically think of relaxing each principal muscular mass of the body; neck, shoulders, back, abdomen, buttocks, and legs.

    b. A relaxed muscle does not fatigue and tremble.

    c. Rest arm after an unsuccessful effort to shoot.

4. <u>Deliver the Shot</u>. In order to accomplish a high degree of control, the shooter and coach must be aware of every step in the shot sequence.

    a. The shooter must be deliberate and forceful in the delivery of the shot.

    b. The shot must be delivered exactly as planned.

    c. Record the value of the shot in your scorebook after analysis.

5. <u>Shot Analysis</u>. Shot analysis is a must after every shot. Keep the shooter's self-respect at a high level. Do not hesitate to praise a good performance. Maintain a positive attitude. If a shooter's score is not what you expect, let your approach to analysis of his performance be that of

"room for improvement" instead of "you blew it." It is as important to analyze good shots as bad ones. If there are no errors detected, a thoughtful appraisal of the technique used to keep the shot under control is in order. The steps in shot analysis are:

    a. Follow thru check.

    b. Call shot (describe sight alignment, not the sight picture).

    c. Compare hit with call.

    d. If shot or call is bad, determine cause.

    e. Watch for error pattern to form (repetition of the same error on two (2) or more shots in a string of shots).

    f. Did shot break in normal pattern of minimum movement?

    g. Did you hold too long? (Looking for a perfect sight picture?)

    h. Did you apply positive trigger pressure? (Comparatively easy when based on stable sight alignment but very difficult when waiting for the fleeting instant of perfect sight picture.)

    i. If you benched weapon after an unsuccessful shot effort, Why?

    j. Did you lose concentration?

    k. Did you get a surprise break?

    l. Were you worried about results?

6. <u>Shot Correction (If Necessary)</u>. If you find an error in your delivery, you must take corrective action. Here the control cycle starts again. In planning the delivery of your next shot, pay special attention to those steps in the shot sequence that you may have failed to employ correctly. Stick with the sequence that you know will produce a good shot for you. For best results, there must be:

    a. Agreement between coach and shooter.

    b. Prompt application-include in plan for next shot.

    c. A prevention of recurrence of error.

    d. Corrective measures on every shot if necessary.

Systematic application of these factors in all line coaching will prevent the haphazard approach and will properly influence the shooter. Systematic guidance will result in the shooter organizing his thoughts and actions and thereby becoming better able to exert progressively better control over his shooting performance.

B. COACHING THE CHAMPIONS

1. <u>The shooting champion</u> is the cutting edge in achieving a competitive marksmanship performance beyond the highest level of present human accomplishment. The coach helps to fashion this instrument and assists in guiding it true and steady toward its objective: a record breaking performance.

2. <u>If you are going to coach a champion, you should be able to recognize one.</u> Usually, it is the simple matter of reading the scoreboard at a pistol match. The champion's name is on top. But as a pistol coach, you have to recognize a champion in the making. The potential champion is the shooter who maintains a calm attitude when the going gets tough, whose scores get better when the competition gets stronger. The shooter who displays these qualities is a potential champion if he

receives proper coaching. Your most important task in coaching a champion is instilling confidence in his ability to perform at a consistently high level. You teach him all you can in order to increase his knowledge of shooting. Remind him of techniques he may forget to employ. Knowledge builds confidence.

3. <u>The Delicate Ego</u>. The better the shooter, the more confident he seems. It is probable however, that the confidence of the better shooter is the most vulnerable. Sometimes, on one uncontrolled shot, his ego will crumble. He is surprised and horrified that he could possibly fire a shot with apparently a complete loss of control. If this happens again within a day's workout, there is a possibility that his confidence may suffer further. If there is an absence of adept coaching at this point and the trend continues, even at infrequent intervals, the confidence may shatter. It is shattered by what the shooter thinks is confirmation of a nagging suspicion that his inherent ability to perform like a champion is somewhat lacking. Shattered confidence is despair. The coach must be on his toes to spot the first symptom, the first vestige of flawing in a basically good system of control.

4. <u>Sustain Confidence and Remove Self-Doubt</u>. Top shooters are some of the most insecure people in the world. The important half of the coaches job is to impart assurance and sustain a confident attitude in the minds of his shooters. The basic requirement here is that the coach be someone who can for example, tell the shooter, "Yes, that match score was satisfactory. We will accept that performance because it was your best for now. You will improve your performance in the next course of fire." The coach, for his part must be extremely careful during practice and competitive situations, not to allow the shooter any reason for selfdoubt. The shooter needs to feel that his ability, his ideas and opinions are highly valued. He must never be given the impression that he has really failed, only that there is room for improvement. In correcting some small part of a faulty performance, never give orders, suggest. Suggest to the shooter, in an informal, man-to-man approach, that perhaps he could get a better result by extensive repetition of the technique in his practice sessions or by trying it another way if experience proves conclusively that the technique is unsound.

5. <u>Self-Respect Needs Tender Care</u>. The good shooter's self-respect needs tender care. His self-esteem, precious and fragile, often benefits from the sincere expressions and attitudes of high regard from his coach. If the coach wants the shooter to feel confident that he can win, he has to tell him that, you, the coach, thinks you, the shooter, are championship material. If the shooter knows you hold him in high regard, it makes it easier for your coaching to influence him favorably when he is experiencing trouble or when he needs a slight nudge back on the track of flawless performance.

6. <u>A Good Coach Can Improve the Champion's Performance</u>. Care, high regard, tact, a meeting of minds, not even all of these things combined can necessarily bring about the "greatest performance." Greatness in any endeavor requires strong talent. If the shooter doesn't have it, the coach can't give it to him. But a good coach can release the innate ability, to whatever degree it exists. A good coach can improve any shooter's performance. Jointly plan the course of action and the shooter will inject his ideas and force into the proceedings and the combination will generate superb accomplishment. If something wrong or flawed appears, make suggestions, always encouragingly. If you keep working at it together, when it turns out well, the coach will have been the prod that insisted on trying it over and over until perfection is achieved. As a result, the coach furnished the spark of confidence the shooter needed to project himself forward to new heights.

7. <u>The Formula</u>. What is the formula, what are the carefully couched words that a coach should use to get the sterling response that brings forth the marvelous results? As an example, there was a noted coach in action at a match and one of his team shooters had just succeeded in putting together a performance that rescued his team from sure defeat. Even after a fair start, his team was behind. After leaving the firing line, one of the shooter's teammates asked him how had he managed to complete the match without losing any more points. "The coach just told me the magic words: "I want you to do better."

C. HEAD COACH TECHNIQUE

In successful training, rapid and unbroken progress depends in a large measure on the head coach. It is no accident that many leading shooters, in looking back over the path they have traveled, say that they owe their achievements to their good luck in having a highly qualified coach from the very first time they began to shoot. This is how it should be. From the very first days of training, a coach must start a shooter off correctly so that the latter may avoid forming harmful habits. It is much harder to reteach an incorrectly taught person than to start initially teaching him correctly.

When you hold the title of Head Coach of a pistol team, you will find that there are special responsibilities and techniques that apply to training and motivating a larger group of pistol shooters.

1. <u>Selection of Needed Training Subjects</u>. To determine training requirements, a coach should determine the shooting ability of each of his shooters. This ability varies greatly not only between the All-Army level and the Post, Camp, or Station level, but, of more interest to the coach, between the individual shooters on the same squad. This is a difficult but very important job. What the coach selects as the needed training subjects may well determine the later success of the team. The Head Coach must take into account the individual needs and preferences of his team members in planning a training program. This does not mean separate training programs for all shooters, but it does mean that your training program must be flexible enough to allow the shooters to develop at their own pace. If you schedule extra rapid fire training, you need not require every member of the team to shoot rapid fire. You may well have a couple of people who have no problem with rapid fire. Forcing them to shoot additional rapid fire would be a waste of training time. They may need more slow fire training. The training program should be tailored to the needs of the individual.

2. <u>Scheduling and Supervising Training</u>. Publication of a training schedule will not suffice to insure progress in training. Careful observation of individual members and correcting faults during the training is one of the primary duties of the head and line coach. The coach should observe the manner in which the scores are fired in practice. These scores will help him in initial selection of team members and in selection of training subject for the immediate future. Match performance will become the basis of training subjects selected later in the training program.

3. <u>Limit Record Shooting</u>. A coach should not overdo record shooting. Shooters have a tendency to lose motivation and become stale when they have too many record shooting sessions. The fact that these scores may be useful in putting together a team is no excuse. Too frequent record shooting is very exhausting for shooters and can result in a shooter coming to a competition unable to force himself to do his best because of his previous nervous and physical exhaustion. Therefore, the wisest course to follow in a monthly cycle of training sessions is to have no more than half of them record shooting sessions.

If a shooter feels that his place on the team would be jeopardized by a bad score, he may hesitate to try a new technique during a record match. The technique change may be just what he needs for improvement but he won't try it for fear of losing his place on the team. Allow some time for free practice so your shooters can experiment. Record practice shooting is necessary to keep shooters on their toes, but don't overdo it.

4. <u>Develop a Spirit of Cooperative Effort on Your Team</u>. If you have more than one team, every man should be doing his best to shoot his way onto the first team. However, there should be a spirit of willing helpfulness within your group. Each man should be eager to do what he can to improve the team's performance. Look at the successful pistol teams. Observe the team members. Listen to their conversations and you will detect a cooperative attitude. This is the sort of team spirit you should develop on your squad.

5. <u>Train for a Specific Match</u>. It is a good idea to have the training period prior to a given match duplicate the procedure expected in the record match. If the team matches will be fired in one relay, practice shooting team matches in one relay. If you expect to complete a 900 aggregate before coming off the line, practice shooting a 900 aggregate without a break. A team's performance is handicapped if they enter a match in which a 2700 aggregate is fired in one day if in their experience they have never fired more than a 900 aggregate in one day. An abrupt change-over may create psychological problems because of the necessity to alter complex habits of coordination. This is the shooter's "edge" that must not be disrupted.

6. <u>Weapons and Equipment Checks</u>. The head coach should periodically check the equipment of each of his shooters. This check serves a quadruple purpose:

a. Detailed inspection of match pistols serves to assure the proper mechanical functioning of each weapon. Many times a shooter does not bring a weapon in for repair because it malfunctions very seldom. Every time a weapon malfunctions, it should be brought to the attention of the coach and the armorer.

   b. Weapons in constant use tend to lose accuracy over varying periods of use. The handicap of a weapon that shoots groups larger than the shooter's ability to hold will undermine even the performance of a champion. Thus, when a shooter's declining scores indicate a possible loss of mechanical accuracy in the pistol, that weapon should be checked on a static testing device and refitted if tests indicate it is necessary.

   c. The coach should check on the cleanliness of each weapon. Most shooters get lazy about cleaning their weapons, particularly if they are fired every day. Unfortunately, many shooters think that cleaning a dirty, though functioning weapon will somehow cause it to start malfunctioning and to lose accuracy.

   d. The security of government issue weapons is paramount. Regular serial number checks of weapons issued to individual shooters and security storage inspection insure proper accountability and security of all weapons and equipment.

   7. <u>Periodic Written and Oral Tests</u>. The coach should conduct periodic written or oral tests in order to check how much the shooters have learned. This means how much of the fundamentals, of the NRA Match Rules, of general match procedure and how well they have mastered techniques. The results of these tests should become part of the shooter's record.

   8. <u>Constantly Maintain a Current Evaluation</u> and an estimate of potential of each shooter regularly assigned to the All-Army team. Also, an evaluation and an estimate of potential should be made concerning outstanding Army shooters in major command level competition. This estimate must include: an analysis of the rate of progress; individual morale and attitudes, on and off the range; the degree of team effort exercised; all of which should be current. A good coach must distinguish between who is shooting for the team to win and which shooters are merely hanging on for the free ride.

   9. <u>Refine Doctrine and Raise Performance Standards</u> by constant review of training manuals, materials and methods. This research should affirm new ideas and techniques proven to be sound and reliable and weed out unsound doctrine.

   10. <u>Improve the Team Potential</u> by conducting periodic, organized, group instruction.

   11. <u>Improve the Individual Shooter's Potential</u> by personal and private interview and conducting individual coaching sessions.

   12. <u>Supervise the Team Preparation</u> for match participation.

   13. <u>Evaluate the Coaching Technique</u> of individual line coaches.

   14. <u>Assist in Preparation</u> of courses of marksmanship training.

   15. <u>Assist in Rehearsal</u> of instructors.

   16. <u>Participate in All Registered Competition</u>. A shooting coach is a person who appreciates the problems faced by the competitive shooter and is not inclined to be arbitary in his judgment.

   17. <u>Influence the Morale, Attitudes, Enthusiasm</u> of the shooter and stimulate his will to win by exhibiting individual consideration. Good coaching should promote a feeling of confidence and create an atmosphere of inevitable success. This influence will assure a favorable response if the guidance in personal habits and activities results in measurable improvement in performance. During the shooting season, placing certain limitations of a personal nature on his living habits will help the individual to build and maintain excellence. There must be no use of tobacco, alcohol, coffee or unprescribed use of certain drug preparations that may adversely affect shooting performance. Physical fitness is enhanced by avoiding both late hours and overindulgence in rich or unaccustomed foods. The shooter must be encouraged not to change his normal living routine. Pride and esprit de corps go hand in hand with an attitude that reflects the will to win.

18. __Correction and Recommendation.__ During the match, the head coach should keep notes on any mistakes made by team members or himself. Then he must put into his training program the necessary subject matter to bring about a correction of the errors. He should make recommendations to the team on anything that will cause the group to perform better in future team matches. The coach is the director of the team. If he fails, the team fails; if he is a success, the team will be a success.

19. __Individual Approach.__ When a coach is training a team, he should not adhere to unchanging standardized schedules. He must not make their provision equally applicable to all shooters. He should not forget that adept coaching provides an individual approach to each shooter. Therefore, he should exercise special care that every shooter accomplish that which he is potentially capable. Demand a performance level that corresponds to his mental and physical attributes, and peculiarities. Special Hours should be set aside for individual work during which time the shooters should concentrate on improving the weaknesses in their shooting. If a coach discovers signs of over-fatigue or signs that shooter is approaching a state of over-training, he should recommend a rest period. However, it should be borne in mind that signs of over-fatigue for one or two shooters is not ample cause for lightening the work load for the others.

20. __Training Methods Should Be Flexible.__ Training should be flexible enough to permit change in form and method, so that a shooter in some cases will be allowed to decide for himself what type of training is best for him. A coach should avoid tying a shooter down to static routine. He should teach his shooters to think and to experiment. He should unobtrusively bring them around to drawing their own conclusions as to what they should do instead of always setting his conclusions before them. He should remember that too much spoon feeding in the training period discourages a shooter's initiative and independence as well as other qualities associated with the development of will power.

21. __Study Shooter's Behavior During Shooting.__ In order that a coach may have the opportunity to study a shooter's shooting peculiarities and behavior during shooting, he should not try to work with more than two shooters at the same time. An experienced head coach should check all targets periodically with binoculars while shooting is going on, accompanying his inspection with observations and instructions. However, a coach must from time to time give some attention to personal assistance. While he is correcting a shooter's shooting technique, the coach can study the shooter's traits and assess his personality. When guiding the shooter's technique of employing the fundamentals, observe his temperament. See to it that he takes enough rest and the right kind of rest between shots. Also, as he is following a shooter's reaction to good and bad shots, a coach has the opportunity of studying his discipline, or lack of it. Work should never stop during the training period, on instilling the self discipline which will be required of a shooter when he shoots for record scores when participating in tournaments.

22. __Recognition of the Shooter.__ A coach's recognition of a shooter's work is very important. A shooter's successes and his conscientious work should not go unobserved by his coach. Excellence should be publicized and encouragement rendered. At the same time, a coach should show some reserve in such matters. Such shooters become conceited with being praised too much, begin to think too highly of their abilities. They may slack up on the hard work necessary to sustain the ability to shoot championship scores. Praise for the team shooter is based on direct comparison of his attributes and performance with those of his team mates and competitors. A coach ought to announce the results of practice and record shooting by posting scores, bulletins, etc. The shooters should achieve sense of responsibility for their own scores. An impartial totaling of the scores should be made in record shooting according to the rules of competition. Take notes of all the reasons for certain lower scores. It is only after due account has been taken of all these observations that any conclusion will be reached regarding the relative proficiency of a shooter.

23. __Esprit De Corps.__ In addition to his concern for a shooter's development of a high degree of control and the strengthening of his willpower, a coach should devote the most serious attention to the state of mind of each of the shooters on a team. Bolstering their pride in themselves, removing any cause for worry or dejection is the path to a happier team. In implanting and fostering a healthy state of mind i.e., a feeling of duty and obligation toward the group establishes a feeling of friendship and camaraderie. The coach should make a careful study of the character of each shooter from a psychological point of view. Wage unrelenting campaign against any show of self-seeking. If a shooter starts to think only of a champion's laurels from a personal standpoint and enters a tournament with these thoughts in his head, he may prove to be a burden to his team. Such shooters, thinking of themselves instead of the team, generally become extremely nervous due to the striving for individual

honors. This may cause their team performance to decline. The other members of the group quickly sense this selfish attitude on the part of a teammate and usually react negatively.

24. <u>Fighting the Fear of Failure</u>. The demand that a shooter show a sense of responsibility to the team for his own score is a reasonable one. He should not be allowed however, to develop an overpowering, oppressive sense of his individual responsibility for the team's success. Threats should never be used. It should be borne in mind that a common cause of poor scores in a tournament is a lack of confidence resulting from fear that one may let his team down by shooting poorly. The responsibility for success is equally divided between the five team members, i.e., four shooters and the coach.

25. <u>Rest Period Before Each Match</u>. At least a full day before a tournament starts, a coach should stop all training and give the shooters an opportunity to have a good rest. Following such a rest, a shooter is more apt to shoot willingly and attentively, and thus he will give more effort to controlling each shot. However, it is certainly not out of place for a coach to advise the shooters to practice "dry firing." Such sessions aid the shooter in maintaining the finely tuned "edge" he has attained in diligent training.

D. LINE COACH TECHNIQUE

Teaching and coaching in marksmanship must be logical and methodical. In order to receive the necessary response and results from the average shooter the coach must be able to analyze mistakes, dispel false notions, be sympathetic, be encouraging, and be honest. As each shooter is an individual, the coach must be quick to determine the individual characteristics that affect him, so that he may apply his coaching psychology to establish a satisfactory relationship. The initial contact between a coach and shooter will be the training period. During this time the coach will call upon his experience to impart knowledge and to correct shooting habits. This training period is of as much benefit to the coach as the shooter. As the training progresses a feeling of mutual confidence must come into existence.

One of the principal problems facing a line coach in a record match is to maintain the shooter's shot group in the center of the target. However, the line coach's role is not limited to the technical aspects of shooting. For example, if a shooter is nervous during a tournament, the line coach should help him to control his nervousness.

1. <u>The Shooter's Nervousness</u>. A shooter's nervousness during a pistol match is not limited in scope to minor symptoms. For example, his pulse beat is faster and his coordination is upset. It is also the result of a change in his behavior and of those characteristics of his personality which ordinarily do not manifest themselves. A shooter becomes irritable, touchy, less restrained, etc. Sometimes a shooter, who under ordinary circumstances, is very calm and complacent, becomes suddenly unpredictable because of his nervousness. Everything annoys him on a very hot day, his jacket is uncomfortable. Someone laughs too loudly or someone else is staring at him. He is annoyed by the fact that the line coach is displaying indifference to his difficulties and shown no interest in how he is doing. Irritation, dissatisfaction, and probably insults follow. Another shooter may become exceedingly absent-minded. You say something to him, he nods his head to show he understands and does exactly the opposite. A line coach must be ready for these deviations in conduct. He must display tact and the greatest patience toward this behavior so that he can keep his shooter, in spite of all diversions, shooting well and help him to shoot a winning score.

A line coach should understand that a shooter's nervousness and state of mind should not be thought of as fixed once it has changed. For example, a shooter may be nervous at the start of shooting, later regain his composure, and once more get the shakes. A shooter may be calm at the beginning of a match and suddenly become nervous, settle down and on the later shots once again have to fight to overcome a nervous tremor. The changing conditions of competitive shooting cause emotions to fluctuate, one variation sometimes replacing its opposite. This means that a line coach should know his shooter, have some knowledge of his character, and the trend of his state of mind at all times. Thus, if, after a poor start, a shooter regains mastery over himself and begins to shoot with boldness and assurance, there is no reason why the line coach should interfere for a while at least. If, however, the line coach feels that the shooter's control is again slipping away and that the latter wants to keep on shooting even after he has shown obvious signs of disturbance - which usually results in some poor shots - the line coach must interfere before it is too late, to prevent additional lost points. Force the shooter to take a short break. A shooter's state of mind changes very quickly when he is shooting poorly. What was a healthy frame of mind may disappear without a trace and be replaced by feelings of irritation and anger. A line coach must sense this sort of thing very clearly and his behavior should be governed by the character of his shooter. Also pistol shooters react differently to bad shots. On some, they have a sobering effect, causing them to strive harder and to be more observant in making the next shot. On another shooter, a poor shot may have the effect of urging him on in desperation to firing the next shot as soon as possible in order to compensate for a bad performance by a quick ten ring hit. This usually results in his committing the same error over again. In this case, the line coach's duty is to stop the shooting and insist that the shooter take a short break and check his stance, position, etc. Still others may be rattled by poor shooting and lose their self-assurance and exhibit a reluctance to continue. If the line coach thinks that the cause of the bad shots can be corrected promptly, he should suggest to the shooter a course of corrective action and insist that shooting continue in a suitable tempo. If, however, the shooter is unable to continue because he is fast losing his confidence in himself, the line coach should have him unload his pistol and shoot "dry." A reemphasis of good shooting habits will stimulate the conditioned reactions so painstakingly developed during training.

In order to get a shooter to behave as calmly as possible, the line coach must be a person in whom the shooter has trust, in whom he believes, and who he feels has his best interests at heart.

2. <u>Unintentional Harm to Shooter's Performance</u>. A good performance is not always assured by a complacent character and a benevolent attitude on the part of the line coach toward the shooter. Sometimes a line coach may make things difficult for a shooter without knowing it. The person who acts as line coach should have had considerable experience in tournaments and should be knowledgeable in shooting matters. He should understand the technical information about the weapons and ammunition; he should have practical experience and knowledge concerning the influence of weather conditions on shooting accuracy. He must understand the character of his shooters from the point of view of their habits, their disciplinary and psychological peculiarities, their competitive capabilities, and their method of operation. Choose your words carefully, exhibit a friendly, winning, cooperative attitude. Maintain your composure, radiate enthusiasm as regards the possibility of winning the match, don't reveal anxiety and avoid unwarranted changes in procedure.

3. <u>Control Tempo of Shooting</u>. Part of the line coaches job is to keep close watch on the tempo and rhythm of shooting. If a shooter is shooting smoothly and doing well in a particular rate of firing, he should continue at this rate and not slow down or speed up. Also, the coach should know what to do when a poor shot is fired. If, when a shooter is shooting well and shooting with assurance, a poor shot occurs, there is no need to suggest to the shooter that he stop or change his tempo. If two poor shots are fired consecutively, however, and the shooter is showing signs of losing his self-control, the line coach ought to take a decisive stand and insist that he take a short break so that an analysis of performance can be made and the shooter can make an effort to regain his composure. Check out his stance, position, grip or any of the other fundamentals. Only then should he continue to shoot, attempting to do so at his former rate of fire. Meanwhile, the line coach should observe the time closely and regulate the tempo according to the remaining time. If the shooting is at a fast pace, and under control, the line coach should not take him off his rate of fire for the rest of the stage. But, if the shooter is shooting with difficulty and if there is some danger that he may run out of time, the line coach should suggest to the shooter that he time his shots with a stop watch at exactly even intervals as a means of getting him to work with greater speed. It is especially important to watch the time if there is a gusty wind. Under such conditions, the line coach must insist on a somewhat stepped-up pace so that the shooter will be certain to have some time to spare in case the gusts are so strong that the shooter cannot shoot and is in danger of having his time run out before he has finished shooting.

4. <u>Reestablishing Coordination by Dry Firing</u>. If a shooter loses his control and can't get back to a semblance of order, if he becomes nervous and shows signs of indecisiveness and fear of making a bad shot, the line coach must insist that he unload his pistol and make a few "dry fire" shots. This should reestablish his coordination, overcome his indecisiveness; he may regain his temporarily lost ability to sight accurately and press the trigger with coordination and smoothness.

5. <u>Limitations of Dry Fire Practice</u>. Do not overestimate the value of dry fire training as compared to shooting with live ammunition. The two types of training complement each other. Nothing can replace the psychological and nervous processes that a shooter experiences during actual shooting. When practicing with "dry firing", no loud noise accompanies a shot and there is no recoil. The nervous system, and one's reflexes in particular, are in an altogether different state. Consequently,

muscle tension is not changed as much as it is just before firing a real shot. Moreover, the dry shooter seldom feels as responsible for the quality of his shooting and may not work at it as carefully because there are no bullet holes and no regrets. There is not the same feeling that a mistake is irrevokable and that poor shots can't be redeemed as in live practice. Every shooter should, therefore, find a suitable balance in his training for these two approaches without overestimating or underestimating the value of either one.

      6. <u>The Right Word at the Right Time.</u> The winning shooter experiences tremendous nervous tension in a pistol match. He must have an inordinate amount of will power to force himself to fire a large number of controlled shots with the greatest of care while at the same time trying to overcome his nervousness. He must use every means to take his mind off distractions. The existence of such conditions gives considerable significance to those words which can calm and those which can excite him. In the difficult conditions which demand great efforts of will from a shooter, overcoming some knotty problem is sometimes accomplished by the right word spoken at the right time: a word of encouragement, of assurance, of sympathy or guidance. Words of conviction stand for real things, when the shooter is under conditions of stress.

      7. <u>Cooperative Attitude.</u> Line coaches must keep the shooter in a good frame of mind so that he will shoot with assurance and boldness. They must try as hard as possible to bring about such a situation during shooting. If the shooter senses in every word, every gesture, that the coach's attitude toward him is friendly, that the coach is taking a vital interest in his effort, that he understands the stresses and strains which a shooter undergoes during a match, the shooter will have the incentive to excel.

      8. <u>Deciding the Order of Relays.</u> In a two relay team match, the relays should be made up of compatable pairs of shooters. In a four relay set-up, the shooter who has extensive experience in match shooting and who is less apt to be nervous is put on the first relay. Such a stronger-willed shooter should also be put on the last relay when the competitive pressure is at its highest pitch and when the scores of the last relay may determine the outcome of the match for the whole team. The relays which fall between are assigned to shooters of less experience and those whose abilities the coach has some minor doubts. However, in deciding upon the order of the relays, a coach ought to take into consideration the desires of his shooters to shoot on a particular relay. He should bear in mind that some shooters, because of their psychological nature, prefer to shoot first, some want to shoot later, when the results of preceding relay are already known. Therefore, a coach must, on the one hand, pay attention to the desires of the shooter and do what he can to satisfy them. On the other hand, when there are two or more who want to shoot on the same relay, he must make his decision based on the best interest of the team.

      9. <u>Team Selections.</u> Sometimes, a coach who has studied a shooter both by the results of his shooting and by observing him in action, decides to place him on a lower ranking team. The coach may have doubts as to whether or not he would have done as well under the stresses and strains that go with shooting on the first team. The coach should not, under such circumstances, become perplexed or show any weakness of will. He should always be firm in his decision and stand by his standards of selection.

      10. <u>Encourage Thoughtful Analysis.</u> A shooter must develop the ability to approach each shot thoughtfully and critically and by careful analysis find out what factors are affecting his accuracy. Without a careful study of his actions, without an analysis of his shooting, especially of his poor shots, it will be impossible to become a better shooter. It is necessary to impart to the coach an accurate report of what transpired during the firing of the shot and to search out the reasons for a poor performance. Some shooters have an excellent memory for their good training scores. They carry them about with them, and brag about them. But their poor record scores - which most shooters have in amply supply - are tossed in the mental ash can, not being regarded worth the thought it takes to figure out why they are bad. In such cases, there will be no solid improvement because such a shooter cannot extract from his training that which is useful to him. Since he does not find out the causes of mistakes, he will experience only slow progress, if any.

      Close communication must be maintained between the shooter and the line coach. After each shot is fired, the coach and shooter should discuss the nature of the shot, good, bad or unknown. An accurate shot call is mandatory because if a shooter cannot call his shot accurately, he does not have control of his shooting. To arrive at comprehensive shot analysis, the coach and shooter must be in close liaison. Everything the shooter saw during the shot delivery must be revealed. Everything he heard, even the most insignificant distractions must be mentioned. The innermost thoughts that occupied the shooter's mind, even fleetingly, are of tremendous importance to a coach when he is making a determined effort to pin-point the reasons for even a slight loss of control. When shot analysis and positive correction seem to fail in improving control, the coach should start asking leading questions. An example of some dilemma breaking questions appear on both the slow fire and rapid fire work sheets.

Improvement in performance may or may not be spectacular. Impress on the shooter that you are interested in his progress and will feel justly rewarded by even the smallest evidence of improvement or understanding. By reducing the frequency of error by one per cent, measurable progress is being made. Slow, steady progress is indicative of firm improvement of control. Flashy, fantastic performance followed by average results is of no consequence unless the shooter should happen to learn why and how his shooting suddenly improved.

E. ANALYZING AND IMPROVING COACHING TECHNIQUE

1. <u>Why Some Shooters DON'T SHOOT WELL.</u>

a. Lack of know how. A symptom is vacillation. Switching equipment, employing gadgets and not using an established technique in a desperate attempt to offset his lack of knowledge and skill.

b. Lack of incentive. Satisfied with present level of proficiency. Indifferent to need to perform even at actual ability level. Will fire only well enough to make the cut-off in team selection and will not be concerned about further improvement or being a member of winning team.

c. Lack of team spirit. This type of shooter is a glory seeker. During individual matches he is adept and hard working. During team matches, he lacks the cooperative will to win. His team performance and results may be less than excellent.

d. Lack of sense of responsibility-seeking excuses for bad performance. This type of shooter is sensitive to any conceivable situation or act on the part of his teammates or fellow competitors upon which he can transfer blame for a marginal performance. Examples are: derogatory statements or attitudes on the part of the coach or his teammates; upset by innocent remarks he believes to be of a slighting nature.

e. Lack of ability to analyze and correct errors. This situation applies to the shooter who is not showing any lasting improvement.

f. Lack of proper guidance. Shooter is subjected to a minimum of intelligent coaching technique, know-how and inspiration. Most shooters need comprehensive direction to achieve improvement. Steady improvement is the basis of most shooters interest in competitive marksmanship. It is much easier for the coach to motivate an interested shooter.

g. Performance hurt by pessimism or overconfidence. If a shooter does not expect to shoot well in a match, he probably won't. If he becomes so confident in his ability that it begins to look easy, the coach must expect a sickening crash.

h. Closed mind. Usually will not try new approaches to improving control. When a new method is tried and spectacular results are not immediately forthcoming, the closed mind forthwith settles back and resumes the old habits without allowing time for conditioning the nervous system to the demands of the new method.

2. <u>Why Some Shooters DO SHOOT WELL.</u>

a. Has that extra edge of confidence and is busy at planning a positive course of action.

b. Utilizes all available skills-assisted by confidence in good equipment.

c. Willing to sacrifice certain activities and practices that hurt his shooting.

d. Enthusiasm generated by taking part in the team effort.

  e. Has the will to win by his own resources and determination.

  f. Will to win is stimulated by the enthusiasm and cooperative attitude of his team coach and team mates.

  g. Enjoys above average coordination and timing. He wasn't born with these attributes. This type shooter has trained long and hard to develope these qualities.

  3. <u>Though They Are in a Sense Competing</u>, each trying to perfect a team performance of a higher caliber than all other teams on the pistol squad, coaches must never forget that an important technique of successful organization is cooperation. What team coach "A" knows has helped his team or what he can do to assist team "B" or "C" becomes his official and moral obligation to transmit. Further, he must know the problems faced by the other coaches and how they resolved them. Achievement develops out of unity of action.

  4. <u>A Good Coach Knows When it is Important Not to Coach.</u> He, of course, is obligated to do all for the shooter that he knows but he must exercise less direct supervision for certain periods to see if the shooter can control all the factors that add up to good shooting. The coach must not be a crutch.

  5. <u>If a Coach Tries to Correct Too Many Mistakes</u> on the part of a shooter in one practice session, it serves only to confuse. He must correct each fault after full analysis and explanation to the shooter as to the nature of the fault.

  6. <u>If Team Shooters Desire to Coach Each Other</u>, all the established functions of coaching must be adhered to. It is not the perogative of individual shooters to dispense with certain functions of team preparation and control and include only those actions that come to mind. It is mandatory that all coaching responsibility be superimposed upon the already considerable burden of responsibility presently guiding the actions and thoughts of the successful individual competitors. An organized, systematic routine rather than the haphazard approach must be followed as in successful individual competition. The established responsibilities of a team coach are in danger of not being carried out completely or reduced to a confused state when the responsibility is shared equally by four team members. This condition would automatically necessitate the chief responsibility for coordination to become the function of the ranking member of a military team. This person then becomes a de facto coach and this results in the watering down of his attention to his individual efforts. The team can't affort to have one team member performing at less than the peak of his potential.

  7. <u>Equipment and Ammunition Selection.</u>

  a. The coach should be instrumental in making certain that each member of his team is in possession of weapons of the highest degree of inherent accuracy and that each weapon is mechanically reliable. Each weapon should be functionally adapted to the particular desires of each individual shooter.

  b. He should make certain that each member of his team has available to him the highest quality, accurate ammunition possible to procure and have it test fired to establish its superiority of accuracy in each shooter's weapon. The advanced pistol marksmanship coach should endeavor to develop the shooter from the level he receives him and instill in him the desire to progress. The will to win and be a part of a winning team serves to encourage the shooter to achieve excellence. Never let the shooter forget, slide by or detour around the proper employment of the fundamentals.

F. EVALUATION OF THE TEAM SHOOTER

Proper evaluation of a shooter's potential ability is based on: knowledge of his personal traits; his understanding and retention of instruction; his ability to analyze his shooting faults and the trend of his shooting; his reactions to external conditions; the outstanding shooting problems he has overcome and those problems he still faces as a challenge to his improvement.

Accurate evaluation of each shooting member of a pistol or rifle team is a requirement made necessary by the failure of guess work. Intuition, prayer and hope will not bring about a grouping of the four best shooters on your team by accident. Accidents are inherently destructive and a team coach should not willfully court disaster.

A dynamic system of operation engages all of the participating individuals, measures them against exacting standards, tests them for constant progress and relegates each of them to a level of potential commensurate with their assessed capabilities. If a person's will can be demoralized in the face of competitive stress, his category is with those who have abject natures. An unskilled person does an unskilled job. The audacious and the resourceful forge ahead over a path made soft by the unprotesting, yielding bodies of those who conveniently play dead.

To be efficient, the system must evaluate unerringly. It must remove the veneer of less important attributes and penetrate deeply to the core. At the core is the inherent skill and intellect. Toughness, tenacity and initiative weld these essentials of character into a driving, relentless force for victory in competition.

The evaluation of the shooter by the line coach, the head coach, and the officer in charge is made up of the following: A line coach evaluation, a graphic progress record and an individual aggregate and team aggregate record which reflects the shooter's average for the complete training period; an examination, oral or written; the shooters individual score book and an analysis of his personal traits or attributes. An example of each of these evaluation factors follows:

G. INDIVIDUAL INFORMATION AND EVALUATION SHEET

1. <u>Line Coach Evaluation</u>:

<u>NOTE</u>: Information recorded here will by its very nature be considered confidential and will be only for the use of coaches and team officials. Each coach will make his remark in such a manner as to reflect his considered OPINION and what action he has taken to orient the shooter as to his particular problem if a problem is noted. Extreme care is indicated here and it is suggested that this portion of the file be filled in after the shooter has been under observation for at least a month or just prior to completion of training period.

2. <u>Graphic Progress Record</u>: The graphic record of an individual's shooting must furnish the following information: The level of individual performance as compared to team performance in practice shooting and in match shooting. It is well known that the hot individual shooter can maintain an aggregate that is equal to or better than a good team shooter who is unable to post equally high individual scores. The graph furnishes this important knowledge at a glance where a running aggregate has to be studied in detail. The graph also shows the all important trend of a shooter's performance. A hot starter and weak finisher can have the same total aggregate as a slow starter and a hot finisher. The best team member is the one who is firing the best scores <u>now</u>. One more important factor that the graph shows is consistency. The prospective performance of a shooter is somewhere between the peaks and valleys of the graph. The wider the span from best score to worst score, the less the accuracy of predicting the approximate score a certain shooter will get in the next team match. Are the peak performances attained during practice and a somewhat poorer performance attained during competition? This feature is immediately apparent on even a short term graph. What is John Doe's highest potential score?

EXAMPLE OF LINE COACH'S EVALUATION FORM

NAME _____ RANK _____ SN _____ ORG _____

  a. Personal Traits:
    (1) Mental Attitude.
    (2) Behavior.
    (3) Team Spirit.
    (4) Physical Condition.
    (5) Personal Appearance.
    (6) Care of Equipment.

  b. Specific Problems:
    (1) Weapons.
    (2) Slow Fire.
    (3) Timed Fire.
    (4) Rapid Fire.
    (5) Other.

  c. Reaction to Weather Conditions:
    (1) Wind.
    (2) Rain.
    (3) Cold.
    (4) Very Hot.

  d. Response to Coaching:
    (1) Preliminary Preparation.
    (2) Position and Grip.
    (3) Sight Alignment versus Sight Picture.
    (4) Trigger Control.
    (5) Mental Discipline.
    (6) Use of Slow, Timed and Rapid Fire Technique.
    (7) Shot Analysis.
    (8) Use of Corrective Measures

  e. Individual Aggregate Average.
    Competitive: .22____ C.F.____ .45WC____ AGG____ .45HB____
    Practice: .22____ C.F.____ .45WC____ AGG____ .45HB____

  f. Team Average:
    Competitive: (NMC) .22____ C.F.____ 45WC____ .45HB____
    Practice: (NMC) .22____ C.F.____ 45WC____ .45HB____

  g. Test Grades: I____ II____ III____ IV____ V____ VI____ VII____

  h. Line Coach Evaluation: _____

                                               Signature of Line Coach

  i. Head Coach Evaluation: _____

                                               Signature of Head Coach

                       Noted by _____
                                                 OIC Pistol Division

(EXAMPLE)

Graphic Progress Record (Individual Average, Solid Line--Team Average, Dotted Line)

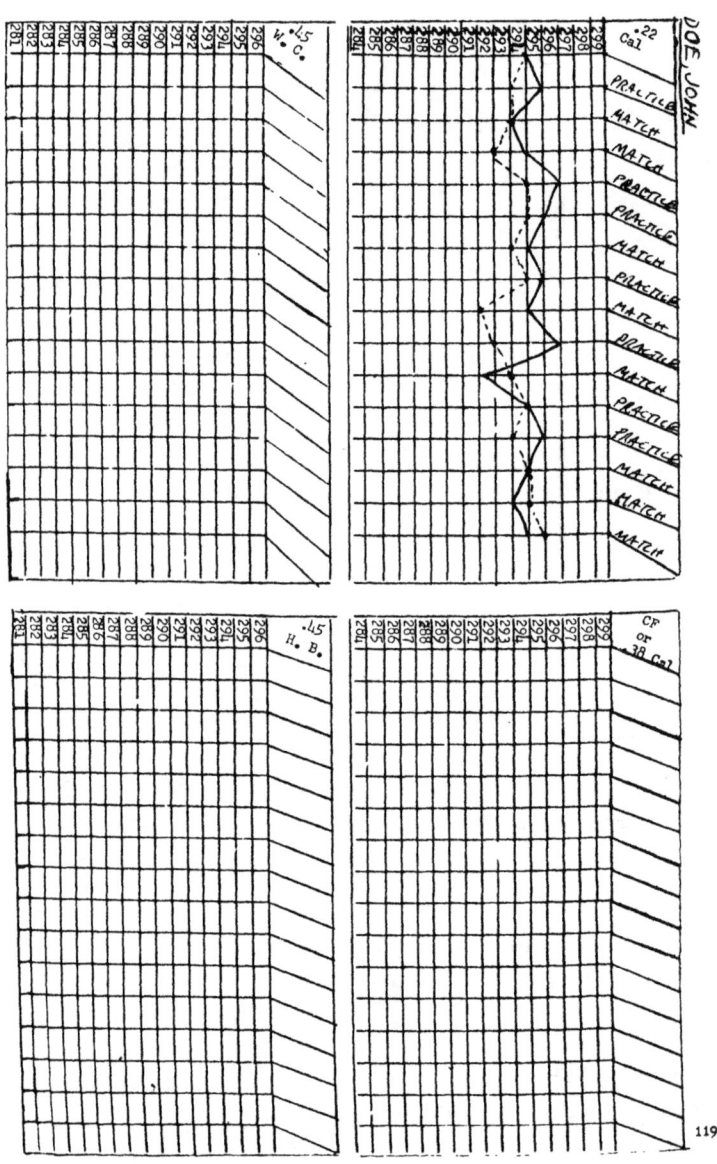

3. Aggregate Record.

   a. The dependable team shooter is the primary objective in the US Army Marksmanship Training Program.

   b. It is readily apparent in this system that team shooting has the ratio of 50-50 instead of 75-25. Also apparent is the immediate comparison of individual versus team performance. This arrangement places the team effort in its proper focus. The high scoring individual shooter has a difficult time compiling a running aggregate that will cause him to be chosen over a team member who has higher team scores. For example, team member number 8 above, might possibly be selected as one member of best four-man team. Malfunction of a weapon may have caused low individual score.

148

(EXAMPLE)

INDIVIDUAL AGGREGATE AND TEAM AGGREGATE RECORD FOR ONE CALIBER WEAPON IN THREE MATCHES

Ranking	Name		Indiv Agg	Tripled Team Agg	Grand Running Agg
1	SFC	SHOOTER	2610	2592	5202
2	"	"	2606	2589	5194
3	"	"	2582	2604	5186
4	"	"	2579	2607	5186
5	"	"	2589	3595	5184
6	"	"	2575	2604	5179
7	"	"	2592	2583	5175
8	"	"	2548	2622	5170
9	"	"	2576	2586	5162
10	"	"	2592	2568	5160
11	"	"	2561	2595	5156
12	"	"	2577	2568	5145
13	"	"	2579	2544	5123
14	"	"	2569	2547	5116
15	"	"	2552	2562	5114
16	"	"	2555	2541	5096
17	"	"	2533	2559	5092
18	"	"	2546	2526	5072
19	"	"	2539	2526	5065
20	"	"	2531	2526	5057

NOTE: 1st place individual aggregate score is average of 870 per match.
1st place team aggregate is average of 288 x 3 or 864 points per match.

 c. Practice scores should not be combined with match scores in a running aggregate. Use match scores only. If a practice score aggregate is kept initially, it should be discarded when sufficient match scores are available for accurate evaluation.

 d. There are other enormous handicaps in selection of the four best pistol shooters on your pistol squad. The four best .22 caliber shooters are not necessarily the best .38 caliber shots and vice versa. Total aggregates of all three weapons are of limited use. Base your selection on separate caliber aggregates and graphs. Consult the line coach's evaluation of each shooter on his performance problems with each weapon. Further, direct consultation with the line coaches on the day of the team matches may reveal conditions such as mild illness, family problems, changes in normal living habits, depression or elation over current performance that have a profound effect on team scores. If the wind is blowing today, there are certain members of your pistol squad who will perform immeasurably better than others. (The line coach's evaluation sheet for each shooter should have a complete list of pertinent information and other items needed by the coach in making an intelligent decision. It is difficult to remember all these facts concerning a pistol squad of many shooters.)

 e. To foster incentive and avoid activities and formation of habits that are detrimental to the team effort, certain guidelines are necessary for instilling discipline among the individual shooters:

  (1) No overeating during the shooting day.

  (2) No smoking during duty day or at matches.

  (3) No coffee during duty day or at matches.

  (4) No alcohol or drugs (medicine) in the training period before or immediately prior to or during record or match firing.

  (5) Daily physical training (necessary for body tone).

(6) No gambling (creates nervous tension).

(7) No late hours (2300 hours is the latest bed time acceptable in the training period and during the days of match firing).

(8) No night driving (be off the highways between sundown and sunup).

(9) Keep your weapons secure, clean, lubricated and periodically tested for accuracy.

(10) Keep a record of every round fired with each caliber weapon in your scorebook.

(11) Never score yourself during record firing.

    f. (1) Par score for record practice 3 gun indiv aggregate - 2640x2700.
           Par score for record practice 3 gun team aggregate - 880x900.

       (2) Par score for record practice 45 H. B. indiv aggregate 876x900.
           Par score for record practice 45 H. B. team score 292x300.
           (Par scores will be modified as weather conditions dictate)

H. THE SHOOTERS SCORE BOOK

The Shooters Score Book is a valuable aid to the competitive pistol shooter. It is an individual shooters record of all firing by stage, such as slow, timed, and rapid fire, plus National Match Course and all shots fired should be recorded. Practice scores should be kept separate from Match Score.

The Score Book is valuable only if it is kept accurately and up to date. Record the bad scores as well as the good ones. By keeping the scorebook accurately it will be an aid in finding weak points in the shooters performance. The Score Book will show over a period of time a trend in the shooters overall progress. It will reflect progress in the different stages of fire. Averages may also be kept in the score book for each stage of fire. Record of ammunition used, sight adjustment, windage, and elevation, weather, light, wind, temperature, may be an aid to the shooter at a later day. There are many type score books that will do the job well, but choose the ones that are simple and will furnish the data you need quickly and enable you to start firing under existing conditions and eliminate the handicap of not knowing where to set your sights. An example of a Score Book page appears below:

DATE			LOCATION									AMMO CALIBER						WIND			LIGHT	
1	2	3	4	5	6	7	8	9	10	TOTAL		1	2	3	4	5	6	7	8	9	10	TOTAL
											SL											
											TM											
ELEV			WIND			SF TOTAL					RP											
1	2	3	4	5	6	7	8	9	10			NMC TOTAL										
												TOTAL AGGREGATE										
												NMC TEAM MATCH										
ELEV			WIND			TF TOTAL						1	2	3	4	5	6	7	8	9	10	TOTAL
1	2	3	4	5	6	7	8	9	10		SL											
											TM											
											RP											
ELEV			WIND			RF TOTAL						NMC TOTAL										

I. ATTRIBUTES OF A TEAM SHOOTER (for the coaches' analysis).

A popular fallacy that competitive marksmanship scores are engendered by a blank or nerveless mind has been proven completely erroneous. A quick look at the nation's outstanding marksmen will show you individuals of higher than average level of intelligence. The necessity for intense concentration and strict adherence to a multitude of sometimes unnatural but correct shooting fundamentals and techniques, quickly eliminates those of lower intellect.

A good team shooter needs to have many favorable attributes. However, if he lacks the ability for intense concentration and the intestinal fortitude to make up his mind to adhere to fundamentals regardless of match pressure, adverse weather conditions or any other conceivable distraction, he is useless to himself and the team.

1. <u>Consistency</u>. Consistency is a most useful and important attribute if it is coupled to performance at a high level of efficiency. If you, as a coach, can place four consistently high scoring shooters on the line for a team match, your probability of winning is considerably better than that of a less fortunate team. If you have a man who shoots 297 part of the time and 287 at other times, and a man who shoots scores ranging between 290 and 294, which of these shooters would you prefer to have on your team? Naturally, you would choose the 292 average shooter with the smallest score spread. An inconsistent or erratic shooter can ruin a good team in the long run. The other members never know whether he is going to "hit it hard" or have an "OFF" day.

2. <u>Compatability</u>. Compatability, as it pertains to team shooting, is the ability to get along with your teammates. You are working as a team, in the highest sense of the word. All of your long hours of practice and hard work can be ruined if one of your team members makes a remark which you interpret as a personal reproach, or which may cause you to form a lasting dislike for him. Your mind may dwell on this remark instead of concentrating on your shooting. You must also be able to get along with shooters other than your teammates. If some of these fine competitors find that you don't have a sense of humor they may antagonize you by making you the butt of a joke of a practical nature. It helps to be an easy going person while shooting on a pistol team because with the pressure on during the team match, the rest of the team doesn't need a grouch or a sore head among them to complicate things. Smile and the world smiles with you. Grouch and the world tries to ignore you.

3. <u>Eagerness</u>. Eagerness or love of shooting on the part of the team shooter will apply not only to competitive match shooting, but to the long and grueling periods of practice with no other goal except self-improvement. You have no doubt seen the avid shooters, when a lull in practice occurs, gather up their powder horns, shot measures and experiment with that old muzzle loader. Some of them may be shooting an old junker of a .22 rifle or pistol which was old when they were born. A good percentage of them are hunters, or outdoorsmen of other calibre. Many of our better pistol shooters are part time gun collectors of one degree of proficiency or another. In short, they enjoy guns and shooting. They know they are capable shooters and they look forward to matches with anticipation, knowing they are going to turn in a superb performance.

4. <u>Sure-Footed and Deliberate</u>. The good shooters always seem to be standing around telling jokes when everyone else is hastily preparing to shoot. Why? The next time you go to a match, watch one of these top shooters, he does everything ahead of time and more or less by the numbers. When the range officer says "IS THE LINE READY," that shooter is ready. He is deliberate in everything he does, and he is thorough. Where does this sure-footed business come in? If our sure-footed

shooter is shooting a group in the 9 or 10 ring at 3 o'clock, he immediately moves his sights, not relying on hope that the next shot will be an X. If he is sure and positive in his actions and has confidence in his ability, he will move his sights enough in one bold adjustment to put his next shot in the center of the X ring.

     5. <u>Confidence</u>. It is impossible to shoot consistently good scores if you don't have confidence in your ability. A point concerning confidence in ability is the four minute mile. For years it was thought that it was impossible to run a mile in four minutes, however, as soon as one man did it, several others also accomplished it. The mental block was removed and people knew it was possible and they were able to duplicate it. Where confidence enters the picture in shooting is when the first relay has fired a good score in a team match. The thoughts in each team member's mind at that time is a good indication of a confident attitude or the lack of it. Gloomy Gus will say, "I hope they don't goof up," or "Good lord, what do I do now?" The confident shooter will say, "We're bound to have a good score. I have been able to carry out my plan for control of each shot on a 97% basis and I expect to improve the average in this match."

     6. <u>Good Equipment</u>. Good equipment being considered as an attribute of a team shooter may sound a little unusual, but upon careful consideration, it will mean a lot more. The team shooter builds confidence in good personal shooting equipment. This is of primary importance in a winning team performance. It is immaterial whether the equipment is issued to the shooter or whether it belongs to him, it should be the best that is possible to obtain. Not only is it necessary to have the best obtainable but is important to have his equipment complete and checked out in good working order. A shooter can be just as easily upset by not having an accurate weapon or a malfunctioning weapon as by having buck fever.

     7. <u>Good Health.</u> Good health is another attribute of a team shooter that is easily perfected. If a team shooter is to give a good performance he must be in good physical condition in order to have the stamina on the firing line to make each shot or string of shots, the best he is capable of firing. Physical conditioning is imperative for the reason that it gives resiliency to the muscles and better nerve control. A shooter should strive to build better general health and he should keep himself in the best possible condition, prior to and during matches. This conditioning is aided by the knowledge of the effects that certain foods have on the system. In subsequent instruction, the effects of some of these foods and certain other detrimental items such as coffee, alcohol, drugs, and tobacco will be discussed.

     8. <u>Open Mind</u>. Be always on the alert to help a teammate. Accept and give constructive criticism in the spirit of being helpful and of pulling for the common goal. Accepting the little peculiarities or personality quirks of his teammates as part of the days work, will greatly add to the shooter's ability to remain calm, serene and ready to concentrate on the job at hand: control each shot to the best of his ability. Your teammates have to have an open mind to be able to put up with you, you can at least return the favor.

     9. <u>Sportsmanship</u>. <u>There is no room on any team anywhere for a poor sport</u>. This is especially true of an Army marksmanship team. Any time you wear the Army uniform to a match you are representing the US Army. Considering your rank and pay, you may think differently, but in the eyes of all civilians at the match, YOU are the Army. What is more, each civilian will look at you with the idea that he alone, with his taxes, bought your weapons and equipment and even paid your entry fees. This behooves you to conduct yourself in a manner that will bring no criticism on the Army, the marksmanship program and finally, but most assuredly, YOU. Any complaining letter that is written or remark that is made, will, with the slow certain steps of death and taxes gradually flow down hill to YOU. BY YOUR CONDUCT make sure this is a letter of appreciation and not one of complaint. Our illustrious leader has said that there is no such thing as a good loser, however, you don't want to be a bad loser either.

     10. <u>Mental Attitude</u>. A philospher once said, "You are what you think." What you tell yourself or what you convince yourself you can do, you more than likely can accomplish. How many pistol firers have ever stood on the firing line when all of a sudden a shooter slams his weapon down and says "D/#G*()&&%'''''/@1/46$#''', I just jerked a seven". It happens frequently and it may shake you up. The shooter that operates under such a thin emotional veneer that he easily loses his temper is the victim of a monumental lack of self-control. Such a lack is a serious handicap in an endeavor that places great importance on controlled concentration.

Your mental attitude governs your complete performance. How many have ever gone into the last string of rapid fire knowing exactly what you had to get for the NMC to beat the existing record. If you failed, you were probably saying to yourself the whole time "If I just don't goof this string up." Control is based on careful planning and the coordination that comes from extensive match experience. Hoping, praying or wishing, will not bring home championship records.

11. <u>Dependability</u>. The team shooter that is on time for a team briefing before the Big Team Match, that shows up with all of his equipment and ammunition, that knows what target and relay he is assigned to, who voluntarily scores the adjacent team during his off relay without the necessity of having the coach look for him, sets up methodically and thoroughly on the firing point without undue supervision, is the kind of dependable man who can also do a better than average job of shooting. It is surprising how much can be accomplished when there is no worrying about who gets the credit.

12. <u>Honesty</u>. It is sufficient to say that honesty, like virtue, is its own reward.

13. <u>Ability to Shoot</u>. Without the ability to shoot, the foregoing attributes are pointless unless you are planning to be a salesman. For example, eagerness to participate in a sport you haven't mastered, possessing a compatible personality and bursting with good health, are obviously not the primary requirements necessary to produce winning teams. The ability to shoot is the foundation on which the other attributes are used to fashion a winning team shooter.

CHAPTER XII

SERVICE RIFLE COACHING

To clarify this coaching guide, one must understand the relationship of the various terms used for coaching competitive rifle marksmanship. These are outlined as follows:

RIFLE GROUP: All personnel selected from a given unit to participate in competitive rifle marksmanship.

SQUAD: A subdivision of a Group which contains from one to several teams trained for a specific type of competition.

TEAM: The base competitive firing unit consisting of a team captain, team coach, alternates, and a number of shooters as established by the ground rules of the match.

INSTRUCTOR: This individual is primarily concerned with the teaching of marksmanship fundamentals. He should be a qualified marksman and a trained teacher.

HEAD COACH: This individual has the responsibility of the organization, training, selection, and performance of the unit or organizational rifle squad. The individual selected for this duty is generally the most experienced instructor-shooter of the squad. In a sense, he is the commander, supervisor, instructor, and coach. The head coach supervises the team coaches to insure implementation of proper firing and coaching techniques.

TEAM CAPTAIN: This individual has the responsibility for all administrative duties of his team during the conduct of a team match. During the conduct of the Infantry Trophy or the Combat Rifle Match he also performs as a team coach.

TEAM LEADER: This individual has the same responsibilities as listed above for Team Captain during the conduct of the Combat Rifle Match.

TEAM COACH: This individual has the responsibility for the organization training and performance of his team under the guidance of the head coach. For small units or squads, the team coach may also be the head coach.

ASSISTANT TEAM LEADER: This individual has the same responsibilities as listed above for Team Coach during the conduct of the Combat Rifle Match.

INDIVIDUAL COACH: This individual is generally a shooter who during training is required to assist another shooter under the direction of the team or head coach. He is primarily concerned with insuring the proper application of marksmanship fundamentals and with the detection and correction of errors.

It must be understood that there is no substitute for knowledge and experience in marksmanship. An individual assigned the specific duties of teaching marksmanship fundamentals or coaching must be completely sold on the importance of marksmanship. A winning team is the pride of any commander--whereas a losing team, while performing to the best of its ability, a personal desire to excel. Good marksmanship and good coaching has never been conveyed quickly, nor can it be gained in a short period of time. Good marksmanship is the result of good coaching combined with mental and manual dexterity after long practice and hard work. A good instructor and coach will call on experience to aid him in his job and will give his men the advantage of everything he has learned. With proper application of the techniques discussed in this guide, the United States Army will continue producing championship teams.

Section V. TEAM COACHING

Figure 12-1.

A. Coach - Shooter Relationship.

Teaching and coaching in marksmanship must be logical and methodical. In order to receive the necessary response and results from the average shooter, the coach must be able to analyze mistakes, dispel false notions, by sympathetic, encouraging, and honest. As each shooter is an individual, the coach must be able to determine the characteristics that affect him. Knowling these, he may apply his coaching psychology to establish a satisfactory relationship. The initial contact between a coach and shooter will be the training period. During this time the coach will call upon his experience to impart knowledge and correct bad firing habits. This training period is of as much benefit to the coach as it is to the shooter. As the training progresses, so does the mutual confidence that is so necessarty to their relationship.

There are many characteristics and attributes of a good coach which contribute to the success fo a firing team. The most important, however, is the coaches ability to establish the proper "Coach - Shooter" relationship. This is nothing more than a mutual understanding which creates conditions that allow the coach and the shooter to produce the highest scores possible. If there is a conflict in personalities between the coach and shooter, the overall team is adversely affected. This has been and will continue to be a problem for all coaches.

B. Attributes of a Coach.

The following attributes are by no means the key to success for establishing the coachshooter relationship. If a shooter cannot benefit from training, correction, or practice, all these points have little value. The psychology and techniques of a coaching are still a personal thing and the developing of the coach-shooter relationship can only be the product of experience, practice, and the diligent application of these attributes.

    1. The coach must be an experienced marksman. The coach's previous experience as a shooter will enable thim to understand the problems that his shooters have.

    2. He must approach his team assignment with a cheerful and understanding attitude.

    3. He should insist on proper applciation of fundamentals.

    4. He should be thoroughly grounded in the principles of detection and coorection of errors.of errors.

5. He must have the ability to read the wind and mirage consistently from day to day. This is an important aspect in establishing zeros. This ability increases with practice and experience.

6. He must promote team spirit and the will to win.

7. He must apply the correct coaching techniques.

8. He must insist on correct, prompt execution of instructions and commands.

9. He must never overlook the fact that each shooter is an individual with a personality of his own.

Figure 12-2.

C. Technique of Team Coaching and Conduct of the Match.

1. Techniques of Coaching.

The technique employed by the team coach during the conduct of team firing is the single factor responsible for the attainment of winning scores. The coach's skill, patience, and enthusiasm are all directed toward this one goal, and he must acquire the best possible performance from his shooters. But how does he accomplish this?

After a period of training on the principles and fundamentals of competitive marksmanship, the coach ends up with the trained shooter. He is then faced with the task of selecting his firing team. He accomplishes this based on his system of evaluation and is now ready to train his team in preparation for competition.

The success of his team is based on many factors; however, the selected team generally consists of the best shooters, guided by the best coach, consistent with the correct application of firing and coaching techniques. Competitive teams are generally equal as pertains to talent and the winning team is the one that makes the fewest the application of his coaching technique. The technique can best be described as <u>"a process of detection and correction of errors insuring the proper application of marksmanship fundamentals and the guidance of his shooters with commands and information to produce winning scores."</u> The techniques outlined herein are used throughout the Army and were developed as a result of years of experience and evaluation.

Throughout the team training phase and during the competitive matches, it is important to remember that there is no substitute for the coach-shooter relationship. A Coach who has gained the respect and confidence of his shooters; who knows their capabilities and characteristics; and has the ability to evaluate their performance while employing the correct coaching techniques will produce winning scores.

2. Conduct of the Match.

In the discussion of the conduct of the match the National Trophy Rifle Team organization will be used as a guide. This team is composed of six shooters, two alternates, a non-firing coach, and a team captain. Keep in mind however, that the techniques discussed are equally applicable to any team organization or course of fire for competitive firing or training with the service rifle.

 a. One of the first tasks of the coach is to group the six shooters into pairs or firing partners. This is an important aspect of team firing. He considers the shooters sight picture, personal characteristics, performance, etc. For example; two shooters into pairs or firing partners. This is an important aspect of team firing. He considers the shooters sight picture, personal characteristics, performance, etc. For example; two shooters with the same physical characteristics should be paired together providing they fire at approximately the same speed and are mutually compatible; just as two shooters who use the same sight picture may be paired together.

 b. The coach must consider all of the many characteristics of his shooters and as a result of the pairing, he expects the maximum performance and resultant scores under any given condition. These many characteristics are brought to light as a result of training evaluation, , and maintenance of the "Coach's Plotting Sheet." (Figure 12-8).

 c. Before continuing with the discussion of coaching techniques we should review the "Coach's Performance Checklist" (Figure 12-12).This checklist is step by step procedure for the coach to follow without a detailed discussion of technique.

 d. Being familiar with the Coach's Performance Checklist we can now relate the checklist to the firing of the National Match Team Course; also examining Coaching Techniques and duties of the Team Captain.

  (1) First stage of the National Match Course.

   (a) Standing or offhand firing is possibly the most difficult of all firing stages, and pair rather than individual firing is normally required. During this stage, the coach must rely basically on a shooter's ability and conditioning, in addition to the correctness of his calls. He will advise and encourage the shooters while analyzing their "Hits" vs "Calls". Through this analysis he keeps their groups as close to the center of the bull's-eye as possible.

   (b) Careful pairing of shooters is particularly important. While considering the previous pairing requirements which apply to all pairing, the coach wants a strong, fast firing pair on the line to start the match, with the most experienced and reliable shooter on the right. These are pace setters, and their success has a definite psychological advantage. The last, or anchor, pair should be capable of performing well under extreme pressure of time or score.

RANGE COMMAND: TEAM CAPTAINS AND COACHES YOU MAY MOVE YOUR EQUIPMENT TO THE FIRING LINE.

NOTE: Targets are exposed for inspection.

   (c) At this time the team captain and coach move their equipment to the firing line and check their target for proper facing and appearance.

RANGE COMMAND: COACHES MOVE YOUR FIRST PAIR TO THE FIRING LINE FOR THE FIRST STAGE OF THE NATIONAL TROPHY TEAM MATCH. PAIRS WILL (SPLIT THE STAKE) (FIRE TO THE RIGHT OF THE NUMBERED STAKE). YOUR THREE MINUTE PREPARATION PERIOD STARTS NOW. (During this time the targets are exposed). AMMUNITION HANDLERS PASS OUT 60 ROUNDS TO EACH TEAM CAPTAIN.

    (d) The three minute preparation period is timed by the coach. As the pair is moved to the firing line, the coach will assist the shooters in finding a suitable level piece of ground. When the shooters are in position, the coach accomplishes the following, starting with the shooter on the right; confirm the sight setting and observe the shooters set their sights for elevation and windage (200 SF zero); observe them as they assume the standing position; remind them of the target number; caution them on assuming a natural point of aim; and finally, have them dry to check trigger control and hold.

    (e) During this preparation period the coach may distribute ammunition which is again checked by the shooters. Upon completion of the check of the shooters, the coach sets up the team scope between them (Figure 12-3). He observes the wind condition and mirage; however, their effect on the shooter and the bullet will be of little consequence unless blowing quite strong. If the wind is quite strong, the coach may correct for it, giving the shooter sufficient clicks into the wind.

    (f) The team captain positions himself to the rear of the scorer where he is best able to verify the value of each shot scored and still communicate with the coach.

Figure 12-3. COACHING TECHNIQUE, STANDING

RANGE COMMAND: YOUR PREPARATION PERIOD HAS ENDED. THIS IS THE FIRST STAGE OF THE NATIONAL TROPHY TEAM MATCH. 10 ROUNDS PER FIRING MEMBER, SLOW FIRE, STANDING POSITION. TOTAL TEAM TIME 66 MINUTES. WITH ONE ROUND LOCK AND LOAD, IS THE LINE READY? THE LINE IS READY. READY ON THE LEFT. READY ON THE FIRING LINE. YOUR TIME COMMENCES WHEN YOUR TARGETS APPEAR.

(g) As the targets appear the coach starts the stop watch to check the team time and give any necessary windage to both shooters. He reminds the shooter constantly of the target number and tells the number one shooter to fire when he is ready.

(h) After the right shooter fires, he relays his call to the coach who generally watches the shot as it hits the target. The coach should be able to immediately relate the shooter's call to the hit and if a correction is needed, he is ready to relate this to the shooter. As the target is disked, he checks the location and value, plots the hit on the plotting sheet, and gives any correction to the shooter. For example; "Jones, your hit was left of call, come right two clicks." Number two man fire." After the number two man fires, he calls a "ten" at 6 o'clock but the coach saw the man fire an eight at 6 o'clock. As the target is disked, the coach plots the shot on the plotting sheet and cautions the shooter to come up. The coach does not normally give a shooter elevation changes, this is the responsibility of the shooter. But if a shooter fails to correct, the coach may caution or even give a shooter elevation changes.

(i) Individual coaching, as indicated, may be a necessity at the start of the string of slow fire; however, once the shooters are centered and synchronized, they should fire without word as soon as the value of the last shot is given by the scorer, or unless the coach directs the shooters to stop. It is important to remember that if the target is not pulled immediately upon being hit, the coach should call for a mark; after disking, the target cannot be fired on until the value is relayed by the scoer. If the value as recorded is incorrect the team captain must initiate a challenge.

(j) This firing procedure is continued for each pair. The changing of pairs should be completed quickly and orderly. As a pair is about to compelte their string, the team captain signals to the assembly area for the next pair, indicating the number of rounds remaining to be fired. This signal is normally given when each shooter has two shots remaining. The next pair will proceed to the ready line completely prepared except for the zero sight setting. As one pair completes their string, they check with the coach as to their elevation, windage, and score. Their weapons will be cleared by the block officer or coach. As the pair leave the line, the next pair will select their positions on the line, place their equipment, and wait for instructions from the coach.

(k) Frequently during the slow fire phase, the coach checks the time to insure that sufficient time remains without pressuring the shooters unnecessarily.

(l) It was previously mentioned that during firing, the coach will advise and encourage the shooters while analyzing their "Calls" vs "Hits". How well a shooter does offhand on any particular day is a measure of his conditioning, both mental and physical. There are times when a shooter may stand to fire and for some reason cannot get off a round. This may be caused by pressure, wind, or many other factors. If the condition arises where the shooter cannot get a round off in approximately 10 seconds, the coach should command the shooter to sit down, unload, relax, breath deeply, and start over again. He may have the shooter dry fire or execute any combination of acts to calm him.

(m) A technique for the coach to remember in using the scope is to watch the shooter until he fires, then shift the eye focus to the target to pick up the hit before the target is pulled. The performance of the shooter is of most importance during the 200 yard slow fire stage. To avoid mistakes, observe the shooter while he makes sight changes to insure correct manipulation of the sights and frequently remind the shooter of the target number.

Upon expiration of the allotted time the targets are pulled and the Range Officer commands:

RANGE COMMAND: CEASE FIRE. LOCK AND CLEAR ALL WEAPONS. IS THE LINE CLEAR? THE LINE IS CLEAR. IS THERE ADDITIONAL TIME REQUIRED? NO ADDITIONAL TIME IS REQUIRED.

(n) At the completion of firing the line will be policed by the shooters as the team captain checks the score and signs or initials the scorecard.

(2) Second and Third stages of the National Match Course.

(a) The rapid fire stages, standing to sitting or kneeling (200 yards, 50 seconds) and standing to prone (300 yards, 60 seconds), are a real test of the shooter's ability; and when fired under a good coach will result in the highest attainable score. Here a coach can direct his shooters with sight changes in elevation and windage, favors (Figure 61), and with information to keep their groups in the center.

(b) The coach will fire his team individually in any sequence he desires. Usually he will start with an experienced shooter who best knows his zero. This will help the coach in determining subsequent windage for the following relays. The last shooter should be experienced and capable of performing well under pressure.

RANGE COMMANDS: RELAY ONE TO THE READY LINE FOR STAGE TWO OF THE NATIONAL TROPHY TEAM MATCH. AMMUNITION HANDLERS PASS OUT 60 ROUNDS TO EACH TEAM CAPTAIN.

(c) At this time the shooter moves to the ready line, adjusts his sling, checks his equipment, and studies his scorebook. The team captain should move to the Ready Line, obtain the shooter's magazine, then return to the firing line and load them.

RANGE COMMAND: RELAY ONE TO THE FIRING LINE FOR STAGE TWO OF THE NATIONAL TROPHY TEAM MATCH. YOUR THREE MINUTE PREPARATION PERIOD STARTS NOW. (During this time the targets are exposed)

(d) As the shooter moves onto the firing line he will acquire his loaded magazines from the team captain.

(e) The preparation period and actions of the coach are the same as in the slow fire stage with the following additional requirements. The position differs and the coach may assist the shooter in the check of the position. When the shooter has assumed his position, the coach may strike the handle of the operating rod to simulate recoil and cock the rifle. This is accomplished in cadence and allows the shooter to dry fire to check position balance, natural point of aim, cadence, and trigger control. He may also have the shooter rise and reassume his position to check markings and return to natural point of aim.

(f) After checking the shooter, the coach positions the team scope. It should be positioned as close to the shooter as possible without interfering with, or touching the shooter and with the barrel of the scope pointing directly over the axis of the rifle barrel to the target. (Figures 59 and 60) This positioning allows the coach to see the trace or shock wave of the bullet to the target. The focus of the scope is on the target for both rapid fire stages. With the scope focused, the coach checks the target condition and the mirage. As previously mentioned, the wind or mirage must be quite strong at 200 yards to affect the bullet. At 300 yards the mirage must be checked closely and the initial windage must be accurate. The primary use of the scope in rapid fire is to see the shots hit the target, or if the holes are not visible, to see the shock wave of the bullet as it moves through the air to the target. In this manner the coach can move the shooters shots or group in any direction with the use of simple voice commands. If the scope is not positioned correctly as stated, the coach will not be seeing the true path or trace of the bullet. It is also recommended that the scope be positioned at the lowest possible level above the barrel of the rifle. The coach's position behind the scope is important and the following is recommended. The coach should kneel at the rear of the scope and be ready to shift forward to give the shooter commands In any case the coach must be able to quickly leave the scope and move forward and observe the shooter when necessary.

Figure 12-4. COACHING TECHNIQUE, SITTING RAPID FIRE

Figure 12-5. COACHING TECHNIQUE, PRONE RAPID FIRE

When the preparation period ends the targets are pulled and the Range Officer commands:

RANGE COMMAND: YOUR PREPARATION PERIOD HAS ENDED. THIS IS THE SECOND (THIRD) STAGE OF THE NATIONAL TROPHY TEAM MATCH. TEN ROUNDS PER FIRING MEMBER, RAPID FIRE, FROM STANDING TO SITTING OR KNEELING (PRONE), 50 SECOND (60 SECOND) TIME LIMIT. SHOOTERS RISE. LOCK AND LOAD. IS THE LINE READY? THE LINE IS READY. READY ON THE RIGHT, READY ON THE LEFT, READY ON THE FIRING LINE.

(g) During the commands, the coach stands facing the shooter to insure proper application of loading, sight manipulation, and to advise the shooter if necessary. If windage is needed, the coach transmits this to the shooter at the last possible moment to eliminate a change or correction due to any late wind shift. As the target appears, the shooter assumes his position. The coach starts the stop watch, checks the position, and reminds the shooter of the target number. As the shooter starts his aiming process the coach moves to the scope to observe the initial shots.

(h) The coach must see the trace of the initial shot or the location of the hole in the target. The location of this shot is on what the coach must base his subsequent commands to his shooter. If the shot is centered, the coach calls "good" to the shooter. This is important as the shooter is mentally calling his shots. If the first shot is not centered, due possibly to an incorrect zero, additional wind, or as a result of a bad shot by the shooter, the coach must give the shooter additional commands to center the group in the black. These commands are favors. "Favor right"; "Favor left"; "Take white"; and "Hold closer". They are termed full favors and are as indicated below, using the bull's-eye and front sight (figure 61). These favors are generally six inch favors in the direction indicated. These favors are further reduced by the commands "Favor a little right"; "Favor a little left"; etc. Any time the coach calls for a favor, the next shot fired should be called by the coach. This is also the case after firing the 6th shot, regardless of a change. If good, the coach should call "good".

If not good, give a subsequent favor, but as in slow fire, the less said the better. A shooter has much to concentrate on and should only be spoken to when necessary. After a few shots fired with a favor it may be necessary to cancel the favor. This is accomplished by telling the shooter to fire "straight away".

        (i) At the completion of the first magazine the shooter relates his calls to the coach while reloading. The coach watches the shooter and upon completion of reloading, if a windage or elevation change is needed, he gives this to the shooter.

NOTE: All changes are given verbally and visibly with the use of the fingers.

        (j) The shooter will repeat all changes given to him by the coach. Changes are given in a positive manner and by elevation, then windage. For examples: "Up one; right two." If no change is necessary, the coach says "good". It may be necessary at times when a strong wind is blowing, and after a change, or if no change is given, to favor the shooter again. The general rule to follow, in most cases, after firing the first magazine is as follows: <u>If a change is needed, give a sight change and never a favor</u>. Also, watch that the shooter correctly manipulates the sights. As the shooter reassumes his position, the coach again checks the shooters position and reminds him of his target number. He then returns to the scope to observe the remaining shots. The sixth shot is important, and the coach must again give a subsequent command to the shooter. The coach must be constantly aware of the cadence and the normal time needed for firing each round, reloading, etc. While this is developed during training, a mistake or a slip in a match may cause a delay that will make the time factor extremely important. Any saved rounds are a coach's responsibility. If the shooter is firing slowly, the coach, after the third or fourth shot, should tell the shooter to "speed it up". He repeats this for subsequent shots as necessary. If, when five seconds are remaining, the shooter has more than one round to fire, the coach commands "shoot em". The meaning of this command is: Fire your rounds, even if they don't hit the black. This is the only situation in which accuracy is sacrificed for speed. Two eights are better than a ten and a miss.

        (k) Individual characteristics of the shooter are important during rapid fire. The same may be said for some weapons. The coach will become aware of these characteristics during training and should be prepared to cope with them during a match. The following examples are cited: Some shooters by virtue of position will constantly fire the first five shots of a string to the left or right, then without a change will fire the remaining five rounds well centered. Some weapons characteristically fire the first shots out of the group, however, magazines should be checked for similarity if different groups are apparent. At the completion of the string the coach checks the shooter's time. The shooter remains in position and the team captain checks the exposure time of the targets. The Range Officer will command:

RANGE COMMAND: CEASE FIRE. LOCK AND CLEAR ALL WEAPONS. ARE THERE ANY ALIBIS? (Alibi targets will be pasted and personnel re-squadded) THE TIME WAS CORRECT. IS THE LINE CLEAR? CLEAR ON THE RIGHT. CLEAR ON THE LEFT. THE FIRING LINE IS CLEAR. SHOOTERS MOVE YOUR EQUIPMENT TO THE REAR OF THE SCORER AND STAND BY TO RECEIVE YOUR SCORE.

        (l) As the coach clears the shooter's weapon and the line is cleared the shooter will move back to the team captain where he checks his sight setting. When the targets are spotted and ready for scoring, the Range Officer is informed.

RANGE COMMAND: RELAY TWO TO THE FIRING LINE FOR STAGE TWO OF THE NATIONAL TROPHY TEAM MATCH. YOUR PREPARATION PERIOD STARTS NOW.

Figure 12-6. COACHING TEAM FAVORS

RANGE COMMAND: STAND BY, YOUR TARGETS ARE COMING UP FOR SCORING.

(m) During the scoring the team captain makes sure that the correct score is disked and recorded. The coach plots the group and discusses the string with the shooter. The team captain loads the magazines of the next shooter and the coach, using the same techniques he used for the first relay, prepares the succeeding relay for firing.

As soon as scoring is completed the Range Officer commands:

RANGE COMMAND: HAS THE SCORING BEEN COMPLETED ON THE RIGHT? ON THE LEFT? SCORING IS COMPLETED. WITH THE EXCEPTION OF TARGETS BEING CHALLENGED OR RE-DISKED, PULL AND PASTE.

NOTE: When scoring has been completed the pits are informed and the targets are pasted. As soon as all targets are pasted they will be exposed again for a few seconds for the shooters to see.

(n) When all relays are completed, the line is policed and the team captain signs or initials the scorecard.

(3) Fourth stage of the National Match Course.

(a) The 600 yard slow fire stage is difficult by virtue of the range at which fired, and of all the stages of the NMC, requires the utmost in precision firing by the shooter and mirage reading by the coach. The coach must keep abreast of the wind and sight changes made by the shooters under his direction. His use of the coach's plotting sheet becomes extremely important. Here again pair firing is normally required. The coach will analyze the shooter's "Hits" vs "Calls" to keep their groups centered in the bull's-eye.

(b) Careful pairing of the shooters is again important. A strong, fast firing pair should start this stage with the most experienced and reliable shooter on the right. The last, or anchor, pair should be capable of performing well under pressure of time and score.

RANGE COMMAND: TEAM CAPTAIN AND COACHES YOU MAY MOVE YOUR EQUIPMENT TO THE FIRING LINE.

NOTE: Targets are exposed for inspection.

(c) At this time the team captain and coach move their equipment to the firing line and check their target for proper facing and appearance.

RANGE COMMAND: COACHES MOVE YOUR FIRST PAIR TO THE FIRING LINE FOR THE FOURTH STAGE OF THE NATIONAL TROPHY TEAM MATCH. PAIRS WILL (SPLIT THE STAKE) (FIRE TO THE RIGHT OF THE NUMBERED STAKE.) YOUR THREE MINUTE PREPARATION PERIOD STARTS NOW. (During this time the targets are exposed.) AMMUNITION HANDLERS PASS OUT 120 ROUNDS TO EACH TEAM CAPTAIN.

(d) The actions during the preparation period are the same as the slow fire standing, except that the shooters bring to the firing line individual scopes, which are used for plotting purposes only.

(e) After checking the shooter, the coach again sets up his scope as close to the shooters as possible without touching or disturbing them. While there is no required placement of the scope for this stage, it is recommended that it be placed at sitting height between the shooters (Figure 62).

(f) The scope should be equipped with a 24 power eye-piece for most conditions, as this provides the magnification most coaches accept as best. Variable eye pieces are only available with the larger team scopes. The scope is initially focused on the target for a serviceability check and then clearly focused on the 300 yard line. This focus gives a mid-range mirage reading that is of most value to the coach. Regardless of the position of the scope, the coach should be as close to the shooters as possible without touching them. This gives the coach close control of the pair, and the physical proximity of the three helps to promote team spirit.

(g) As long range scoping is fatiguing, for long periods of time it is best to use the scanning technique. Staring at one place will cause fatigue in the "scoping" eye and will make it difficult to pick up changes quickly. If the coach gazes through the scope, shifts his focus from one extreme to the other in the field of view, and watches the mirage as it moves across the numbered boards and the targets, he can make a more accurate comparison of changes. Periodically he should look away from the scope to check the flags and other wind speed and direction indicators. This technique will produce the most accurate "wind doping" with the least fatigue.

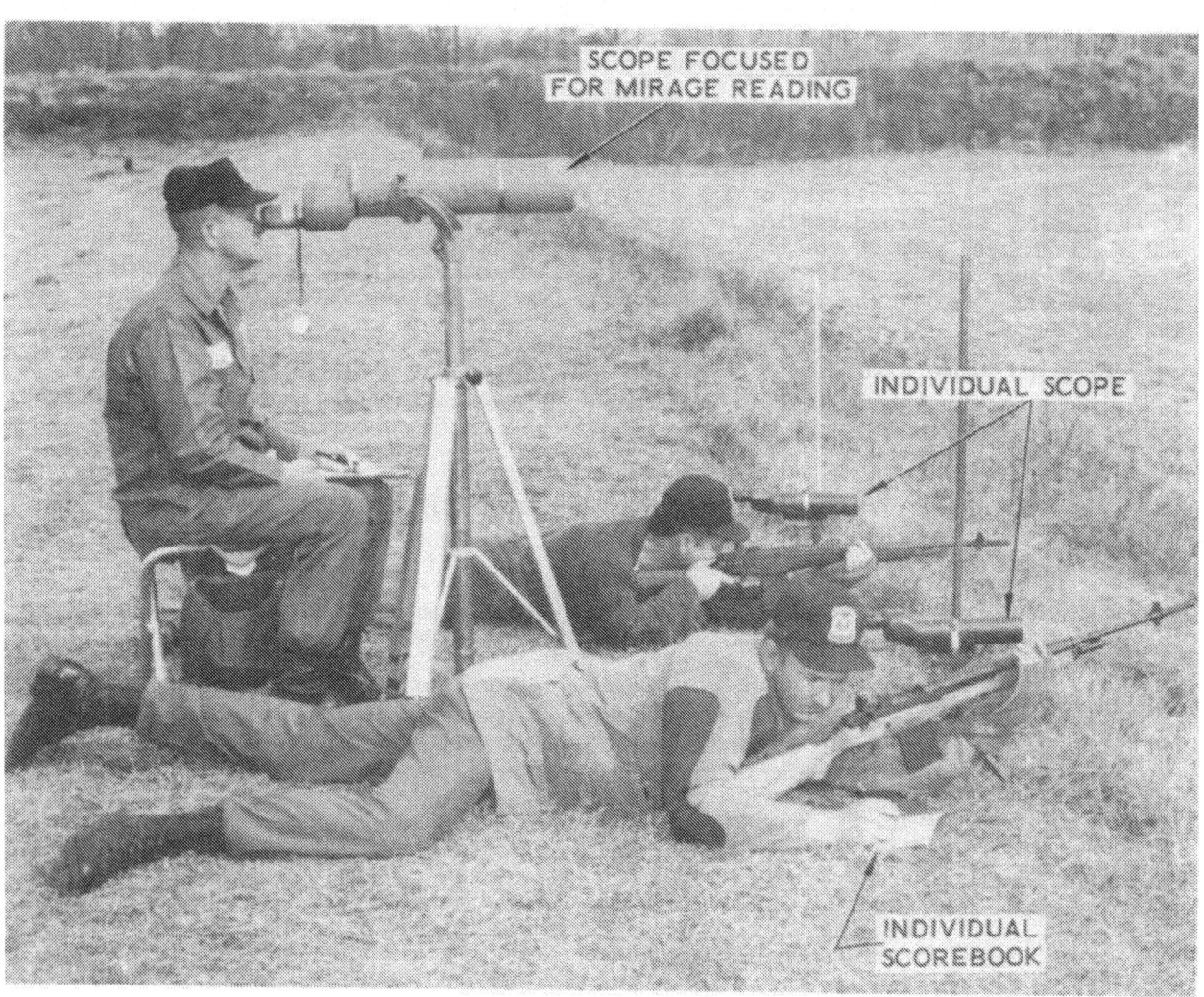

Figure 12-7. COACHING TECHNIQUE, PRONE-SLOW FIRE

When the preparation period has been completed the Range Officer commands:

RANGE COMMAND: YOUR PREPARATION PERIOD HAS ENDED. THIS IS THE FOURTH STAGE OF THE NATIONAL TROPHY TEAM MATCH. 20 ROUNDS PER FIRING MEMBER, SLOW FIRE, PRONE POSITION. TOTAL TEAM TIME 126 MINUTES. WITH ONE ROUND, LOCK AND LOAD. IS THE LINE READY? THE LINE IS READY, READY ON THE RIGHT. READY ON THE LEFT. READY ON THE FIRING LINE. YOUR TIME COMMENCES WHEN YOUR TARGETS APPEAR.

    (h) After the preparation period is completed, and the firing commands given, the coach will have had sufficient time to study the conditions and give a windage adjustment to both shooters for the first shot. If the shooters are ready at the command to commence firing, he directs the right man to fire. If the shooter knows his zero and the coach correctly estimates the wind, the first shot should be in.

    (i) Having fired, the shooter calls the shot, plots his shot when marked, and awaits his turn to fire his second shot. Based on the location of the hit as marked, and related to the call, the coach determines the correctness of his initial windage adjustment. If a correction is needed he will relay this to both shooters by stating, "Both guns right (or left) one click." "Number 2 man fire." As previously mentioned, if the shooters zeros are synchronized, any windage adjustment will apply to both shooters. The coach may start off using the synchronized system, but must be flexible enough to revert to individual coaching if one shooter's zero should suddenly change. Under the individual system the shooter's sight setting is based on his last shot, and last call, disregarding his partners last shot and call. For the purpose of team firing, the sooner the shooters are synchronized, the better for the team. This will speed up the firing and minimize the possibilities of firing through many wind changes. Regardless of the employment of either the individual or the synchronized technique, subsequent shots will be good if the shooters perform and the coach gives proper wind changes.

    (j) Elevation changes are again the responsibility of the shooter. However, as previously mentioned, if the coach notices a group build up, either high or low, he should caution the shooter to correct for it.

    (k) During this stage of slow fire it may be necessary to utilize favors. Under most conditions it is best and recommended that sight changes be used to move a group or shot. However, if firing in a fish tailing or shifting wind of such frequency that it becomes fatiguing and distracting to continually hold up the shooters and have them come out of position to make changes, it is recommended that windage favors be used (Figure 61). The most common favors used at 600 yards are negative favors. Example: "Stay off the right (left)".

    (l) The time checking, change overs, sight checks, critique, checking scores, policing brass, etc., are the same as with the 200 yard slow fire stage.

    e. A well trained team, that applies the fundamentals as discussed, will be a winning team.

CARTOON COACH'S PLOTTING SHEET

Fig. 12-8

D. Use of the Coach's Plotting Sheet.

   1. The plotting sheet is to the coach what the scorebook is to the shooter. In this respect it is a record of the shooter's performance, the coach's performance, the rifle, ammunition, weather, etc. Because of its importance, the coach maintains a plotting sheet for each firer, for each match or course of fire, and uses it as a guide in critiquing the shooter in his performance. These sheets are maintained as part of the shooter's evaluation file; and whenever the shooter comes under the direction of a new coach these sheets will accompany him.

   2. The plotting sheet, as mentioned above, is a measure of the coach's performance Through remarks, plotting of hits and calls, and recording sight changes and favors, the coach can analyze his coaching ability. His ability to dope the wind and read the mirage will be evident when reviewing the horizontal spread of a group at 600 yards. The placement of the groups during rapid fire will clearly indicate, in relation to the calls and favors given, whether the coach gave the proper and necessary favors. The importance of recording the conditions that affect the shooter from day to day will be evident if properly evaluated.

   3. A sample plotting sheet, as revised, is illustrated in Figure 63. This revision was accomplished to enable the coach to more completely record and evaluate the shooters performance. It also facilitates the ease of handling and recording. A block is provided for each stage of firing. The plotting sheet is divided into two sections to facilitate pair firing during slow fire. Cut the sheets in half for filing purposes.

   4. The plotting sheet may be modified by the coach to meet individual requirements, however, the following is its generally accepted usage.

      a. <u>General Notes</u>. Standard locations are adopted for recording weather conditions and other general information. In each instance, the information should be recorded in a concise and complete manner with the use of descriptive words or phrases. The recording of the light and wind direction is with the use of arrows. In using arrows, we indicate the direction from which the wind is blowing or from which direction the light is coming. For example; utilizing the clock, with the shooter at the center, and 12 o'clock being the target, a 9 o'clock wind, or a wind blowing from the left, would be recorded with the arrow passing through 9 o'clock pointing to the opposite side of the clock. The same is true for the light. For each stage of firing, a zero or sight change block is provided. This is used to record the shooter's rifle zero, the elevation and windage used for a shot or string, and the correct zero. For the slow fire stages, additional blocks are used to record subsequent changes due to wind changes. A block is provided in each stage to record the shooter's aiming technique. A plotting or hit target is suitable for each stage of firing with an additional call target for the slow fire stages. The target number is recorded in the triangle provided, as this is the standard outline of the target number backings. The recording of the time is important, as light conditions vary depending on the time of day. For specific information as to each stage the following should be noted.

    b. <u>200 yard, Slow Fire</u>. This block is utilized to record the conditions and performance of the shooters for the standing or offhand firing. The call and plot targets are located side by side for ease of use. As previously mentioned, the coach requires the shooter to call each shot. When a shot is called, the coach indicates its location on the call target. When the target is marked, the exact location of the hit is plotted on the plot target. To record these hits and calls, the numbers 1 through 20 are used. Sighting shots are recorded by utilizing the prefix S to the number of sighters used. The numbers should be small so as not to clutter the target. In this manner the hits and calls can quickly be analyzed to determine if the shooter is calling properly, if he has his proper zero, or if the coach has properly compensated for weather conditions.

    c. <u>200 and 300 yards Rapid Fire</u>. These are identical blocks, and noted is the absence of any call target. However, a remarks section is labeled "calls" for the coach's use. Here he would indicate the call of the shooters first shots and any bad shots indicated at the conclusion of the string. Favors given to the shooter are recorded in the space provided. The group, as fired, is plotted utilizing small x's to indicate each hit, with the exception of the 1st, 2d, and 6th shot where numbers are used. Numbers are also used to indicate the location of any bad shot or shots out of the group.

    d. <u>600 yard, Slow Fire</u>. This stage is possibly the most difficult for the coach with respect to maintaining it. Yet it must be maintained accurately. Pair firing is normally required and, depending on the match, sighting shots may be authorized. Here the coach must record each shot fired in the same manner as for the 200 yards slow fire stage. An additional remarks section has been added to indicate the mirage whether light heavy, etc. Keeping abreast of each shot under the varying conditions of the wind is difficult due to the coach's continual observation of the mirage down range. However, every effort should be made to record all sight changes and windage corrections given to the shooters. There are sufficient blocks provided to record the elevation and windage for each shot fired. It is not necessary to repeat elevation or windage entries. A recording is made only upon a change. The shooter's initial zero is placed in the block provided. If the coach has the shooter compensate for the wind, the actual number of clicks is placed in the wind column. For example, a three o'clock wind may be worth six clicks right, this is recorded as R6. The shooter's windage zero is not recorded except as mentioned above. In this manner, any subsequent changes or corrections right or left are merely added or subtracted to the initial windage adjustment. Using this system, the coach will at a glance know the correct windage adjustment the rifle is using at all times. Elevation changes are recorded from the shooters zero. Since elevation is a shooter's responsibility, he should be instructed to relay any changes made. At the completion of the string, the shooter, prior to leaving the firing line, is checked for his specific elevation and windage to verify the rifle's zero and also the windage value at the completion of the string.

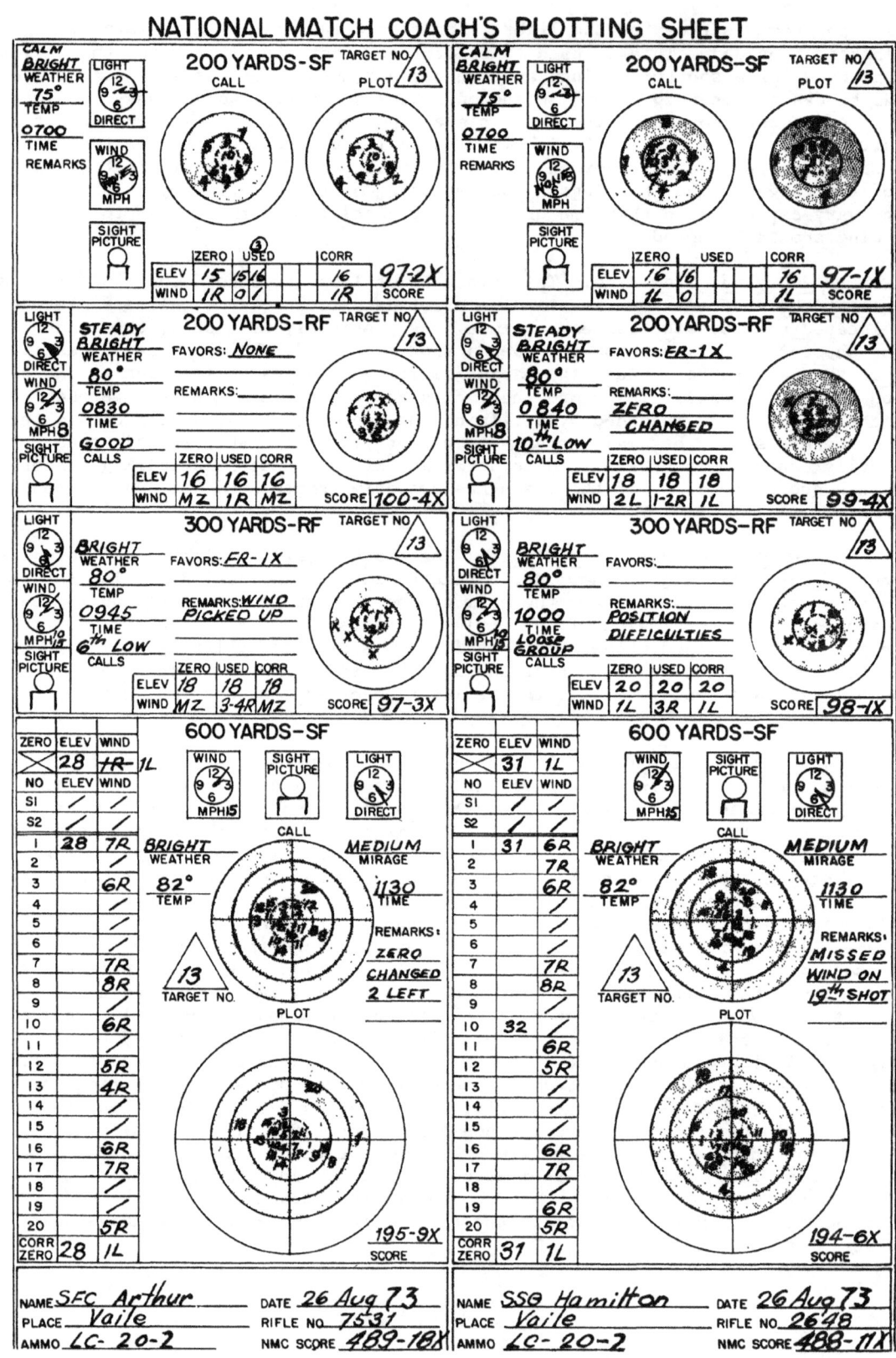

Figure 12-9. NM COACH'S PLOTTING SHEET

```
┌─────────────────────────────────────────────────────────────────────────┐
│              COACH'S PERFORMANCE CHECK LIST AND EXAMINATION             │
│ NAME:_____  UNIT:_____                 │
│                                                                         │
│ A.  COACH'S PERFORMANCE CHECK LIST  (Students not graded, for information only.) │
│                                                                         │
│     1.  BEFORE REPORTING TO RANGE        Did Coach:                     │
│         a.  Receive instructions from the head coach:                   │
│             (1)  Team member's names (Evaluation File)?        _____  │
│             (2)  Target assignment (Squadding Ticket)?         _____  │
│         b.  Analyze previous Coach's Plotting Sheets?          _____  │
│         c.  Check coaching equipment for completeness and serviceability: │
│             (1)  Team spotting scope?                          _____  │
│             (2)  Stop watch?                                   _____  │
│             (3)  Clipboard and pencil?                         _____  │
│             (4)  Coach's Plotting Sheets?                      _____  │
│             (5)  Stool?                                        _____  │
│             (6)  Lens tissue?                                  _____  │
│             (7)  Ear plugs?                                    _____  │
│             (8)  Allen wrench, screwdriver, and combination tool? _____ │
│                                                                         │
│     2.  ASSEMBLY AREA                    Did Coach:                     │
│         a.  Check that team is present and ready?                       │
│         b.  Fill in the appropriate entries of the day's Coach's Plotting Sheets? _____ │
│         c.  Minimize mental anxiety and human error by talking to and checking │
│             the team on:                                                │
│             (1)  Course of fire?                               _____  │
│             (2)  Firing order?                                 _____  │
│             (3)  Equipment serviceability (rifle, sights, etc.)? _____ │
│             (4)  Scorebook analysis?                           _____  │
│             (5)  Weather conditions?                           _____  │
│             (6)  Correction of individual errors?              _____  │
│                                                                         │
│     3.  READY LINE              Did Team Captain or Coach:              │
│         Notify the next relay to move to the ready line in time to relax and think │
│     of what they are going to do when the preceding relay has finished firing. │
│                                                                         │
│                                    (1)                                  │
└─────────────────────────────────────────────────────────────────────────┘
```

Figure 12-10. COACH'S PERFORMANCE CHECKLIST (Page 1 of 3)

B. COACH'S PERFORMANCE EXAMINATION (One point per question, except c. AFTER FIRING 200 and 600 yards slow fire, half point per question.)

1. FIRING LINE 200 AND 600 YARDS (SLOW FIRE)

   a. PREPARATION PERIOD   Did Coach: 200 Yds   600 Yds

   (1) Help shooters find suitable location on firing line?   _____   _____
   (2) Confirm sight setting to be used?   _____   _____
   (3) Check that sights are blackened and set properly?   _____   _____
   (4) Remind shooter of his target number?   _____   _____
   (5) Have shooters dry fire and check natural point of aim?   _____   _____
   (6) Check that ammunition is protected from weather?   _____   _____
   (7) Position scope properly?   _____   _____
   (8) Check that target is suitable?   _____   _____
   (9) Focus scope properly?   _____   _____

   b. DURING FIRING   Did Coach:

   (1) Start stop watch and record each relay's starting time on Coach's Plotting Sheet?   _____   _____
   (2) Wait until shooters are ready?   _____   _____
   (3) Determine the value of the wind and have shooter set sights?   _____   _____
   (4) Call off target number before firing?   _____   _____
   (5) Require shooter to call each shot?   _____   _____
   (6) Require shooter to maintain scorebook properly?   _____   _____
   (7) Give correct sight changes?   _____   _____
   (8) Maintain Coach's Plotting Sheet properly?   _____   _____
   (9) Check on time?   _____   _____
   (10) Insure scores are recorded correctly and have team captain challenge when necessary?   _____   _____

   c. AFTER FIRING   Did Coach:

   (1) Insure that the weapon is cleared?   _____   _____
   (2) Prepare firing line for next relay?   _____   _____
   (3) Confirm elevation and windage used?   _____   _____
   (4) Analyze performance and confirm or establish zeros?   _____   _____

2. FIRING LINE 200 AND 300 YARDS (RAPID FIRE)

   a. PREPARATION PERIOD   Did Coach: 200 Yds   300 Yds
   (1) Help shooter find suitable location on firing line?   _____   _____
   (2) Confirm sight setting to be used?   _____   _____
   (3) Check that sights are blackened and set properly?   _____   _____
   (4) Remind shooter of his target number?   _____   _____
   (5) Help shooter dry fire and check natural point of aim?   _____   _____

(2)

Figure 12-11. COACH'S PERFORMANCE CHECKLIST (Page 2 of 3)

	200 Yds	300 Yds

    (6) Check that ammunition is clean and serviceable and magazine properly loaded?

    (7) Position scope properly?

    (8) Check that target is suitable?

b. <u>DURING FIRING</u>   Did Coach:

    (1) On command "Shooters Rise", move to side of shooter?

    (2) On command "Load", insure safety is engaged, magazine is latched, and top round chambered?

    (3) On command "Ready on the Firing Line", insure safety is disengaged?

    (4) Give estimated wind adjustment?

    (5) Tell shooter when flag is up, waving, and down; and to watch for the targets?

    (6) Start stop watch when targets appear?

    (7) Remind shooter of target number when targets appear?

    (8) Give appropriate commands?

    (9) Move from scope to shooter's side after 1st five shots?

    (10) Insure weapon is reloaded before giving sight changes?

    (11) Give necessary sight changes correctly?

    (12) Remind shooter of his target number after reloading?

    (13) Check shooters time periodically during firing with stop watch?

    (14) Insure that target exposure time is noted?

c. <u>AFTER FIRING</u>   Did Coach:

    (1) Insure that weapon is cleared if shooter does not have an alibi?

    (2) Have shooter call bad shots?

    (3) Prepare firing line for next relay?

    (4) Complete entries in Coach's Plotting Sheet to include plotting of group?

    (5) Make sure shooter fills out scorebook properly?

    (6) Analyze performance and confirm or establish zeros?

    (7) Insure scores are recorded correctly and challenge when necessary?

(3)

Figure 12-12. COACH'S PERFORMANCE CHECKLIST (Page 3 of 3)

SECTION II - SELECTION OF SQUAD MEMBERS

In the selection of squad members it is important to utilize the best talent available insuring that sufficient new talent is developed and utilized as required by the competitive new shooter rule as outlined in FORSCOM Suppl 1 to AR 350-6. The identification of marksmanship talent is directed by HQ FORSCOM, as is the career advancement of personnel engaged in the competitive program.

a. Selection at the lowest level.

Selection of shooters at the company level or at the lowest level of participation is generally accomplished as follows:

1. The person responsible for organizing the team will generally ask for volunteers These personnel may be experienced or may only have professed an interest in marksmanship. In the absence of volunteers, personnel firing the highest scores during the annual qualification should be used. Also, a review of individual qualification records will indicate personnel with prior competitive experience as well as those individuals who may possess natural ability.

2. All personnel considered should be interviewed to determine if they possess those personal characteristics or qualities that are important to become a good shooter. Some of the most important attributes are:

ATTRIBUTES OF A SHOOTER

a. Interest in marksmanship.

b. Previous experience.

c. Eligibility (as outlined in FORSCOM Suppl 1 to AR 350-6.)

d. Availability.

e. Physically qualified.

f. Cooperative, Honest, Ambitious, and Reliable.

g. Competitive Spirit.

3. Those individuals selected should then be given instruction and practice to determine potential.

B. Selection at Higher Levels.

If faced with the task of selection of a squad subsequent to the selection at the lowest level, such as at the completion of inter-unit or inter-post matches, the problem is somewhat minimized and is accomplished generally as follows:

1. A senior member from among the selected individuals or an experienced shooter or coach will more than likely be assigned to organize the squad. This assignment should be accomplished prior to the completion of the subject match to enable the individual to contact team coaches of individuals firing in the match, as well as give him the opportunity to actually observe shooters on the firing line during the competition.

2. As a result of his observation, contact with team coaches, and his utilization of the match bulletins, the squad coach will select his squad members. He will insure that he has sufficient old and new shooters for subsequent organization of his team(s) for the next level of competition. (As mentioned previously, refer to the new shooter rules outlined in CON Suppl 1 to AR 350-6).

3. While not indorsed, it is conceivable that selection of a certain individual might be based on potential displayed, or a specific need or requirement of the squad. Therefore an individual, an old shooter with past experience, or a new shooter with certain potential, may be selected over the proven competitor.

4. Selection of any squad, should as a rule, be based on proven ability in keeping with the spirit of the competitive directives.

Fig. 12-13

C. Team Selection and Evaluation File.

1. After the selection and training of the shooters, the head coach is responsible for choosing those individuals who will fire in team matches. In order to select those shooters with the greatest capability, the coach must have a thorough knowledge of every individual in the squad. To accomplish this, it is necessary to rely on the observation of the team coaches.

2. To provide the team coach with a systematic method of recording data concerning each shooter, an evaluation file should be initiated for every individual. This file, if properly maintained, will not only reflect pertinent information to be used in selecting team members, but will assist in the planning of the training program.

3. The Evaluation File consists of a Personal Information Sheet, a Shooter's Graph and Daily Log Section. The Evaluation File used by the Army Team is printed on a manila type folder. The cover sheet (Figure 1) contains the instructions for the use of the file. The inside left of the folder (Figure 2) contains the Shooter's Graphs for 200 and 600 yards slow fire and 200 and 300 yards rapid fire. The inside right of the folder (Figure 3) is the National Match Course Graph followed by the Daily Log, which allows for quick reference between the Shooter's Graphs and the initial training period of the individual shooter. The file is completed with a continuation of the Daily Log on the back, and a mimeographed Personal Information Sheet (Figure 4) which is retained by the team coach for his own information. A sample evaluation file with examples inserted follows:

<u>SHOOTER'S EVALUATION FILE</u>

<u>INSTRUCTIONS</u>

Each competitor should have an evaluation file throughout his competitive season. The file contains a PERSONAL INFORMATION SHEET and a SHOOTER'S GRAPH and DAILY LOG SECTION. This file will be in the custody of the team coach at all times, except when needed by squad officials. The coach will make appropriate entries as events occur. He will also insure that all plotting sheets resulting from firing under his supervision are associated with this file before passing it on the the next coach.

PERSONAL INFORMATION SHEET: Is considered confidential and should be only for the use of the coach and squad officials. Each coach will make his remarks in such a manner as to reflect his considered OPINION and what action he has taken to orient the shooter concerning his particular problem if a problem is noted. Extreme care is indicated here, and it is suggested that this portion of the file be filled in after the shooter has been under observation for at least a week. Because of the subjective nature of these comments, they should be used only for corrective action by the coach who maintains the file. Critical remarks are worthless unless they are followed up with appropriate counseling. If improvement is noted, these remarks should be destroyed. This sheet should be destroyed when the shooter changes coaches.

SHOOTER'S GRAPH AND DAILY LOG SECTION:

    1. SHOOTER'S GRAPH: Is used for the entry of all practice, match, and average scores, whether they are fired as a member of a team or as an individual. All other factors being equal, this is the information which will provide the basis for the final selection of team members. The following color code is used to record and qualify scores.

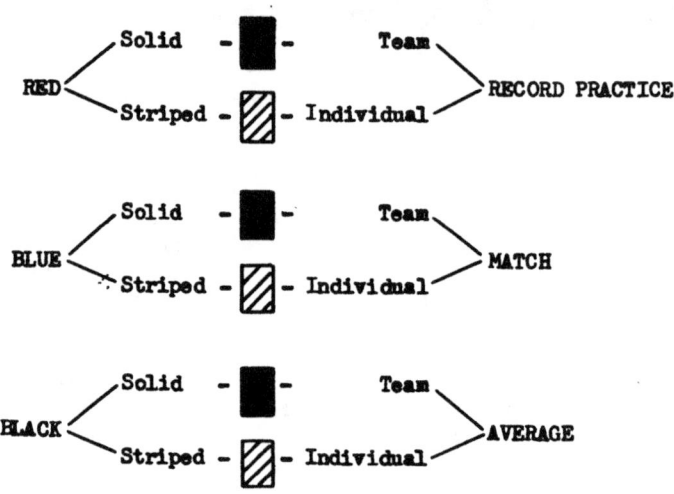

Fig. 12-14

    2. DAILY LOG: Is used to record individual strengths and weaknesses of the shooters. Comments must be of a factual nature and used to qualify each score on the SHOOTER'S GRAPH. Examples are: effects of weather, coaching of firing errors, equipment failure, etc. These comments present a picture of the shooters progress and are particularly useful in programing instruction and remedial training.

    USAMTV Form
FB   33
    15 Nov 66

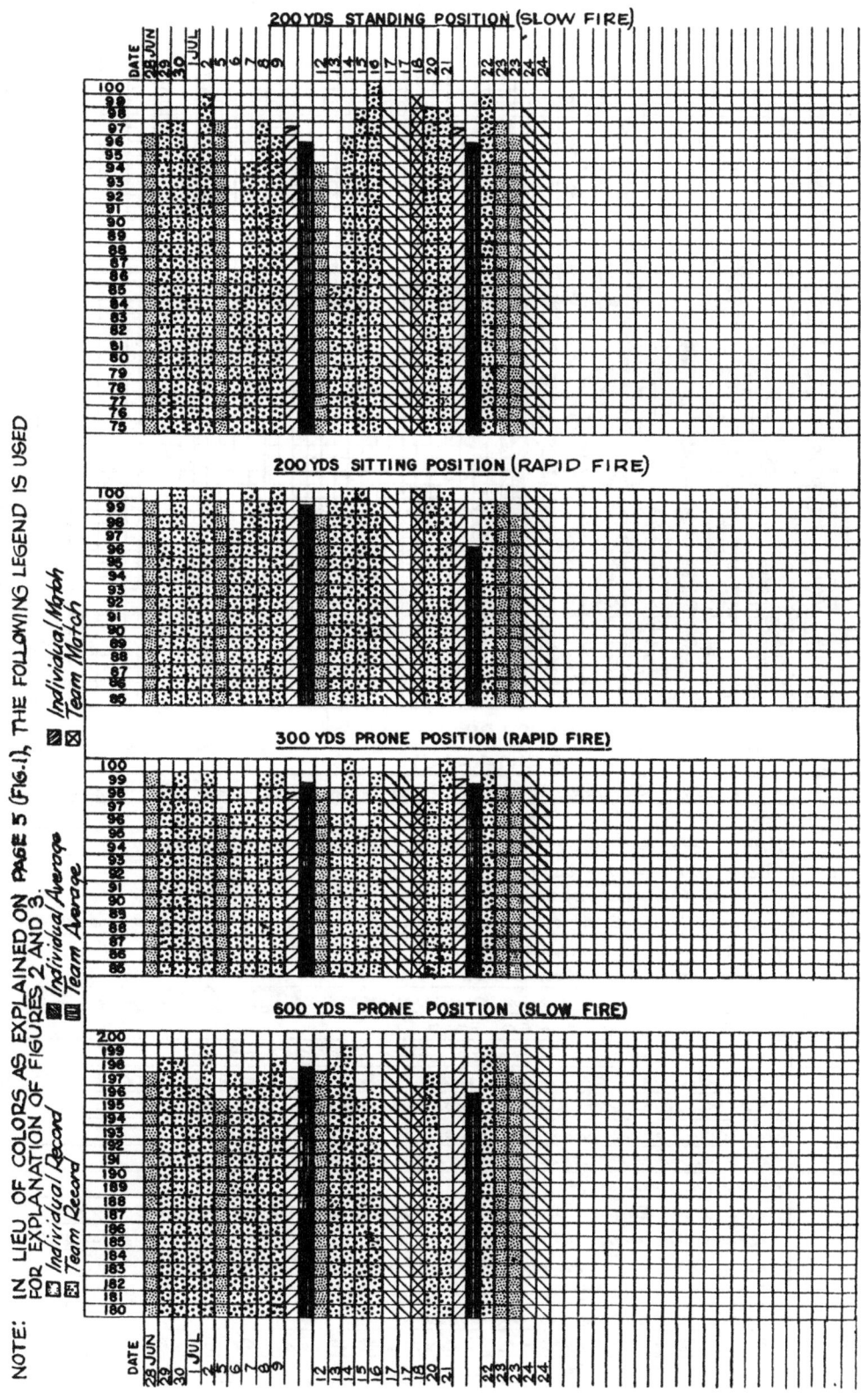

Figure 12-14. EVALUATION FILE (SHOOTER'S GRAPH)

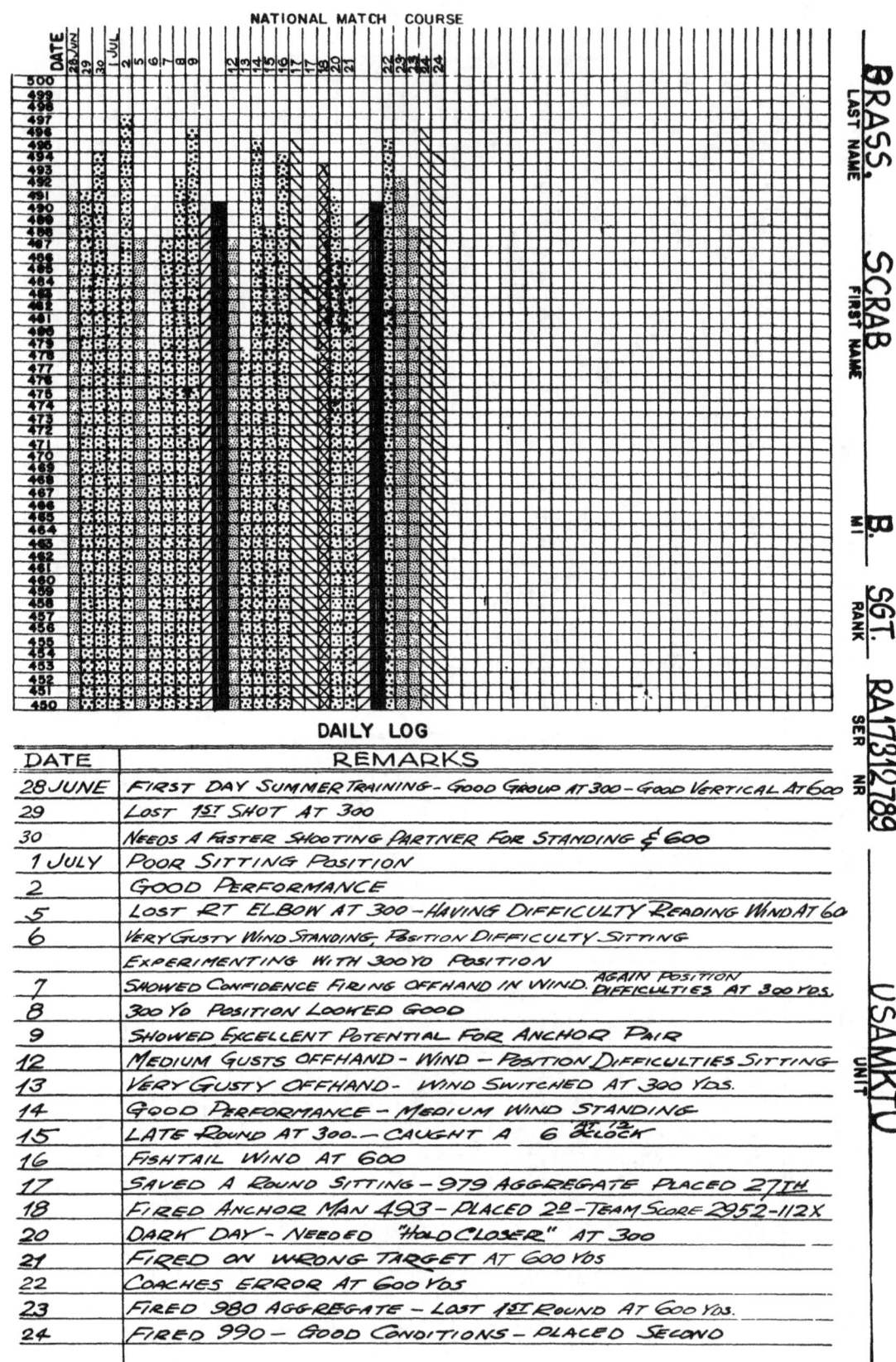

DATE	REMARKS
28 JUNE	First day summer training - Good group at 300 - Good vertical at 600
29	Lost 1st shot at 300
30	Needs a faster shooting partner for standing & 600
1 JULY	Poor sitting position
2	Good performance
5	Lost Rt elbow at 300 - Having difficulty reading wind at 60
6	Very gusty wind standing, position difficulty sitting
7	Experimenting with 300 yd position. Showed confidence firing offhand in wind. Again position difficulties at 300 yds.
8	300 yd position looked good
9	Showed excellent potential for anchor pair
12	Medium gusts offhand - wind - position difficulties sitting
13	Very gusty offhand - wind switched at 300 yds.
14	Good performance - medium wind standing
15	Late round at 300. - Caught a 6 at 12 o'clock
16	Fishtail wind at 600
17	Saved a round sitting - 979 aggregate placed 27th
18	Fired anchor man 493 - Placed 2nd - Team score 2952 - 112X
20	Dark day - needed "hold closer" at 300
21	Fired on wrong target at 600 yds
22	Coaches error at 600 yds
23	Fired 980 aggregate - Lost 1st round at 600 yds.
24	Fired 990 - Good conditions - Placed second

Figure 12-15. EVALUATION FILE (NMC GRAPH AND DAILY LOG)

PERSONAL INFORMATION

NOTE: INFORMATION CONTAINED IN THIS PAGE IS CONFIDENTIAL

MENTAL ATTITUDE: Reflects the individual's ability to perform under the pressure of competition. In order to determine the quality of a shooter's mental attitude, answer these questions and explain affirmative answers.

1. Is he easily perturbed?
   No
2. Does he give up easily?
   No
3. Is he easily discouraged by unfavorable conditions?
   Yes Sgt Gross needs a "pep talk" during bad weather, otherwise he will not concentrate on his firing.
4. Is he susceptible to rumors?
   No
5. Is he disturbed by scores fired by other competitors?
   No
6. Does he worry about equipment?
   No
7. Does he lack the will to win?
   No
8. Other

PERSONAL BEHAVIOR: Indicates other traits of character which measure the capabilities and shortcomings of a potential team member. While thus evaluating the shooter, the wise coach will also evaluate himself, for he, as well as the shooter, will be a member of the team. Answer these questions and explain negative answers.

1. Is he cooperative?
   Very cooperative except when pushed. Recommend a "soft sell" approach
2. Is he ambitious?
   Yes
3. Is he aggressive?
   Yes
4. Is he honest?
   Yes
5. Is he reliable?
   Yes
6. Is he neat in appearance?
   Yes

REMARKS: Describe the shooters abilities, characteristics and potential.
Much potential - makes "new-shooter" mistakes but will soon overcome these misfortunes. He fires best when allowed to sleep between relays

EVALUATION FILE (INSERT)
Figure 12-16

SECTION III - SQUAD EQUIPMENT

To facilitate training and development of rifle teams for ultimate participation in competition, Headquarters, Department of the Army has published a <u>Table of Allowance, 60-18</u>, that prescribes the equipment authorized, and a <u>Table of Allowance 23-100</u>, that prescribes the authorized annual match grade ammunition allowance. In addition to procurement, the coach and shooter must know how to select and care for this equipment.

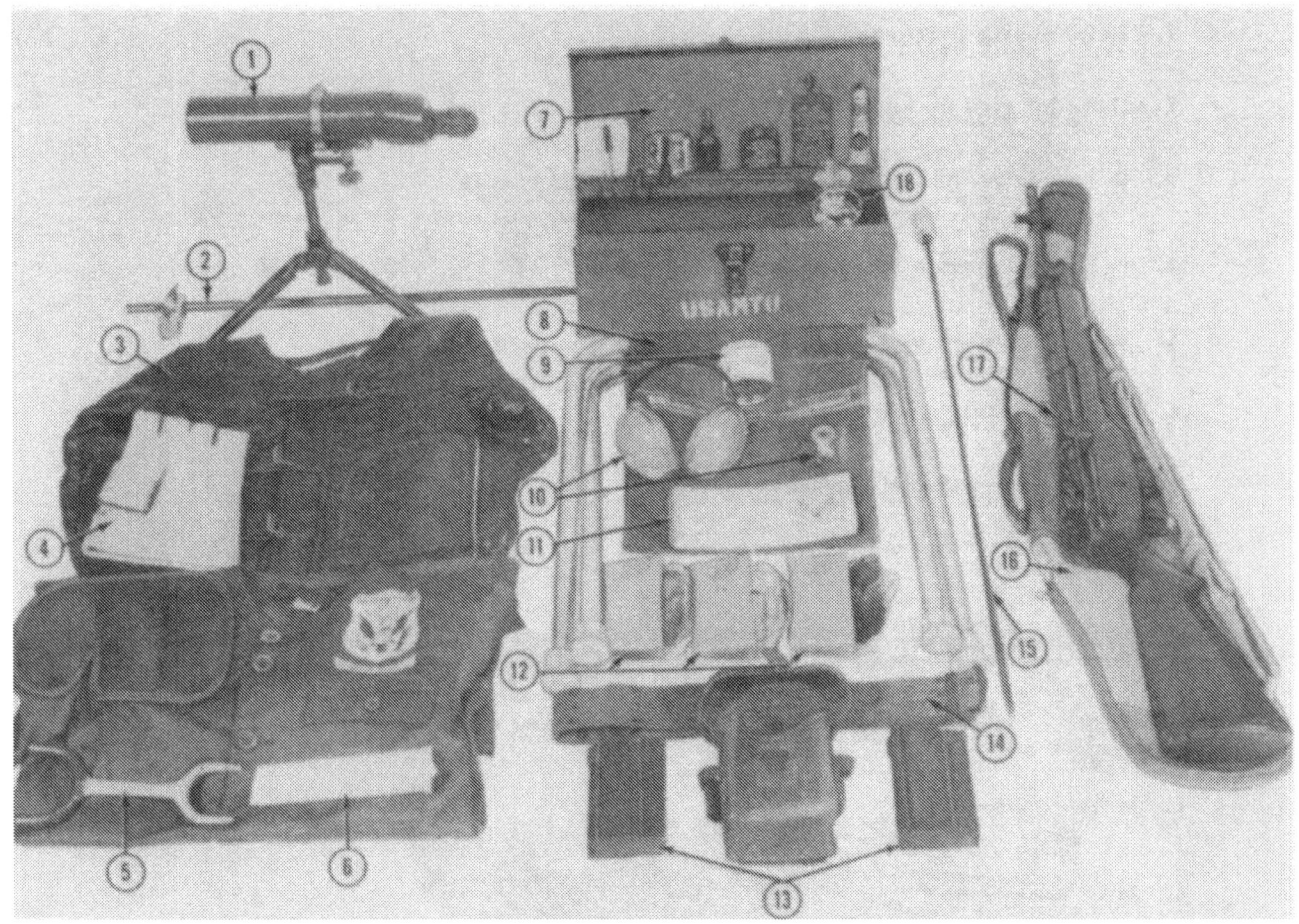

Figure 12-17. INDIVIDUAL EQUIPMENT.

A. Individual Equipment:

    1. Spotting telescope

    2. Telescope extension

    3. Shooting jacket - (cloth or leather)

    4. Shooting glove or mitt

    5. Rifle fork

    6. Sweat band

    7. Cleaning equipment and box

    8. Stool

    9. Firm grip

10. Ear plugs or protectors

11. Scorebook

12. Shooting glasses

13. Magazines

14. Cleaning rod

15. Rifle case

16. Rifle with sling

17. Carbide lamp

   It is becoming increasingly apparent that fitting and selection of equipment has much to do with performance. A shooter having a jacket that is loose and sloppy will not produce his best scores. Where possible, items of wearing apparel should be issued on the basis of two per man. This allows for proper care and cleaning. For safety reasons it is recommended that all shooters wear glasses, however, certain color glasses are not particularly suited for some shooters and if used, will result in eye strain or poor scores.

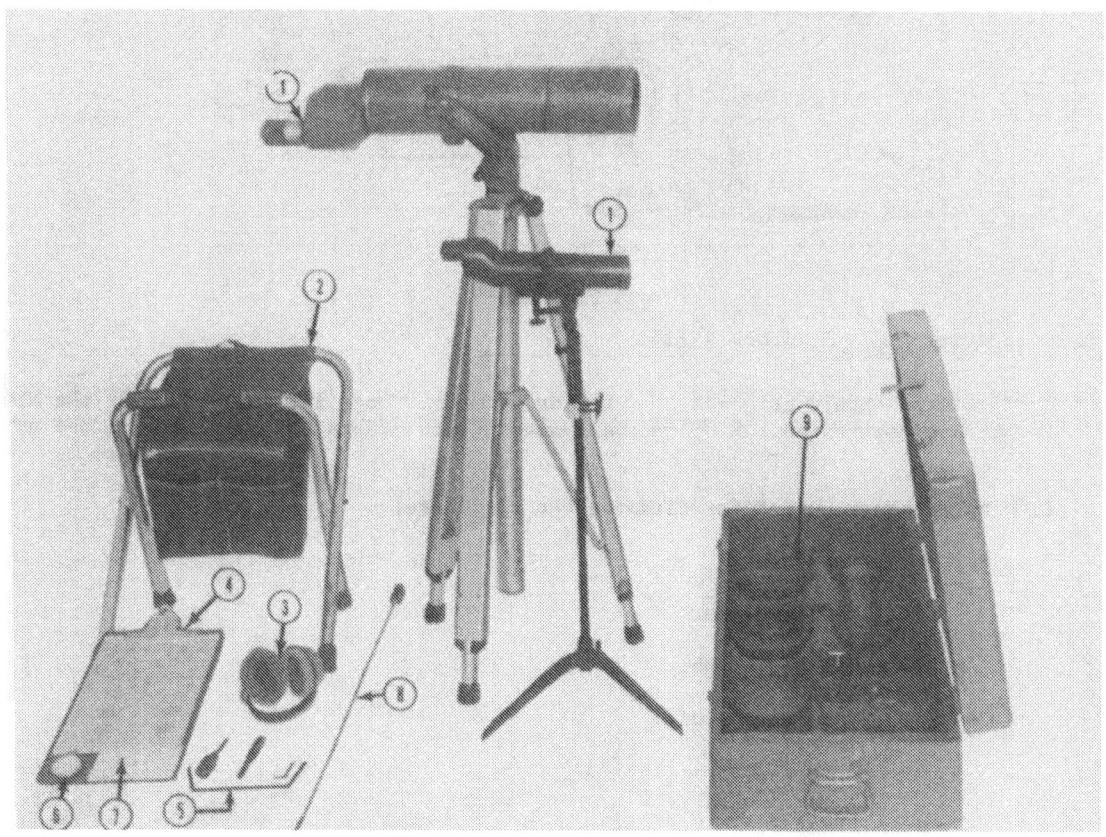

Figure 12-18. COACH'S EQUIPMENT.

B. Coach's Equipment:

   1. Team spotting (100 MM) or individual scope (20-25 power) with extension

   2. Stool

   3. Ear plugs or protectors

   4. Plotting sheets

   5. Allen wrench, screwdriver, and combination tool

   6. Stopwatch

   7. Lens tissue

   8. Cleaning rod

   9. Scope carrying case with dust covers and alternate lenses

Fig. 12-19.

C. Care and Cleaning of an M14 Rifle.

   Extreme care should be devoted to the daily maintenance and inspection of the M14 whenever it has been fired. The materials and recommended method for cleaning are as follows:

   1. Materials required for cleaning the rifle are:

      a. Military Issue.

         (1) Cleaning rod.

         (2) Chamber brush.

         (3) Lubricating oil.

         (4) Bore cleaner.

         (5) Patches.

         (6) Bore brush.

         (7) Linseed oil.

- (8) Lubriplate grease.
- (9) Neat's foot oil.
- (10) Rag.
- (11) .45 cal. cleaning rod.
- (12) .45 cal. bore brush.

b. Purchased.

- (1) Plastic coated cleaning rod.
- (2) Satin silicone compound for stock.
- (3) Rust inhibitive grease.
- (4) Tooth brush.
- (5) Shaving brush.
- (6) Sight cleaning brush or rag.
- (7) Artist brush.
- (8) Plastic grease.

2. Cleaning the Match M14 rifle is accomplished in the following manner: Extra care must be taken to insure that the rifle is cleaned thoroughly. The first step is to clean the sights and wipe off all external dirt. The bore should be swabbed four or five times with a brass brush dipped in bore cleaner. Run the brush all the way through and all the way out to insure complete cleaning and to avoid crimping the bristles. The chamber is also swabbed with bore cleaner using a chamber brush. The bore cleaner should be left in the bore and chamber until the rest of the rifle is cleaned. The throat of the chamber, receiver, and other interior areas may be cleaned with a patch or a piece of cotton on the end of a stiff wire or thin stick. To remove the hardened carbon from the interior of the flash suppressor a .45 cal rod with a .45 cal bore brush and patches are used. Use the rod and brush to remove heavy carbon build-up. Follow this up with patches saturated with bore cleaner, being sure to clean between the prongs. At this point the bore and chamber are swabbed with clean patches. As the final step in the cleaning process, four or five drops of bore cleaner are placed in the lower gas port in the gas cylinder. Then elevate the muzzle, close, then open the bolt so the piston drops all the way to the rear; repeat this several times to insure complete saturation of the interior of the cylinder.

NOTE: The rifle will never be disassembled by the shooter for cleaning or lubrication. Disassembly should be performed only by a qualified gunsmith and should be cleaned thoroughly when it is disassembled for repairs.

3. Lubrication is performed in the following manner: lubricant must be removed and replaced everyday. All surfaces that have become shiny from metal to metal contact should be lubricated. The parts that must be lubricated are:

a. Lip of receiver.

b. Locking lugs.

c. Operating rod guide groove.

    d. Operating rod.

    e. Operating rod guide.

    f. Bolt camming lug.

    g. Bolt camming recess.

    h. Bottom, right side of barrel immediately forward of the receiver.

   4. Magazines must be inspected daily for cleanliness and damage as either may cause a malfunction. When damaged, replace the magazine; when dirty, clean by disassembling or by depressing the follower and removing all foreign matter with a cotton swab. A light coat of oil should then be applied to the interior walls of the magazine.

   5. When transported, the M14 rifle should be inclosed in its case and placed in the firing position (sights up), and when stored, it should be hung muzzle down, bolt forward, hammer released, and flash suppressor free from contact. If stored for an extended period, the trigger guard should be disengaged, and the bore coated with a rust inhibitive grease.

   6. Any M14 rifle used in competition must be handled with care and properly maintained if accuracy is to remain consistent. Protective care on the range is of the utmost importance because it is here that the rifle is exposed to the careless handling of the shooter and the effects of weather. The shooter must be careful not to drop or excessively jar his weapon. When not used, it should be placed sights up on the rifle fork and protected from the elements of weather (rain, dust, and sun) as exposure to the weather will result in malfunctions and decreased accuracy. Prior to firing, the shooter must perform a visual inspection and clean or lubricate his weapon when necessary It has been proven by testing that care and cleaning has a direct relationship to accuracy.

SECTION IV - SQUAD TRAINING

Fig. 12-20

A. Training Program.

Once the selection of individuals has been made to comprise a squad, an organized training program should be laid out to mold the shooters into individual and team competitors. The goal of the program being to prepare each individual to adequately represent himself and the team at a given level of competition.

1. After selection of individuals for the squad, at least two weeks, time permitting, should be devoted to the instructional training phase. This instructional training should include physical conditioning, mental conditioning, rules and regulations, safety, fundamentals of marksmanship, detection and correction of errors, dry firing, and range firing.

2. The primary goal for the squad OIC and Head Coach is to win the team match at any level of competition. Therefore, at the conclusion of the instructional training phase, and based upon the progress of each individual as recorded in the evaluation file, teams should be formulated and a concentrated effort placed on team firing. However, time must be reserved for individual firing, since a good portion of all competition is of an individual nature.

    a. Having formed the team(s) in reference to ability and organizational requirements, a training schedule which has proven successful over the last few years includes the following:

       (1) <u>Physical Conditioning, Limbering Up Exercise, and Organized Athletics:</u> These exercises should be performed daily during practice and discontinued prior to and during competitive firing.

       (2) <u>Individual Firing:</u> Monday and Thursday of each week should be devoted to individual firing. During this phase of training, those deficiencies noted by the coaches on team days, can and should be corrected.

       (3) <u>Team Firing:</u> Tuesday and Wednesday should be devoted to team practice Fridays should be devoted to team record firing. The shoulder to shoulder team record competitions on Friday will aid in the mental conditioning of the team shooter.

       (4) <u>Match Firing:</u> All available competitive matches should be entered. There is no substitute for this type of training which offers the true competitive spirit of marksmanship. Here the individual is afforded the opportunity to build and develop the mental conditioning and discipline so necessary to the shooter.

b. When a weakness is noted, a review of the technique, fundamental, or subject is recommended. A detailed explanation of the training techniques used to successfully develop the individual and team, follows in this guide.

Fig. 12-21

B. Physical Conditioning.

The objective of physical training in the marksmanship program is to condition the muscles, heart, and lungs; thereby increasing the shooters capability of controlling the body and rifle for sustained periods without experiencing fatigue. The ultimate level of endurance must be determined by the individual shooter. A general state of good health is beneficial for all shooters. This guide is designed to develop those areas of physical proficiency necessary to achieve championship form.

1. The following physical condition characteristics are desirable for the competitive rifleman.

   a. Control of an adequately developed muscle system.

   b. Endurance to fire over long periods of time without perceptible lowering of scores.

   c. Highly efficient heart and lung system.

   d. Good reflexes and coordination.

2. Physical training exercises, practice firing, and sports are the best means of developing these qualities in a shooter and will also aid in developing self control and confidence.

   a. The shooter must understand that a physical training program is long range in nature and may not show immediate results. To benefit from such a program the shooter must diligently adhere to this program. Once he is satisfied with his level of physical proficiency he then must maintain this level.

   b. Physical Training Exercise.

      (1) Isometrics: Contraction of muscles under tension without moving the various parts of the body. These exercises will improve muscle tone and, to some extent, strength. Prolonged practice of Isometrics will also tighten and shorten ligaments that support joints and make them more susceptible to injury.

186

(2) Weight lifting: Beneficial as warmup exercises but does not materially strengthen the heart and lung systems. Being over-muscled is as detrimental to shooting as being overweight.

(3) Calisthenics: Beneficial as warmup exercises, but again does not materially strengthen the heart and lung systems.

(4) The following exercises are useful to develop strength and coordination in areas needed to aid a shooter and as warmup exercises. The number of repetitions are best determined by the individual needs of the shooter.

    (a) BEND AND REACH

        1. Starting Position--Side straddle, arms overhead, palms in.

        2. Cadence--Moderate.

        3. Movement:

            a. Bend trunk forward and downward. At the same time swing arms between the legs, touching fingers to ground between and behind the heels. Knees are bent. Touch fingers as far behind heels as possible.

            b. Recover to starting position.

            c. Repeat count (1).

            d. Recover to starting position.

START   (1)   (2)   (3)   (4)

Fig. 12-22

    (b) SQUAT BENDER

        1. Starting Position--Standing with feet slightly separated, hands on hips.

        2. Cadence--Moderate.

        3. Movement:

            a. Do a full knee bend, and thrust arms forward. Keep fingers extended, palms down, and trunk erect.

- b. Recover to starting position.
- c. Bend trunk forward, keeping knees straight, touch ground in front of toes.
- d. Recover to starting position.

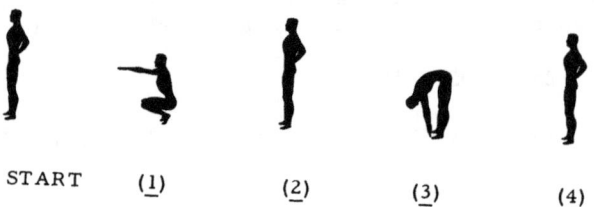

Fig. 12-23

(c) SIDE BENDER

1. Starting Position--Side straddle, arms overhead, thumbs interlocked.

2. Cadence--Slow.

3. Movement:

- a. Bend sidewards sharply to the left, bending the left knee. Bend straight to the side without twisting the trunk or shoulders.
- b. Recover slightly and repeat with a bounce.
- c. Repeat count (2).
- d. Recover to starting position. Repeat on right side for counts (5), (6), (7), and (8)

Fig. 12-24

(d) TRUNK TWISTER

1. Starting Position--Side straddle, fingers laced behind head, elbows backward, chin in.

2. Cadence--Slow.

3. Movement:

   a. Keeping knees straight, bend forward sharply, with a **slight** bouncing movement that causes slight recovery from the bend. This is a vigorous movement.

   b. Bounce downward, and simultaneously turn the trunk sharply to the left so that the right elbow swings downward between the knees.

   c. Repeat count (2) to the right. This time the left elbow swings down between knees.

   d. Recover to starting position, pulling head backward and chin inward strongly.

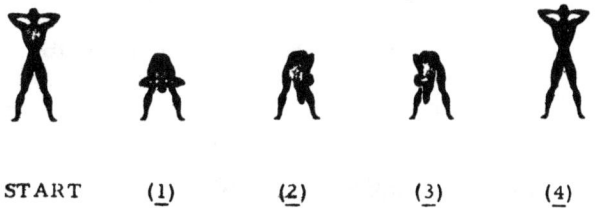

START   (1)   (2)   (3)   (4)

Fig. 12-25.

(e) PUSH UP

1. Starting Position--Front leaning rest, body straight from head to heels, weight supported on hands and toes.

2. Cadence--Moderate.

3. Movement:

   a. Bend elbows and touch chest to ground keeping body straight.

   b. Straighten elbows, and recover to starting position.

   c. Repeat count (1).

   d. Repeat count (2).

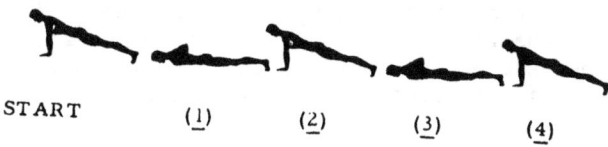

START   (1)   (2)   (3)   (4)

Fig. 12-26.

(f) BODY TWIST

1. Starting Position--On back, arms on ground and extended sideward, palms down, legs vertical, feet together, knees straight.

2. Cadence--Slow.

3. Movement:

    a. Lower legs to the left, twisting trunk and touching ground next to left hand. Keep knees straight, and both shoulders on ground. Legs must be lowered, not dropped.

    b. Recover to starting position without bending knees.

    c. Lower legs to right, twisting trunk, and touching ground near right hand.

    d. Recover to starting position.

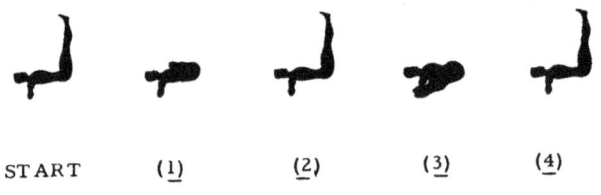

START   (1)   (2)   (3)   (4)

Fig. 12-27

(5) Dash events, such as those used in performance sports, require sub-maximal to maximal effort for periods of less than three or four minutes. When interspersed with frequent rest periods, such as speed running for one or two minutes, these exercises do not allow the body to attain a "steady state" or a "level off" period; and therefore lack endurance developing capabilities.

(6) Running, fast walking, swimming, and cycling require continual sub-maximal to maximal effort for periods of more than four minutes. Tests have shown that shooters so trained have higher developed heart and lung systems than those trained with other types of exercises. A shooter whose heart and lung systems are in good condition has better control of competitive pressure by being able to more fully relax in a shorter period of time. This type of exercise also develops self-control by forcing the shooter to endure long periods of body stress. If the maximum benefit is to be derived from conditioning, an endurance program should be followed. Care must be exercised not to overtrain, as this can defeat a positive Physical Training Program.

c. Practice Firing.

The actual practice of taking up the firing positions, whether for live or dry firing, exercises the firing muscles. To attain top firing performance, no matter how well conditioned the rest of the body is, the shooter must maintain the muscle tone of the muscles used primarily to sustain a stable firing position.

d. Sports.

Tennis, Vollyball, Softball, Handball, and Bowling offer the opportunity to develop mobility, reflexes, and precision. These exercises also offer a break in the body conditioning portion of the marksmanship training program and improve the esprit de corps of the shooters.

Fig. 12-28.

C. Mental Conditioning.

Mental conditioning can be defined as the developing or disciplining of a shooters emotions to prepare him for the act of firing. No matter how well a competitor has physically mastered the basic fundamentals of marksmanship, he will not do well on the competitive firing line if he has failed to develop mental discipline. Lack of mental discipline as evidenced by certain undesirable emotions, is manifested in certain physical reactions such as shaking, rapid breathing, etc. The coach can recognize these emotions or physical reactions by conversing with or observing the shooter; or studying the evaluation file.

1. All too little emphasis is placed on teaching and training the student of marksmanship on how and what to think. Of course nothing adds to mental conditioning and confidence more than <u>practice and more practice</u>. Everyone who has fired in a match is aware of the effects of emotional or mental strain. Those who have progressed to the higher levels of competitive firing have found that they must always strive to improve their ability to control these reactions.

    a. The primary emotion felt by most competitors is fear or anxiety. To a certain degree, depending on the individual, this results in mental and physical reactions which are natural and involuntary. Of these effects, the most detrimental are rapid pulse, rapid breathing, muscular tension, and impairment of the reasoning power. Some of these symptoms may be suppressed to a certain degree. By breathing slowly and deeply, both the rapid breathing, and to some extent, the muscular tnesion will be reduced. Unfortunately, since the mind has lost some of its reasoning power, it has also lost a portion of its control over the body, making it very difficult to overcome these reactions Therefore, the competitor must discipline his mind to enable it to control these otherwise involuntary reflexes. Several techniques which have been found to be successful in accomplishing this are as follows:

    (1) Prior to and during the match, the competitor should avoid having thoughts that create anxiety. Many shooters have trained themselves to doze between relays in order to eliminate all conscious thoughts, (CAUTION: If a competitor uses this method he must first be certain that he will be awakened in time for his next relay). Another method is to think about some subject other than the match. A very effective

technique is to establish a system of preparing equipment prior to the match and between relays. Disturbing thoughts have no chance to get started. This method has the added advantage of insuring that the equipment is in the best possible condition.

(2) If a shooter finds that he is easily disturbed by range rumor and other person's scores, he should make every effort to avoid hearing them. If necessary, he must keep himself separated from other shooters during the match, and he must stay away from the scoreboard.

(3) Muscular spasms caused by tension are best controlled by relaxation. Again, dozing between relays will promote this condition of relaxation. Also, by concentrating the full attention of the mind on each part of the body, a complete state of relaxation can be reached.

(4) While on the line in a match, the experienced shooter may find that the impairment of his reasoning powers may be to an advantage. If he has succeeded in relaxing and controlling his breathing, he will find that his well trained body will do the right thing at the right time without any conscious help from his mind. On the other hand, if his mind becomes fully aware of its surroundings, anxiety may appear with all its undesirable effects.

b. Other related emotions which are detrimental to the shooter are negative thinking and overconfidence.

(1) Negative thinking on the part of a shooter can, and usually does, affect his performance. It demonstrates that though the shooter has mastered all the physical skills of marksmanship, he has not accomplished the principle of mental discipline. For instance, if we think that we are going to be sick, we usually find ourselves being sick. By the same token, if we think that we may not do so well on the firing line, usually we don't. Because we do not discipline ourselves mentally, we permit ourselves to be too pessimistic about our firing.

(2) On the other hand, overconfidence on the part of the competitor has played havoc with many a would-be fine score. The competitor must strive to reach a happy compromise in developing the proper state of mental discipline. His state of conscious thinking must fall between that of overconfidence and negative thought. He should not, however, push the positive element to the extent that it is converted into overconfidence.

2. In order to accomplish mental discipline, the shooter must know what situations and habits are good for him and avoid situations and habits that are bad for him. Following is a list of habits that the average competitor finds himself doing in most matches. These are a few of the many. You must understand yourself in order to ascertain whether they are good for you, or bad for you. They are listed as "Don'ts", but they may not all be so, depending on your individual characteristics.

a. Don't be perturbed. The very nature of the habit makes it a bad one, and it should be considered a don't for every shooter. It is certain that everyone will agree that a marksman cannot perform on the firing line if he is experiencing the emotional sensation of anger and bitterness. Further, it is by no means in the best interest of safety for a competitor on the firing line with a weapon in his hands to even approach the point of losing his temper.

b. Don't give up after making a bad shot. Many competitors have given up in disgust after firing a bad shot, but if he had not quit he might have won the match or at least posted a much higher score. Mentally discipline yourself to never give up or quit. A quitter never wins.

c. Don't get "shook up" by adverse weather or range conditions. Remember that all other competitors are under the same handicap. Practice mental discipline by positive thinking to the effect that you welcome an opportunity to try your skill under adverse conditions. Chances are that you will beat your competitors because you are not thinking in negative and pessimistic terms and they probably are.

d. Don't believe in boast and rumors uttered by other competitors. By your very nature, you may be susceptible to the influence exerted by another competitor's "Flappin Jaws" and be pressured out of firing a good score. Practice mental discipline, do not allow yourself to believe or even listen to how good the other fellow is. The only thing you want to believe is what you see on the official bulletin board after the last round has been fired and the smoke has cleared away.

e. Don't watch the score board during the match. Of course this is one of the big thrills of the game, and may not bother some people. However, if you are one of those who cannot successfully cope with the pressure and anxiety that might build up within you by checking the scoreboard, then wait until after you have fired your last round. Then you may enjoy the pleasure and satisfaction of seeing how you compare with the rest of the competitors.

f. Don't score any target other than your own while firing. By their own admission, many top competitors have lost an important match simply by letting their curiosity get the best of them. If you are disciplining yourself to think in positive terms about your performance while on the firing line, you have no time to concern yourself with how well or how poor the other competitor is doing.

g. Don't worry about your total or aggregate score. The most important shot that you fire is always the one that you are in the process of firing, therefore, while you are in position on the firing line, that single shot is all that you should be consciously concerned with. Discipline yourself along that line and the score will take care of itself.

h. Don't dismay when competing against superior equipment. Many a match has been won as a result of superb human performance employing equipment which was considered to be inferior compared to that used by other competitors.

3. The most important phase of mental conditioning is accomplished during training. It is in this period that the body is taught to function automatically. The shooter develops a high degree of confidence in himself and his equipment. He learns that he is capable of performing well even in bad weather. In general, he becomes physically and mentally fit.

Fig. 12-29.

D. Rules and Regulations.

Rules and regulations may vary from one match to another. A good guide for instructional purposes is the current NRA Rule Book and AR 920-30. These rules and regulations are supplemented by ground rules established by the Match Executive Officer's Bulletin.

1. In early stages of training it is recommended that the boredom of position drills and cadence exercises be broken with scheduled periods of instruction on rules and regulations. All of the hard work that goes into the training of a rifle squad may be of no avail if an individual or team is disqualified. It is the competitor's responsibility to understand and obey match rules and regulations. As a guide, the following outline can be used to instruct personnel. Those applicable portions of the NRA High Power Rifle Rules and AR 920-30 should be used.

    a. Purpose of rules and their application during matches.

    b. NRA Official Referee-duties and responsibilities.

    c. Competitors duties and responsibilities.

    d. Equipment and ammunition authorized for match competition.

    e. Legal positions for match competition.

    f. Competition regulations and range operations.

    g. Range control and commands.

    h. Marking and scoring system.

    i. Procedures for initiating challenges and protest.

2. It is important for squad officials to be familiar with the regulations that affect the formation and operation of a squad. Some are:

    a. FORSCOM Suppl 1 to AR 350-6 "Army-Wide Small Arms Competitive Marksmanship

    b. "NRA High Power Rifle Rules."

    c. AR 920-30, "Rules and Regulations for National Matches."

    d. AR 350-6, "Army-wide Small Arms Competitive Marksmanship."

    e. AR 672-5-1, "Awards."

    f. TA 60-18, "Army Rifle and Pistol Team."

    g. TA 23-100, "Ammunition for Training."

    h. FM 23-5, "U.S. Rifle Cal .30 M-1."

    i. FM 23-8, "U.S. Rifle 7.62 MM, M14."

    j. FM 23-71, "Rifle Marksmanship."

    k. FM 23-9, "Rifle, 5.56-MM, M16E1.

Fig. 12-30

E. Safety.

Regardless of the degree of experience of the shooters on a team, they must all be oriented on safety regulations prior to their marksmanship training. Safety must be a primary consideration at all times, and even experienced competitors must be reminded periodically of their responsibilities on and off the range. The rules of safety can be divided into three groups: Those regulations which apply behind the firing line, those which apply on the firing line, and general safety regulations which apply to any situation on the range.

1. Rules to be observed on the range <u>behind the firing line</u> are as follows:

    a. Rifle will be cleared at all times. The bolt will be open, magazine removed, and the safety locked.

    b. When handling rifles, keep the muzzle pointed up in the air and down range.

    c. There will be no aiming, dry firing, or position work behind the firing line.

    d. Draw ammunition according to ground rules.

    e. When called to the ready line, inspect the bore to insure that it is clear.

    f. Do not run on the range.

2. Rules and precautions to be observed <u>on the firing line</u> are as follows:

    a. Inspect ammunition for cleanliness, serviceability, and proper caliber.

    b. Keep rifle clear and pointed down range until the range has been declared safe for firing.

    c. Load only on the command of the range officer.

    d. After firing, remain in position until the rifle is cleared and permission to leave the firing line has been granted. In many matches, shooters must remain in position until the firing line is clear.

    e. After firing, each shooter, while still in position, must open the bolt, remove the magazine, inspect the chamber, and if appropriate, lock the safety.

    f. No one will go forward of the firing line before the firing line is cleared by the Range Officer.

    g. Upon hearing the command "Cease Fire", shooters will immediately lock and clear all rifles; except for those people who need alibis verified.

h. Everyone should wear shooting glasses and ear plugs while firing.

i. In addition, Range or Match SOP's will be followed.

3. <u>General safety</u> regulations and precautions are as follows:

a. Any person who observes an unsafe condition on or in front of the firing line <u>will</u> give the command "Cease Fire". Also, he <u>will</u> correct any unsafe condition observed behind the firing line.

b. Before use, dummy rounds must be inspected to insure that no live rounds are present. Dummy rounds should be stored separately from live rounds.

c. All rifles should be inspected before conducting dry firing exercises.

d. Before firing on any range, the Range Officer must insure that Range Regulations are observed which pertain to range fans, range guards, range flags, and range clearance.

Accidents result from violations of "common sense" safety rules. Any violation can result in the disqualification of the shooter or his team. Safety consciousness can only be developed if all squad officials and experienced shooters set the example and <u>insist</u> that everyone adheres to these same rules.

Fig. 12-31

F. Detection of Errors.

As mentioned earlier the most difficult type of errors to detect are those caused by improper trigger control, especially with the experienced shooter. When it is suspected or observed that a shooter's poor firing is caused by improper trigger control, as manifested by flinching, bucking, or jerking the trigger, the coach should utilize the ball and dummy method.

a. The main points to consider in determining the number of dummy rounds to use on a shooter or the frequency of the exercise are as follows:

(1) The shooter's experience and needs.

(2) For the experienced shooter it may be sufficient to prove to him that he is flinching, bucking or jerking the trigger.

(3) For the new shooter it may be necessary to repeat the exercise until he produces a few, then more and more good shots.

(4) Do not use the exercise unnecessarily.

(5) Do not use the exercise if errors in aiming or position are manifested

b. Some suggested exercises are listed below:

(1) Ball and Dummy in the standing position.

(2) Ball and Dummy in rapid fire. During our discussion of rapid fire, three drills were discussed. The first shot, reloading, and the ten shot drill. These drills are excellent to utilize with the ball and dummy method.

(3) Blank target firing exercises will be beneficial since they remove the impulse to jerk the trigger when the perfect sight picture is attained.

(4) Any combination of the above with the coach applying pressure on the trigger to show the shooter that trigger control is the most important fundamental of marksmanship.

Ball and Dummy exercises are considered to be an effective training technique to increase the shooter's power of concentration and to assist him in overcoming normal reactions to the explosion and recoil.

G. DESCRIPTION OF THE NATIONAL MATCH COURSE

Section 113 of the National Defense Act of 1916, as amended, provides that there shall be held an annual competition, known as the National Matches. The course fired during these matches became known as the National Match Course. From this course the Excellence in Competition and Distinguished badges are awarded. The National Course is fired both individually and as a team.

The National Match Course, individually, is fired in four stages.

FIRST STAGE: 200 yards, slow fire, standing, 10 rounds in ten minutes, Target "A".

SECOND STAGE: 200 yards, rapid fire, sitting or kneeling from standing, 10 rounds in fifty seconds, Target "A".

THIRD STAGE: 300 yards, rapid fire, prone from standing, 10 rounds in sixty seconds, Target "A".

FOURTH STAGE: 600 yards, slow fire, prone, 20 rounds in twenty minutes, Target "B".

The National Match Team Course is essentially the same as the individual; with a team normally consisting of four or six shooters, team captain, and a team coach. Each team is assigned a target and all shooters will fire at this target.

During slow fire stages, two shooters will be on the firing line simultaneously, alternating shots; with the team allotted one minute per round and three minute preparation periods for each pair of shooters. For example, at the 200 yards first stage, six shooters have 66 minutes and at the 600 yards fourth stage they have 126 minutes.

Rapid fire stages are identical, with the team shooters following in rotation on this same target.

CHAPTER XIII

INTERNATIONAL RIFLE COACHING

A. PURPOSE

This chapter is written for the benefit of those individuals in International shooting who may find themselves facing the job of coaching an international rifle team with little or no international marksmanship background to draw upon. This chapter should also be of value to those persons who are in a position to appoint coaches. You may also be called upon to coach a champion shooter.

The champion shooter is the cutting instrument that penetrates through and beyond the bounds of what is presently considered the best possible shooting performance. The coach is the tool that hones this superb instrument and keeps it razor sharp. The coach can keep the champion shooter continually striving to break the existing records. The score that equals or breaks a previous record is never good enough to stand unbroken for the years to come.

A coach exists for the benefit of his shooters, and not the shooters for the benefit of the coach. The coach's job is to direct the shooter in his own development. His most important function is to make the shooter think.

B. PRINCIPLES OF COACHING

Coaching clinics have revealed that very few individuals are familiar with the principles involved in coaching a free-rifle shooter.

Many of the shooter's ideas must be influenced by personal and individual coaching. In general, positive influences can best be made by an individual coach in whom the shooter has a great deal of confidence.

1. First we must recognize that the basic principles and fundamentals of marksmanship do not change. However, the application of those principles and fundamentals will differ from shooter to shooter, and from one weapon to the next.

2. We must recognize also that the style and techniques of coaching will differ as we move from service rifle and pistol to the more individualistic free-rifle type of shooting. This is because the free rifle shooter is trained for individual performance; he never fires as a team member in the same sense that service rifle and pistol shooters do. Consequently, one of the foremost responsibilities a free-rifle coach is to instill self-reliance and confidence in his shooters.

3. A difficult coaching task is to create an atmosphere in which each individual shooter can experiment with and refine his own techniques. The progress made by a shooter in advancing his score is in direct proportion to his thinking about and analyzing his own performance. He must have, as well, the desire to be a World or Olympic Champion.

4. The coach is aided in creating confidence in the team as a whole if he carefully uses the performance of his shooters for purposes of research and analysis. He should constantly watch the performance of those shooters who are "on top", and he should seek the advice and counsel of those top shooters. Then, armed with a knowledge of the techniques employed by the best shooters, he should acquaint each new shooter with these techniques. Teach him to adopt the techniques that work best for him, and to discard the methods that are unsound. The coach must be careful to prevent his shooters from using "gimmicks"--easy solutions in the form of novel equipment or trick techniques. Inevitably some shooters will begin to rely on gimmicks and fail to concentrate on basic improvement

of their performance. The use of gimmicks may become the basis for the development of alibis. Excuses for poor performance will cause confidence to dwindle away. On the other hand, if the coach keeps his shooters concentrating on performance, and aids their progress by displaying a sound knowledge of shooting techniques, then he is building confidence. A coach bears the responsibility of creating the right amount or maximum of confidence in each shooter.

5. All successful coaches, in any form of shooting, have one thing in common; they have some attributes in their personality that induces excellence of performance from those under their guidance. Minor shortcomings of personality do not negate this art. Of course, there are no perfect coaches because there are no perfect men. Because of this fact, we are accustomed to over-looking minor flaws in one who possesses true leadership qualities. A coach or a shooter is only parading his own inferiority if he scorns or refuses to cooperate with his coach or one of his teammates because of some small personality trait he dislikes.

6. If a coach is not compatible with the members of his team he may become a source of friction, no matter how much knowledge or talent he may possess. The ability to get along with people is not an inherited talent; it must be cultivated. Usually the person who fails to get along with others, fails to make an effort to get along. In each coach there must be strong traits of human understanding, tolerance and patience.

7. A coach must be temperate in all things. He must have the will power, the intestinal fortitude and the character to deny himself those things that will compromise his standing as a leader. The members of his team will respect him as a coach only if they can respect him as a man.

8. It is not necessary for a coach himself to be a champion shooter; indeed, in many instances a champion may possess too strong a drive toward personal achievement to be able to coax achievement out of others. There are, of course, exceptions to this rule. A good coach must not necessarily be the equal of a champion in his ability to shoot high scores, but he must be the equal of a champion in the depth of his shooting knowledge.

9. A coach must be dedicated to his work. He must constantly keep in mind that his job is to get the maximum performance out of his shooters, and he must be willing to make personal sacrifices to that end.

10. He must at all times and in all things exercise patience and self-control. Irreparable damage can be done if a coach allows his loss of temper to antagonize a shooter. He will erect an unbreachable wall between them, and further constructive coaching will be severly impaired or made impossible. For this reason, it is wise to have one individual in charge of the team to execute disciplinary measures, and another individual to act as coach. In this way the coach avoids friction between himself and the shooters.

11. Patience, tolerance, and self-control must extend not only to problems of discipline, but to problems of performance as well. A coach may point out to a shooter some error in his performance, only to see the shooter continue to repeat that same error again and again. To become discouraged or dismayed at this point would solve nothing; it would only destroy the shooters' confidence and desire. The coach must work patiently and display a confidence in the shooter's ability to overcome his problem.

12. When the problem is solved, then it is the coach's duty to give the shooter credit for his success. Unselfishness is necessary in match performance. Also, the coach must remember that even if the shooters could not have won the match without him, still it is they who should receive the credit for winning. He himself must be content to blend into the background when the laurels are passed out. If he accepts the laurels for himself, he may thereafter get only a grudging performance from his team.

13. A coach must instill in a shooter the ability to confidently analyze his own performance. He does this by showing a respect for the individual's intelligence and by paying due attention to his ideas. To do this, a coach must maintain an open and progressive mind. He must accept new ideas, and remember that most new ideas will come from the shooters themselves. Complacency with old techniques is fatal to progress. However, all new ideas are not sound ones and after due consideration, a coach must firmly reject ideas that are worthless or harmful. He must also be alert to recognize the difference between a shooter's conduct of a useful experiment and a shooter's adoption of a gimmick.

14. A coach must at all times be objective and observant. Many times he will be able to detect irregularities that affect performance even before the shooter is aware of them. One of the coach's responsibilities is to establish a routine for his shooters. He must then learn how this routine can be upset by matches on unfamiliar ranges and how a change in routine upsets the individual shooters. He must consider unfavorable match conditions as a deterrant to good scores. He must then train his shooters to accept and adjust to such irregularities.

15. During practice, a coach must set exacting standards for himself and for his shooters. He must demand that each individual exert as much effort in practice as he exerts in an actual match. The concentration and determination required to produce a winning score cannot be turned on and off at will; it must be developed. It can be sustained only through continuing intensive training and practice.

    a. It is in this area of practice that the coach will have his greatest difficulty. Some shooters invariably feel that practice performance is not as important as match performance, and therefore does not require as much effort. They must be made to recognize that practice is not only a physical exercise, but also an exercise of the will.

    b. Practice sessions are also the time to teach the shooter to rely upon himself to analyze and critique his own performance. He must learn to diagnose a symptom and prescribe a treatment. This training is necessary preparation for that time when the shooter is firing a match and must be his own coach.

16. A coach should at all times be alert for a bad rifle. He must recognize and accept that occasionally a rifle will be the cause of poor performance. Most good shooters are reluctant to blame a rifle for a poor score. The coach, however, must constantly consider this possibility and take action to see that the rifle is maintained to permit maximum performance on the part of the shooter.

Never, under any circumstance, should a man be allowed to continue to shoot a rifle that is not performing properly. A poor rifle may destroy a shooter's confidence in himself, or become an excuse for a poor performance. Detecting a bad rifle and insuring that the fault is corrected is the responsibility of the coach; otherwise, some shooters will become amateur gunsmiths and unintentionally alter the accuracy of otherwise good rifles.

17. The coach should keep a constant check on the condition of all equipment in addition to rifles. This need not be a formal, organized inspection, but rather a continuing process during training. Setting a standard of maintenance not only helps to preserve equipment, it also gives the shooter confidence in the reliability of his equipment. Never allow a shooter to arrive at a match site with faulty equipment. Even if an excellent replacement item is available, the shooter is handicapped to a degree by entering the match with an unfamiliar item.

18. A coach must never be satisfied with an average performance from a shooter nor should he allow the shooter to be satisfied. He should inspire the shooter to strive for perfection at all times. However, standards of performance must be flexible enough to comply with the abilities of both the novice and the expert. When each shooter has reached the level of performance set for him, the coach must convince the shooter that he

must set a new, higher standard for himself. It must be remembered that behind any poor performance there must be a valid reason. Identification and correction of errors is the only solution to preventing repetition, and this must be worked out by proper rapport between the coach and shooter.

19. A training program must be designed to meet the requirements of each individual rather than the requirements of the group as a whole. Training requirements will normally vary from one individual to the next because of differences in physical condition shooting experience, etc.

20. Training must be planned, scheduled and supervised. Simply publishing a training schedule is not enough. Supervision is necessary to insure that standards of performance are met by each individual shooter. The coach must be objective in the evaluation of his training program, and he must be prepared to modify his program if it is not achieving the desired results.

21. In order to remain objective about each shooter's performance, some sort of record must be maintained. Most coaches prefer a graph scale. Generally, an examination of such a scale will reveal things that neither coach nor the shooter was aware of. It will serve as an indicator of progress, enabling the coach to evaluate the state of training. It will point out deficiencies in training. It is best to keep a graph of each position, and an aggregate graph. This is the best way to evaluate over-all progress and to detect specific weaknesses. The shooter nor the coach should rely on memory for recollection of performance statistics over an extended period of time. This would negate much of training effort, and cause improper emphasis to be placed in certain areas.

22. Generally, a coach's duties during a match will differ greatly from his duties during practice. In training, the coach will have stressed the development of self-reliance and will power. The shooters will have learned to analyze and correct their own performances.

    a. This self-reliance must continue during a match, but the coach must remember that during competition a shooter is under tremendous pressures and stress.

    b. During the match, the shooter will be helped immeasurably by the knowledge that his coach is nearby to help, should help be needed. The coach's presence will exercise a profound influence on the morale, attitude, enthusiasm and confidence of a shooter. He may even help in the physical preparation for the match. This will help the shooter feel that everything possible is being done to create the best conditions for his performance.

    c. During the match, the coach must be constantly alert for irregularities in procedure or match conditions that will affect the shooters. He must intercede in the interest of his shooter in any argument, and protest in his behalf when any rule infraction occurs. The shooter's only responsibility should be to perform as best he can.

23. An important phase of coaching should always occur immediately after a match. Simply telling the shooters they did not do well is not enough. The coach must make them realize that there are reasons for both good and bad performances. He must help them determine the reasons why their own performances were good or bad, as the case may be. If a shooter discovers the reasons for a poor performance, then that performance was not an entirely wasted effort. Listed below are a few of the more frequent reasons for poor performance.

    a. Lack of knowledge. The shooter with this difficulty is blindly groping for an answer to his shooting problems. He is constantly changing equipment and methods in his search for a satisfactory new technique.

    b. Lack of incentive. The individual is satisfied with his level of performance and has no desire to be a champion unless all of his competitors' performances will decline sufficiently to allow him to be a winner.

    c. Lack of team spirit. The individual with this difficulty may not be fully in accord with the team effort because of personal differences with other team members or the coach.

    d. Inability to make corrections. This individual cannot admit that he has committed a mistake but is always ready with an excuse for a poor performance.

    e. Lack of proper guidance. The individual has the ability, but may be handicapped by overconfidence or pessimism. He needs more objective coaching where the emphasis is on performance rather than on probable results.

    f. Lack of ability. This reason is the most difficult for a shooter to accept.

  24. We have, by design, refrained from giving specific solutions to hypothetical or actual problems. We feel that solutions will come of themselves with the development of an individual coaching style and technique. While shooters A and B have identical problems, their solutions will be approached in a different manner, by various coaches. Both coaches may obtain different results by hard work on the part of themselves and their shooter. The coach who knows as much as possible about fundamentals and basic principles, who persuades his shooters to employ these fundamentals with enthusiasm, and who does not alienate himself from his shooter--that coach has the qualifications for success.

SECTION THREE. U.S. ARMY MARKSMANSHIP TRAINING PROGRAM

CHAPTER XIV

ARMY-WIDE SMALL ARMS COMPETITIVE MARKSMANSHIP TRAINING

FORSCOM/TRADOC Supplement 1   22 January 1975
to AR 350-6 (Condensed)

Appendix   A.   General

                      B.   Service Rifle and Service Pistol Competition*

                      C.   Combat Rifle Match

                      D.   Precision Combat Rifle Match

                      E.   Combat Pistol Match

                      F.   Loading and Reduction of Malfunctions

                      G.   Smallbore Rifle and Pistol Competitions

                      H.   M60 Light Machine Gun Competition

                      I.   National Matches

                      J.   Omitted

                      K.   Reserve Components Program

*Subject to revision by letter HQ FORSCOM, dated 1 June 1975, attached.

APPENDIX A (EXTRACTED)

GENERAL

1. PURPOSES.

   a. Provides direction for all phases of the US Army Small Arms Marksmanship Training Program as delegated to HQ FORSCOM in accordance with this regulation.
   b. Prescribes the objectives, responsibilities, policies, and procedures applicable to the selection and training of individuals and teams for Army-wide competitions, from unit level through the US Army Championships, Interservice Championships, National Matches, and Army participation in tryouts for membership on the US International shooting teams.
   c. Provides additional guidance for security of marksmanship weapons while in the possession of team members.

2. OBJECTIVE. Provides authority and guidance for conducting small arms training and competitive marksmanship activities within the Army, as delegated to HQ FORSCOM in accordance with this regulation, that will achieve:

   a. Outstanding marksmen for the Army in order to attain and maintain US Army leadership in the Armed Forces national and international competition.
   b. An overall marksmanship training program that will improve the readiness condition of the Active Army and Reserve Components.
   c. An opportunity for the US Army to further domestic relations and increase prestige by assisting marksmanship training activities of ROTC personnel in colleges and universities and law enforcement agencies that engage in marksmanship training activities.
   d. A cadre of skilled marksmanship instructors.

3. POLICIES.

   a. Command interest and emphasis are essential to accomplish the objectives of the competitive marksmanship program.
   b. Official International Shooting Union (ISU) or international small arms competitions such as Olympic and Pan-American Games, World Championships, and Council of International Military Sports (CISM) Shooting Championships are classified as sporting events.
   c. Competitions up to and including post/division, ARCOM/GOCOM, and NG championships within CONUS are encouraged. Smallbore competitions outlined in Appendix G Are also encouraged.
   d. Installation commanders are encouraged to organize marksmanship training units as provisional units (supported by a small permanent cadre and TDY/SD personnel) in accord ance with paragraph 7, AR 220-5. All Active Army installations with <u>4000</u> or more assigned personnel strength will provide one or more rifle and pistol teams for competition in their respective CONUSA area matches. <u>When more than one team is selected for rifle or pistol competition, each team will be assigned a specific command designation and will be entered in competition as a separate entity. Team composition will be in accordance with guidance established by this supplement.</u>
   e. Competitive marksmanship programs at all levels will include combat-type courses of fire prescribed in Appendixes B through E. Competitors will use weapons and equipment prescribed by the appropriate appendix and this regulation.
   f. The US Army rifle and pistol championships will be conducted annually. Course of fire will be as prescribed in paragraph 6b, Appendix B, this supplement. The host installations will provide competition M16 rifles.
   g. Commanders authorized the M14 rifle by CTA 60-18 will retain these rifles to permit personnel to practice long range precision firing, Army-wide familiarization with the weapon, and to support contingency requirements for sniper training. Major oversea commanders are authorized to retain the M14 rifle for these purposes.
   h. Personnel engaged in off-post marksmanship activity may transport small arms ammunition and weapons by Government or privately owned vehicles provided that:
      (1) Weapons and ammunition are issued in accordance with current regulations.
      (2) Ammunition and weapons are properly safeguarded during travel and stops enroute and at the destination.

(3) The expenditure and disposal of ammunition comply with current procedures.

i. Off-duty marksmanship practice will be encouraged to supplement scheduled training. Policies relating to weapons use and security will be reviewed to insure ready availability of weapons, ammunition, targets, equipment, and ranges. Special attention will be taken to insure that weapons issued to members of Reserve Components will be readily accessible to the users under the provisions of AR 190-11. Individuals as members of USAR units/team members are considered to be on extended periods of duty/daily activities and therefore may, when necessary, retain weapons in their personal possession to accomplish daily practice providing the security requirements of this paragraph 4-2c and c(1), AR 190-11 are met. Travel orders issued will contain appropriate instructions (Appendix B, AR 310-10) authorizing team or individuals conducting marksmanship activities to transport weapons, ammunition, both Government and privately owned, in connection with marksmanship activities.

* * * * * * *

APPENDIX B

SERVICE RIFLE AND SERVICE PISTOL COMPETITIONS

1. GENERAL. Service Rifle and Service Pistol Competitions consist of combat and precision type competitions designed to accomplish the objectives listed below. In cases not covered by this supplement, these competitions will be conducted in accordance with NRA rules.

2. OBJECTIVES.

a. To raise the standards of marksmanship proficiency with the Service Rifle and Service Pistol within the US Army by improving methods and supervision of Marksmanship Training.

b. To instill confidence in the individual soldier to the degree that he will engage an armed enemy instinctively with confidence and without hesitation.

c. To develop skilled riflemen and pistolmen who will be available to unit commanders as marksmanship instructors.

d. To generate interest and accelerate self-improvement in rifle and pistol marksmanship by creating a desire to excel in competition.

3. PARTICIPATION.

a. Participation by major command/CONUS Armies and higher level teams in interservice; interpost/interdivision; and local, state, and regional/sectional NRA matches, and in NRA gallery rifle and pistol leagues is encouraged.

b. Participation by installation/division level MTU personnel in civilian and interservice competition at Government expense is authorized within funding limitations. This participation will be controlled and supervised by installation/division level commanders.

c. Participation in interservice matches sponsored by other services is limited to squads representing major command/CONUS Armies and higher level activities.

d. Participation in the National Matches is limited to the teams eligible to enter teams in the National Trophy Team Matches under the provisions of AR 920-30.

e. Squads entering major command/CONUS Army and US Army competitions should provide as much participation as possible. Each team is authorized a team captain and coach who may be firing members.

f. The composition of squads entered in CONUS Army competitions through the US Army championships will be governed by the following:

(1) There must be sufficient new shooters on the squad at each level of competition in order to field an eligible team(s). New shooters are those individuals who have not previously fired in a Service-Rifle or Service-Pistol Team Match (National Match Course or Combat Match Course) at the particular level or higher and in that weapon category (rifle or pistol). However, one half of the firing squad members in the All-Army Matches must be eligible new shooters for the National Trophy Team Match at the National Matches.

(2) Individuals who previously represented the ROTC or Service Academy on teams at any level of Army, interservice, or national competition are considered new shooters until they compete as members of Army teams described in this supplement. National Guard personnel who have competed in the National Guard Championships (Winston P. Wilson Matches) as new shooters are considered new shooters in major command matches.

g. Each MTU/CONUS Army will enter a minimum of one rifle team and one pistol team representing each component of the Army in the US Army Rifle and Pistol Championships.

h. These teams will be allowed a minimum of 32 rifle shooters, 21 pistol shooters representing each component, and 5 administrative personnel from the Active Army.

i. Each command participating will notify the Commander, USAIC, of the number of individuals on his squad and the date of their arrival at Fort Benning, Georgia, not later than 45 days prior to the start of the US Army championships.

j. The officer in charge of each Active Army squad entering CONUS Army matches and US Army Rifle and Pistol Championship Matches will present to the Match Director, not later than the designated closing date for match entry, a roster of their respective squads to include the official title designation of administrative personnel, competitor assignment by weapon, the complete name, grade, present service number (SSN), organization of assignment with unit identification code, home address, old or new team classification, individual classification (for US Army Rifle and Pistol Championship Matches only), team membership eligibility for the National Trophy Matches at the National Matches, and certification that all members of their squad are eligible for retention to represent the US Army through the National Matches. Sample format is shown at Figure B-1. Each match director will forward one copy of each squad roster to the Commander, USAMU, ATTN: S3, Fort Benning, Georgia 31905.

k. Selection of Active Army squads for participation in the National Matches will be made by the CO, USAMU, at the conclusion of the US Army Rifle and Pistol Championship Matches. Personnel Selected for the USAR and NG squads will be selected by their respective commanders.

l. TDY orders for Active Army personnel will contain, as a minimum, the following information:

(1) Name, grade, service number, MOS, unit identification code, and the unit to which the individual is permanently assigned.

(2) The purpose of TDY shall be indicated in the orders (e.g., "participation in the US Army Rifle and Pistol Championships and the National Matches").

(3) An authorization to carry personal or Government weapons and ammunition while enroute or during TDY period away from home station.

(4) An individual selected for further competition upon completion of the US Army Championships will be further attached to the USAMU, Fort Benning, Georgia, for rations, quarters, and administration until the National Matches are completed or he is released.

(5) An authorization for 30 DDALV upon completing the TDY and before returning to the home station. Commander, USAMU, will notify appropriate commanders of number of days' leave, destination, and date of return of individual concerned.

(6) An authorization for the Commander, USAMU, or the designated team OIC to amend orders if necessary for the individual to complete the mission.

(7) Oversea tour credit for TDY in excess of 60 days is authorized (Table 1-2, AR 614-20).

m. Individuals shall be provided 50 copies of their orders.

n. Each participant will carry a complete military pay record, with DD Form 1588 (Record of Travel Payments), personnel records, shooting equipment, and adequate utility and summer uniforms. An enlisted man to be discharged and reenlisted during the summer training period will have a record of his reenlistment physical examination included with his medical records.

o. If an individual is selected for membership on a higher level squad eligible to participate in the National Trophy Matches, he will be placed on further TDY by the commander who selected him.

p. Each enlisted man scheduled for a proficiency pay test must have a completed DA Form 2166-4 (Enlisted Efficiency Report) in his 201 file.

q. Any person alerted for oversea shipment shall have an approved deferment, covering him through the National Matches, in his 201 file when reporting for the US Army championships.

4. TYPES OF COMPETITIONS.

a. Combat-type matches.

(1) The Combat Rifle Match will be used for competition through installation/division level (Appendix C).

(2) The Precision Combat Rifle Match (M16A1) combined with the National Match Rifle Course (M14) will be conducted at CONUS Army matches and the US Army Championship (Appendix D).

(3) The Combat Pistol Qualification Course (paragraph 58-67, FM 23-25) or the Combat Pistol Match, Appendix E, will be used for competition at company/battalion/brigade level.

(4) The Combat Pistol Match (appendix E) will be used from installation/division level through the US Army Championships combined with National Match Pistol Courses of fire.

(5) Competitions from company through installation, division, ARCOM, GOCOM, or State NG level will use standard-issue service rifles and pistols without alteration or modification.

b. National Match-type competitions. Rifle and pistol matches fired over the National Match Course will be in accordance with AR 920-30 and NRA Rifle/Pistol rules with the following exceptions:

(1) Weapons authorized.

(a) Competitions from company through installation, division, ARCOM, GOCOM, State NG level will use the standard issue service rifle and pistol without alteration or modification.

(b) CONUS Army and US Army championships will use the service rifle or pistol as described in AR 920-30.

(2) Ammunition. Service grade ammunition will be used from company through installation, division, ARCOM, GOCOM, State NG level. Match grade ammunition is authorized at the CONUS Army level and US Army Rifle and Pistol Championship Matches.

(3) Safety. Loading and reduction of malfunction procedures for the .45 caliber service pistol will be as specified in Appendix F.

c. Excellence-in-Competition Matches at installation, division, ARCOM, GOCOM, State NG (4 point leg match) level.

(1) This match will be fired using the M16 service rifle and the M1911 service pistol with service ammunition only. Match weapons and/or ammunition are strictly prohibited.

(2) Course of fire.

(a) Rifle. The Precision Combat Rifle Match as outlined in this supplement.

(b) Pistol. The Combat Pistol Match as outlined in this supplement.

(3) Personnel with "distinguished" credit using the same type weapons will not participate in this match. The purpose of this match is to provide an opportunity for the soldier, using his individual weapon, to enter the competitive program and to stimulate interest in marksmanship at the unit level through use of individual weapons.

(4) Deviation from these requirements will be made only upon approval of HQ FORSCOM. Requests for deviation will be forwarded to HQ Forscom, ATTN: AFOP-TS.

d. NRA procedures for challenges and protests will apply. Fees collected for challenges not sustained will be donated to central post and major command welfare funds for use in connection with marksmanship activities.

5. AWARDS.

a. Awards and prizes for major commands/CONUS Army competitions will be established locally. Awards for the US Army Championship Matches will be established by Commander, USAIC.

b. Commanders at all levels are encouraged to present appropriate individual and team awards.

c. A system of classification in individual matches shall be established for major commands/CONUS Army competitions to promote the interest of the novice who is competing for the first time at that level.

d. Competitors will be classified for individual matches at the US Army Championships as follows: (For team classification upon which eligibility of squad membership is based, see paragraph 3f, above.)

(1) Open Class - Competitors who have previously fired with the same arm in a US Army Championship or the National Matches.

(2) Novice Class - Competitors who have not previously fired with the same arm in a US Army Championship or the National Matches.

f. Closing ceremonies will be held for presenting awards to recognize the achievements of participants and the contribution of support personnel.

6. COMPETITIVE MATCH PROGRAMS. Match programs for competition will be as follows:

    a. Company through brigade/installation level.

        (1) One individual combat rifle or precision combat rifle match with accompanying team match.

        (2) One individual combat pistol qualification course as prescribed in paragraph 59-67, FM 23-35, Pistols and Revolvers, or one each of individual and team combat pistol match as described in Appendix E.

    b. The program for major commands (CONUSA) and the US Army rifle and pistol championships will be for one week. The schedule will be as follows:

MAJOR COMMAND AND ALL ARMY FIRING SCHEDULE
PISTOL MATCHES

MONDAY:	.22 caliber individual 900 aggregate and team match, national match course.
TUESDAY:	.45 caliber 900 aggregate and team match, national match course.*
WEDNESDAY:	Service pistol individual 900 aggregate and team match, national match course.
THURSDAY:	Individual combat pistol and combat pistol team match.
FRIDAY:	Excellence-in-Competition Match; awards ceremony.

*Center fire may be substituted for the .45 caliber match at the annual All-Army championship matches. All team matches will be fired using the new shooter rule. Individual aggregate award will encompass all individual matches and comprise a 3600 aggregate. Team aggregate will encompass all team matches and comprise a 1200-point aggregate. The commander's trophy will be presented to the winning Service Pistol team.

RIFLE MATCHES

MONDAY:	M16 rifle individual precision combat rifle match.*
	M16 rifle precision combat rifle team match (fired by team members designated prior to starting the match).
TUESDAY:	M14 rifle individual off hand-200 yards-20 shots.
	M14 rifle individual rapid fire-200 yards-2 strings.
	M14 rifle individual rapid fire-300 yards-2 strings.
WEDNESDAY:	M14 rifle individual slow fire-600 yards-40 shots.
THURSDAY:	M14 rifle team match; national match course.
FRIDAY:	M14 rifle individual Excellence-in-Competition Match; awards ceremony.

*M16 rifles will be furnished at match site by host installations; issued on firing line and returned after firing. Rifles will be "as issued" and no modifications permitted. The M16 match will not be included in aggregate matches.

All team matches will be fired using the new shooter rule. The individual aggregate will not include M16 matches or the excellence-in-competition matches and will comprise 1000 points. The commander's trophy will be presented for the winning M14 rifle team.

7. WEAPONS AND EQUIPMENT SECURITY. Personnel in possession of Government-issued weapons and equipment will comply with the provisions of AR 190-11.

APPENDIX C

COMBAT RIFLE MATCH

1. GENERAL. The Combat Rifle Match fired with the M16 rifle is designed to increase the proficiency of the individual soldier. The Combat Rifle Match or the Precision Combat Rifle Match will be used for competition from company level through installation/division level in all service rifle competition. This match is a superior marksmanship training vehicle and is highly recommended for training and familiarization.

2. AWARDS.

    a. Commanders are encouraged to present appropriate individual and team awards. Separate awards for stages of the match are not recommended.

    b. CONUSA commands will prepare official bulletins announcing the top 10 teams and the top 10 individuals from their command matches. Three copies of the bulletin will be submitted to HQ FORSCOM not later than 30 April of each year (RCS exempt report, paragraph 7-2p, AR 335-15). As a minimum, the bulletin will indicate the stage scores and bonus points for the individuals and teams named in the bulletin. HQ FORSCOM will announce the top three teams and individuals from each major command. Scores for teams or individuals that were not fired on the standard course described herein will be indicated by an asterisk.

3. TEAM COMPOSITION AND ELIGIBILITY.

    a. A team is designed to approximate the rifle firepower of a squad which will consist of six firing members, two alternates, one team leader, and one assistant team leader.

    b. One half of the team firing members must be new shooters. All members will be eligible to compete in the individual portion of the matches in which the team is entered.

4. MATCH CONDITIONS.

    a. Clothing and equipment. Only clothing and equipment normally associated with combat/field duty will be used by the competitors. The match director will prescribe a field uniform appropriate to the season and will insure compliance with the following:

        (1) A steel helmet with liner will be worn by all participants. The chin strap will be fastened.

        (2) Load-bearing equipment will be worn, to include a pistol belt with suspenders; full canteen with a cup and carrier; first aid packet and pouch; magazine pouch; and, if issued, a bayonet.

        (3) Issue binoculars, not exceeding 7 X 50 power, may be used by the team leader and assistant team leader to direct firing during team matches.

        (4) The team leader and assistant team leader will wear the equipment required of firers and will carry a rifle, unloaded and locked.

        (5) No individual or team member will be allowed to fire unless the equipment prescribed above is worn or carried in the manner intended. Failure to meet this requirement will result in disqualification of the individual for the match involved.

            (a) An individual discovered to be incompletely equipped will be cleared from the firing line and will not be allowed to continue until his equipment has been retrieved.

            (b) Equipment may be recovered by the team leader or assistant team leader; or an exchange of equipment made with the individual and the dropped equipment recovered by the team leader or assistant team leader.

    b. Weapons. The only rifles used in the competition will be the standard issue service rifle (M16A1 or M14) as described in TM 9-1005-223-series, and TM 9-1005-249-14.

    c. Ammunition. Only issue service-grade ammunition may be used.

        (1) Ammunition for individual matches will be issued by the umpire to the firers during the preparation period for each stage of the match.

        (2) In team matches, 300 rounds of ammunition will be issued to the team leader to be issued to the members of his team. An additional 60 rounds will be given the team leader for use in stages 2 and 4 of the team match.

    d. Positions.

        (1) The firing positions used in the course of the matches will be as specified in FM 23-71, unsupported. The use of the hasty sling is optional.

        (2) Prior to any movement prescribed in the match, the firer must have his abdomen and one elbow touching the ground. The rifle will be held at the balance when carried during prescribed movements in the match.

    e. Safety.

        (1) Appropriate range safety rules and regulations will be followed.

        (2) During stages of fire requiring movement, rifles will be unloaded, locked, and the muzzle of the weapon will be pointed down range. The rifle can be loaded and unlocked only when the firer has assumed the proper firing position on the firing line.

        (3) Personnel assigned the duty of umpire for the match will, at the direction of the range officer or his assistant, check competitors for safety and completeness of uniform and equipment; record scores; and, in individual competitions, issue ammunition as required for each stage of the match. One umpire will be assigned to each firer, both in individual and team competition. In addition, in team events there will be an extra umpire assigned for each team to keep the team score. Firers may be used as umpires; however, if a firer must subsequently participate in the match, he should be given time to prepare himself for firing.

(4) Participants shall be subject to disqualification for violation of safety rules.

(5) A participant shall be disqualified for failure to follow the prescribed course of fire.

(6) If a competitor falls during the matches and the umpire has no doubt that the bore of the rifle is still clear, the firer may continue. If, however, the umpire has any doubt about the condition of the bore, he will halt the firer and inspect the rifle to insure that the bore is clear before allowing the firer to continue. In team matches, the firer may use a leader's weapon if his own cannot be used.

5. CONDUCT OF THE MATCH.

    a. General. The Combat Rifle Match is composed of four stages of fire: slow fire, rapid fire, quick fire, and fire and movement. The maximum possible score for the individual competition is 250 points; for a team - 2056 points.

    b. Organization of the range. The known-distance range will be used and organized in a manner that will allow the stages of the match to be fired consecutively and with minimum delay. Equipment such as range towers, benches, rifle racks, and ammunition tables, which would be an obstacle to rapid movement from one firing line to another, will not be situated forward (down range) of the 350-yard line.

    c. Targets. In stages 1, 2, and 4, the standard D prone silhouette target (FSN 6920-922-7450) is used with modification; the present white background will be changed to olive drab using locally available resources. In stage 3, the F prone silhouette target is used.

    d. Individual matches. When the course is fired as an individual match, the firers will be squadded to allow a minimum of one unused (unexposed) target between each firer.

    e. Stages of fire.

    (1) Stage 1. Two sighter rounds and 10 rounds for record, slow fire, one minute for each shot, from the prone position at 400 yards. The range officer will call each relay to the firing line and allow the firers one minute to prepare for each relay of the match by commanding: "RELAY_____ TO THE FIRING LINE FOR STAGE 1 OF THE COMBAT RIFLE MATCH. YOUR ONE-MINUTE PREPARATION PERIOD STARTS NOW." The targets will be raised at this time. At the end of the preparation period, the targets will be lowered. The range officer will determine that the firing line is ready and will start the firing by commanding: "YOUR PREPARATION PERIOD HAS ENDED. WITH A MAGAZINE OF 12 ROUNDS LOCK AND LOAD, YOU WILL HAVE 12 MINUTES TO FIRE 12 ROUNDS SLOW FIRE FROM THE PRONE POSITION. IS THE LINE READY? THE LINE IS READY. UNLOCK. YOU MAY COMMENCE FIRING WHEN YOUR TARGET APPEARS." The targets will be raised. After each shot, the target will be pulled and marked with a spotter. The target will be raised and the value of the shot will be disked. The first two shots fired will be sighting shots and will not count in the total score. When the firer completes firing, he will remove the magazine, the bolt will be left open, and the rifle will be locked. When the last firer has finished or 12 minutes have expired, the range officer will command: "CEASE FIRE. CLEAR ALL WEAPONS. (Umpires will clear the weapons.) IS THE LINE CLEAR? THE LINE IS CLEAR." The total score possible is 50 points.

    (2) Stage 2. This will be a movement and fire exercise of 10 rounds in 75 seconds. Starting from the prone position, at 400 yards as the targets appear, the firer must rise, move to the 300-yard line, assume the prone position, load, unlock, and fire 10 rounds within the time limit. The range officer will allow the firer one minute to prepare for the exercise at the 400-yard line by commanding: "RELAY_____ PREPARE FOR STAGE 2 OF THE COMBAT RIFLE MATCH. YOUR ONE-MINUTE PREPARATION STARTS NOW." The targets will not be raised during the preparation period. At the completion of the preparation period, the range officer will determine that the line is ready and will start the firing by commanding: "YOUR PREPARATION PERIOD HAS ENDED. THIS EXERCISE IS 10 ROUNDS RAPID FIRE IN 75 SECONDS FROM THE PRONE POSITION AT 300 YARDS. IS THE LINE READY? THE LINE IS READY. YOUR WEAPON WILL NOT BE LOADED UNTIL YOU HAVE ASSUMED THE PRONE POSITION AT THE 300-YARD LINE. YOU MAY RISE AND MOVE FORWARD WHEN YOUR TARGET APPEARS." As the targets appear, the firers will rise, move to the 300-yard line, assume the prone position, load, unlock, and fire 10 rounds. The targets will remain up for 75 seconds, then they will be pulled and marked with spotters. As the targets are pulled,

the range officer will command: "CEASE FIRE, CLEAR ALL WEAPONS, IS THE LINE CLEAR? (Umpires will clear the weapons.) THE LINE IS CLEAR. STAND BY FOR YOUR SCORE." When the targets are ready for scoring, they will be raised and disked; after the range officer determines that scoring is complete, he will command: "REMAIN IN PLACE FOR STAGE 3." The total possible score is 50 points.

(3) Stage 3. This will be 10 rounds for record, quick fire, 5 points for each hit, 2 rounds for each 6-second exposure, fired from the prone position at 300 yards. The range officer will allow the firer one minute to prepare for the exercise by commanding: "RELAY_____PREPARE FOR STAGE 3 OF THE COMBAT RIFLE MATCH. YOUR PREPARATION PERIOD STARTS NOW." The target used will be the F prone silhouette attached to a pole raised and lowered by the pit personnel. It will not be raised during the preparation period. The targets used in other stages of the match will be kept in the pits and will not be exposed during this exercise. At the end of the preparation period, the range officer will determine that the line is ready and will start the firing by commanding: "YOUR PREPARATION PERIOD HAS ENDED. WITH A MAGAZINE OF 10 ROUNDS, LOCK AND LOAD." The range officer will then command: "THIS EXERCISE IS FOR RECORD, FIVE 6-SECOND EXPOSURES, FIRING TWO ROUNDS FOR EACH EXPOSURE. UNLOCK. WATCH YOUR TARGET." The targets will be raised for 6 seconds and lowered for 6 seconds. After the fifth exposure, the range officer will command: "CEASE FIRE. CLEAR ALL WEAPONS. (Umpires will clear the weapons.) IS THE LINE CLEAR? THE LINE IS CLEAR. STAND BY FOR YOUR SCORE." Spotters will be placed in each hit on the target and the line will be informed that the targets are ready for scoring. The range officer will command: "WATCH YOUR TARGET TO RECEIVE YOUR SCORE." At this time the targets will be raised for 30 seconds to display the targets to the firers and then lowered, then either a white disk will be raised above the butts, once for each hit on the target, or the score will be called to the firing line by telephone. When scoring is complete, the range officer will command: "RELAY_____MOVE BACK TO THE 450-YARD LINE FOR STAGE 4 OF THE COMBAT RIFLE MATCH." The total possible score is 50 points.

(4) Stage 4. This will be an assault firing exercise requiring movement from the 450-yard line to the 100-yard firing line; 20 rounds will be fired in the exercise; 2 rounds from the prone position at 400 yards; 8 rounds from the sitting or squatting position at 300 yards, and 5 rounds each from the kneeling position at 200 yards and the standing position at 100 yards. Targets will remain up for a 30-second exposure for movement to and firing from the 400-yard line. They will be withdrawn for 15-second intervals between exposures thereafter. The target will be exposed for 65 seconds for the remainder of the exercise in Stage 4. All movement must start from the prone position. The range officer will move each relay to the starting line and allow the firer one minute to prepare for the exercise by the command: "RELAY_____TO THE STARTING LINE FOR STAGE 4 OF THE COMBAT RIFLE MATCH. YOUR ONE-MINUTE PREPARATION PERIOD STARTS NOW." The targets will not be raised. At the completion of the preparation period, the range officer will determine that the line is ready and will give the firing commands: "YOUR PREPARATION PERIOD HAS ENDED. WEAPONS WILL NOT BE LOADED DURING MOVEMENT. WHEN YOUR TARGET APPEARS, YOU WILL HAVE 30 SECONDS TO RISE AND MOVE TO THE 400-YARD LINE, ASSUME THE PRONE POSITION, LOAD, UNLOCK YOUR WEAPON, AND FIRE TWO ROUNDS. WHEN YOUR TARGET DISAPPEARS, LOCK YOUR WEAPON, UNLOAD, REMOVE THE MAGAZINE, CLEAR YOUR WEAPON, THEN PREPARE TO RISE AND MOVE TO THE 300-YARD LINE. WHEN YOUR TARGET APPEARS, YOU WILL HAVE 65 SECONDS TO RISE AND MOVE TO THE 300-YARD LINE, ASSUME A SITTING OR SQUATTING POSITION, LOAD, UNLOCK, AND FIRE EIGHT ROUNDS. WHEN YOUR TARGET DISAPPEARS, LOCK YOUR WEAPON, UNLOAD, REMOVE THE MAGAZINE, CLEAR YOUR WEAPON, ASSUME THE PRONE POSITION, AND PREPARE TO MOVE TO THE 200-YARD LINE. WHEN YOUR TARGET APPEARS, YOU WILL HAVE 65 SECONDS TO RISE AND MOVE TO THE 200-YARD LINE, ASSUME THE KNEELING POSITION, LOAD, UNLOCK, AND FIRE FIVE ROUNDS. WHEN YOUR TARGET DISAPPEARS, LOCK YOUR WEAPON, UNLOAD, REMOVE THE MAGAZINE, CLEAR YOUR WEAPON, ASSUME THE PRONE POSITION, AND PREPARE TO RISE AND MOVE TO THE 100-YARD LINE. WHEN YOUR TARGET APPEARS, YOU WILL HAVE 65 SECONDS TO RISE AND MOVE TO THE 100-YARD LINE, ASSUME THE STANDING POSITION, LOAD, UNLOCK, AND FIRE FIVE ROUNDS. WHEN THE TARGET DISAPPEARS, LOCK YOUR WEAPON, REMOVE THE MAGAZINE, AND CLEAR YOUR WEAPON. IS THE LINE READY? THE LINE IS READY. WATCH YOUR TARGET." As the targets appear, the firers will start the exercise as specified in the commands. Umpires must be particularly alert to insure that all movement is made with rifles locked and with the muzzle pointed down

range. Any firer not moving out at any particular stage as the targets appear will be disqualified. Strict adherence to safety is required. At the completion of the 100-yard stage, the range officer will command: "CEASE FIRE, CLEAR ALL WEAPONS. IS THE LINE CLEAR? (Weapons will be checked by an umpire.) THE LINE IS CLEAR. STAND BY FOR YOUR SCORE." When the targets are marked by spotters and ready for disking, the line will be informed and the range officer will command: "WATCH YOUR TARGETS." The targets will be scored. After disking and scoring is completed, he will command: "IS THE SCORING COMPLETE? SLING YOUR RIFLES AND MOVE TO THE REAR OF THE RANGE." The senior umpire will form and march the firers and umpires to the rear of or off the range. The total possible score is 100 points.

 f. Team matches. The team (composed as required by a above) will fire the match described above for individual matches, with the following exceptions:

  (1) A team will fire as a unit on the line. When the number of teams in a match exceeds the range capacity, the teams will draw for the relay on which they will fire. In no case will teams representing the same competing command fire on adjacent targets.

  (2) In team matches, the range will be organized to provide banks of eight targets, with each bank separated by an unexposed target. In Stages 1 and 3, only the six middle targets of each bank will be exposed and in Stages 2 and 4, all eight targets in the bank will be exposed.

  (3) The firing members of the team may take positions on the line, in the space allotted their team, without regard to target numbers on the line. This is to allow the leaders of the team to achieve better control and assist the team.

  (4) Sighting rounds will be fired by each team member as in the individual match.

  (5) Prior to commencement of Stages 2 and 4 of the team match, the team leader may distribute, as he deems appropriate, the extra 60 rounds of team ammunition provided in order that a firer may engage more than one target and the team may thus take advantage of scores and bonuses allowed by hits on the additional targets in the team's assigned bank.

  (6) The team leader will distribute his team's ammunition. The team leader and his assistant may help the firing members by giving commands, providing sight changes, assisting with dropped equipment, and redistributing ammunition. In the interest of safety, the team leader and his assistant may not touch a firer during the firing, nor may they advance forward (down range) of an imaginary line created by the butts of the firer's rifles.

  (7) In Stage 4 of the match, each of the six firers must fire at least two rounds from the 400-yard line, eight rounds from the 300-yard line, five rounds from the 200-yard line, and five rounds from the 100-yard line.

  (8) No outside assistance may be given to the team during the conduct of the match.

 g. Match program. Company through brigade level championships should be a course of fire consisting of a minimum of one firing of an individual match and one team match. Installation/division level championships will consist of at least three individual matches and one team match.

6. SCORING.

 a. The score recorded by the umpire on the line will become the official score after it has been signed by the individual firer, the team leader, and the umpire. In case of improperly totaled scores, the statistical officer may change the total score based upon the recorded individual shots.

 b. Scores will be signaled by disking from the pits to indicate the value of each shot for Stage 1, and by telephone or disking in Stages 2, 3, and 4. If the score is questioned or challenged, the score given by telephone or radio will be considered official.

 c. In Stages 1, 2, and 3 of the individual matches, only the 10 hits of highest value will be counted on each target. In Stage 4 of the individual match, only the 20 hits of highest value will be counted.

  (1) If during individual matches an individual fires fewer than the prescribed number of shots and there are more hits on his target than shots fired, he will only be scored the number of shots of highest value equal to the number of shots fired.

(2) In individual matches, no credit will be given for shots fired on the wrong targets.

d. On the D prone silhouette target, if the edge of a shot hole comes in contact with the outside of a higher value scoring ring, the value of the higher scoring ring will be scored. On the F prone silhouette target (Stage 3), any hit or nick in the edge shall be counted toward the total score allowed.

e. All shots fired by an individual after he has taken his position on the firing line will be counted in the score. Shots fired before permission to fire has been given or after the command "CEASE FIRE" will be scored as misses.

f. Bonus points will be allowed in team competition as follows:

(1) Stages 1 and 3: No bonus.
(2) Stage 2: A bonus will be awarded for those targets with a score of 35 or more points.
(3) Stage 4: A bonus will be awarded for those targets with a score of 70 or more points.
(4) The bonus will be computed separately for Stages 2 and 4. For each of the stages, the sum of the square of the number of targets with 35 or more points (Stage 2) or 70 or more points (Stage 4) will be doubled and added to the total fired score of the team for that stage. (For example, a team firing Stage 2 has seven targets with a score of 35 or more. The bonus is computed as follows: 7x7x2=98 bonus points.) The maximum bonus allowable at each stage is 128 points.
(5) If a team has qualified for the maximum bonus (128 points), it will be awarded the value of all hits in excess of 10 that are on the team's bank of targets. (Note to scorer: When a team has achieved the maximum bonus, the range officer will be notified immediately so that the pits can ascertain the value of any excess hits for addition to the team score).

g. Ties will be broken in both individual and team matches in the following order:
(1) By the highest score in Stage 4.
(2) By the highest score in Stage 3.
(3) By the highest score in Stage 2.
(4) By the highest score in Stage 1.

h. To break ties within a stage, NRA High-power Rifle Rule 15.4(c)(1), (c)(2), (c)(3), and (c)(4) will be applied.

i. Ties not resolved by procedures outlined in g and h above will be resolved by NRA High-power Rifle Rule 15.13(b).

j. Team matches will be scored by recording the scores in individual cards in Stages 1 and 3 and transferring these scores to the team card. In Stages 2 and 4, scores will be disked for each target; however, the score for each target will be telephoned or radioed to the team umpire, announced by him to the team leader, and entered on the team score card. Formats for the score cards, to be locally reproduced on 5 by 8-inch cards, are at Figs. 14-1 and 14-2.

7. MISFIRES, STOPPAGES, AND ALIBIS.

a. No refires will be authorized for stoppage due to faulty ammunition, rifles, or competitor errors.

b. A refire (range alibi) may be authorized in case of faulty target operation or improper range management if, in the judgment of the chief range officer, the firer was penalized by the failure, and the firer concerned protests the faulty procedures before the target is exposed for scoring. When an alibi has been allowed, the protested target and/or score will not be shown or indicated to the firer.

SCORECARD -- INDIVIDUAL COMBAT RIFLE COMPETITION												
NAME			GRADE	SSN		ORGANIZATION				RELAY		TARGET
STAGE 1												TOTAL
	SS	1	2	3	4	5	6	7	8	9	10	
STAGE 2												TOTAL
STAGE 3	Number of hits ________ X 5											TOTAL
STAGE 4												TOTAL
		1 / 11	2 / 12	3 / 13	4 / 14	5 / 15	6 / 16	7 / 17	8 / 18	9 / 19	10 / 20	
SIGNATURE OF COMPETITOR						SIGNATURE OF UMPIRE (SCORER)					MATCH TOTAL	

FORSCOM Form 246-R Jul 73 CONARC Edition of Feb 72 may be used until exhausted.

Figure 14-1

SCORECARD -- COMBAT RIFLE TEAM COMPETITION										
NAME OF TEAM					TEAM LEADER			ASST TEAM LEADER		
	TGT NO	TGT NO	TGT NO	TGT NO	TGT NO	TGT NO	TGT NO	TGT NO		TOTALS
STAGE 1	■									
STAGE 2										
BONUS STAGE 2	NO OF TARGETS WITH 35 OR MORE POINTS ___X___=___X 2 = (EXAMPLE: 7 X 7 = 49 X 2 = 98)										
STAGE 3	■							■		
STAGE 4										
BONUS STAGE 4	NO OF TARGETS WITH 70 OR MORE POINTS ___X___=___X 2 = (EXAMPLE: 7 X 7 = 49 X 2 = 98)										
SIGNATURE OF TEAM LEADER				SIGNATURE OF UMPIRE (SCORER)				TEAM TOTAL		

FORSCOM Form 247-R CONARC Edition of Feb 72 may be used until exhausted
Jul 73

NAME OF TEAM			ORGANIZATION		
TEAM MEMBERS	GRADE	SSN	MILITARY ADDRESS	HOME ADDRESS	
TEAM LEADER					
ASST TEAM LEADER					
1.					
2.					
3.					
4.					
5					
6.					
ALT					
ALT					
DATE	PRINTED NAME OF TEAM LEADER			SIGNATURE OF TEAM LEADER	

(Reverse Side)

Figure 14-2

APPENDIX D

PRECISION COMBAT RIFLE MATCH

1. GENERAL.
    a. The precision Combat Rifle Match fired with the M-16 rifle is designed to increase the proficiency of the individual soldier. The Precision Combat Rifle Match may be used for competition from company level through installation/division level in all service rifle competition. This match is a superior marksmanship training vehicle and is highly recommended for training and familiarization. The Precision Combat Rifle Match is the prescribed course of fire for the installation/division/ARCOM/GOCOM/State NG level Excellence-in-Competition Match.
    b. All rules and regulations that pertain to the National Match type competition will govern the Precision Combat Rifle Course (AR 920-30 and current NRA Rules).
2. TEAM COMPOSITION.
    a. Teams will consist of not more than a team captain, a team coach, six firing members, and, if desired, two alternates.
    b. Team captain and coach may be firing members.
3. MATCH CONDITIONS.
    a. Uniform. The uniform for major command/CONUSA and US Army rifle and pistol championships will be the duty uniform.
    b. Weapons. The only rifle used in this competition will be the standard issue M16A1 service with web or leather sling optional, as described in TM 9-1005-249-14.
    c. Ammunition. As authorized by CTA 23-101.
    d. Positions. As prescribed in current NRA rules.
    e. Safety. Appropriate range safety rules and regulations will be followed.
4. CONDUCT OF MATCH.
    a. General. The Precision Combat Rifle Match is fired in four stages: sustained fire - 100 yards, rapid fire - 200 yards, rapid fire - 300 yards, and slow fire - 400 yards. The maximum possible score for individual competition is 250 points - and for a team, 1500 points.
    b. Targets. The current Rifle Target "D" FSN 6920-922-7450 with replacement center FSN 6920-922-7541.
    c. Stages of fire.
        (1) Stage 1. Fired from the 100-yard line. Two sighters (to be fired during the three-minute preparation period) and 10 rounds sustained fire, standing position, two-minute time limit, possible score - 50 points.
        (2) Stage 2. Fired from the 200-yard line. Ten rounds rapid fire from the standing to the sitting position, 50-second time limit, possible score - 50 points.
        (3) Stage 3. To be fired from the 300-yard line. Ten rounds rapid fire from the standing to the prone position, 60-second time limit, possible score - 50 points.
        (4) Stage 4. To be fired from the 400-yard line. Two sighters and 20 rounds slow fire, prone position, 22-minute time limit, possible score - 100 points.
    d. Team matches.
        (1) Course of fire. Same as for individual match per firing member except the 100 through 300-yard stages will be fired individually and by relays, with assistance of the coach as in National Trophy Rifle Team Matches. The fourth stage, 400 yards, will be fired in pairs.
        (2) Two sighter rounds will be fired by each firing member of the team at the 100-yard line only. No sighter rounds are authorized for the remainder of the match.
    e. Scoring. Targets will be scored in the same manner as the National Match Course (i.e., slow fire stages, targets will be pulled, marked, and scored after each shot); rapid fire at the completion of a 10-round string.

APPENDIX E

COMBAT PISTOL MATCH

1. GENERAL.
    a. The Combat Pistol Match fired with the service pistol is designed to increase the proficiency of the individual soldier. The Combat Pistol Match will be used for

competition from company through installation/division level in all service pistol competition. This match is a superior pistol marksmanship training vehicle and is highly recommended in training and familiarization. The Combat Pistol Match is the only course of fire authorized for the installation/division/ARCOM/GOCOM/State NG level Excellence-in-Competition Match.

    b. Competition at the company, battalion, and brigade levels may be conducted over the Combat-Pistol Qualification Course as prescribed in paragraph 59-67, FM 23-35, or the Combat-Pistol Match Course as described herein.

2. TEAM COMPOSITION. Pistol teams entering team matches will be composed of four firing members, a coach, and a team captain and, if desired, two alternates. The coach and team captain may be alternate firing members of the team. At least two of the individual firing members must not have previously fired at that level of competition. All team members will be eligible to compete in the individual portions of the matches.

3. MATCH CONDITIONS.
    a. General. In events not covered by these rules the NRA Pistol Rules are suggested as a basis for decision.
    b. Clothing and equipment.
        (1) Clothing and equipment normally associated with combat/field duty will be worn by each member of the team through major command competitions. The match director will prescribe a field uniform appropriate to the season and will insure compliance with the following:
            (a) Issue binoculars, not exceeding 7 x 50 power, may be used by the team captain and coach to direct fire during team matches.
            (b) The team captain and coach will wear the equipment required of firers.
        (2) Participation in the US Army championships will be in duty uniform.
    c. Weapons.
        (1) Company through installation/division. The only caliber .45 pistol allowable is that listed under LIN N96741, SB 700-20. Internal or external modification or alteration of parts is not permitted.
        (2) Major command and US Army championships. National Match Grade caliber .45 pistol listed under LIN N96878, CTA 60-18, or the same type and caliber of commercially manufactured pistol as described in AR 920-30 is authorized.
    d. Ammunition. .45 caliber service grade ammunition will be used through installation/division level. Match grade ammunition is authorized at the FORSCOM Army/major command and US Army Championship Matches.
    e. Target. Target, Silhouette: Paper, Standing (FSN 6920-713-5385) which may be trimmed and mounted on a Target, Silhouette, Pasteboard, E, Kneeling (FSN 6920-795-1806) (Fig E-1). When a full silhouette is used, the Target, Silhouette, Pasteboard, Trapezoidal (FSN 6920-600-6881) is added to the bottom of the E-type silhouette. Repair centers with an "X" ring center may be used. Target, Indoor Silhouette, B-24, NRA approved for 50 feet gallery shooting may be used as a substitute only in the US Army Pistol Championships. Target must be pasted to Standard American target backing for use on standard 25-yard NRA National Match Course Pistol Range (Turning Targets).
    f. Range. 25 yards.
    g. Timing. When rotating targets are used, time is measured from the time the target starts to move to face the shooter until it starts to move to the edge position.
    h. Positions. The STANDING, PRONE, KNEELING, and CROUCH positions used in this course of fire will be as follows:
        (1) PRONE: The firer will assume the prone position on command before the command "LOAD" is given and will start firing when the targets face. The shooting hand must not touch the ground or rest on the nonshooting hand that is touching the ground. The firer may use a ground cover that renders no artificial support.
        (2) KNEELING: In the kneeling position, either knee may be grounded for support but not both knees simultaneously. Firing with a two-handed support is permitted. The position of the foot used as main support of body weight, i.e., under the buttocks, must be vertical with the body weight resting on heel and toes.
        (3) CROUCH: In the crouch position, the firer must face the target with the feet parallel to the target line with the feet placed in relation to each other not exceeding half the length of the firer's foot to the front or rear. Firing with a two-handed support is permitted provided a crouching position low enough to place the nonshooting hand on the knee is used.

(4) STANDING: In the standing position, the firer may either assume the stance permitted by NRA Pistol Rules for participating in National Match type competition or may face the target using the two-handed supported stance.
    (5) READY POSITION: The firer is in a STANDING POSITION with the pistol held in one hand with the mechanism cocked and his arm stretched downward so that his arm forms a maximum angle of 45 degrees to the vertical or so that the pistol or his hand touches the bench. The pistol must not be raised until the COMMENCE FIRING signal is heard or the targets start to move to face the firer.
NOTE: The two handed grip is not used in the READY POSITION.
  i. Refires.
    (1) In the event of a defective cartridge, disabled pistol, or malfunction before a string is completed, the firer will not attempt to clear the malfunction but will fire another five-shot string, except in the PRONE POSITION where he will be allowed 15 seconds per unfired shot, provided he signals the range officer by holding up his hand at the end of the time period. Failure of the firer to notify the range officer of the malfunction, or touching the pistol in an attempt to reduce the malfunction before the range officer inspects the pistol, forfeits the right to refire.
    (2) Two refires per table will be authorized due to a defective cartridge, disabled pistol, or malfunction. However, two consecutive refires will not be allowed in a 10-shot string or stage of fire. The firer will not attempt to reduce a failure to fire in a refire string.
  j. Scoring.
    (1) All shots fired by the firer count and are considered in his score, even if his pistol is not directed toward the target or is accidentally discharged.
    (2) All shots fired on the wrong target shall be scored as misses.
    (3) If the firer should have excessive hits on his target, he may elect to be scored the value of the lowest hits or refire that stage of the course.
    (4) Unfired shots in the refire string shall be scored as misses.
  k. Decisions of ties. When two or more firers have the same numerical score, ties will be ranked in the order listed below:
    (1) By the greatest number of hits.
    (2) By the greatest number of X's (If X ring is used).
    (3) By the greatest number of 10's.
    (4) By the highest total score in Table 3.
    (5) By the highest score in the 10-second stage of Table 3; if still tied, the highest score in the 15-second stage.
    (6) By the highest total score in Table 2; if still tied, the highest last 10-shot string in Table 2.
  l. Penalties.
    (1) A firer in the READY POSITION who raises his arm too early or does not lower it sufficiently shall first be warned that his position is incorrect by the range officer. The firer may be penalized the value of the string fired should the fault be repeated.
    (2) Unsafe handling of the pistol will be grounds for disqualification.
    (3) No individual or team member will be allowed to fire unless the equipment prescribed in b, above, is worn in the manner intended. If equipment is lost by the firer during the conduct of the course, he will not be allowed to fire until the equipment is replaced in its proper position without delaying the match or creating a safety hazard. Failure to meet this requirement will result in disqualification of the individual for the match involved.
  m. Safety.
    (1) Appropriate range safety rules and regulations will be followed.
    (2) Loading and reduction of malfunction procedures for the .45 caliber service pistol will be as specified in Appendix F.
4. CONDUCT OF THE MATCH.
  a. General.
    (1) The Combat Pistol Match is divided into three tables with 30 rounds fired in each table. Ninety rounds are fired in individual competition with the maximum score of 900 points. One hundred and twenty rounds are fired in 4-man team competition with the maximum score of 1200 points.

(2) Firing may be conducted so that each firer completes each table before starting the next table or completes the entire course while on the line.

(3) Team members shall be squadded so that they will fire as a unit on the line. In the event that all teams cannot fire at the same time, relays will be assigned so that two members of each team will fire together. Team captain/coach may direct the team's firing.

(4) The Team Match will be an abbreviated course, i.e., 10 shots prone position, 10 shots duel stage and 10 shots military rapid fire (with each 5 shots in 10 seconds) for a total of 30 shots.

  b. Course of Fire.

(1) Table 1 - Combat Position Firing. This table tests the firer's ability to engage a target with accurate fire from a position he may use in combat: prone, kneeling, or crouch. Thirty shots, 10 shots fired from each position.

(2) Table 2 - Duel. This table tests the firer's ability to rapidly engage a target that is exposed for a limited period of time. Thirty shots total.

(3) Table 3 - Military Rapid Fire. This table tests the firer's ability to rapidly engage a target and place accurate sustained fire on the target under decreasing time limits. Thirty shots total.

  c. Firing line commands. Firers will be called to the firing line and be given a 3-minute preparation period. At the conclusion of the preparation period the following commands will be given to control the conduct of firing:

(1) "THE PREPARATION PERIOD HAS ENDED."

(2) "THIS WILL BE THE_____STAGE OF THE COMBAT PISTOL MATCH. TAKE YOUR POSITION ON THE FIRING LINE."

(3) "WITH 5 ROUNDS LOAD."

(4) "IS THE LINE READY?" (Firers must assume the READY POSITION on this command.)

(5) "THE LINE IS READY." (This command will not be given until the firers assume the READY POSITION)

(6) The targets will be exposed or the signal to "COMMENCE FIRING" will be given in approximately 5 seconds after the command "THE LINE IS READY."

(7) "CEASE FIRING - MAGAZINES OUT - SLIDES BACK - CLEAR ALL WEAPONS." (In Table 1, PRONE POSITION, the firer is commanded, "FIRERS RISE."

  d. Firing tables

(1) Table E-1 - Combat Position Firing

POSITION	DESCRIPTION
PRONE	Ten shots fired in two 5-shot strings, reloading second magazine of five rounds without command, 2½ minutes per 10 shots. Firer loads on command while in the prone position and starts firing when the targets face. The firer will remain in the firing position until commanded to rise. CAUTION: The firer must extend the arms far enough to prevent the pistol slide from striking the face during recoil.
KNEELING	Ten shots fired in two 5-shot strings, 20 seconds per string. Firer starts in the READY POSITION and assumes the kneeling position when the targets face. CAUTION: The arms must be extended far enough to prevent the slide from striking the firer in the face during recoil.
CROUCH	Ten shots fired in two 5-shot strings, 20 seconds per string. Firer starts in the READY POSITION and assumes the crouch position when the targets face.

NOTE: The construction of shooting tables on some competitive ranges may make the firing of Table 1 impractical from the 25-yard firing line. Table 1 may be fired from in front of the shooting tables at a distance of approximately 22 yards.

(2) Table E-2 - Duel

POSITION	DESCRIPTION
STANDING	Thirty shots fired in six 5-shot strings, 5 seconds per shot. Targets are faced 5 seconds and edged 5 seconds. Firers start in READY POSITION and shoot one shot each time target appears, returning to READY POSITION after each shot.

(3) Table E-3 - Military Rapid Fire

POSITION	DESCRIPTION
STANDING	Ten shots in two 5-shot strings, 20 seconds per string. Firer starts in the READY POSITION and shoots 5 shots at one target during the time limit.
STANDING	Ten shots in two 5-shot strings, 15 seconds per string. Firer starts in the READY POSITION and shoots 5 shots at one target during the time limit.
STANDING	Ten shots in two 5-shot strings, 10 seconds per string. Firer starts in the READY POSITION and shoots 5 shots at one target during the time limit.

APPENDIX F

LOADING AND REDUCTION OF MALFUNCTIONS

1. PROCEDURES WHEN LOADING. To prevent accidental firing of the caliber .45 service pistol when loading, the following procedures will be followed:

    a. With the slide locked to the rear, grip the pistol firmly with the right hand as when actually firing.

    b. With the trigger finger, depress the trigger (engaging the disconnector).

    c. Extend the right arm, wrist straight, elbow locked, arm muscles flexed (as when firing) to form approximately a 45-degree angle to the body.

    d. With the left hand, insert the magazine, insuring that the magazine catch is engaged.

    e. Placing the thumb on the left hand on the hammer spur, depress the hammer to it most rearward position, and while maintaining this position, with the index finger on the left hand, depress the slide catch, allowing the slide to travel forward loading a round into the chamber.

    f. Maintaining the hammer in its most rearward position, remove the left index finger from the slide catch and place it between the hammer and the firing pin.

    g. Slowly lighten the thumb pressure on the hammer, allowing it to rotate forward to the full cocked position.

    h. When the hammer stops in the full cocked position, release the trigger and remove the left index finger from its position between the hammer and firing pin. The weapon is now safely loaded and ready to fire. At this point, left handed shooters will engage the safety and changeover to the left handed mode.

    i. If when releasing the pressure on the hammer (g above) the hammer should pass the full cock notch and come to rest against the index finger, the weapon is defective. In this situation, allow the hammer to ease forward to the half cock notch, release the trigger pressure, and remove the left hand from the weapon. Notify the range safety officer by raising the left arm.

    j. Under the supervision of the safety officer, unload, clear, and remove the defective piece from the firing line.

2. MALFUNCTION. To prevent accidental firing of the .45 caliber service pistol when a malfunction occurs (failure to feed, failure to eject), the procedures below will be followed:

    a. The firer will remain in the firing position, arm and pistol extended toward the target, wrist straight, elbow locked, and arm muscles flexed.

    b. The trigger will be depressed (engaging the disconnector).

    c. The pistol will be lowered to form approximately a 45 degree angle to the body. Do not rest or lay weapon on bench.

    d. The nonfiring arm will be raised to signal the safety officer.

    e. The safety officer will supervise the unloading and clearing of the malfunction

    f. Only after the weapon has been cleared will the trigger be released and the firing hand (grip) and firing arm be relaxed.

3. PROCEDURES WHEN UNLOADING. To prevent accidental firing while unloading the caliber .45 service pistol, the following procedures will be adhered to:

    a. When unloading at the end of a string, it is not necessary to maintain solid arm control of the piece. If all rounds have been fired and the slide has stopped in the rearward position, simply remove the magazine and inspect the chamber to insure that the weapon is clear. The weapon may be rested on the bench during this procedure.

b. To unload when the weapon is functioning normally but with the slide forward and the hammer in the full cock position, the procedures below will be followed:
      (1) Remove the trigger finger from the trigger.
      (2) With the nonfiring hand, engage the safety.
      (3) Depress the magazine catch and remove the magazine.
      (4) Disengage the safety.
      (5) Draw the slide to the rear extracting the chambered cartridge and engage the slide stop.
      (6) Inspect the chamber to insure that the pistol is clear.
4. REDUCTION OF MALFUNCTION.
   a. If necessary, to facilitate employment of the left hand while loading or the nonfiring hand during reductions of malfunctions, the firer may shift his position to face the firing line squarely (90 degrees).
   b. Extreme caution must be exercised by the shooter during loading and reducing malfunctions to prevent the weapon, firing hand, or arm from coming in contact with the bench. Only after the weapon is safely loaded (slide fully forward and hammer at full cock), or unloaded in case of malfunction, may a firer relax his firing arm and rest the pistol on the bench.
   c. In the interest of safety, the firer should exercise extreme caution by assuring that the pistol is always pointed down range toward the target during loading or reducing malfunctions.

APPENDIX G

SMALLBORE RIFLE AND PISTOL COMPETITIONS

1. PURPOSE. The smallbore rifle and pistol match outlined herein established an annual Army-wide company level competition designed to emphasize the maintenance of basic marksmanship skills. It affords company commanders an inexpensive and excellent means of supplementing and achieving marksmanship training prescribed in AR 350-4 and enhancing unit morale and prestige.
2. GENERAL.
   a. Each CONUS Army and major command may conduct an annual smallbore rifle and pistol match during the period between 1 October and 31 March. The USAR and ARNG are invited and encouraged to participate. Respective CONUS Army area MTU's will sponsor and administer these matches.
   b. Firing will be conducted on locally available outdoor/indoor ranges under rules stated herein.
   c. This program is not intended to limit the conduct of other smallbore programs.
3. BASIS FOR ENTRY.
   a. Entries will be limited to one rifle team and one pistol team for each company-size unit. Team members must have been assigned to the unit for 30 days before the date of record firing, except persons arriving on PCS from another installation. Provisional organizations approximating the strength of a rifle company (TOE 7-18E) may be established for units with an assigned strength of less than 50 personnel. When it is necessary to combine units, it is desirable to provide a basis for the entry of as many teams and as much individual participation as possible. Individual entries are encouraged, and individuals need not fire with a team to be eligible for individual awards.
   b. Personnel assigned or on special duty to TD or provisional marksmanship detachments will not be eligible to compete in these matches, either with teams or individually.
4. ISSUING AND STORING TARGETS.
   a. MTU's will establish control procedures for issuing and scoring targets that will preclude:
      (1) Interchanging assigned targets.
      (2) Issuing duplicate targets.
   b. A disinterested officer witnessing the firing of the match will execute a statement certifying that all match conditions were met as specified herein. Targets will not be "plugged" by the witnessing officer.

c. The decision of the MTU commanders on scores will be final.

d. Participating commands will submit three copies of official bulletins, to reach this headquarters by 1 May of each year. The bulletins will include the stage scores of at least the top four teams and top ten individuals. In addition, the bulletin will indicate the total number, broken down by the weapon fired, of teams and individuals that competed in the match (RCS exempt report, para 7-2p, AR 335-15).

5. AWARDS.

a. All command levels are encouraged to present appropriate individual and team awards.

b. Announcement of the US Army team and individual winners and presentation of FORSCOM trophies will be made in conjunction with the annual US Army Rifle and Pistol Championships at Fort Benning, Georgia. Winners will be determined by FORSCOM after review of MTU official bulletins. Scores fired in the team match will be used as the basis for individual awards, except in the case of individual entries (para 3a, above).

6. RIFLE RULES.

a. In cases not covered by these rules concerning the conduct and scoring of the competition, NRA smallbore rifle rules will apply.

b. A team will consist of six firing members, a team captain, and a coach. The team captain and coach may also be firing members.

c. The international 50-meter target, reduced for 50 feet (A-36 NRA international target), will be used for record firing. Targets will be marked plainly and numbered for each stage.

d. Firing positions will be as described in NRA Smallbore Rifle Rule.

e. Following are the courses of fire:

(1) Ten shots for record in each of three positions - prone, kneeling, and standing - will be fired in any order desired. There will be one target for each position, with one shot fired at each record bull's eye. Sighters and foulers will be permitted. A maximum of four sighters may be fired in each position and must be fired before the 10 record shots. For misplaced sighting shots, NRA Smallbore Rifle Rule 9-8 will govern.

(2) A 60-minute time limit will be allowed for each match. Sighters and foulers must be taken within the overall time limit.

(3) A team will fire as a unit. If range facilities cannot accomodate six individuals at one time, no more than 10 minutes will elapse between relays.

f. The rifle authorized for use in smallbore rifle matches is the .22 caliber rim rifle chambered for cartridges commercially cataloged as .22 short, .22 long, or .22 long rifle. There are no restrictions on barrel length. Overall weight of the rifle and accessories may not exceed 17.6 pounds. The rifle must be in a safe operating condition. The competitor may change rifles during a match if he so desires. Rifles may be equipped with metallic sights only. A single lens may be attached to the rear sight as a substitute for prescribed spectacles. Scheutzen-type buttplates and palm rests are authorized.

g. Rim-fire cartridges, commercially cataloged as .22 short, .22 long, or 22 long rifle, that have an overall length of not more than 1.1 inches and are loaded with a lead or alloy bullet not heavier than 40 grains are authorized. Hollow point, tracer, incendiary, and explosive bullets are specifically excluded from ammunition authorized for match use.

h. Rests or artificial support will not be used. Kneeling rolls meeting the requirements of NRA Smallbore Rifle Rule 3.14.1 are authorized.

i. The range will measure 50 feet from the firing line to the face of the target hung in position for firing.

7. PISTOL RULES.

a. In all cases not covered by these rules concerning the conduct and scoring of the competition, NRA pistol rules will apply.

b. A team will consist of four firing members, a team captain, and a coach. The team captain and coach may be firing members.

c. The firing position will be standing, with the pistol held in one hand only, the other hand and arm used in no way to assist. All portions of the shooter's clothing, body, and weapon must be clear of artificial support.

d. The course of fire is as follows:

(1) Three stages: Ten shots, slow fire, 1 minute for each shot; 10 shots, time fire, 2 strings of 5 shots each, 20 seconds for each string; and 10 shots, rapid fire 2 strings of 5 shots each, 10 seconds for each string.

(2) A team will fire as a unit. If range facilities cannot accommodate four individuals at one time, no more than 10 minutes will elapse between relays.

APPENDIX H

M60 LIGHT MACHINE GUN COMPETITION

1. PURPOSE. The Light Machine Gun (LMG) competitions prescribed herein are designed to emphasize the importance of the LMG as a base of fire and to enhance the prestige of the LMG Team.

2. OBJECTIVES. The objectives of these competitions are to:

    a. Raise the standards of proficiency of the LMG team.

    b. Improve the combat effectiveness of weapons squads by increasing the accuracy of LMG fire.

    c. Develop highly skilled LMG teams that will be available to unit commanders as instructors.

    d. Accelerate interest and self-improvement in LMG marksmanship that stem from a desire to win in competition.

3. UNIT COMPETITIONS.

    a. CONUSA/major oversea commanders may direct the conduct of competitions from company level through post/division level for all units that employ the LMG.

    b. Maximum participation in company competitions is authorized.

4. CONUSA/MAJOR COMMAND COMPETITIONS.

    a. Dependent upon the number of units armed with the LMG and the number of posts having competitions in a CONUSA/major command area, the major commander concerned may designate a post/division level competition as the major command championship. This will preclude unnecessary travel and expense.

    b. Official bulletins for each match and for the 1800-point aggregate will be prepared and sent to HQ FORSCOM, ATTN: AFOP-TT (RCS exempt report, para 7-2p, AR 335-15).

    c. After review of all post/division bulletins, HQ FORSCOM will announce the high-scoring teams.

5. RULES AND REGULATIONS. The following rules and regulations apply to LMG competitions:

    a. Team composition and eligibility.

    (1) Participation in this program is limited to competitors representing units that employ the LMG as a primary weapon (i.e., infantry, airborne infantry, armored or mechanized infantry, and armored cavalry units). The grade of team members must be within the grade authorization specified in the applicable TOE.

    (2) Each team will consist of two firers equipped with one LMG (the same weapon to be fired by both members).

    (3) Competitors selected by any commander for participation at the next higher level must consist of the two-man teams available within his command. Specifically, the two members of a team selected to compete in any match above battalion level must be the same two individuals used to form that team by the battalion commander. The identity of the parent unit, such as company and battalion, will follow a successful LMG team through all levels of competition.

    b. Courses of fire.

    (1) The commander responsible for the competition may prescribe substitute courses of fire at installations lacking 400-yard firing points. Scores fired on substitute courses should be indicated in the official bulletins.

    (2) The course outlined in this appendix will be used in the LMG matches at post/division level. The post/division championship will be determined by firing three times over this course.

    c. Positions.

    (1) During firing, no part of the weapon except the bipod and sling may touch the ground.

(2) Prior to firing the weapon, the firer will be in the prone position.

(3) Prior to any movement prescribed in the match, the firer must have his stomach and forearm and hand touching the ground. The other hand will grasp the carrying handle of the weapon.

(4) Prior to firing, the number 2 or assistant will be in a prone position, stomach on the ground, with both elbows touching the ground. During firing he may move to the rear of the firer, assume the kneeling or sitting position, and adjust the direction of fire on the target. Standing will not be permitted.

(5) No form of rest for the weapon except the official mounting is permitted. The number 2 or assistant may not sit astride or support the firer, or hold the weapon in any way to steady it during firing. The number 2 man may assist in changing belts and rectifying stoppages, and he may give advice.

    d. Dress and equipment.

        (1) Competitors will wear the combat/field uniform, to include the following:

            (a) Steel helmet.

            (b) Field uniform appropriate to the season.

            (c) Pistol belt with suspenders, first-aid packet, full canteen, canteen cup and cover.

        (2) Tinted glasses are permitted.

        (3) Government-issue binoculars, not exceeding 7x50 power may be used during firing. Telescopes are not permitted on the firing line.

        (4) Any item of waterproof clothing may be worn, provided that it is a normal item of issue.

        (5) A poncho may be worn when needed.

    e. Weapons. The condition of all weapons used in these competitions will be within approved ordnance standards. Modifications or alterations of the weapon or its parts not authorized by applicable manuals (TM 9-1005-224 Series) are not allowed.

    f. Ammunition. Only standard-issue service ammunition will be used at all levels of competition with the LMG.

    g. Targets and scoring.

        (1) The standard rifle "D" target will be used at ranges of 200, 300, and 400 yards.

        (2) The standard methods of scoring, marking, and disking for rapid fire will apply.

    h. Safety lever. In any match involving movement, the safety lever must be set at "safe" and will not be unlocked until the competitor arrives at the firing point and is in the actual firing position.

    i. Misfires, stoppages, and mishaps.

        (1) No allowances will be made for misfires, stoppages, or faulty target exposures.

        (2) If a target or frame breaks during a match, the competitor may request permission to refire the stage, provided that he does so before the targets are scored for that stage.

    j. Penalties and disqualifications.

        (1) A team will be disqualified and its score annulled if during a match:

            (a) The team breaks the minor rules regarding position twice.

            (b) Either member breaks the rule about supporting the gun or steadying the firer.

            (c) Either member behaves dangerously on the firing line or fails to make his weapon safe.

            (d) Either member has any form of communication with an outside observer.

        (2) Penalties for firing on a wrong target, or for failure to adhere to the course of fire, are as follows:

            (a) If more than the required number of hits appear on a target, the competitor will be given a choice of accepting the required number of hits or lowest value appearing on the target or firing a complete new score. If all hits are of equal value, the score will be recorded as the required number of hits of that value.

(b) All shots fired by a competitor after he has taken his position on the firing point will be counted in the scores.

(c) All shots fired before the signal to commence fire, or after the signal to commence fire, or after the signal to cease fire, will be scored as misses.

(d) If a competitor fires fewer than the prescribed number of shots, and there should be more hits on the target than shots fired, the number of shots of highest value equal to the number he fired will be scored, and each unfired round will be scored as a miss.

(e) A competitor will not be given credit for shots fired on a wrong target and he will be scored with a miss for each shot so fired.

k. Challenges and protests. Rules and procedures for challenges and protests may be established by the commander sponsoring the competition. NRA High Power Rifle Rule 16 is suggested as a guide.

l. Organization for firing.

(1) Squadding will be the same as that used for service rifle matches, except that only alternate firing points will be used.

(2) Teams from the same command or organization should not be squadded next to one another during match competition.

(3) Classification of the firer does not affect these matches.

m. Small-arms firing school. Small-arms firing schools for the LMG should be conducted at an early level of competition to insure that firers employ proper firing techniques and understand the rules and range procedures concerning this competition.

6. CONDUCT OF LIGHT MACHINE GUN MATCHES.

a. 400-Yard Rapid-Fire Match. Two belts of 20 rounds for score, and one belt of 10 rounds as warmers, will be issued to each team.

(1) Each team will assume the prone position on the 400-yard firing line and will be allowed five minutes in which to fire 10 rounds into the butts as warmers, with targets lowered. After firing warmers, competitors will remain lying in their action positions, bolt to the rear, safety on SAFE, and cover raised. About 30 seconds before the first exposure, the range officer will give the command to LOAD - "READY ON THE RIGHT; READY ON THE LEFT; READY ON THE FIRING LINE." Safety catches may be put off and the firers may come into the aim. The targets will be raised for 24 seconds, and number 1 will fire his belt of 20 rounds. After the targets have been lowered, number 1 will clear the gun, set the safety on safe, and change places with number 2 on orders from the range officer.

(2) When all are ready, the targets will be raised for another 24 seconds, and number 2 will fire his 20-round belt, using same range commands as for number 1. After the second exposure the gun will be cleared and set at safe, and competitors will remain on the ground awaiting orders from the range officer. Binoculars may be used and advice exchanged between members of any pair.

(3) The highest possible score is 200.

b. Midrange Rapid-Fire Match. The movement will be from 400 to 300 yards. To start the match, each firing team of two men will face the targets on the 400-yard firing line. Two belts of 20 rounds each will be issued to each team.

(1) The competitors will assume the prone position at the 400-yard firing point with weapons half loaded, no round in the chamber, sights set and the safety catch applied. One belt may be carried by number 1 of each pair. Number 2 of each team will carry the spare parts kit, barrel and second belt. The range officer will order "READY ON THE RIGHT; READY ON THE LEFT; READY ON THE FIRING LINE" about 30 seconds before the targets are due to be exposed. The target will be raised for 60 seconds. During this time the teams will run to the 300-yard firing point and number 1 will fire the first belt. No team may open fire until both its members are lying down in their correct positions on the firing point. When the targets disappear, number 1 will clear the gun, pull bolt to the rear, and set the safety on safe.

(2) The targets will be down for 60 seconds; during this time numbers 1 and 2 will change places. Number 2 prepares the weapon for firing (half load) on the second exposure of the targets, which will last 60 seconds, competitors will run to the 200-yard firing point, and number 2 will fire the second belt. When the targets go down, number 2 will unload, clear the gun, apply the safety, and all will remain lying down awaiting an order from the range officer. While awaiting the next event, competitors will have at least the stomach, elbow, and hand on the ground. Binoculars may be used and advice exchanged between members of any pair.

    (3) The highest possible score is 200.
  c. 200-Yard Rapid-Fire Match. Two belts of 20 rounds each will be issued to each team.

    (1) Each team will lie on the 200-yard firing line. Number 1 will take the firing position, load the weapon, set sights, and place weapon on SAFE. About 30 seconds before the first exposure, the range officer will order "READY ON THE RIGHT; READY ON THE LEFT; READY ON THE FIRING LINE" - safety catches to fire position and then firers may come into the aim. The targets will be raised for 24 seconds, and number 1 will fire his belt of ammunition. After the targets have been lowered, number 1 will clear the gun, set the safety on SAFE, change places with number 2, and load the weapon on command of the range officer.

    (2) When all are ready, the targets will be raised for another 24 seconds, and number 2 will fire his belt of ammunition. After the second exposure the gun will be cleared and set on SAFE and competitors will remain on the ground awaiting orders from the range officer. Binoculars may be used and advice exchanged between members of any pair.

    (3) The highest possible score is 200.
  d. CONUSA Match. The CONUSA Match will consist of three firings of each match outlined in this appendix. The highest possible score is 1800 points.

7. CHECKLIST FOR SAFETY OFFICER AND RANGE OFFICERS.
  a. Positions.

    (1) The only parts of the weapon that may touch the ground during firing are the bipod and the sling.

    (2) A hole may not be dug in the ground for steadying the gun.

    (3) Prior to any movement, the gunner must have his stomach, one forearm, and one hand touching the ground.

    (4) The gunner may not commence firing until number 2 is in position
  b. Weapon. Using the sling for support with the LMG is permitted.
  c. Assistant gunner. This assistant is only permitted to help the gunner to change belts, rectify stoppages, and give advice.
  d. Uniform and equipment.

    (1) The following clothing and equipment will be worn.
      (a) Steel helmet.
      (b) Field uniform.
      (c) Pistol belt with suspenders, first aid pouch and packet, full canteen and cover, and poncho.

    (2) Shooting jackets, elbow pads, and shooting gloves are not permitted.

    (3) Telescopes are not permitted on the firing line.

    (4) Tinted glasses are permitted.

    (5) Issue binoculars with a magnification no greater than 7x50 are permitted.

    (6) Issue waterproof clothing is permitted.
  e. Disqualification. A team is disqualified if:

    (1) The team breaks the rules concerning position twice.

    (2) The team commits a breach of rules concerning support of the gun or the firer.

    (3) The team members commit unsafe acts on the firing line, to include failure to set the safety of a weapon when an exercise is terminated.

    (4) A member has any communication with an outside observer.

    (5) The weapon is carried in movement by a person other than the person who is to fire.

    (6) The weapon used in the match is not within the standards set forth by the appropriate TM.
  f. Penalties.

    (1) When more than the required number of shots appear on the target, the competitor may elect to accept the lowest value or refire.

    (2) No alibis will be allowed for accidental discharges.

    (3) Shots fired before the signal to commence firing will be scored as misses.

    (4) When a gunner fires fewer than the prescribed number of shots and his total shows more hits than shots fired, he will receive the highest values equal to the number of rounds he fired and be scored a miss for each unfired round.

APPENDIX I

NATIONAL MATCHES

1. PURPOSE. This appendix prescribes the requirements for squads to be eligible for the National Matches.

2. PARTICIPATION.

   a. Participation in the annual National Matches is limited to squads eligible for entry in the National Trophy Team Matches from the US Army, the Reserve Components, and members of the ROTC (AR 920-30).

   b. Details pertaining to the US Army Reserve Components are contained in Appendix K.

   c. Estimated billeting requirements for the National Matches will be submitted by the CO, USAMU, to the Executive Officer, National Matches, Camp Perry, Ohio, not later than 15 June of each year. The requirement will indicate billet needs for rifle and pistol entrants and indicate the total number for each weapon. This estimate will include personnel competing with squads representing the US Army eastern and western regions (RCS exempt report, para 7-2c, AR 335-15).

3. NATIONAL TROPHY TEAMS.

   a. Army entries in the National Trophy Team Matches and the National Trophy Infantry Team Match are restricted to one rifle team and one pistol team representing each of the following:

       (1) The US Army at large.

       (2) The US Army eastern region composed of personnel from MDW, First US Army, 193d Inf Bde, USAREUR, and USAMU.

       (3) The US Army western region composed of personnel from Fifth and Sixth US Armies, 172d Inf Bde and USASCH.

   b. Sponsoring, supervising, selecting, organizing, and training the regional squads will be the responsibility of the CO, USAMU.

   c. Personnel for membership on squads from which National Trophy Teams are formed will be selected at the conclusion of the US Army Rifle and Pistol Championships. Selection of personnel for the US Army squads will have priority.

       (1) The US Army Squads will be limited to 38 members for rifle and 26 members for pistol in the National Matches.

       (2) The eastern and western region squads will each be limited to 28 members for rifle and 15 members for pistol in the National Matches.

   d. Sufficient members of rifle squads must be new shooters who have not previously fired in the National Trophy Rifle Team match, and sufficient members of pistol squads must be new shooters who have not previously fired in the National Trophy Pistol Team Match (para 16a, AR 920-30).

   e. Rifle and pistol squads will train as a single group during preparation for and participation in the National Matches. Training will be conducted in accordance with the following:

       (1) The Commander, USAIC, will provide facilities and station support, as required, at Fort Benning, Georgia, for squads to train in conjunction with the US Army rifle and pistol squads.

       (2) The USAMU will:

           (a) Establish procedures for selecting, training, and fielding one US Army rifle squad and one US Army pistol squad.

           (b) Provide weapons, shooting equipment, coaching, technical and maintenance assistance, and match-grade ammunition to National Trophy squads. Squad identity for regions will be established before departure for the National Matches.

       (3) Funding arrangements are contained in Appendix J.

4. ASSISTANT INSTRUCTORS TO THE SMALL ARMS FIRING SCHOOLS. Competitors from all components of the Army are subject to duty as assistant instructors at rifle and pistol small arms firing schools (SAFS) at the National Matches. HQ FORSCOM, through CO, USAMU, will announce instructor requirements to participating commands and units as required.

5. ROTC PARTICIPATION.

a. Each ROTC Region in coordination with their respective MTU is encouraged to enter one ROTC 10-man rifle squad and one 6-man pistol squad in the annual National Matches. Authority for institutions to develop individuals for membership on the Army area ROTC squad is contained in paragraph 5-23, AR 145-2. Entry in the National Trophy Team Matches is authorized by AR 920-30.

b. ROTC students from units in Alaska and Hawaii are authorized to compete as members of the Sixth US Army ROTC squads. ROTC students from units in Puerto Rico are authorized to compete as members of the First US Army ROTC squads.

c. Expenses, including travel, subsistence, clothing, and equipment, at Army area matches, NRA regional matches, and National Matches are a proper charge to the OMA appropriation under Budget Program 8.

d. Appropriated funds are not authorized to cover expenses incident to participation in local competitions approved by public or private sources; e.g., state fairs.

e. Costs of awards and trophies will be chargeable to funds as set forth in AR 230-1, and AR 350-6.

f. ROTC squad members participating in National Matches will be furnished travel allowances at the rate of five cents a mile and subsistence allowances at the rate of $1.50 per day.

g. ROTC Regions will submit data concerning eligible individuals and schools to supporting marksmanship training units not later than 1 May annually.

APPENDIX K

RESERVE COMPONENTS PROGRAM

1. US ARMY RESERVE (USAR) PROGRAM. Each CONUSA will establish within its command a competitive small arms marksmanship program for individuals and units of the USAR. USAR commanders will provide an equitable opportunity for all personnel from ARCOM/GOCOM units and Individual Ready Reserve (IRR), regardless of rank, job, or station to prepare for and participate in small arms competitions.

2. SERVICE RIFLE AND SERVICE PISTOL COMPETITION. Commanders at each level (company through division or comparable size unit) are encouraged to conduct interunit matches to determine the best qualified service rifle and service pistol marksmen in their respective units to form teams for the next level of competition. Provisional organizations may be established, as required to conduct unit level competitions.

    a. Competitions will be conducted in accordance with the instructions and procedures outlined in Appendix A.

    b. ARCOM/GOCOM will conduct an annual match to select qualified individuals to form squads for participation in the Army Commander's Match.

3. PARTICIPATION IN MAJOR COMMAND/CONUSA MATCHES. Each ARCOM/GOCOM will enter at least one rifle team and one pistol team to represent the ARCOM/GOCOM in the appropriate major command/CONUSA match.

4. PARTICIPATION IN THE US ARMY RIFLE AND PISTOL CHAMPIONSHIPS. Each CONUSA is authorized a USAR 32-man rifle squad and a 21-man pistol squad for participation in the US Army Championships. One administrator is authorized to supervise the activities of both squads. Additional administrative support and maintenance services will be provided by the sponsoring CONUSA contingent.

5. PARTICIPATION IN THE NATIONAL MATCHES.

    a. Outstanding competitors will be selected from CONUSA USAR squads for the USAR rifle and pistol squads based on their performance in the US Army Championships for participation in the National Matches. Squad size is limited to 35 rifle and 30 pistol competitors. All USAR team OIC will be selected by HQDA (DAAR) in coordination with Commander, FORSCOM. USAR National Trophy squad members will assist CO, USAMU, in the conduct of the SAFS as required.

    b. All USAR teams will conduct practice for the National Matches at sites and times determined by each team OIC with approval of HQDA (DAAR). Training plans will be announced each year. USAR squad members will be attached to HQDA (DAAR) for training.

    c. Entry arrangements and payment of entry fees for USAR squad members in the National Matches will be the responsibility of the squad OIC.

6. SUPPORT OF US ARMY RESERVE TEAMS. Weapons, equipment, and ammunition will be made available by the appropriate Marksmanship Training Unit to USAR personnel selected for participation in the US Army Championships as members of the CONUSA squads at the conclusion of the major command/CONUSA matches to facilitate training for the US Army Championships. USAR squad members will be provided equipment and ammunition at the conclusion of the US Army Championships by the CO, USAMU, to facilitate training. Requirements will be submitted by HQDA (DAAR) to Commander, HQ FORSCOM, for the following FY by 30 September annually.

7. AUTHORITY FOR USAR MARKSMANSHIP TRAINING. Troop unit personnel may participate in this program in lieu of AT with the consent of their unit commanders. Personnel selected to compete in higher level matches are authorized an additional special tour not to exceed 30 days. When feasible, unit competitions and major USAR command matches may be accomplished as multiple training assemblies. Entry fees will be paid from Operation and Maintenance Army Reserve funds.

8. REPORT. Each CONUSA will furnish one copy of the Marksmanship Training and Competition Report (RCS CSRES-198) required by AR 140-125 to HQ FORSCOM, ATTN: AFOP-TS.

9. SMALLBORE RIFLE AND PISTOL COMPETITIONS. Smallbore rifle and pistol firing is an excellent means of stimulating interest and improving marksmanship skills. USAR units are encouraged to participate in the Army-wide, company level smallbore rifle and pistol competitions outlined in Appendix G and in the Army Reserve/National Rifle Association Rifle and Pistol Postal Matches (AR 140-125).

10. ARMY NATIONAL GUARD PROGRAM. The basic Army National Guard competitive marksmanship program is prescribed in NGR 44-1.

Letter from: Department of the Army
HQ, United States Army Forces Command
Fort McPherson, Georgia 30330

SUBJECT: FY 76 Competitive Marksmanship Program

1. References:

    a. AR 350-6.

    b. FORSCOM/TRADOC Supplement 1 to AR 350-6.

    c. FORSCOM message, Rogers Sends, 291155Z April 75.

2. Reference 1a assigns responsibility to Commander, FORSCOM, to conduct the Army-Wide Competitive Marksmanship Program. The major objective of the Competitive Marksmanship Program is to "shoot-to-win" and thereby provide unit commanders with a vehicle to enhance individual marksmanship proficiency through competition and recognition.

3. The FY 76 program gives the soldier an opportunity to progress in levels of competition to the limit of his ability to excel. The program is designed to:

    a. Improve overall Army marksmanship proficiency through individual and unit competition.

    b. Retain team-unit identification to the highest level of competition.

    c. Emphasize combat shooting in matches.

FOR THE COMMANDER:

1 Incl   SS: John P. Irving
as   JOHN P. IRVING III
                                              MAJ, AGC
                                              Asst AG

REVISED COMPETITIVE MARKSMANSHIP PROGRAM

1. General. These instructions outline phases of competition for the revised competitive marksmanship program. The program consists of three phases, or levels of competition as described below:

2. Phase 1. Installation/ARCOM/GOCOM/state level competition.

    a. Each Active Army installation of 4,000 or more total military strength, each ARCOM, each GOCOM and each state National Guard will conduct commander's matches to select winning teams in the following categories:

        (1) Best TOE battalion combat rifle team (8 man team).

        (2) Best TOE battalion combat pistol team (6 man team).

        (3) Best TOE light machine gun team (2 man team).

    b. Additionally, each commander conducting these matches will select a command composite combat rifle and pistol team to participate in the next higher level. This team will be formed from the best shooters (8 rifle, 6 pistol) on losing teams participating at the commander's matches. The 50 percent new shooter rule applies.

    c. Each USAR command, state National Guard, and Active Army installation, regardless of strength, not having TOE battalions assigned, will hold commander's matches to determine the best battalion equivalent team. A battalion equivalent is any nonstandard unit of not more than 1,000 military strength. Additionally, those commands will form a composite team of the best shooters from the losing teams.

    d. All of the above teams will represent that command at the CONUS regional matches or in the case of OCONUS commands, EUSA, and USAREUR commander's matches.

3. Phase II. OCONUS command/CONUS regional matches.

    a. Commander, EUSA and Commander, USAREUR, will conduct command matches to determine best MG teams winners, and to select best rifle and pistol teams for participation in the all Army matches in CONUS.

    b. Commander, FORSCOM will hold three regional matches at CONUS installations. Normally, Region I will be held at Ft Meade, MD, Region II matches will be held at Ft Riley, KS, and Region III at Ft Ord, CA. CONUS region areas will approximate current CONUSA areas; however, CONUSA Commanders will have no responsibility for these matches other than USAR participation. Adjustments to region boundaries will be made in revised FORSCOM Supplement 1 to AR 350-6 to insure each participant will attend matches as close as possible to his home station. FORSCOM MTU's will assist the designated host installation commanders in conducting the CONUS regional matches.

    c. Categories of competition at OCONUS command/CONUS regional matches are as follows:

        (1) TOE battalion combat rifle and combat pistol team matches.

        (2) TOE 2-man light machine gun team matches.

        (3) Command composite combat rifle and pistol team matches.

    d. A winning team in each of the above categories plus a regional composite trophy team will be selected for each Army component participating in regional matches. The composite trophy team members will be selected from winning command composite teams and from losing team in other categories (50 percent new shooter rule applies). These teams will fire competitive weapons and will represent OCONUS command/CONUS region in the All-Army matches.

    e. At the CONUS regional matches, FORSCOM MTU's will select the Active Army composite trophy team and will assist the CONUSA marksmanship coordinators and NGMTU in making selections for the USAR and NG composite teams.

4. Phase III. All-Army matches.

    a. All components (Active, NG, and USAR) compete for top honors in the following categories:

        (1) Best TOE battalion combat rifle and combat pistol teams.

        (2) Best battalion equivalent combat rifle and combat pistol teams.

        (3) Best major command trophy rifle and pistol teams.

    b. Machine gun competition will not be held at All-Army level. Personnel from all teams participating will be the major sources of old and new shooters to field U.S. Army teams for inter-service, national and international competition.

5. Special rules.

    a. Team membership. An individual competing as a unit team member must have been assigned to the unit at least 60 days prior to conduct of the phase I matches. Exceptions will be made for PCS transfer from outside the installation.

    b. Weapons. All competition except the composite trophy team matches at All-Army matches will be with unmodified M16 rifles and M1911A1 Cal 45 pistols and standard ammunition. Trophy teams will use match grade weapons and ammunition. Machine gun competition at phase I and II will be with unmodified M-60 light machine guns.

    c. Team Organization. Combat rifle teams in all phases will consist of eight personnel; combat pistol teams in all phases, six personnel and machine gun, two personnel. Machine gun teams must be TOE assigned gunner and assistant gunner. Unit/Installation composite combat rifle and pistol teams in phases I and II, will also consist of eight and six personnel respectively. Composite trophy teams formed for competition at all Army, will have 25 rifle personnel and 15 pistol personnel. Composite trophy teams formed by CONUS for competition at All-Army, will have 25 rifle shooters and 15 pistol shooters. OCONUS command trophy teams will have a minimum of 15 rifle shooters and 10 pistol shooters and a maximum of 25 rifle shooters and 15 pistol shooters.

    d. Funding. Phase I and II; TDY/TRAVEL expenses incurred by competitors will be borne by the team's parent unit/installation. Phase III; CONUS Active Army Competitors' TDY/TRAVEL expenses will be borne by the AREA FORSCOM MTU. OCONUS commands will fund for TDY/TRAVEL expenses incurred by their competitors. CONUSA commander's will fund for TDY/TRAVEL expenses incurred by USAR competitors. National Guard Bureau will fund participation of National Guard Competitors.

CHAPTER XV

PROGRAMS OF INSTRUCTION

A. ARMY AREA PISTOL MARKSMANSHIP INSTRUCTORS AND COACHES CLINIC (40 HOUR PROGRAM OF INSTRUCTION)

Apr 1975

FILE NR	SUBJECT	HOURS	TYPE OF INSTRUCTION	SCOPE	REFERENCES
124	Orientation to Clinic (US Army Marksmanship Training Program)	1 Hr	Lecture	To familiarize the student with the overall Army Marksmanship Program to include the structure of the Marksmanship ladder, Distinguished Rifleman and Pistol Badge, competitive type shooting, International shooting, Sniper Program and the clinic program of instruction to include introducing the instructors to the students.	AR 350-6, FORSCOM/TRADOC Suppl 1 to AR 350-6 and POI for 40 hour Army Area Rifle and Pistol Marksmanship Instructors' and Coaches' Clinic
301	Fundamentals I (Attaining Minimum Arc of Movement)	1 Hr	Lecture and Demonstration	As a result of this instruction, students must be able to DEFINE minimum arc of movement, EXPLAIN the effect of arc of movement on accurate shooting, DESCRIBE the steps necessary in achieving minimum arc of movement and EMPLOY stance, position, grip, breath control to attain minimum arc of movement.	Chapter I, the USAMU Pistol Marksmanship Guide.
302	Fundamentals II (Sight Alignment)	1 Hr	Lecture and Demonstration	As a result of this instruction, students must be able to DEFINE sight alignment, EXPLAIN the effect of sight alignment on accurate shooting, DESCRIBE technique for maintaining sight alignment, and EMPLOY sight alignment in a method of shooting.	Chapter II, USAMU Pistol Marksmanship Guide.
303	Fundamentals III (Trigger Control)	30 Min	Lecture and Demonstration	As a result of this instruction, students must be able to DIFFERENTIATE between the types of trigger control and COORDINATE trigger control with the other fundamentals.	Chapter III, USAMU Pistol Marksmanship Guide.

FILE NR	SUBJECT	HOURS	TYPE OF INSTRUCTION	SCOPE	REFERENCES
304	Technique of Fire I (Establishing a System)	1 Hr	Lecture and Demonstration	As a result of this instruction, students must be able to LIST the steps in establishing a system, DESCRIBE a firing system to include the prescribed method of zeroing.	Chapter IV, USAMU Pistol Marksmanship Guide.
305	Technique of Fire II (Slow Fire)	1 Hr	Lecture	As a result of this instruction, students must be able to DESCRIBE the slow fire shot sequence and EMPLOY proper slow fire techniques under ideal and adverse conditions.	Chapter V, USAMU Pistol Marksmanship Guide.
306	Techniques of Fire II (Sustained Fire)	30 Min	Lecture	As a result of this instruction, students must be able to DESCRIBE the sustained fire shot sequence and EMPLOY proper sustained fire techniques under ideal and adverse conditions.	Chapter VI, USAMU Pistol Marksmanship Guide.
307	Mental Discipline	1 Hr	Lecture	As a result of this instruction, students must be able to DEFINE mental discipline, DESCRIBE the causes and effects of match pressure, LIST the steps in the development of mental discipline, EXPLAIN the essential nature of mental discipline to champion shooters and PRACTICE mental discipline techniques of reducing the effects of tension.	Chapter VII, USAMU Pistol Marksmanship Guide.
308	Coaching I (Attributes, Aids and the Duties of a Pistol Coach)	1 Hr	Lecture	As a result of this instruction, students must be able to LIST the attributes of a coach, LIST and DESCRIBE the duties of a Line Coach.	Chapter X, USAMU Marksmanship Instructors' and Coaches Guide.

FILE NR	SUBJECT	HOURS	TYPE OF INSTRUCTION	SCOPE	REFERENCES
309	Coaching II (Technique)	1 Hr	Lecture and Demonstration	As a result of this instruction, students must be able to EXPLAIN the technique of coaching a pistol team, LIST the factors in systematic coaching techniques, INCORPORATE suggestions for improvement of coaching techniques.	Chapter XI, USAMU Marksmanship Instructors' and Coaches' Guide.
310	Physical Fitness I (Physical Conditioning)	1 Hr	Lecture, Demonstration and Practical Exercise	As a result of this instruction, students must be able to EXPLAIN, DEMONSTRATE and PERFORM each of the Pistol Shooters' Daily Dozen exercises and DESCRIBE the other exercises, sports and activities beneficial to the competitive pistol shooter in attaining physical fitness.	Chapter VIII, USAMU Pistol Marksmanship Guide.
311D	Physical Fitness II (Diet)	1 Hr	Lecture	As a result of this instruction, students must be able to EXPLAIN the need for having a well balanced diet and NAME the essential nutrients and the foods in which they are found.	Chapter IX, USAMU Pistol Marksmanship Guide.
311	Physical Fitness III (Effects of Alcohol, Coffee, Tobacco and drugs)	1 Hr	Lecture	As a result of this instruction, students must be able to DESCRIBE the effects of Alcohol, Tobacco, Coffee, and Drugs on shooting performance.	Chapter X, USAMU Pistol Marksmanship Guide.
600	Review of Pistol Fundamentals	1 Hr	Conference	As a result of this instruction, students must be able to CITE HIGHLIGHTS of all previous instruction on pistol fundamentals. Students request explanation of all information not clearly understood.	Chapters I, II & III, USAMU Pistol Marksmanship Guide.

FILE NR	SUBJECT	HOURS	TYPE OF INSTRUCTION	SCOPE	REFERENCES

313	Pistol Marksmanship Examination	1 Hr	Practical Exercise	The Instructor will issue test instructions as follows:
				1. Announces number of questions and pages.
				2. Announces changes to test as printed and have students make any necessary corrections.
				3. Explain test subdivisions:
				   a. Multiple choice questions have only one correct answer.
				   b. True-False. A true answer is true only if it is 100% correct.
				   c. Fill in the blank with word(s) which are a proven fact or have been included within a statement made by an instructor(s) during course.
				4. If there is any doubt as to meaning of a question, assistance may be obtained from an AI by raising hand.
				5. Give the instructions for filling out heading of examination answer sheet.
				6. Announce starting time and time allotted to complete test.
				7. At completion of test, turn in test to _____. Answers to the test will be posted at _____.
					All previous references and Chapter VIII, USAMU Marksmanship Instructors' and Coaches' Guide.

315S	Technique of Fire IV (International Slow Fire) Annex I (International Air Pistol)	1 Hr	Lecture and Demonstration	As a result of this instruction, students must be able to EXPLAIN the International Slow Fire Course and DESCRIBE the special techniques and weapons used in International Air Pistol and course of fire and EXPLAIN the special techniques of firing the Air Pistol in International competition.	Chapters XI, USAMU Pistol Marksmanship Guide. Annex 1 to Lesson Outline AAPMI&CC 315S.

FILE NR	SUBJECT	HOURS	TYPE OF INSTRUCTION	SCOPE	REFERENCES
315R	Technique of Fire VI (International Rapid Fire)	1 Hr	Lecture and Demonstration	As a result of this instruction, students must be able to EXPLAIN the International Rapid Fire Course and DESCRIBE the special techniques and weapons used in the course of fire.	Chapter XII, USAMU Pistol Marksmanship Guide.
315C	Technique of Fire V (International Center Fire)	1 Hr	Lecture and Demonstration	As a result of this instruction, students must be able to EXPLAIN the International Center Fire Course and DESCRIBE the special techniques and weapons used in the course of fire.	Chapter XIII, USAMU Pistol Marksmanship Guide.
315SP	Technique of Fire VII (International Standard Pistol)	1 Hr	Lecture and Demonstration	As a result of this instruction, students must be able to EXPLAIN the International Standard Pistol Course and DESCRIBE the special techniques and weapons used in the course of fire.	Chapter XIV, USAMU Pistol Marksmanship Guide.
701	Competitive Regulations I (NRA Pistol Match Rules)	1 Hr	Lecture, P.E.& Conference	As a result of this instruction, students must be able to EXPLAIN the provisions of pistol match rules used in the conduct of an NRA Registered Pistol Match.	NRA Pistol Rule Book.
701A	Competitive Regulations III (National Trophy Pistol Match Rules) Competitive Regulations IV (Earning Distinguished Pistol Badge)	1 Hr	Lecture, P.E.& Conference	As a result of this instruction, students must be able to EXPLAIN the provisions of the National Trophy Pistol Match Rules, and describe the steps in earning a Distinguished Pistol Shot Badge.	Rules for conduct of National Trophy Pistol Match, AR 920-30 and Chapter XXVI, USAMU Marksmanship Instructors' & Coaches' Guide.

FILE NR	SUBJECT	HOURS	TYPE OF INSTRUCTION	SCOPE	REFERENCES
701B	Competitive Regulations V (International Shooting Union Rules for Pistol)	1 Hr	Lecture, P.E.& Conference	As a result of this instruction, students must be able to EXPLAIN the provisions of ISU Pistol Rules as concerned with weapons, position on firing line, time limits and scoring.	ISU Pistol Rules.
702	Competitive Regulations II (Range Procedure and Safety)	1 Hr	Lecture, P.E.& Conference	As a result of this instruction, students must be able to EXPLAIN the provisions of NRA Pistol Rules as concerned with the procedures for conducting NRA type pistol matches. STATE the safety regulations to be observed during pistol competition, and DESCRIBE the proper method of loading a caliber 45 service pistol.	NRA Pistol Rule Book and FORSCOM/TRADOC Suppl 1 to AR 350-6.
703	Panel Discussion	1 Hr	Conference	As a result of this instruction, students must be able to RESOLVE any questions that may have arisen as a result of instruction received.	Chapters VIII and XVII, USAMU Marksmanship Instructors' and Coaches' Guide.
704	Graduation (Award of Certificates & Closing Ceremonies)	1 Hr	Lecture	OIC, USAMU Pistol Instructor Team will present awards and certificates as appropriate.	Chapter XVII, USAMU Marksmanship Instructors' and Coaches' Guide.
705	Instructor Training I (Methods of Instruction)	1 Hr	Lecture, Demonstration, Conference and practical exercise	Given the assignment of instructing Pistol Marksmanship, the student must be able to PERFORM the following Methods of Instruction: Lecture, Conference, Demonstration and Performance; USE effective speaking and CONTROL the interest of a student group; PREPARE and UTILIZE Lesson Outlines and DESIGN and USE Training Aids.	Chapters II, III, IV, V and VII, USAMU Marksmanship Instructors' and Coaches' Guide.

FILE NR	SUBJECT	HOURS	TYPE OF INSTRUCTION	SCOPE	REFERENCES
706	Instructor Training II (Control of interest and effective speaking)	1 Hr	Lecture and Conference	Given the assignment of instructing Pistol Marksmanship, the student must be able to PERFORM the following Methods of Instruction; Lecture, Conference, Demonstration and Performance; USE effective speaking and CONTROL the interest of a student group; PREPARE and UTILIZE Lesson Outlines and DESIGN and USE Training Aids.	Chapters II, III, IV, V and VII, USAMU Marksmanship Instructors' and Coaches' Guide.
707	Instructor Training III (Lesson Planning and Use of Training Aids)	1 Hr	Lecture, P.E. and Conference	Given the assignment of instructing Pistol Marksmanship, the student must be able to PERFORM the following Methods of Instruction: Lecture, Conference, Demonstration and Performance; USE effective speaking and CONTROL the interest of a student group; PREPARE and UTILIZE Lesson Outlines and DESIGN and USE Training Aids.	Chapters II, III, IV, V and VI, USAMU Marksmanship Instructors' and Coaches' Guide.

FILE NR	SUBJECT	HOURS	TYPE OF INSTRUCTION	SCOPE	REFERENCES
INSTRUCTIONAL FIRING					
400	Zeroing 22 Cal & 45 Cal at 25 Yds and 50 Yds	2 Hrs	Practical Exercise	As a result of this instruction, students must be able to EMPLOY the techniques of zeroing taught in Problem 304, Technique of Fire I.	NRA Pistol Rule Book, Chapter IV, USAMU Pistol Marksmanship Guide, and FORSCOM/TRADOC Suppl 1 to AR 350-6.
402	Practice Timed and Rapid Fire (22 Cal Pistol and 45 Cal Pistol)	1 1/2 Hrs	Practical Exercise	As a result of this instruction, students must be able to FIRE two rapid fire strings at 25 yards using both the 22 cal pistol and the 45 cal pistol.	NRA Pistol Rule Book, Chapter VI, USAMU Pistol Marksmanship Guide, and FORSCOM/TRADOC Suppl 1 to AR 350-6.
403	Practice International Slow Fire (1/2 Course)	1 Hr	Practical Exercise	As a result of this instruction, students must be able to fire an International Slow Fire Match (1/2 Course) at 50 meters, using a 22 cal free pistol or a suitable match type 22 cal target pistol.	ISU Rules and Regulations for firing the International Slow Fire Course.
404	Practice International Center Fire (full course)	2 Hrs	Practical Exercise	As a result of this instruction, students must be able to fire an International Center Fire Match (full course) at 25 meters, using a center fire (not to exceed 38 cal) pistol.	ISU Rules and Regulations for firing the International Center Fire Course.
407	Practice International Standard Pistol (full course)	1 Hr	Practical Exercise	As a result of this instruction, students must be able to FIRE International Standard Pistol Match (full course) at 25 meters, using a 22 cal int'l standard pistol.	ISU Rules and Regulations for firing the International Standard Pistol Course.

FILE NR	SUBJECT	HOURS	TYPE OF INSTRUCTION	SCOPE	REFERENCES
408	Practice International Rapid Fire (1/2 course)	1 Hr	Practical Exercise	As a result of this instruction, students must be able to fire an International Rapid Fire match (1/2 course) at 25 meters, using a 22 cal rapid fire pistol.	ISU Rules and Regulations for firing the International Rapid Fire Course.
410	Practice NMC 22 cal. Pistol	1 1/2 Hrs	Practical Exercise	As a result of this instruction, students must be able to FIRE a National Match Course at 25 yds and 50 yds, using the 22 cal. Pistol plus ten extra rounds slow fire and ten extra rds rapid fire.	AR 920-30, Chapter IV, USAMU Pistol Marksmanship Guide and NRA Pistol Rule Book.
411	Practice NMC 45 cal. (Service Pistol)	1 1/2 Hr	Practical Exercise	As a result of this instruction, students must be able to FIRE a National Match Course at 25 and 50 yards, using a caliber 45 pistol plus ten extra rounds slow fire and ten extra rounds rapid fire.	Same as for Problem 410 and FORSCOM/TRADOC Suppl 1 to AR 350-6.
000	Administrative Processing	1 1/2 Hrs	Practical Exercise	Weapons maintenance, turn in issued equipment, clear billets, sign out. (Resv personnel submit pay vouchers) USAMU Instructor Personnel remain until next day. Obtain attendance report (Army, Navy, Air Force, USMC, USCG, Reserve, Police, Collegiate and Civilian.)	USAMU SOP 1975.
510	Combat Pistol Course (Alternate to Problem 411, NMC with cal. 45 Service Pistol)	1 1/2 Hrs	Practical Exercise	As a result of this instruction, students must be able to FIRE a Combat Pistol Course at 25 yards, using a caliber 45 pistol.	FORSCOM/TRADOC Suppl 1 to AR 350-6, Appendix E or FM 23-35, Pistols and Revolvers.

B. ARMY AREA RIFLE MARKSMANSHIP INSTRUCTORS AND COACHES CLINIC

(40 HOUR POI AA/RMCC)
1 April 1975

FILE NO	HOUR	SUBJECT	TYPE OF INSTRUCTION	SCOPE	REFERENCE
124	1 Hr	Orientation to Clinic	Conference	To familiarize the student with the overall Army Marksmanship Program to include the structure of the marksmanship ladder, competitive type shooting, and International shooting. Discuss the clinic program of instruction and training schedule to include the introduction of the instructors to the students.	AR 350-6, FORSCOM/TRADOC Suppl 1 to AR 350-6 POI and Training Schedule for the 40 Hour AA/RMCC.
125	1 1/2 Hr	Squad Selection, Organization and Training	Conference	To teach the student the proper methods and techniques used to select a rifle team for participation in competitions at all levels. The use of the TDA and AR's in requesting authorized individuals and team equipment. The techniques used to organize and conduct individual and team training in conjunction with the use of the evaluation file to assemble the strongest possible team for team competition.	USAMU Service Rifle Marksmanship Guide, Chapter 1, Sect. II & IV.
126	1 Hr	Selection of a Rifle and Care and Cleaning	Conference	To familiarize the students with the duties of a gunsmith and the major difference between a National Match and standard issue M-14 Rifle, to include the parts selection process, modification of component parts, assembly, techniques, how to select an M16A1 Rifle, the assembly and disassembly steps and care and cleaning of each type weapon.	USAMU Service Rifle Marksmanship Guide, Chapter 1, Section III, USAMU Accurized National Match Rifle Guide, and FM 23-9.
128	30 Min	Mental and Physical Conditioning	Lecture and Conference	To teach the student the necessity for attaining and maintaining a good physical condition to include the development of a proper mental attitude.	USAMU Service Rifle Marksmanship Guide, Chapter 1, Section IV.

FILE NO	HOUR	SUBJECT	TYPE OF INSTRUCTION	SCOPE	REFERENCE
131.1	4 Hr	Combat Rifle Match (P.E.)	P.E.	Given the assignment as Combat Match Team Coach, the student must be able to explain the method used for the selection of squad members, itemize the squad equipment, and conduct a squad training program emphasizing the fundamentals of Combat Rifle Match firing.	USAMU Service Rifle Marksmanship Guide, Chapter 4.
226	4 Hr	National Trophy Team Coaching Techniques (Examination)	P.E.	As a student in the Coaching Clinic, the student must be able to demonstrate the National Match Team Coaching Techniques as outlined in Lesson File Number 132. The students are divided into four man groups with one USAMU Instructor/Coach as a grader. The students rotate the duties until all have performed as coach for each stage of the National Match Course. During this period the Coach's Performance Check List and Examination may not be used as a reference by the student. The Instructor-Coach will utilize the Coach's Examination as a grading sheet.	USAMU Service Rifle Marksmanship Guide Chapter 1, Section V.

133	1 1/2 Hr	Rifle Marksmanship Exam.	P.E.	The principal instructor will issue test instructions as follows:
				1. Announce number of questions and pages; have students examine test material and make corrections if necessary.
				2. Announce changes to test as printed.
				   Explain test subdivisions.
				   a. Multiple choice questions have only one correct answer. Where there is doubt, mark the most correct.
				   b. True-False-A true answer is true only if it is 100% true.
				   c. Fill in the blanks - with word(s) which are a proven fact or have been included within a statement made by the instructor(s) during the AA/RMCC.
					All previous references.

FILE NO	HOUR	SUBJECT	TYPE OF INSTRUCTION	SCOPE	REFERENCE
133		(Rifle Marksmanship Examination Continued)		4. If any doubt, assistance may be obtained from AI by raising hand. 5. Give instructions for heading completion. 6. Announce starting time, and time allotted to complete test. 7. At completion of test, move to the rear of the classroom, turn in the test to _____. Answers to the test will be posted at _____.	
134	1 Hr.	Panel Discussion	Conference	During this panel discussion, the clinic OIC introduces the instructors, and the classes presented by them during the coaching clinics. The OIC conducts a question and answer session between the students and the panel. Questions should be directed so the principal instructor who presented that subject may provide the answer with additional comments supplied by other panel members.	All preceding references.
135	1 Hr.	Graduation	P.E.	To introduce the guest speaker, make presentation of certificate and closing remarks in recognition of student accomplishments.	AARMCC Lesson Plan NR. 135
130	1 Hr.	Precision Combat Rifle Match Principles	Conference, Demonstration	As a team coach, the student must be able to: Explain the conduct of the match, the individual and team course of fire, team composition, targets used, scoring system, amount of ammunition, authorized firing positions, selection of squad members; itemize the squad equipment; and conduct of a squad training program.	FORSCOM/TRADOC Suppl 1 to AR 350-6, AR 350-6 CTA 23-101.
130.1	5 Hrs.	Precision Combat Rifle Match (P.E.)	Conference, Practical Exercise	Given the assignment as a team coach PERFORM correctly all of the duties of a shooter and team coach.	FORSCOM/TRADOC Suppl 1 to AR 350-6, Chapter 2, Section I and V USAMU Service Rifle Marksmanship Guide.

FILE NO	HOUR	SUBJECT	TYPE OF INSTRUCTION	SCOPE	REFERENCE
132	2 Hrs	National Trophy Team Coaching Techniques	Lecture Demonstration	As a team coach or captain, the student must be able to select, train and organize a National Trophy Rifle Team. The student must apply the principles of a coach-shooter relationship and Coaching Techniques at each stage of the National Match Course; organize the shooters and their equipment; and prepare and use the coach's plotting sheet. For the Practical Exercise portion of this class, see File No. 225; National Match Team Coaching Techniques. (Practical Exercise)	USAMU Service Rifle Marksmanship Guide, Chapter 1, Section V.
225	4 Hrs	National Trophy Team Coaching Techniques (P.E.)	P.E.	As a student in the Coaching Clinic, the student must be able to put into practice the National Match Team Coaching Techniques as outlined in Lesson File Number 132. The Students are divided into four man groups under the supervision of a USAMU Instructor-Coach. The duties are rotated until each student has performed as a coach for each stage of the National Match Course, with the aid of the Coach's Performance Check List and Examination.	USAMU Service Rifle Marksmanship Guide, Chapter 1, Section V.
131	1 Hr	Combat Rifle Match Principles	Lecture & Demonstration	As a Team Leader in the Combat Rifle Match, the student must be able to: Explain the conduct of the match, the Individual and Team course of fire, team composition, targets used, & scoring system, amount of ammunition issued for individual and team competition, authorized firing positions, and fire plans. Also, the student must be able to explain the duties of the Umpire and the Assistant Team Leader.	USAMU Service Rifle Marksmanship Guide, Chapter 4.

FILE NO	HOUR	SUBJECT	TYPE OF INSTRUCTION	SCOPE	REFERENCE
129	30 Min	Rifle Range Safety	Lecture	To teach the student to practice and enforce proper safety habits in the conduct of firing and handling firearms and to familiarize the student with the importance of enforcing Range Safety Regulations.	USAMU Service Rifle Marksmanship Guide, Chapter 1, Section IV.
224	10 Hrs	Shooting Techniques	Lecture, Conference, Demonstration & Pract exercise.	To teach the student how to properly demonstrate to his shooters with the M-14 and M-16A1 the prone slow fire, prone rapid fire, sitting rapid fire, standing slow fire, and standing sustained fire, to include adjusting and applying the loop sling, natural point of aim, correct eye relief, sight alignment, sight picture, rapid fire cadence, trigger control, and the breathing and aiming process. Additionally, the student will receive instruction on how to organize and conduct P.E.	USAMU Service Rifle Marksmanship Guide, Chapter I, Section IV.
127	1 Hr	Effects of the Weather, and use of the scorebook and telescope.	Conference, Demonstration & Prac Exer.	The student must be able to compute and adjust the sights of the M-14 rifle and compute and adjust the aiming point of the M16A1 Rifle to compensate for the effects of: 1. The wind; by use of the wind formula, mirage, and wind diagram. 2. The effects of light; direct or angular light on the target and front sight. 3. The effects of temperature. Also, the student must be able to adjust and make use of the telescope, and utilize the scorebook in his final determination of a zero and sight adjustment needed for one shot slow fire or a rapid string. In addition, the student must be able to organize and conduct practical exercises using the telescope for reading the mirage.	USAMU Service Rifle Marksmanship Guide, Chapter 1, Section IV.

C. INTERNATIONAL RIFLE MARKSMANSHIP CLINIC
3 HOUR PROGRAM OF INSTRUCTION (INTERNATIONAL RIFLE)

1 April 1975

FILE NO	SUBJECT	HOURS	TYPE OF INSTRUCTION	SCOPE	REFERENCES
800	Equipment, procedures and techniques of training	1 Hour	Lecture	Given the requirement to properly equip an International Rifle Team, select the proper equipment; faced with the problem of proper breath control, aiming and exercising trigger control, apply the proper methods; master the problem of assuming stable shooting positions by applying the general requirements fundamental to all positions; develop an adequate training program that will meet the needs of all shooters and stress the importance of retaining mental composure during competition.	Chapters II, III, and IV USAMU International Rifle Marksmanship Guide.
801	International Prone Kneeling & Standing Positions	1 Hour	Lecture and Demonstration	Given the factors of the prone position, assume the positions and explain the technique of steadying the hold and accomplishing position orientation, (1) assume proper leg positions, (2) correctly position the arms, (3) stabilize the head position (4) gain proper rifle/shoulder contact. Given the factors of the kneeling position, correctly use the (1) kneeling roll, (2) right foot & knee, (3) body and head, (4) left arm, elbow and knee, (5) left leg and foot, (6) right arm and hand, and (7) sling and accessory adjustments. Given the factors of the standing position, assume a free rifle type standing position by correct placement of the (1) feet, (2) knees, (3) hips, (4) back bend, (5) body twist, (6) left arm, (7) right arm, and (8) head.	Chapters V, VI, & VII, USAMU International Rifle Marksmanship Guide.

FILE NO	SUBJECT	HOURS	TYPE OF INSTRUCTION	SCOPE	REFERENCES
802	International Rifle Match Program	1 Hour	Lecture & Conference	Discussion of the participation in International Rifle Competition according to the ISU Rules and according to the modified NRA Rules for domestic competition. Familiarize the student with the problems involved in sponsoring and conducting domestic International Rifle Matches. Relate to the International Rifle student the record of accomplishment of United States and US Army International Rifle Shooters in International Rifle competition worldwide over the last 10 years. Reveal the opportunities for world competition available to the new US Army International Rifle Shooter.	Chapter I and Annex 1 to USAMU International Rifle Marksmanship Guide.

D. US ARMY INTERNATIONAL SHOTGUN MARKSMANSHIP CLINIC
   3 HOUR PROGRAM OF INSTRUCTION (SKEET & TRAP)

10 April 1975

FILE NO	HOURS	SUBJECT	TYPE OF INSTRUCTION	SCOPE	REFERENCES
900	1	Introduction to International Skeet & Trap Marksmanship	Lecture	To familiarize the International Skeet and Trap student with the United States International Skeet and Trap Program, the mission and goals of the US Army Marksmanship Unit in regard to International Clay Target shooting, the performance of US International Skeet and Trap Teams in recent International competition, cover the major differences between American style and international style Skeet and Trap, and reveal the opportunities for international competition by those US Army personnel who participate in the US Army International Skeet and Trap Program.	Chapters I and VI, USAMU International Skeet and Trap Guide.
901	1	International Clay Pigeon (Trap) Marksmanship	Lecture, P.E., and Demonstration	To enable the International Trap Marksmanship Student to exhibit a knowledge of the history of trap shooting, describe the International Trap field layout, to explain shooting match procedures according to ISU rules of International Trap, to select the best of quality in available trap shooting equipment and to understand the proper shooting techniques to win in International Trap competition.	Chapter V, USAMU International Skeet and Trap Guide.
902	1	International Skeet Marksmanship	Lecture, P.E., and Demonstration	To enable the International Skeet Marksmanship Student to exhibit a knowledge of the history of skeet shooting, describe the International Skeet Field layout, to explain the shooting match procedures according to ISU rules of International Skeet, to select the best of quality in available skeet shooting equipment and understand the proper shooting techniques to win in International Skeet competition.	Chapter IV, USAMU International Skeet and Trap Guide.

E. US ARMY INTERNATIONAL RUNNING TARGET MARKSMANSHIP CLINIC
   3 HOUR PROGRAM OF INSTRUCTION (RUNNING TARGET)

FILE NO	HOURS	SUBJECT	TYPE OF INSTRUCTION	SCOPE	REFERENCES
903	1 Hr	Running Target Marksmanship Program	Lecture & Demonstration	To familiarize the International Running Target Marksmanship student with the United States International Running Target Program and the origin of the sport, the goals and accomplishments of the US Army Marksmanship Unit in International Running Target competition; to enable the student to describe the International Running Target Range layout, explain the match procedures according to ISU Rules of International Running Target, to select the proper specialized equipment available to the Running Target Shooter and understand the proper shooting techniques to produce the winning score in International Running Target Competition.	Chapters I, V and VI USAMU International Running Target Guide.
000	2 Hrs	Running Boar Practical Firing Exercises	Practical Exercise	Fire 1 course Slow Run, 1 course Fast Run	Same as Problem 903

CHAPTER XVl

TRAINING STANDARDS AND SCHEDULING COURSES OF INSTRUCTION

A. GENERAL

It must be understood that there is no substitute for knowledge and experience in teaching marksmanship. An individual assigned the specific duties of teaching marksmanship fundamentals, techniques of fire or coaching techniques must also be completely convinced of the importance of Marksmanship training to the mission of the Armed Forces of the United States. A winning team is the pride of any commander---as it should be---whereas a losing team, while performing to the best of its ability, may possibly lose by a single point and never receive the recognition it deserves. This fact in itself necessitates a program of instruction that will increase the skill and provide the knowledge that will motivate in a military unit or service, a continuing desire to excel. Good marksmanship and good coaching has never been conveyed quickly, nor can it be gained in a short period of time. Good marksmanship is a combination of good coaching, a commanding knowledge of fundamentals, mental and manual dexterity, a desire to excel and a willingness to undergo long practice and hard work. A good instructor and coach will call on experience to aid him in his job and he will give his men the advantage of everything he has learned. With proper application of the techniques discussed in the training program, the resulting proficiency will continue producing championship teams.

B. IMPLEMENTING THE TRAINING MISSION

To properly implement the training of an Army marksmanship team, a comprehensive training program has been devised. The basis of this training program, of course, is the knowledge and experience of the instructor-shooters and the coaches who work with them, past, present, and future. The wealth of information on shooting skill present among these personnel must be made available to other competitive shooters in the nation. The marksmanship instructor-shooter-coach must become, during his duty tour, a highly qualified instructor. The pistol or rifle coach must not only assist in training the shooter to be an instructor and a highly qualified marksman but he must also be the main tool in implementing the training program. Program organizing, instructor rehearsing, setting up of facilities, arranging for fabrication and proper use of training aids, studying refinements of various techniques and presenting instructional matter in an exemplary and interesting manner is but part of the function of a coach. To accomplish this latter mission more effectively, the instructor coach must be familiar with advanced instructional techniques, be acquainted with the physiological-psychological processes of the human body, must learn to influence the developing shooter toward a mentally conditioned state so that he will properly respond and coordinate his thoughts and actions for control of his shooting.

C. STAGES OF DEVELOPMENT OF THE PISTOL OR RIFLE SHOOTER

Certain factors and specific stages of development govern the progress of a shooter and the instructional techniques should reflect these degrees of development and learning.

These degrees of learning are, in a sense, phases of advancement: one, exploratory; two, articulation or conversation; three, organization or resolution; four, accomplishment or consummation.

First, a comparatively new shooter usually finds that he is involved in exploring the skill of marksmanship when his interest in shooting is induced by other shooters or a coach. This period is characterized by the sensing of the important factors of a skillful endeavor. In his limited experience, he has learned that certain vital and dynamic qualities are present and his curiosity is aroused. His interest in shooting can be sustained if the one endeavoring to expand the beginner's knowledge confines his teaching to the simpler aspects.

Second, effective instruction proceeds by making sure that exploration is followed by extensive, inquisitive conversation. A shooter's learning possibilities are limited if he continues to ponder by himself. Beyond a certain point, he must become articulate, begin to ask questions and look things up. Real conversation begins when the developing shooter does not like the answers he is getting, and offers rebuttal. A meeting of minds results if there ensues an exchange of ideas and information between coach and shooter or shooter and shooter. Open and free communication is essential to this learning process. During this period the shooter may zig toward excessive self-doubt and zag toward excessive self-confidence but the skillful coach-instructor brings him back to the proper bearing. No longer is the shooter merely probing, again and again he convinces himself that he is in a firm position of logic from which he can contribute materially to the expansion of his knowledge of marksmanship.

Third, intelligent coaching hurdles an important challenge by leading the shooter to an organization and full grasp of all the fundamentals, principles and factors that contribute to the controlling of his skill as a marksman. The unanswered questions are being answered satisfactorily, the problems are being resolved. There eventually comes a time when a technique of operation is established. This phase of organization or resolution is marked by the shooters' decision to adopt an acceptable method of approach. It is also a characteristic of this period that the shooter has evolved a system of operation peculiar to him and figuratively speaking, he strikes out on his own. There has come into being a personal technique for accomplishment.

Fourth and last, the phase of consummation. Teaching becomes a sustaining force when the inducement to accomplishment is reached. Coaching and instructing is now counseling and advisement, a keeping on the track, so to speak. The shooter has assimilated knowledge, accepts a method of operation, exercises skillful judgement, can control vagaries of the will and has correlated various courses of action into a coefficient of the kind of shooter he now is, how he will accomplish his end. After this he emerges as a master of pistol or rifle marksmanship. His degree of accomplishment depends upon the peculiar nature of this entity. The limitations if any, lie in the ambition, force of character, completeness of knowledge, enthusiasm and being associated with group incentive. Whether he will eventually fulfill his desire to become a champion is ultimately only his business.

D. THE SUCCESSFUL MARKSMANSHIP INSTRUCTOR

Marksmanship instruction involves presenting ideas and doctrine to students in a concentrated training schedule. To carry out this assignment successfully, the instructor must demonstrate certain positive qualities. Many of the necessary characteristics are the same as those which make a successful marksman. The instructional traits listed approximately in the order of importance are: knowledge of subject; demonstrated ability to communicate ideas; ability to organize lesson materials; knowledge of instructional methods and skill in techniques of presentation and an ability to grow professionally. Instructors must also have a favorable attitude towards teaching and exercise a fair degree of personal imagination.

The marksmanship instructor must have a deep, full knowledge of his subject. An outstanding instructor does not parrot other instructor's examples and phrases inherited from the vault file. Rather, he constantly searches for current, meaningful examples or appropriate events to support his teaching points. The doctrine or teaching points may remain the same but the presentation has been personalized by his own research. So rich is his background of teaching experience and training that he can use a variety of presentation methods to maintain student interest. No two of his presentations are identical even though the teaching points may be the same. The ebb and flow of student questions and discussion condition each presentation so that there is always something new and stimulating. He is aware of correct conference and demonstration techniques and is flexible enough to adapt them to each specific teaching situation.

A good instructor realizes that securing rapport or a sense of close communication between himself and the student body is difficult. Nevertheless, he is quick to appreciate the attitude of his class and he adapts to the student point of view in presenting his material. It is this sensitivity to student reaction which makes him a successful instructor. He quickly realizes when his class is having difficulty in understanding the subject matter and attempts to clear up this difficult by encouraging student participation through discussion. He is aware of the significance of students' questions and skillful at answering them. He is adept at summarizing student comments in language understandable to the class. By proper handling of student questions, comments, and responses he creates a rapport within the class which facilitates learning. His thinking and expression are in terms of ideas, not words.

The lesson plan of a capable instructor reflects a good sense of organization. The instructor has carefully researched his subject and outlined, planned and rehearsed his instructional material. He has timed his presentation and every orderly element of it is an integral and necessary part. Since his students are working within a training schedule, it is imperative that his presentation start and stop on time. Yet his primary concern is to ensure student learning; hence, he must allow sufficient time for his students to understand the teaching points and to work the practice exercises which were designed to reinforce learning or to practice the application of the principles learned.

An instructor must display a great deal of personal force. Sometimes he may be presenting his instructional material out-of-doors, on the range, where distractions are commonplace. Here he must hold the attention of his audience and maintain their interest in the subject matter. Under all circumstances, he must be convinced of the truth and significance of what he is saying and his task is to persuade and convince his students. As the principal instructor, he is the acknowledged authority on the subject and he takes his stand firmly so there is no doubt left in the student's mind.

Although his diction and vocabulary may not be so polished as that of a network announcer or university professor, he nevertheless uses acceptable grammar and has the vocabulary and fluency of speech that enables one man to communicate effectively with another. He can think on his feet and give competent answers to student's questions. He gains the respect and wholehearted cooperation of his class because of his sincerity, his sense of humor and his enthusiasm.

The instructor is sold on the worth and value of what he is presenting. His interest in his students and the subject matter is contagious. He is able to project this enthusiasm even to large classes of two hundred or more students so that they too take an active, enthusiastic part in the learning process. It is this active student involvement and the give and take of thought-stimulating questions and ideas which characterizes the teamwork necessary to develop top-notch marksmen.

Part of his task is to help train new instructors. While these beginners are receiving an instructor training course, he may be designated as a student's sponsor. He cheerfully supplies all necessary administrative and logistical support. He rehearses with him conscientiously. He is careful not to dominate the students learning process by insisting that the student parrot his words or mimick his actions. He knows that instructor individuality is essential for effective teaching.

The instructor's appearance plays a key part in his presentation. His voice must be firm, positive and convincing. His manner should be poised, friendly and personal

E. SCHEDULING COURSES OF INSTRUCTION

There follows an example of a five day period of instruction for a pistol marksmanship clinic. This is to be utilized mainly for accelerated courses of instruction with a minimum of practical work, that is, range practice and coach-shooter teamwork. The ideal period of time on which to base this cycle of training is approximately four (4) weeks. The expanded schedule would include an increase in the hours of practical work and repetition of certain basic phases of instruction that need reiteration. An example of a weekly training schedule is found in Chapter XXII, "Marksmanship Squad Administration

Annex I to Training Standards and Scheduling Courses of Instruction.

Subject: Example of A Marksmanship Clinic Training Schedule

THE UNITED STATES ARMY MARKSMANSHIP UNIT
Fort Benning, Georgia 31905

Army Area Pistol Marksmanship Coach/Instructor Clinic Training Schedule

Day & Time	Subject	Area	Instructor	Uniform	Remarks
1st Day					
0720-0730	Introduction	Classroom	OIC USAMU Instructor Group	Duty	All students must have weapons and equipment (Distribute student questionnaire) Bring Pistol Manual, paper and pencil to all classes.
0730-0820	Fundamentals I (Minimum arc of movement)	"	USAMU Instructor	"	
0830-0920	Fundamentals II (Sight Alignment)	"	"	"	
0930-0955	Fundamentals III (Trigger control)	"	"	"	(Collect student questionnaire) (Prepare master scoreboard with team relay and target assignment for each student)
1000-1050	Technique of Fire I (Establish a system)	"	"	"	20 rds .22 cal * 20 rds .45 cal (Ser)*
1100-1200	Zeroing .22 cal and .45 cal at 25 and 50 yds (Off relay, Competitive Regulations II (Range procedure and range safety)	Pistol Range "	All Instructors (USAMU Instructor)	" "	*Individual ammunition requirement for each student. (Announce team relay and target assignment for each student)
1300-1400	Zeroing .22 cal and .45 cal at 25 and 50 yards (Off relay, Competitive Regulations II (Range Procedure and range safety)	" "	All Instructors (USAMU Instructor)	"	20 rds .22 cal. 20 rds .45 cal (Ser)
1400-1450	Technique of Fire II (Slow Fire)	"	USAMU Instructor	"	Bleachers needed in range area

Day & Time	Subject	Area	Instructor	Uniform	Remarks
1st Day					
1500–1600	.22 cal and .45 cal service pistol slow fire practice at 50 yards (Off relay, Competitive Regulations III, national trophy pistol match rules.)	Pistol Range "	(USAMU Instructor) (USAMU Instructor)	Duty "	30 rds .22 cal 30 rds .45 cal (Serv)
1600–1700	.22 cal and .45 cal service pistol slow fire practice at 50 yards (Off relay, Competitive Regulations III, National Trophy Pistol Match Rules)	"	All Instructors (USAMU Instructor)	"	30 rds .22 cal 30 rds .45 cal (Ser)
2d Day					
0730–0800	Technique of Fire III (Sustained Fire) .22 cal and .45 cal service pistol sustained fire practice (Off relay sustained fire critique)	"	USAMU Instructor All Instructors	"	40 rds .22 cal 40 rds .45 cal (Ser)
0845–0930	.22 cal and .45 cal service pistol sustained fire practice (Off relay, sustained fire critique)	"	"	"	40 rds .22 cal 40 rds .45 cal (Ser)
0930–1020	Technique of Fire IV (International Slow Fire)	Classroom	USAMU Instructor	"	
1030–1120	Technique of Fire VI (International Rapid Fire)	"	"	"	
1130–1200	Technique of Fire V (International Center Fire)	"	"	"	
1300–1350	Coaching I (Attributes, Aids and Duties)	"	"	"	
1400–1450	Coaching II (Technique)	"		"	
1500–1515	Demonstration of Coaching Technique	"	"		

Day & Night	Subject	Area	Instructor	Uniform	Remarks
2d Day					
1520-1610	.22 cal and .45 cal Service Pistol National Match Course (Practice) (Off relay, half Course International Rapid Fire at 25 meters)	Pistol Range	All Instructors (USAMU Instructor)	Duty	60 rds .22 cal 30 rds .45 cal (Ser) (Students may fire this course with standard .22 cal pistol)
1610-1700	.22 cal and .45 cal Service Pistol National Match Course (Practice) (Off relay, half course International Rapid Fire at 25 meters)	"	"	"	60 rds .22 cal 30 rds .45 cal (Ser) (Students may fire this course with standard .22 cal pistol)
3d Day					
0730-0820	Mental Discipline	Classroom	USAMU Instructor	"	
0830-0920	Physical Fitness I (Conditioning)	"	"	"	Practical Exercises will require outdoor class area
0930-1020	Physical Fitness II and III (Diet and effects of alcohol, coffee, tobacco and drugs)	"	"	"	
1030-1150	Competitive Regulations I (NRA Pistol Match Rules)	"	"	"	Hand out pistol rules questionnaire and NRA Pistol Rule Book
1300-1430	.22 cal and .45 cal Service Pistol NMC (Record) (Off relay, half course International Slow Fire at 50 meters)	Pistol Range	All Instructors (USAMU Instructor)	"	60 rds .22 cal 30 rds .45 cal (post scores on master scoreboard)
1430-1600	.22 cal and .45 cal Service Pistol NMC (Record) (Off relay, half course International Slow Fire at 50 meters)	"	"	"	"
1600-1700	Weapons and Equipment Inspection	"	Unit Armorers	"	Repairs made at option of local unit commander

Day & Night	Subject	Area	Instructor	Uniform	Remarks
4th Day					
0730-0820	Instructor Training I (Method of Instruction)	Classroom	USAMU Instructor	Duty	
0830-0920	Instructor Training II (Control of Interest and Effective Speaking)	"	"	"	
0930-1020	Instructor Training III (Lesson Planning and Use of Training Aids)	"	"	"	
1030-1150	**Instructor Training Practical Exercise**	"	"	"	Hand out lesson plan format
1300-1430	.22 cal and .45 cal Service Pistol NMC Team Match (Off relay, full course International Center Fire at 25 meters, precision and duel)	Pistol Range	All Instructors (USAMU Instructor)	"	(Collect Duplicate team score-cards) 30 rds .22 cal 60 rds .38 cal (Wadcutter) 30 rds .45 cal (Service) (Post scores on master score-book)
1430-1600	.22 cal and .45 cal Service Pistol NMC Team Match (Off relay, rull course International Center Fire at 25 meters, precision and duel)	"	"	"	30 rds .22 cal 60 rds .38 cal (Wadcutter) 30 rds .45 cal (Service) (Collect duplicate team score-cards) (Post scores on master scoreboard)
1600-1700	Critique of Student Performance by Line Coaches	"	"	"	
5th Day					
0730-0920	Combat Pistol Course, Demonstration and Practical Exercise	"	USAMU Instructor	"	Distribute student comment sheets. 30 rds .45 cal (Service)

Day & Night	Subject	Area	Instructor	Uniform	Remarks
5th Day					
0930-1050	Review and critique of coach/instructor course	Classroom	USAMU Instructor	Duty	Coach/Instructor students
0930-1050	Review of Fundamentals of Pistol Marksmanship	Pistol Range	USAMU Instructor	"	Pistol Marksmanship students
1100-1200	Pistol Marksmanship Examination	Classroom	"	"	Designate Honor Graduate
1300-1350	Panel Discussion	"	"	"	Collect student comment sheets
1400-1500	Preparation for Closing Ceremony	"	OIC USAMU Instructor Group	TBA	Brief students on ceremony procedure and designate seating arrangement.
1500-1600	Closing Ceremony - Awards and Certificates	TBA	"	"	Attendance certificates and honor graduate certificate prepared for distribution.
1600-1700	Turn-In of equipment, clear billets and sign out	TDY Unit Area	"	Optional	Load all instructional equipment in designated vehicles. Prepare after-action report to include categories and number students. a. US Army f. Reserve b. US Navy g. Police c. US Air Force h. College d. US Marines i. Civilian e. US Coast j. Total Guard

CHAPTER XVII

LESSON PLANS FOR SERVICE PISTOL MARKSMANSHIP INSTRUCTORS'
AND COACHES' COURSE

A. FUNDAMENTALS I 　　　　　　　　　　　　　　　　　　　　　AAPMI+CC301
    (ATTAINING A MINIMUM ARC OF MOVEMENT)　　　　　　　　　　50 Minutes
    　　　　　　　　　　　　　　　　　　　　　　　　　　　　　MAR 1975

LESSON OUTLINE

I. LESSON OBJECTIVE: To enable the Pistol Marksmanship students to attain a minimum arc of movement by the proper employment of the fundamentals of stance, position, grip, and breath control.

II. STUDENT PERFORMANCE OBJECTIVES: As a result of this instruction, the student must be able to accomplish the following student performance objectives:

　　A. Given the main requirements of a proper stance or shooting posture, DESCRIBE the basic physical characteristics of the human body that affect the attainment of a good shooting stance and ASSUME a stance that provides stability, immobility and correct head position as outlined in Chapter I, USAMU Pistol Marksmanship Guide.

　　B. Given the definition and requirements of a shooting position, DISCLOSE why natural alignment of the shooting arm and body with the center of the target is essential and DEMONSTRATE a method of orienting the body position so that it naturally points the shooters extended arm at the center of the aiming area on the target as outlined in Chapter I, USAMU Pistol Marksmanship Guide.

　　C. Given the requirements of a proper grip on the pistol, DEMONSTRATE and STATE the steps in attaining a proper grip and EXPLAIN the method of checking the grip to assure the attainment of natural sight alignment as outlined in Chapter I, USAMU Pistol Marksmanship Guide.

　　D. Given the objective of breath control in shooting, EXPLAIN briefly the physiological process of breathing and REGULATE the breathing by proper breath control prior to and in conjunction with firing as outlined in Chapter I, USAMU Pistol Marksmanship Guide.

III. ADVANCE ASSIGNMENT: USAMU Pistol Marksmanship Guide, Chapter I, (Attaining a Minimum Arc of Movement.)

IV. INTRODUCTION:

　　A. Gain Attention: Have your pistol scores improved enough in the last year to enable you to win a pistol match? If not, do you know why?

　　B. Orient Students:

　　　　1. Lesson Tie-In: Advanced pistol marksmanship is the highly skilled and trained part of competitive pistol shooting that enables the master pistol shooter to fire winning scores. This period of instruction will be devoted to the fundamentals that are the foundation of advanced pistol marksmanship.

　　　　2. Motivation: Pistol Champions are not born that way. They are the ultimate evolution of the competitive pistol shooter. Through intense, continuous hard work they have gained the ability to control the employment of the fundamentals in practice and in matches. Winning scores do not come easy. Concentration on learning the fundamentals is the first important consideration in readying oneself to master the pistol.

NOTE: SHOW SLIDE #1 - ATTAINING A MINIMUM ARC OF MOVEMENT

        3. Scope: You will be better able to attain minimum arc of movement by applying the fundamentals of pistol marksmanship if you can:

            a. Choose a stance that provides stability, immobility and correct head position.

            b. Select the position which naturally orients the body and shooting arm with the center of the target area.

            c. Employ a uniform grip on the weapon that achieves natural sight alignment.

            d. Regulate your breathing by proper breath control.

V. BODY:

NOTE: SHOW SLIDE #2 - ILLUSTRATION OF A MINIMUM ARC OF MOVEMENT.

TRANSITION: QUESTION: How would you explain to a beginner if he asks "What is the minimum arc of movement?

ANSWER: This is a phrase used to describe the movement of the arm and body of the shooter during aiming - the shooter's ability to hold.

        A. <u>First Student Performance Objective</u>: Given the main requirements of a proper stance or shooting posture, DESCRIBE the basic physical characteristics of the human body that affect the attainment of a good shooting stance and ASSUME a stance that provides stability, immobility and correct head postion as outlined in Chapter I, USAMU Pistol Marksmanship Guide.

           1. <u>General</u>.

                a. There is a direct relationship between accuracy in shooting and the degree of body movement. Therefore, the shooter should fire only during the period when the arc of movement is at a minimum.

                b. In pistol marksmanship the words stance and position have been used to describe both the orientation of the shooters body in relation to the target and the posture of the body while shooting. For the purpose of this instruction, stance is defined as: The arrangement or posture of the body only.

                c. The stance is a major factor in the shooters system of control. Individual body configuration (height, weight and muscle development) permits no all-purpose stance. The shooter may develop whatever variation is necessary to conform to his needs after careful analysis. There are, however, some basic rules which should apply in selecting the stance.

QUESTION: What are the requirements of a shooter's stance that contribute to attaining a minimum arc of movement.

ANSWER: 2. <u>Main requirements of the stance</u>:

                a. Greatest possible degree of stability or balance without muscle strain.

                b. Greatest possible degree of immobility, i.e., the absence of an independent movement of any part of the body. The body must be like a statue.

     c. Correct head position for best use of eyes and the preservation of body equilibrium.

TRANSITION: These are the requirements. To arrange the body into a stance, the shooter should be acquainted with certain important features of the human body.

    3. <u>Characteristics of the Human Body</u>: The human body is divided into two main systems, active and passive.

     a. <u>The Passive Apparatus</u> includes the bones and ligaments which exert an influence on muscle stamina when the body assumes a static pose. The bones and ligaments must bear most of the burden of the shooting stance because the muscles tire relatively quickly.

     b. Active: The active portion of the body includes the muscles and nervous systems.

      (1) There are over 600 muscles in the body. The use of the muscles should be just enough to hold the body in a static pose.

      (2) The nervous system controls the flexing of muscles due to various stimuli. One of these stimuli is the impulses sent from the central nervous system to control body balance.

      (3) Neck tendons and skin in the area of the neck also effect balance.

      (4) The eyes perform an important function in maintaining the balance by also transmitting impulses toward restoring balance anytime the eye ball axis is not situated so the vision will be straight forward and level in relation to the horizon

TRANSITION: These are some of the factors that control the body. The shooter should become familiar with assuming the proper stance in accord with these control systems and practice getting the same stance each time.

NOTE: SHOW SLIDE #3 - STANCE.

NOTE: DEMONSTRATOR TO HIS POST AND IS DIRECTED BY INSTRUCTOR IN ASSUMING THE STANCE.

    4. <u>The Stance</u>:

     a. In the position of attention or natural standing posture the body has a tendency to rock and sway. This is due to the small support area provided by the feet and area in between. Some shooters may feel that the legs spread widely to an extreme, will provide a larger support area and reduce body movement. The placement of the feet increases the load on the inner arch of the foot and places a greater strain on holding the leg muscles in tone.

QUESTION: How far apart should the shooter place his feet?

ANSWER: b. Placing of Feet: shoulder width apart, toes pointed out at a slight angle.

     c. Weight load: equal on both feet with balance shifted slightly forward on the toes. This is to reduce the flexing of the lower back, leg and abdominal muscles caused by the corrective action of the balance reflexes.

     d. Legs: no tension on the muscles except to remain upright.

     e. Knees: firmly straight, not locked.

     f. Hips: as the weapon is held away from the body's center of gravity by muscle support, the hips may be shifted forward to counterbalance the weight of the weapon while keeping the center of gravity over the support area.

     g. Abdomen: relaxed.

     h. Shoulders: level with the non-shooting side relaxed.

     i. Head: erect in an upright position with no inclining to either side.ide.

     j. Non-shooting arm: relaxed with no tension and the hand in the pocket or hooked on the belt.

     k. Shooting arm: solid arm control. Elbow locked and wrist firm.

NOTE: DEMONSTRATOR IS GIVEN COMMAND "AT EASE".

TRANSITION: Having assumed a stable stance, the shooter must be able to place his body in natural alignment to the target center.

NOTE: SHOW SLIDE #4 - POSITION.

  B. <u>Second Student Performance Objective</u>: Given the definition and requirements of a shooting position, DISCLOSE why natural alignment of the shooting arm and body with the center of the target is essential and DEMONSTRATE a method of orienting the body position so that it naturally points the shooters extended arm at the center of the aiming area on the target as outlined in Chapter I, USAMU Pistol Marksmanship Guide.

   1. <u>Position</u>:

    a. Position is defined as the relationship of the shooter's <u>body</u> to the target. Before each shot or string, it is necessary to check this relationship.

NOTE: DEMONSTRATOR IS DIRECTED BY INSTRUCTOR IN THE METHOD OF ORIENTATION.

    b. Method of Orientation:

     (1) Face to left of target 40-50 degrees.

     (2) Turn head to face directly to target.

     (3) Raise arm to target and close eyes. Swing body and arm from ankles, come to rest with natural point.

     (4) Open eyes - check alignment of arm to target.

     (5) Move rear foot in direction of error to change angle of body relation to the target. Pivot on forward foot.

     (6) Recheck.

NOTE: DEMONSTRATOR EXCUSED.

    c. Devote serious attention to perfecting stance and position. Do not blindly copy another individual. Some shooters shoot well in spite of faulty stance and position. Make an intelligent approach to attaining good stance and position by careful analysis of your particular needs.

TRANSITION: Position provides for natural alignment of the body with the target during periods of firing. It is during these firing periods that the shooter must also maintain control of the pistol by proper grip.

  C. <u>Third Student Performance Objective</u>: Given the requirements of a proper grip on the pistol, DEMONSTRATE and STATE the steps in attaining a proper grip and EXPLAIN the method of checking the grip to assure the attainment of natural sight alignment as outlined in Chapter I, USAMU Pistol Marksmanship Guide.

<u>NOTE</u>: SHOW SLIDE #5 - GRIP (SIDE VIEW LEFT)

  1. The proper grip provides maximum control in maintaining sight alignment while applying pressure to the trigger.

  2. The most important single factor in attaining the grip is uniformity. There cannot be any variation in the character of the grip each time the weapon is placed in the hand.

  3. The proper grip on a pistol is one that will meet the following requirements:

    a. Natural sight alignment.

    b. Independent movement of the index or trigger finger.

    c. Force of recoil must travel straight to rear into the shooting arm and shoulder.

    d. A firm grip to prevent the weapon from shifting in the hand without a change of grip pressure.

    e. Comfortable during periods of firing.

<u>TRANSITION</u>: Your ability to consistently place the strike of the bullet in the center of the target depends to a great extent on a proper grip on the pistol.

<u>NOTE</u>: SHOW SLIDE #6 - GRIP (SIDE VIEW RIGHT)

<u>NOTE</u>: DEMONSTRATOR POSTS AND IS DIRECTED BY INSTRUCTOR IN A SEQUENCE OF PLACING THE WEAPON IN THE HAND TO OBTAIN THE PROPER GRIP.

  4. The following is a step-by-step sequence that will provide the proper grip:

    a. Hold pistol by barrel with non-shooting hand.

    b. Spread index finger and thumb to form a "V".

    c. Bend wrist down.

    d. Seat pistol in "V" of hand.

    e. Fit pistol firmly into gripping space.

    f. Grasp stock with lower three fingers.

    g. Thumb placed high on side of stock at a higher level than index finger.

    h. Place index finger on trigger.

           i.   Tighten grip to maximum with tremor.

           j.   Relax grip slightly until tremor disappears.

      5.   The shooter should check out the grip to see that it meets the requirements of the grip. He may do this by asking himself these questions or have a coach perform the tests described below.

NOTE: SHOW SLIDE #7 - GRIP (TOP VIEW).

           a.   With shooting arm extended will the sights stay in natural alignment? If not, reposition the pistol in the shooting hand.

           b.   Does the placement of index finger provide for independent movement of the trigger finger? Make a visual check of the index finger action and dry fire to answer this question. Do the sights move during dry firing?

           c.   Will the force of the recoil travel straight back into the shooting arm? This can be checked by using unloaded weapons. Have a coach simulate recoil by pushing abruptly against the shooting hand.

           d.   Will the grip be firm enough to prevent shifting of the weapon during firing? Again, using an unloaded weapon, have the coach apply pressure by a solid blow to the barrel of the pistol.

           e.   Will the grip be comfortable during long periods of firing? The shooter can answer this by assuring himself that he has followed the sequence of placing the weapon in the hand that precludes any folding or pinching of the skin or the palm.

      6.   The proper grip will provide for the maximum control of the weapon during the firing of a controlled shot.

NOTE: DEMONSTRATOR IS EXCUSED.

TRANSITION: Stance, position and grip are fundamentals and most shooters are aware of their importance. However, the value of breath control in attaining a minimum arc of movement is often overlooked as it is an involuntary reflex action.

      D.   **Fourth Student Performance Objective**: Given the objective of breath control in shooting, EXPLAIN briefly the physiological processes of breathing and REGULATE the breathing by proper breath control prior to and in conjunction with firing as outlined in Chapter I, USAMU Pistol Marksmanship Guide.

      1.   Breath control is an essential part of the shooters system of control. The object of breath control is to hold the breath with comfort during firing so that there is no conscious need to breathe.

QUESTION: Why should you not breathe while aiming the pistol?

ANSWER: 2.   Breathing is accompanied by a rhythmical movement of the chest, shoulders and stomach which will enlarge the shooters arc of movement therefore he should not fire while breathing.

NOTE: SHOW SLIDE #8 - BREATH CONTROL.

      3.   One respiratory cycle consists of an inhalation, exhalation and a respiratory pause. The shooter should hold his breath only during respiratory pause.

NOTE: SHOW SLIDE #9 - RESPIRATORY PAUSE.

        4. The shooter can prolong the normal respiratory pause by deeper breathing prior to firing. This will temporarily decrease the carbon dioxide level in the blood and lengthen the period of comfort while not breathing.

        5. Breath control in coordination with firing or with the fire commands aids the shooter to be systematic about respiration during periods of competition. (Demonstrate)

        6. The respiratory pause is the period between inhalation & exhalation.

TRANSITION: The stable foundation provided by the careful selection of a stance, position grip and the means of breath control will permit the shooter to approach the stress of competitive firing with confidence. He will be able to apply the other fundamentals of sight alignment and trigger control with greater precision.

    VI. CONCLUSION:

        A. <u>Retain Attention</u>: Ask a top shooter "How can I improve my shooting performance?" His advice to you will probably be: "Start with the fundamentals."

        B. <u>Summary</u>:

NOTE: SHOW SLIDE #10 - SCOPE AND SUMMARY.

        1. The smallest arc of movement must be attained by choosing of a stance that provides stable balance with the least body movement. The shooter then selects the position that naturally aligns his body to the target, employs a uniform grip on the weapon that gives him natural sight alignment and controls his breathing before and during firing.

        C. <u>Application</u>: Good scores are an indication of how well the fundamentals have been applied. During the school you can prove to yourself that there is no substitute for applying the fundamentals correctly. Fire the top score of your shooting career.

NOTE: DISPLAY THE USAMU PISTOL MARKSMANSHIP GUIDE, HOLDING IT UP IN VIEW OF THE STUDENTS.

        D. <u>Closing Statement</u>: This is your pistol shooting bible. The first 71 pages are devoted exclusively to fundamentals. Throughout the remaining text on various techniques, constant references are made to the fundamentals. If you fail to understand or properly use the fundamental of attaining a minimum arc of movement, you will fail in the application of sight alignment, trigger control and the various techniques of fire.

B. FUNDAMENTALS II
   (Sight Alignment)

AA PMI&CC 203
50 Minutes
March 1975

LESSON OUTLINE

I. LESSON OBJECTIVE: To enable the Pistol Marksmanship student to achieve and maintain perfection in sight alignment by proper employment of the factors of sight alignment.

II. STUDENT PERFORMANCE OBJECTIVE: As a result of this instruction the student must be able to accomplish the following student performance objectives:

    A. Given the factors of establishing perfect sight alignment; EXPLAIN perfect sight alignment as described in Chapter II, USAMU Pistol Marksmanship Guide.

    B. Given the requirement of attaining perfect sight alignment; FOCUS clearly and precisely on a point on the front sight as illustrated in Chapter II, USAMU Pistol Marksmanship Guide.

    C. Required to fire an accurate shot, EMPLOY intense, uninterrupted mental concentration of sufficient duration to maintain proper relationship between the front and rear sight until the shot is fired as shown in Chapter II, USAMU Pistol Marksmanship Guide.

    D. Given the requirement to relate the normal function of the human eye to accuracy in pistol shooting; briefly DESCRIBE monocular and binocular vision, the determination of the dominant eye, and methods of employing the human eye for the best results in perfecting and maintaining sight alignment as shown in Chapter II, Annex 2 USAMU Pistol Marksmanship Guide.

III. ADVANCED ASSIGNMENT: Chapter II, USAMU Pistol Marksmanship Guide.

IV. INTRODUCTION:

    A. <u>Gain Attention</u>: Sight alignment is the most important of all the fundamentals.

    B. <u>Orient Students</u>:

        1. <u>Lesson Tie-In</u>: The class, Fundamentals I. "Attaining A Minimum Arc of Movement," had the purpose of teaching you how to hold the weapon still. This hour of instruction is designed to establish that a hit on the target within the ability to hold is best achieved by proper sight alignment.

        2. <u>Motivation</u>: The next National Champion will be the competitor who holds his weapon still, perfects sight alignment and employs trigger control so as to achieve a surprise shot without disturbing the sight alignment more times during his two hundred and seventy (270) shots than any other shooter entered in the National Championship A large percentage of the shooters presently holding National records believe maintaining sight alignment to be fundamental requiring the most effort to master. They devote the most intense mental and physical effort towards maintaining absolute perfection in the relationship between front and rear sight.

NOTE: SHOW SLIDE 1A - SCOPE AND SUMMARY.

        3. <u>Scope</u>: Every shooter should know the technique of perfecting and maintaining front and rear sight relationship. This information will enable you to:

            a. Control the relationship of the sights in order to insure the accurate placement of the bullet on the target.

     b. It will enable you to achieve a point of focus on the front sight and to understand why this is so necessary to accurate shooting.

     c. It will enable you to develop the ability to apply intense, uninterrupted mental concentration on the vital fundamental of sight alignment and to realize the effect anxiety has on your ability to concentrate.

     d. It will help you to understand the function of monocular and binocular vision find the dominant eye and, to use your eyes to the best advantage in shooting within the physical limitations of the human eye.

V. BODY:

  A. <u>First Student Performance Objective</u>: Given the factors of establishing perfect sight alignment; EXPLAIN perfect sight alignment as described in Chapter II, USAMU Pistol Marksmanship Guide.

<u>QUESTION</u>: What elements do we refer to when we speak of "Aim" in firing a pistol?

<u>ANSWER</u>: 1. "Aim" is holding the pistol with the smallest possible movement about a selected point on the target and achieving the exact alignment of the front and rear sights.

     2. The essence of competitive shooting is the accurate placement of bullets in the center of a comparatively small target.

     3. In order for the shooter to achieve center hits, the barrel of the pistol must be held in a definite direction relative to the target.

For this reason, proper relationship of the front and rear sight to the shooter's eye must be maintained.

     4. We must consider three (3) factors for proper understanding of sight alignment.

<u>NOTE</u>: SHOW SLIDE #1 - STANDARD AMERICAN FIFTY (50) YARD PISTOL TARGET.

     a. First, we must have something at which to aim. This can be the target we use in competition. The target is a stationary element. The bulls-eye in the center is only a reference point.

<u>NOTE</u>: SHOW SLIDE #2 - A REAR SIGHT IN POSITION BENEATH THE AIMING BULLS-EYE WITH A
    PERFECT SIX O'CLOCK HOLD.

     b. The second factor we must consider is the rear sight. It's importance lies in the rectangular notch wherein the front sight must be centered and aligned.

<u>NOTE</u>: SHOW SLIDE #3 - A FRONT SIGHT IN POSITION BENEATH THE AIMING BULLS-EYE WITH A
    PERFECT SIX O'CLOCK HOLD. CENTER THE FRONT SIGHT IN THE REAR SIGHT NOTCH WITH AN
    EQUAL SPACE ON BOTH SIDES AND ASSURE THAT THE LEVEL SURFACE OF THE TOP OF THE FRONT
    SIGHT IS EXACTLY LEVEL WITH THE TOP OR HORIZONTAL SURFACE OF THE REAR SIGHT.

     c. The third factor to consider is the front sight. It is centered in the rear sight notch with equal space on both sides and the horizontal top surface of the front sight is level with the flat top surface of the rear sight. What you see here is a perfect sight alignment superimposed on a target in clear focus. This is referred to as perfect sight picture. It is a physical impossibility for your eye to transmit this picture to the brain when actually shooting.

QUESTION: Why is this a true fact?

ANSWER: Because the eye is like a camera. It cannot focus on two objects at varying distances.

        5. The design of the pistol sight takes advantage of two characteristics of the human eye.

           a. First, is the eyes ability to discern very minute areas, points or objects such as the space on each side of the front sight when viewed through the rear sight notch.

           b. Second, is the eyes natural attraction to any flaw or irregularity in all even or regular surfaces. Note that a straight edge can be placed in contact with the top surface of both front and rear sights. Should this alignment change by a lowering of the front sight in the rear sight notch, your eye is attracted to the irregularity. Should any of you be using what is called a fine bead, understand that this deliberate error in sight alignment is very difficult to maintain uniformly. The sights are properly aligned only when the top of the front sight is level with the top of the rear sight and exactly centered in the rear sight notch, i.e., an equal amount of light space on each side of the front sight.

NOTE: SHOW SLIDE #4 ANGULAR SHIFT ERROR CAUSED BY MISALIGNING THE FRONT AND REAR SIGHT.

        6. Study this chart carefully. Even though you are pointing the weapon very accurately at your aiming area, any error in sight alignment causes an acute angular error in the alignment of the weapon with the target. For every one-hundredth (.01) inch of error in alignment, there will be three (3) inches of error at fifty (50) yards.

NOTE: SHOW SLIDE #5 A FIFTY (50) YARD PISTOL TARGET. SUPERIMPOSE A MOVABLE SIGHT ALIGNMENT DEVICE NEAR THE SIX O'CLOCK AREA OF THE BULLS-EYE.

           a. On the other hand errors in pointing are much nearer parallel so long as the relationship between front and rear sight is correct. Pointing or hold errors cause far less error in the strike of the bullet on the target.

NOTE: DEMONSTRATE THE ARC OF MOVEMENT BY MOVING THE SIGHT ALIGNMENT DEVICE AROUND IN THE AIMING AREA WHILE MAINTAINING SIGHT ALIGNMENT. PASTE ON FIVE (5) SMALL WHITE DISKS AS TARGET HITS.

           b. A rather small dispersion of hits can be expected if you hold anywhere near the center of your aiming area provided the sights remain in perfect alignment during the arc of movement. All shots will hit the target within the shooters ability to hold.

TRANSITION: You must realize the importance of sight alignment and what happens when errors in alignment occur. You cannot have sufficient control over sight alignment to prevent angular shift errors when attempting to see the perfect sight picture.

        B. Second Student Performance Objective: Given the requirement of attaining perfect sight alignment; FOCUS clearly and precisely on a point on the front sight as illustrated in Chapter II, USAMU Pistol Marksmanship Guide.

        1. By definition--sight alignment is the relationship between the front and rear sight as seen by the human eye.

QUESTION: What is the most accurate method of maintaining sight alignment?

ANSWER: 2. Focus the vision clearly on a point on the front sight and be aware of its relationship with the rear sight notch.

3. There are certain limitations in the optical properties of the human eye that make the locking of the focus on one point necessary.

   a. The human eye is capable of focusing on one object at a time and the area of clear vision on that object is limited in size.

   b. The controlling muscles tire very quickly if called upon to make rapid shifts in point of focus over an extended period of time.

4. The type of work we are requiring our eyes to perform in pistol shooting forces us to select one point of focus if we are to avoid misuse of the eyes.

   a. Remember that the target is stationary, and since it does not move in depth or horizontally, there is no need to focus the vision on it. The target may appear to move in relation to the weapon but this illusion is caused by your minimum arc of movement. This fact eliminates the target as a possible point of focus.

NOTE: SHOW SLIDE #6 STANDARD AMERICAN FIFTY (50) YEAR PISTOL TARGET SLIGHTLY OUT OF FOCUS AND APPEARS BLURRED.

   b. The rear sight is located over the pivot point of the wrist and is relatively stable. Eliminate the rear sight as a point of focus.

NOTE: SHOW SLIDE #7 A SLIGHTLY FUZZY OR ALMOST CLEARLY DEFINED REAR SIGHT. IT IS PLACED WITH THE TOP EDGE OF THE REAR SIGHT NOTCH AT SIX O'CLOCK ON THE AIMING BULLS-EYE.

   c. That leaves the last and most important factor, the front sight. This element alone must be our point of focus. The target must remain out of focus if the shooter is to attain perfect sight alignment. The relationship of the front sight to the rear is the primary consideration and to accomplish a precise relationship they must be seen distinctly.

NOTE: SHOW SLIDE #8 A CLEAR DISTINCT FRONT SIGHT PLACED IN THE REAR SIGHT NOTCH WITH THE HAZY BULLS-EYE IN THE BACKGROUND.

5. When you use your eyes properly the front sight will be clear and distinct, the rear sight will be of slightly less clear definition for some shooters, yet it's outline is easily discernable. The target is hazy or fuzzy and indistinct in outline.

6. The front sight is the most unstable element of the three. It is constantly changing its relationship to the rear sight. For this reason complete attention must be devoted to it. For this concentrated attention to result in perfect sight alignment the shooter must constantly strive to keep the front sight clearly in focus and properly positioned in the rear sight notch.

7. Your ability to maintain perfect sight alignment is in direct proportion to the degree of concentration on maintaining the correct relationship of the two sights.

8. Hold your weapon so as to cause the front sight to appear centered in and level with the top of the rear sight. Maintain this relationship while pointing the weapon at the aiming area on the target. Hold the weapon as still as possible thus establishing a minimum arc of movement.

TRANSITION: Keeping your eye focused on the front sight in a point of focus requires a tremendous amount of effort and concentration.

C. <u>Third Student Performance Objective</u>: Required to fire an accurate shot, EMPLOY intense, uninterrupted mental concentration of sufficient duration to maintain proper relationship between the front and rear sight until the shot is fired as shown in Chpater II, USAMU Pistol Marksmanship Guide.

QUESTION: How long can the average pistol shooter expect to be able to concentrate on sight alignment without interruption?

NOTE: SHOW SLIDE #10 FACTORS APPLICABLE TO MENTAL CONCENTRATION ON SIGHT ALIGNMENT.

ANSWER: 1. Detailed testing indicates that a genius can maintain concentration on only one thing to the exclusion of all else for about eight (8) to ten (10) seconds. The average pistol shooter concentrates from three (3) to six (6) seconds without interruption. This limitation dictates that he must fire promptly or desist and bench the weapon. He must then replan, relax and make a new effort to fire the shots.

2. Carelessness or failure to initiate the necessary degree of concentration results in errors in sight alignment remaining uncorrected to direct the bullet to some point on the target other than the ten ring.

3. Any dilution of thought generally results in a momentary lessening of concentration on sight alignment. Random thoughts of anxiety about apparently stationary pressure on the trigger is an example. They tend to generate an impluse to get more pressure on the trigger. The maintenance of perfect sight alignment suffers and the added trigger pressure usually fires the pistol before concentration returns to the front sight.

TRANSITION: Knowing the critical nature of sight alignment, we must use every faculty available to us to insure that we perform to the utmost in applying this fundamental. Techniques for holding the weapon still were discussed earlier. Trigger control will be covered in a later class. Let us devote our attention to the use of the eye. This is the one absolutely vital element in perfecting sight alignment.

D. <u>Fourth Student Performance Objective</u>: Given the requirement to relate the normal function of the human eye to accuracy in pistol shooting; briefly DESCRIBE monocular and binocular vision, the determination of the dominant eye, and methods of employing the human eye for the best results in perfecting and maintaining sight alignment as shown in Chapter II, Annex 2 USAMU Pistol Marksmanship Guide.

1. Monocular and binocular vision.

a. Monocular vision occurs when one eye is closed or blocked from seeing as by a flap or patch over the eye. The remaining eye must assume the total function of producing images to be relayed to the brain. Closing or squinting of the nonfunctioning eye causes tension on the eyelid and focusing muscles of the eye in use. This tension prevents the functioning eye from working at maximum efficiency.

The method of using a black or solid patch to block vision from the eye not in alignment with the sights doesn't work well. If you reduce the amount of light falling on the pupil of one eye, a sympathetic opening of the pupil in the other eye occurs as it seeks to maintain a balance of acuity. This action reduces the sharpness of the eye in use. If a patch is used, it should be opaque material that will permit light to pass through but blocks the seeing of an image.

b. Binocular vision is merely the use of both eyes when aiming. Binocular aiming has a number of major advantages. The shooter does not have to expend additional effort to squint. This can be quite important in a long match. Binocular vision is normally much more acute than Monocular. With Binocular vision, the line of sight is still achieved by only one eye and the visual impressions of the eye not being used for aiming are simply ignored. Eventually, the psychological suppression of the nonshooting eye becomes so effective that the second target image simply fades from notice. You receive all the benefits of Monocular vision as well as all those of binocular.

NOTE: SHOW SLIDE #15 (DOMINANT EYE).

    2. The Dominant eye.

        a. Many of you are already using one eye almost exclusively. This eye is, for some reason, stronger than the other. It is called the dominant eye. Since the dominant eye is the one you use most naturally, it is the eye you should use in aiming. You may discover which eye is dominant by this simple test. Make a circle of the thumb and forefinger. Hold your arm out and look through the small circle at some object. Close the left eye. If the object remains inside the circle of thumb and forefinger, your right eye is dominant.

        b. Hold the weapon in such a position that the dominant eye has a direct line of vision through the rear sight notch. Align the top edge of the front sight with the top edge of the rear sight and center the front sight with equal lines of light down both sides. The target should be dimly visible as a blurred indistinct outline. The other eye will furnish an impression of a blurred, out of focus target but it will not have the sight alignment superimposed.

    3. Strengthening the eyes for shooting. Function of the natural habits of the eye must not be impaired if the shooter is to achieve maximum use of his vision. Indeed, he should make a conscious effort to improve the condition of his eyes in the intervals when they are not actually aiming by allowing the habits of normal sight to function.

        a. Blinking is the first habit of the normally functioning eye. It is an involuntary action and provides lubrication, cleansing and momentary rest for the eye. You may be convinced that blinking prevents you from seeing properly but research indicates that quite the contrary is true. Blinking increases the actual amount of time you can see, since failing to blink constitutes strain and may cause a drastic reduction in the number of visual images transmitted per second by the eye.

        b. The second habit of normal sight is called central fixation. This is the simultaneous focusing of the eye and complete mental attention on the same object. There is a basic structural reason why this must be done. The only part of the eye that sees with perfect clearness, the Macula Lutea, in the center of the retina is no larger than the head of an ordinary steel pin. This dot of perfect sight is placed in the eye like a point at the bottom center of a globe. This one tiny point has a clear, strong vision. The instant you deviate from that point, there is a tremendous reduction in clarity of sight. There is, instead, blurred collateral vision.

        Since only this point--the Macula Lutea has perfect clear vision, only a very small area can be seen clearly at one time. However, the eye shifts its point of focus so swiftly that the illusion of seeing a large area clearly is created. The images falling on the retina are carried in such rapid succession to the cerebral cortex that the brain interprets these images as one complete ever changing picture.

        One comes to believe the eye can see a large area clearly and so misuse slips in. Any attempt to see a large area results in your trying to see without focusing either the eye or the mind.

        c. The third habit of a normal eye is shifting. The normal eye focus hops around on the observed object like a drop of water in a hot skillet. This habit may seem to conflict with central fixation, but actually it complements it.

        If you do not allow your eyes to shift, you will stare--and staring is the worst form of eye strain. Eye strain reduces the number of images transmitted to the mind.

Shifting is both voluntary and involuntary in character. The involuntary shift is continuous, automatic and very slight. This movement is not visible and is believed to correspond in frequency with the rate of image production in the retina. When the eye is relaxed, the involuntary shifting is frequent and the movement is short. The tense, strained eye tends to shift less often; and the shift, when made, is larger and more spasmodic in nature. It requires relaxation and normality for an eye to keep shifting on a very small area. When the eye is strained and vision becomes abnormal, you should deliberately practice shifting the eye. Frequent shifting will invariably give relief from strain and produce improvement in the vision.

VI. CONCLUSION:

    A. <u>Retain Attention</u>: None of us enjoy shooting bad shots. If you are tired of that blah feeling, come up to a new sensation.------Look only at the front sight and shoot tens.

NOTE: SHOW SLIDE #1 (SCOPE AND SUMMARY).

    B. <u>Summary</u>: During this period we covered the factors of proper sight alignment, disclosed how to place eye focus on the front sight, learned how to employ intense mental concentration to maintain proper relationship of front and rear sights and described the natural habits of the human eye, discussed monocular and binocular vision and how to determine and use the dominant eye. Use this information. Force yourself to align the sights. Focus your eyes on the front sight. Concentrate on maintaining perfect sight alignment to the exclusion of all else that may happen or be noticed around you.

NOTE: SHOW CHART #8 FRONT AND REAR SIGHT IN PERFECT ALIGNMENT AGAINST A BLURRED AIMING BULLS-EYE.

When you see this combination, apply the fundamentals of pistol marksmanship Deliver shots into the ten or X ring consistently. You then will know that your eyes are performing as well as the mind and the rest of the body.

    C. <u>Application</u>: Take what you have learned here and go out and shoot your 2650 or you may break the national record up there in the high 2680's.

NOTE: DISMISS ASSISTANT INSTRUCTORS.

NOTE: SHOW SLIDE #1-(A PICTURE OF AN AIMING BULLS-EYE). "This is the target at which you Shoot."

NOTE: SHOW SLIDE #2-(A REAR SIGHT). "This is the rear sight."

NOTE: SHOW SLIDE #3-(A FRONT SIGHT). "Here is your front sight."

NOTE: SHOW SLIDE #8-(A CLEAR FRONT AND REAR SIGHT SUPERIMPOSED ON A BLURRED BULLS-EYE). "Here is what you should see when your vision is focused on the front sight."

    D. <u>Closing Statement</u>: "If you learned nothing else from this class---remember this." "Keep your eye on the front sight."

C. FUNDAMENTALS III
(Trigger Control)

AA PMI&CC 303
30 Minutes
March 1975

LESSON OUTLINE

I. LESSON OBJECTIVE: To enable the pistol marksmanship student to achieve positive trigger pressure by proper employment of the factors of trigger control.

II. STUDENT PERFORMANCE OBJECTIVES: As a result of this instruction, students must be able to accomplish the following student performance objectives:

    A. GIVEN the basic functions of the human nervous system, EXPLAIN the nerve processes involved in trigger control as shown in Chapter III, USAMU Pistol Marksmanship Guide.

    B. PROVIDED with the elements of trigger control, EXPLAIN each of the factors of correct trigger control as described in Chapter III, USAMU Pistol Marksmanship Guide.

    C. REQUIRED to fire a course of slow fire with a pistol, EMPLOY the area type shooting positive trigger pressure as prescribed in Chapter III, USAMU Pistol Marksmanship Guide.

III. ADVANCED STUDY ASSIGNMENT: Chapter III, USAMU Pistol Marksmanship Guide.

IV. INTRODUCTION:

    A. <u>Gain attention</u>: Many times, the hard work devoted to setting up proper stance, position, grip, and perfect sight alignment is ruined at the last instant before shot delivery by faulty trigger control. This error in performance is generally caused by concentration being shifted from sight alignment to the application of additional trigger pressure shortly before the release of the hammer takes place. Positive trigger pressure should be involuntarily applied (conditioned reflex) in conjunction with a minimum arc of movement and perfect alignment of front and rear sights.

    B. <u>Orient Students</u>:

        1. <u>Lesson Tie-In</u>: In the previous classes we discussed the fundamentals of stance, position, grip, breath control, and sight alignment. The successful employment of positive trigger control depends to a great degree on being coordinated closely with these other fundamentals.

        2. <u>Motivation</u>: When trigger control is employed properly in coordination with the other fundamentals, the pistol can be fired without altering the sight alignment or disturbing the minimum arc of movement. This will result in a winning (world record) score if repeated 270 times in a 3-gun aggregate.

        3. <u>Scope</u>: To master trigger control, you must know:

<u>NOTE</u>: SHOW SLIDE #1-TRIGGER CONTROL SCOPE AND SUMMARY.

            a. What your nerve processes are and how to condition them for championship shooting by proper training.

            b. Explanation of the factors providing for the correct control of the trigger.

            c. How to apply area type shooting positive trigger pressure.

V. <u>BODY</u>:

A. **First Student Performance Objective**: GIVEN the basic functions of the human nervous system, EXPLAIN the nerve processes involved in trigger control as shown in Chapter III, USAMU Pistol Marksmanship Guide.

NOTE: SHOW SLIDE #2 NERVE PROCESSES.

QUESTION: Is it possible to learn to press the trigger causing the shot to fire quickly and exactly when you want to fire?

1. There is a definite interval of time between the beginning of the action of the stimulus signal and the beginning of the response movement. (.18 to .25 seconds)

2. From the beginning of the action to the completion of the response movement, it is essential to put into motion certain muscles and restrain the motion of other muscles. This is known as the process of nerve excitation and takes the form of either stimulation or inhibition.

   a. When the shooter is beginning to train himself in the correct control of the trigger action, the muscular activity at first causes a predominance of the process of stimulation in the cerebral cortex. During the initial phase of training, the process of stimulation spreads within the cerebral cortex, encompassing considerable areas of it. This leads to a situation in which muscular groups which do not take direct participation in an action are drawn into it. This is the reason why beginners, instead of pressing on the trigger by merely moving the trigger finger, accompany its movement by the work of many groups of muscles, thus spoiling the aim at the most critical moment of producing the shot.

   b. Subsequently, as the shooter practices and better habits become ingrained, he learns to limit excessive movements. He is able to develop a certain restraint of movements. The shooter, by means of concentrated practice, has intensified the process of inhibition and the widespread predominance of stimulation in the cerebral cortex is followed by a restriction of stimulation in limited areas. Frequently, during this period of training, the shooter notices that he does not always press the trigger at exactly the right moment. Many shooters are aware of the rather unpleasant sensation that arises from the many times that the trigger seems to come to a complete stop, and refuses to fire the pistol. Many a shooter has been heard to complain "the pistol just sits there, and the finger won't move." After an interval, the pistol begins to describe an enlarged arc of movement, and the shooter, even as he realizes that the best moment for firing the shot has been lost, continues pressure on the trigger. Later, after systematic training, the shooter improves the balance between the processes of stimulation and inhibition. As a result, his movements and timing improve, and are accompanied by the coordinated work of only the necessary groups of muscles.

3. Reflexes are the responses to a stimulus: They are divided into two classes.

   a. Unconditioned reflexes are involuntary reactions in response to definite external stimuli such as heat, pain, or sound and recoil.

   b. Conditioned reflexes are temporary reactions developed by special training or by the influence of your immediate environment; such as, catching a ball, eating, boxing. Conditioned reflexes which have been developed can be retained for a long time only in the event that they are reinforced, by repetition of stimuli.

4. To develop the conditioned reflexes needed in marksmanship, you must practice extensively, and/or dry fire the proper procedure until it becomes a well ingrained habit.

5. To control the unconditioned reflex on the firing line, the trigger finger must be made to start its positive pressure based on the external stimulus, which is the coincidence of perfect sight alignment and your minimum arc of movement for that particular time.

     a. <u>Prevent development of detrimental unconditioned or conditioned reflexes</u>.

       (1) If someone makes a threatening motion at another person, or say, claps his hands in front of his face or fires a shot unexpectedly, the person will involuntarily, unconsciously blink his eyes. This reaction is an unconditioned defensive reflex. Shooting requires a person, during aiming and firing to keep his eyes open. Otherwise he will not be able to call the shot. The shooter must learn to suppress this defensive unconditioned reflex.

       (2) If a beginner who previously has never fired a pistol is given practice cartridges, he will calmly load the pistol with them, aim, and press the trigger. As soon as he fires one or two shots and feels the blow in the hand from the recoil, and in addition, hears a loud sound, he will subsequently act completely different. Knowing that recoil of the shot is accompanied by a blow in the hand, the beginner will involuntarily strain his muscles, stiffen his arm and press his shoulder forward in order to counteract the expected blow. Thus, a conditioned reflex to the expected recoil blow has already appeared. If steps are not taken in time, this conditioned reflex which has manifested itself will become stronger and become a harmful habit for the shooter. Consequently, the shooter must learn not only to suppress certain unconditioned reflexes, but also to prevent the development and reinforcement of conditioned reflexes which are detrimental to shooting. Learn how to relax the body and prevent an increase of muscular tension, so as not to spoil the aim.

    6. Since minimum arc of movement is a part of the stimulus for correct timing in trigger control, it must be highly developed so that the shooter requires it and has complete confidence in his ability to fire a good shot within his ability to hold.

TRANSITION: The knowledge of nerve processes must be applied to the factors of correct trigger control.

   B. <u>Second Student Performance Objective</u>: PROVIDED with the elements of trigger control, EXPLAIN each of the factors of correct trigger control as described in Chapter III, USAMU Pistol Marksmanship Guide.

QUESTION: How much initial pressure should be taken up on the trigger?

NOTE: SHOW SLIDE #3-FACTORS PROVIDING FOR THE CORRECT CONTROL OF THE TRIGGER.

    1. Any free movement of the trigger is called slack. Initial pressure is approximately one-fourth of the total trigger pressure required to fire the weapon.

    2. Trigger pressure is applied with a positive increase of pressure on trigger, smooth and even with no interruption.

    3. The grip must be correct to allow the trigger finger to move freely straight to the rear, independent of any muscle movement in rest of hand.

NOTE: SHOW SLIDE #4-CORRECT PLACEMENT OF THE TRIGGER FINGER.

    4. The trigger finger must be placed on the trigger in such a way as not to deflect the precise sight alignment set up for absolute control of the shot. Smooth, straight to the rear movement is necessary to avoid disturbing sight alignment.

    5. The application of positive trigger pressure must be coordinated with peak visual perception (eight to ten seconds) for correct sight alignment, minimum arc of movement (five to six seconds) and the period of maximum concentration (three to five seconds).

TRANSITION: These are the factors providing for the correct control of the trigger. Precise use of these factors is demanded in applying them to positive trigger pressure.

C. <u>Third Student Performance Objective</u>: REQUIRED to fire a course of slow fire with a pistol, EMPLOY the area type shooting positive trigger pressure as prescribed in Chapter III, USAMU Pistol Marksmanship Guide.

NOTE: SHOW CHART #5 CONTROLLING TRIGGER RELEASE.

QUESTION: What is the difference between area shooting and point shooting?

<u>ANSWER</u>: 1. In point shooting you apply trigger pressure only when the sights are in the alignment and the hold is motionless in the center of your aiming area. This method is attempted by only a few of the top shooters. Many try it unsuccessfully. Trigger pressure is increased as long as the sight picture is perfect. If the sight picture becomes less than perfect, trigger pressure is maintained but not increased until conditions again become perfect. To the average shooter, this method will result in trying to make the shot break during the very short period of perfect sight picture and a jerk or a heel shot (unconditioned reflexes) will develop. All conditions must be perfect in this method and even the best shooters rarely have this combination. This is known as the point method using interrupted trigger pressure.

<u>ANSWER</u>: 2. In Area shooting, the shooter endeavors to complete the firing of the shot once the application of the trigger pressure has started. Trigger pressure will not be started unless the optimum combination of sight alignment and minimum arc of movement is present. This is positive uninterrupted trigger pressure. The steps in this method are:

    a. Take up slack and initial pressure.

    b. Settle into aiming area. (Attain a minimum arc of movement).

    c. Pick up sight alignment.

    d. The trigger finger starts the positive pressure as soon as the sights are in perfect alignment and the hold is reasonably good (minimum arc of movement). The trigger finger only is moving and it is an involuntary movement (conditioned reflex). During this time, maximum concentration is placed on a point focus on the front sight and the result will be a surprise shot hitting the target within your ability to hold. Since it will be a surprise shot, the unconditioned reflexes will not destroy the accuracy of the shot at the last instant. If, during area shooting, the sights move out of alignment or the arc of movement becomes erratic, bench the weapon, replan the shot and repeat the attempt to fire.

    e. The cue for applying positive trigger pressure is perfect sight alignment and a minimum arc of movement not sight picture or the absence of movement in the hold. Sight alignment/minimum arc combination is of much longer duration in area shooting than perfect sight alignment in conjunction with a motionless hold as is attempted in point shooting.

VI. CONCLUSION:

A. <u>Retain Attention</u>: Trigger control is very important in the delivery of a good shot. But remember it must be used in coordination with all of the other fundamentals and it must not disturb sight alignment.

NOTE: SHOW SLIDE #1-TRIGGER CONTROL SCOPE AND SUMMARY.

B. <u>Summary</u>:

1. In order for you to attain good trigger control, you must be able to:

    a. Understand the nerve processes involved in positive trigger control.

    b. Utilize factors providing for the correct control of the trigger.

            c.   Practice the application of positive trigger pressure. The area method overcomes the unconditioned reflexes inherent in the point method.

        C.   <u>Application</u>: By the proper application of positive trigger control, better scores are inevitable. Remember, there is no secret to winning a pistol match, only the proper coordination in employing the fundamentals.

        D.   <u>Closing Statement</u>: To become proficient, a shooter must develop the ability to apply all the fundamentals. The shooter must know in his mind that when he sets up a conditioned reflex (positive trigger pressure) based on perfect sight alignment and minimum arc of movement, the trigger will come back involuntarily, automatically, resulting in a surprise shot that hits the target within the ability to hold.

D.   TECHNIQUE OF FIRE I
    (ESTABLISHING A SYSTEM)

AAPMI&CC 304
50 Minutes
Mar 1975

LESSON OUTLINE

I. LESSON OBJECTIVE: To enable marksmanship students to establish a system of operation that will assist them in delivering a controlled shot or string of shots on the target.

II. STUDENT PERFORMANCE OBJECTIVES: As a result of this instruction the student must be able to accomplish the following student performance objectives:

    A.   Given the steps of Preliminary Preparation for shooting in a Pistol Match, PREPARE both himself and his equipment in an organized manner as described in Chapter IV, USAMU Pistol Marksmanship Guide.

    B.   Having developed a plan for firing a shot in training and practice, UTILIZE during match firing a completely detailed sequence of firing a controlled shot as described in Chapter IV, USAMU Pistol Marksmanship Guide.

III. ADVANCE ASSIGNMENT: Chapter IV, USAMU Pistol Marksmanship Guide.

NOTE: HAVE ASSISTANTS PASS OUT PISTOL SHOOTERS WORKSHEET.

IV. INTRODUCTION:

    A.   <u>Gain Attention</u>: During the past year someone in this group reported to the firing line without his ammunition; one of you missed a relay; somebody forgot to take one item or another of his clothing to a match, hat, belt, boots, or jacket. Some competitor who has just paid a big entry fee will load four (4) rounds in a magazine for a timed or rapid fire string. A certain ball of fire will arrive at the range late, shoot on the wrong target and make a mad dash for the mess hall without signing his scorecard. That person is now standing in front of you giving instruction on how to avoid these disasters.

    B.   <u>Orient Students</u>:

        1.   <u>Lesson Tie-In</u>: You have received instruction on the basic fundamentals. You will now receive instruction on how to coordinate all these fundamentals into a system or technique which will enable you to fire a good shot or string of shots. You must not be handicapped by having forgotten to prepare and organize.

        2.   <u>Motivation</u>: Any mission you undertake can be performed better after thorough organized preparation. This applies especially to shooting. In the national matches at Camp Perry you may fire up to five hundred and ten (510) record shots, depending on how many matches you have entered. In order to win a match you must apply all your skill and knowledge to each of these shots. You must develop a system that will allow you to perform effectively during the entire National Matches. Every step or action must be deliberate and planned. When you raise your arm to fire the first shot of a match you must be completely ready.

        3.   <u>Scope</u>: During this period you will have explained to you the factors necessary to a good system and how to organize them into a sequence that will enable you to maintain an acceptable level of control.

NOTE: SHOW SLIDE #1a - CONTROL FACTORS IN ESTABLISHING A SYSTEM.

a. Preparation before shooting

b. Planning of shot

c. Relaxation before shooting

d. Delivery of shot

e. Analysis of shot

f. Correction of shot (if necessary)

NOTE: SHOW SLIDE #4b - ADDITIONAL AIDS IN ESTABLISHING A SYSTEM.

g. Sight adjustment card

h. Preliminary preparation checksheet

i. Shooters work sheet

j. Pistol score book

TRANSITION: Careful and complete preparation for firing the first shot is the key to success. If the first shot is not under control, a serious handicap is set up by the shooter trying to compensate for his original error.

V. BODY:

A. <u>First Student Performance Objective</u>: Given the steps of Preliminary Preparation for shooting in a Pistol Match, PREPARE both himself and his equipment in an organized manner as described in Chapter IV, USAMU Pistol Marksmanship Guide.

QUESTION: When does preparation for a pistol match start?

ANSWER: When the decision is made to enter the pistol match. The program is received and at that time preparation begins. Your entry is sent in, arrangements made for absence from work, family affairs taken care of, and travel arrangements are made.

NOTE: SHOW SLIDE #1 - CONTROL FACTOR, PREPARATION.

1. Preparation

NOTE: SHOW SLIDE #5 - ZEROING

a. A basic zero must be established for each weapon. A basic zero is one that with perfect sight alignment will place the strike of the bullet in the center of the target on what you consider a normal shooting day.

(1) When initially zeroing a new weapon start at a short range, twenty-five (25) yards, and fire a three (3) shot group and adjust your sight setting. Continue firing single shots until your zero is established.

(2) Make bold sight changes. Move the strike of the bullet to center of target with as few sight changes as possible.

NOTE: REFER TO THE SIGHT ADJUSTMENT CARD ON PAGE 29 USAMU PISTOL MARKSMANSHIP GUIDE OR THE PISTOL WORKSHEET FOR THE DISTANCES OF EACH SCORING RING FROM CENTER OF TARGET.

NOTE: SHOW SLIDE #10 - THE ADJUSTABLE PISTOL SIGHT.

(3) Become familiar with your adjustable sight. Experiment and find out just how far one click moves strike of bullet in both elevation and deflection at fifty (50) and twenty-five (25) yards. Fire ten (10) shots for a group on the target. Move the rear sight adjustment twenty (20) clicks in elevation or windage and fire ten (10) more shots. Measure the distance from shot group center to center and divide by 20.

(4) Mark sights and record basic setting on the sight adjustment card in gun box and on the proper space in the scorebook sheet.

NOTE: SHOW SLIDE #8 - SIGHT ADJUSTMENT CARD.

NOTE: SHOW SLIDE #9 - SCORE BOOK PAGE.

(5) When shooting on other than your home range, try to get to the range a day before the match starts and confirm your zero. This may also be necessary under extreme changes of weather and light conditions.

b. In the assembly area, equipment must be made ready which includes weapons and ammunition. Make sure that you arrive at the pistol match with all your equipment. You should have a list of everything that you need during the match. Put a check mark by each item as you load it in the car. On the day of the match, arrive early enough to complete your assembly area checks prior to being called to the firing line.

NOTE: REFER TO THE PRELIMINARY PREPARATION CHECKLIST IN THE PISTOL WORKSHEET OR PAGE 40 OF THE USAMU PISTOL MARKSMANSHIP GUIDE. SHOW SLIDE #7 - SHOOTER'S SLOW FIRE WORKSHEET.

NOTE: SHOW SLIDE #11 - PRELIMINARY PREPARATION CHECKSHEET.

(1) Physical action

(a) Check squadding for relay and target.

(b) Have proper gun and ammo.

(c) Check sight settings.

(d) Carbide lamp w/carbide.

(e) Blacken sights.

(f) Ear plugs.

(g) Screwdriver.

(h) Stop watch.

(i) Scorebook.

(2) Mental action

(a) Stimulate confidence.

(b) Expect to work hard.

(c) Plan actions on the firing line.

(d) Do not be upset by range irregularities.

(e) Think only of shooting.

(f) Mentally review shot sequence.

TRANSITION: Be ready and waiting for your relay to be called to the firing line.

    c. On the firing line:

        (1) Place shooting box on correct firing point.

        (2) Scope the correct target.

        (3) Inspect target for holes.

        (4) Adjust ear protectors.

        (5) Load magazines w/proper ammunition.

        (6) Check sight blackening.

        (7) Locate accessory shooting equipment.

        (8) Take a few deep breaths.

        (9) Assume stance.

        (10) Check position.

        (11) Assume grip on pistol.

        (12) Trigger pressure straight to rear.

        (13) Be ready for the Command "Load" to be given.

    d. After the command "Load"

        (1) Verify stance.

        (2) Load weapon.

        (3) Safety off.

        (4) Recheck grip.

        (5) Recheck for proper target.

        (6) Relax with pistol at bench rest.

        (7) Resume mental process of planning shot delivery.

TRANSITION: If there is a magic formula or a secret to accurate shooting, it is in the planning of the shot sequence.

    B. <u>Second Student Performance Objective</u>: Having developed a plan for firing a shot in training and practice, UTILIZE during match firing a completely detailed sequence of firing a controlled shot as described in Chapter IV, USAMU Pistol Marksmanship Guide.

NOTE: SHOW SLIDE #2 - CONTROL FACTOR, SHOT PLANNING.

    1. Planning the shot sequence. Most AMU shooters use this sequence for slow fire with only minor modifications.

        a. Extend arm and breathe.

        b. Settle into minimum arc of movement.

        c. Pick-up sight alignment in aiming area.

        d. Take up trigger slack - apply initial pressure.

        e. Hold breath.

        f. Maintain sight alignment and minimum arc of movement.

        g. Start positive trigger pressure.

        h. Concentrate point focus on front sight.

        i. Follow thru.

TRANSITION: After planning the shot sequence the shooter must relax.

    2. Relaxation delays fatigue and aids muscular control.

NOTE: SHOW SLIDE #3 - CONTROL FACTOR, RELAXATION.

        a. Methodically think of relaxing each principal muscular mass of the body; neck, shoulders, back, abdomen, buttocks, and legs.

        b. A relaxed muscle does not fatigue and tremble.

        c. Rest arm after an unsuccessful effort to shoot.

TRANSITION: Now you are ready to fire your first shot.

    3. Deliver the shot.

NOTE: SHOW SLIDE #4 - CONTROL FACTOR, SHOT DELIVERY.

        a. In order to accomplish this you must be aware of every step in your shot sequence. Be deliberate and forceful in the delivery of the shot.

        b. The shot must be delivered exactly as planned.

        c. Record the value of the shot in your scorebook.

    4. Shot analysis is a must after every shot. It is as important to analyze good shots as bad ones. The steps in shot analysis are:

NOTE: SHOW SLIDE #5 - CONTROL FACTOR, SHOT ANALYSIS.

        a. Follow thru check.

        b. Call shot.

        c. Compare hit with call.

        d. If shot or call is bad, determine cause.

        e. Watch for error pattern to form.

        f. Did shot break in normal pattern of movement?

        g. Did you hold too long?

        h. Did you apply positive trigger pressure?

        i. If you benched weapon on shot effort. Why?

        j. Did you lose concentration?

        k. Did you get a surprise break?

        l. Were you worried about results?

    5. Shot correction (if necessary).

NOTE: SHOW SLIDE #6 - CONTROL FACTOR, SHOT CORRECTION.

    1. If you find an error in your delivery, you must take corrective action. Here the control cycle starts again. In planning the delivery of your next shot, pay special attention to those steps in the shot sequence that you may have failed to employ correctly. Stick with the sequence that you know will produce a good shot for you. For best results, there must be:

        a. Agreement between coach and shooter.

        b. Prompt application - include in plan for next shot.

        c. A prevention of recurrence of error.

        d. Corrective measures on every shot if necessary.

VI. CONCLUSION:

    A. <u>Retain Attention</u>: Most of you have probably seen the shooter on the line whose technique of fire goes something like this:

        1. Bang.

        2. Slam gun in box.

        3. Cuss.

        4. Stomp ground or kick bench.

        5. Look thru scope.

        6. Cuss some more.

        7. Pick up gun.

        8. Bang.

    Then he repeats the whole process as long as his anger persists. Why? Because he forgot to take the Rapid Fire stage sight setting off from yesterday's shooting and his first shot slow fire was a six (6) at six o'clock.

NOTE: SHOW SLIDES #1a, and 4b - IN SUCCESSION - CONTROL FACTORS IN ESTABLISHING A SYSTEM AND ADDITIONAL AIDS.

    B. <u>Summary</u>: Establish a system for yourself that will allow you to perform well. This system must include the correct zero, complete preliminary preparation, and a sequence for delivery of a controlled shot that will be used repeatedly without change unless your analysis indicates corrective action is necessary.

        1. Preparation.

        2. Planning.

        3. Relaxation.

        4. Delivery of a shot.

        5. Shot analysis.

        6. Shot correction (if necessary).

    C. <u>Application</u>: Think about the shot sequence you are using. Compare it with the one that I have given you. During your periods of instruction on slow and sustained fire techniques think of how all the different steps in a shot sequence fall into place. The end result is the hammer falling without disturbing that perfect sight alignment while the weapon is pointed with stability at the center of the aiming area.

    D. <u>Closing Statement</u>: Once you have shot a ten or an "X" by using a technique that you can duplicate, you will eventually be able to shoot the second, third or fourth ten, etc. This indicates a degree of improvement in control. If you don't get bored before you have shot 270 in a row you will probably end up winning the National Matches this year.

E. TECHNIQUE OF FIRE II 　　　　　　　　　　　AAPMI&CC 305
    (SLOW FIRE)　　　　　　　　　　　　　　　　　50 Minutes
    　　　　　　　　　　　　　　　　　　　　　　　Mar 1975
                        LESSON OUTLINE

   I. LESSON OBJECTIVE: To enable the Pistol Marksmanship Student to fire winning slow fire scores by employing a technique for coordinated control of slow fire.

   II. STUDENT PERFORMANCE OBJECTIVES: As a result of this instruction, the pistol shooter-student should be able to accomplish the following student performance objectives:

       A. Given the limiting time factors in employing each of the pistol fundamentals, DISCLOSE the method of coordinating the employment of the pistol fundamentals that will result in the delivery of an accurate shot as shown in Chapter V, USAMU Pistol Marksmanship Guide.

       B. Given the techniques that contribute to control of a slow fire shot, EXPLAIN the application of these techniques to control of pistol slow fire as shown in Chapter V, USAMU Pistol Marksmanship Guide.

       C. Given the common deficiencies in slow fire control, REVEAL how the pistol shooter can recognize and correct errors in firing the slow fire stage as shown in Chapter V, USAMU Pistol Marksmanship Guide.

       D. Faced with firing a pistol under adverse condition, EXPLAIN how to compensate for adverse conditions while firing a pistol in the slow fire stage as shown in Chapter V, USAMU Pistol Marksmanship Guide.

       E. As coach of a pistol team, DEVISE a comprehensive slow fire training program incorporating the training methods shown in Chapter V, USAMU Pistol Marksmanship Guide.

   III. ADVANCE ASSIGNMENT: Chapter V, USAMU Pistol Marksmanship Guide.

   IV. INTRODUCTION:

       A. <u>Gain Attention</u>: A technique can be defined as a highly specialized method of performing a specific, complex operation. A technique for control of slow fire without a doubt requires specialization and is based on a complex system of organization and coordination for accurate delivery of each shot.

       B. <u>Orient Students</u>:

           1. <u>Lesson Tie-In</u>: Up to this point you have had instruction in stance, position, grip, breath control, trigger control, and sight alignment. It is necessary to learn to employ all these fundamentals in coordination and develop a technique for slow fire.

           2. <u>Motivation</u>: Generally, in any pistol match more points are dropped at the 50 yard line slow fire than in timed or rapid fire combined.

NOTE: SHOW SLIDE #1 - SCOPE AND SUMMARY OF SLOW FIRE.

           3. <u>Scope</u>: The shooter can improve his slow fire performance by:

               a. Coordination of the employment of the fundamentals in firing of each shot of slow fire.

               b. Developing a technique of slow fire control.

  c. Recognition and correction of common deficiencies in control.

  d. Learning to compensate for adverse shooting conditions.

  e. Using comprehensive training methods.

V. BODY:

 A. <u>First Student Performance Objective</u>: Given the limiting time factors in employing each of the pistol fundamentals, DISCLOSE the method of coordinating the employment of the pistol fundamentals that will result in the delivery of an accurate shot as shown in Chapter V, USAMU Pistol Marksmanship Guide.

NOTE: SHOW SLIDE #2 - EMPLOYMENT OF THE FUNDAMENTALS.

  1. The fundamentals of stance, position, grip and breath control, when employed properly, will give the shooter the ability to hold the pistol almost motionless within the center of the aiming area on the target. This is establishing the minimum arc of movement.

   a. Minimum arc of movement governs the basic size of shot group if sights are kept uniformly in perfect alignment.

   b. The duration of optimum hold or minimum arc of movement is approximately three (3) to six (6) seconds.

  2. Sight alignment is the relationship of the front sight and the rear sight to the shooter's eye while aiming.

   a. Any misalignment of the front and rear sights will cause an angular shift error of three (3) inches on the target at fifty (50) yards for each one hundredth (.01) inch of deflection.

   b. The point of focus in the front sight. The duration of optimum visual perception - the ability of the human eye to maintain a clear point of focus, is limited to a period of six (6) to eight (8) seconds. Focus can be renewed quickly by a blink or shift of the eye but this action may happen after the optimum period of one or more of the other control factors has expired.

   c. Intense mental concentration of the degree necessary to think only of maintaining the front and rear sights in exact alignment will function for the average shooter at an optimum level for approximately three (3) to six (6) seconds.

  3. Positive trigger control is the act of committing the pressure on the trigger to continue until completion of the firing of the shot. If any condition arises that causes an interruption of trigger pressure immediately abandon the attempt. Bench the weapon, rest, replan, and try again. This obviously applies only to slow fire.

   a. Trigger slack and initial pressure is taken up before positive trigger pressure is started.

   b. Uninterrupted positive trigger pressure is not applied until the other fundamentals are settled and are as near perfect as the shooter can set them up.

   c. The duration of positive trigger pressure in firing the pistol varies from one (1) to three (3) seconds because of the care needed to maintain smoothness of application of constantly increasing pressure straight to the rear and not disturb sight alignment. The shooter should observe the limitations of time in which the other fundamentals are optimum. Smooth trigger pressure can be consistently applied in as short a period as approximately one second for each shot as in a rapid fire string.

d. The hammer should fall and fire the shot as a surprise. Any anticipation of the release of the hammer will set up reflex muscular action in the shooting arm and hand and spoil the accuracy of the shot by disturbing sight alignment.

4. To exercise maximum control of shooting a pistol, all of the control factors must be coordinated to be at or very near their optimum state simultaneously. Coincidence should come during the three (3) to six (6) second interval when mental concentration is at peak intensity.

TRANSITION: When the factors are in coordination, employment of the fundamentals is under a high degree of control. A method must be found which will assure constant renewal of the ability to control the employment of the fundamentals.

B. Second Student Performance Objective: Given the techniques that contribute to control of a slow fire shot, EXPLAIN the application of these techniques to control of pistol slow fire as shown in Chapter V, USAMU Pistol Marksmanship Guide.

NOTE: SHOW SLIDE #3 - SLOW FIRE TECHNIQUE.

QUESTION: When all conditions are right, how long should it take a shooter to fire a shot of slow fire?

ANSWER: Approximately three (3) seconds.

1. A shooter should form a habit for firing a few "dry" shots before beginning to fire. This action brings into play, if executed properly, all of the control factors that will be utilized in match firing. If the coordination is inconstant or the employment of one or more of the fundamentals is faulty, immediate corrective steps can be taken to smooth out the operation. Mastery of the employment of the fundamentals is a result of constant practice and extensive match firing experience after the correct method is learned.

2. Great care is one of the mainstays of the control of slow fire shooting. Many of the poorly controlled shots are the result of various degrees of carelessness. A habit of an infrequent compromise is a form of carelessness. An important sector of slow fire control lies in accepting no compromise on perfection. A series of good shots is the result of the repetition of a perfect performance.

3. Patience is of extreme importance to sustaining an effective technique for control of slow fire. Without patience the shooter may disrupt an otherwise perfect performance an instant from successful completion.

4. Holding too long is a result of overcaution caused by the shooter's excessive fear of getting a bad shot. Slow pressure on the trigger arises when holding or sighting is difficult. Sometimes the shooter simply will not press positively on the trigger even though the other control factors are optimum. Lack of confidence in his ability to employ the other fundamentals at optimum level long enough to fire an accurate shot is the underlying cause of this reluctance.

5. The most efficient manner in which the shooter can impart control to his shooting is by careful organization of the techniques used to exert maximum control. For this a system is needed. When organized, this system of delivering an accurate shot on the target should be used uniformly for each shot. The detailed method of developing a personal system is outlined in the shooters slow fire worksheet.

6. In determining pace or tempo of shooting, experience has shown that the most effective approach is to shoot rather rapidly or within a maximum of six (6) seconds after settling. This brief time limit is necessary if the technique of shooting follows the premise of coordination of control factors in employment of the fundamentals.

7. The time spent between each shot, preparing and planning is limited only by the time allowed for the ten shot string of slow fire. The shooter's energy must be expended with care by taking sufficient rest time between each shot and each ten shot string by observing a definite pacing and rhythm of operation. If the weather or other conditions are not favorable, the shooter may have to alter his timing to meet prevailing conditions.

TRANSITION: Slow fire control produces successful results only when flaws in performance are at a minimum. The shooter should be familiar with the more common deficiencies in control and also how to avoid and correct them.

    C. **Third Student Performance Objective**: Given the common deficiencies in slow fire control, REVEAL how the pistol shooter can recognize and correct errors in firing the slow fire stage as shown in Chapter V, USAMU Pistol Marksmanship Guide.

NOTE: SHOW SLIDE #4 - COMMON DEFICIENCIES IN CONTROL.

    QUESTION: What is the difference between a jerking and a heeling of the shot?

ANSWER: 1. & 2. Jerk or Heel: The abrupt application of pressure either with the trigger finger alone or in the case of heeling, pushing with the heel of the hand at the same time. Apply pressure to the trigger straight to the rear and wait for the shot to break. Anticipation can cause muscular reflexes of an instant nature that so closely coincide with recoil that extreme difficulty is experienced in making an accurate call. Anticipation is the same as flinching. Jerking a shot is the abrupt application of trigger pressure in forcing a shot to fire instantly. For example, a shooter with a large arc of movement on a windy day may try to get a shot off as the sights move thru the bullseye.

    3. <u>Vacillation</u>: This is a common fault. You experience minor imperfections in your performance which as a result causes you to change your technique of operation frequently. The end result is that you may only hope to get a good shot. The method of correcting this fault is to develop a comprehensive plan and follow it without deviation until you see the need for making improvements in it.

    4. <u>Anxiety</u>: You work and work on a shot, meanwhile building up in your mind doubt about the possibility of the shot being good. Finally you shoot just to get rid of that particular round so you may work on others.

    5. <u>Not Looking at the Sights</u>: This quite frequently is listed as "looking at the target." A shooter may be focusing his eye on neither the sights nor the target, but since he does not see the target in clear focus he assumes he is looking at the sights. You must concentrate on maintaining perfect sight alignment, clearly defined and in focus.

    6. <u>Loss of Concentration</u>: If the shooter fails to involuntarily apply positive pressure on the trigger while arc of movement and sight alignment are good, his prior determination to apply the positive pressure at the time both are simultaneously proper needs to be increased and reemphasized.

    7. <u>Holding too Long</u>: Any adverse conditions that disturb a shooter's ability to "hold" will cause him to delay his positive application of trigger pressure waiting for conditions to better. The disturbing factor about this is that you will do it unconsciously; therefore, you must continuously ask yourself, "Am I trying to freeze <u>all</u> arm and pistol movement momentarily so I can get a shot off quickly before movement is resumed?"

8. <u>Overcorrection</u>: Maintaining control of your shooting is a continuous battle. The battle builds tension. Tension tightens the muscles and finally the abrupt motions made in compensation for errors causes the shooter to go beyond the desired degree of correction and deliver shots in exactly the opposite place from where the error was causing him to shoot originally. Smoothly coordinated actions are best assured by the relaxed, confident and carefully planned approach.

9. <u>Lack of Follow Through</u>: Follow through is the conscious attempt by the shooter to keep everything as it was set up to shoot until after the round is on its way to the target without any reflex of anticipation that disturbs sight alignment and spoils the surprise break of the shot. Lack of follow through is a breakdown of one or more of the factors set up by the shooter to control a good shot. For example, lack of follow through might be caused by thinking about the need for more pressure on the trigger A speed up of trigger pressure results in anticipation and a heeled shot at one o'clock. You must continue to maintain concentration on sight alignment even after the shot is on the way. Follow through is not to be confused with recovery. Merely reestablishing the hold on the target after the shot is fired is no indication that you are following through

10. <u>Match Pressure</u>: If there are 200 competitors in a match, rest assured that there are 200 shooters suffering from match pressure. What makes you think you are different? You should exert all your mental energy toward planning and executing the fundamentals correctly. Your shooting match pressure will become controllable and your competitors will congratulate you on your fine performance. There are many causes for bad shots. We have listed those most frequently found. It is not a complete list nor is it intended to provide the shooter with a convenient list of bad habits. It is, however, intended to assist the shooter in finding the source of his trouble.

<u>TRANSITION</u>: Ability to correct deficiencies in performance under ideal conditions points the shooter toward the more difficult conditions under which shooting must be controlled.

D. <u>Fourth Student Performance Objective</u>: Faced with firing a pistol under adverse conditions, EXPLAIN how to compensate for adverse conditions while firing a pistol in the slow fire stage as shown in Chapter V, USAMU Pistol Marksmanship Guide.

<u>NOTE</u>: SHOW SLIDE #5 - WIND SHOOTING AND ADVERSE CONDITIONS.

<u>QUESTION</u>: If the wind is blowing hard and the shooter cannot hold in the bullseye area, should he continue his effort to shoot?

<u>ANSWER</u>: Bench the weapon and wait for a lull in the wind, time permitting. If there is no lull, keep the sights aligned, tolerate a larger arc of movement and press positively on the trigger.

1. Wind shooting is conducive to jerking the trigger. This is true because as the arc of movement increases, the shooter develops a tendency to relax his trigger pressure. He is waiting for a more stable sight picture. His concentration on sight alignment will diminish and he will make an effort to set the shot off on the move as the sights pass the vicinity of the target center. The obvious answer is to, first wait for a lull in the wind; next, concentrate as one normally does on sight alignment and as a minimum arc of movement is achieved, start a constantly increasing positive pressure on the trigger until the shot is fired. Do not continue the hold during extreme gusts. Always take advantage of a chance to rest. Each attempt to fire a shot should be made with a firm resolve to align the sights and to apply constantly increasing trigger pressure until the shot is fired. The surprise shot continues to be the indicator, even under these conditions, of whether you are applying the fundamentals. Your shot group may be larger as a result of the increased arc of movement but the wild shots resulting

from faulty sight alignment, flinching, jerking and over-correction will be minimized. Extensive practice under wind conditions is not recommended but enough firing should be conducted under those conditions to prevent a stampede to the nearest wind shelter when a wisp of air movement stirs the pine tops.

    2.   Adverse weather conditions such as cold, hot or rainy weather or extreme light conditions pose problems that can be solved in much the manner as in wind shooting. Be determined to adhere to the fundamentals and ignore as much as possible the distractions that are demoralizing to the competition.

    3.   Light conditions varies from extremely bright to very dim. The shooter should keep a record of the light conditions on every range he fires on in his score book. Some competitors are affected more by changes in light than others. A note should be made as to how much his zero changes in the different light conditions. Sights should be blackened with care on bright days. As a part of shooting accessories, you should have both amber and green shooting glasses not only for light conditions but for protection against oil, wind and empty brass. Firing from an uncovered firing line usually requires different sight sittings than the firing from under a shed. Ammunition should be kept out of the sun as its accuracy is affected if it is exposed to the direct rays of the sun.

    4.   The major portion of our accomplishments on the firing line stems from our mental capacity to face up to the out-of-the-ordinary and parlay these conditions into a winning margin. Poor conditions must never become an excuse to quit or compromise and consequently deliver a poor performance. Good scores are produced by hard work in the application of the fundamentals regardless of the conditions. Proper application of the fundamentals remain the most important factor in shooting winning scores under adverse conditions.

<u>TRANSITION</u>: The proper employment of the fundamentals necessitates the following of a comprehensive method of training to analyze and perfect the technique chosen by the shooter.

    e.   <u>Fifth Student Performance Objective</u>: As coach of a pistol team, DEVISE a comprehensive slow fire training program incorporating the training methods shown in Chapter V, USAMU Pistol Marksmanship Guide.

<u>NOTE</u>:   SHOW SLIDE #6 - TRAINING METHODS.

    <u>QUESTION</u>: If match competition is available each week, should the shooter participate?

    <u>ANSWER</u>: Yes, if time and finances permit.

    1.   Frequent shoulder-to-shoulder competition and regularly scheduled practice on the range with shooters who approach the problem of improving their shooting with enthusiasm and a serious, determined attitude is the most effective method of accelerating your development as a top competitive shooter.

    2.   For dry-fire practice to be effective, each practice session must have a goal. You should approach the training period with the idea that you are going to distinctly improve one specific aspect of your shooting technique. Dry firing practice should be conducted with the same careful attention to detail as live ammunition practice, for example:

    a.   To improve your ability to deliver each slow fire shot quickly and accurately, we advise a practice session of about ten rounds delivered in the following manner. Adjust the target turning mechanism to face the target and turn it away after approximately three (3) seconds. Use your normal preliminary preparation with maximum attention on delivering the first shot without hesitation as the target turns. Fire one shot only. Repeat the exercise ten times with sufficient time between shots to allow for analysis and mental preparation.

        b. To improve your ability to maintain a point focus on the front sight, place a target face on the frame backwards so that no bullseye or aiming point is visible. Assume your stance, position and grip with meticulous attention to detail. Without a <u>point</u> at which to aim, you will find that you must trust your stance and position to maintain an acceptable arc of movement. You will find it easier to apply the fundamentals and find that you can deliver the shot with amazing accuracy. Sight alignment can be maintained with a startling degree of control and assurance. This is because the distracting effects of having an exact point to aim have been eliminated. You have no way of knowing when a perfect hold occurs. You simply accept minor errors in hold caused by your minimum arc of movement and go ahead and follow your plan of delivery of the shot.

        c. Avoid training and shooting alone. Use a training program that duplicates as near as possible the competitive atmosphere of a match. Develop and use a comprehensive plan that gives you the ability to employ the fundamentals most reliably under pressure and continually strive for improvement. The shooters' slow fire work sheet will provide this guidance.

    3. Ball and dummy exercises will reveal certain errors in performance not apparent otherwise.

    4. Intensive training in the fundamentals and techniques of pistol shooting is vital to progress.

    5. Frequent review of the fundamentals and techniques will assure constant interest on the part fo the shooter to improve his performance.

VI. CONCLUSION:

    A. <u>Retain Attention</u>: The great slow fire shooters do not hope, pray and wish for Providence to smile on their efforts in pistol competition so they can turn in a high score without knowing exactly how it happened. Every thought and action is planned and executed with care and precision. Their technique of slow fire is a tangible, workable method that enables them to enter competition with confidence.

NOTE: SHOW SLIDE #1 - SCOPE AND SUMMARY.

    B. <u>Summary</u>:

    1. The shooter should be able to coordinate the employment of the fundamentals in firing each shot of slow fire.

    2. The shooter must develop a personal technique of slow fire control.

    3. The shooter must recognize and correct the common deficiencies in control.

    4. The shooter must learn to compensate for adverse shooting conditions.

    5. The shooter must use a comprehensive method of training.

    C. <u>Application</u>: A coach or a shooter who has developed a technique for coordinated employment of the fundamentals will eventually achieve winning scores.

D. **Closing Statement**: There are no gimmicks or special pills that boost your slow fire scores. The answer is a lot of hard work, care, patience, coordination analysis of errors, and positive correction to prevent any past errors from occurring again and lowering your performance level.

F. TECHNIQUE OF FIRE III　　　　　　　　　　　　AAPMI&CC 306
    (SUSTAINED FIRE)　　　　　　　　　　　　　　　30 Minutes
　　　　　　　　　　　　　　　　　　　　　　　　　Mar 1975
                        LESSON OUTLINE

   I. LESSON OBJECTIVE: To enable the pistol marksmanship student to fire winning scores in sustained fire by coordinated control of the employment of the fundamentals in the timed and rapid fire stages.

   II. STUDENT PERFORMANCE OBJECTIVES: As a result of this instruction, students must be able to accomplish the following student performance objectives:

       A. Given the fundamentals and factors that are essential to sustained fire control, EXPLAIN how these fundamentals and additional factors are coordinated in controlling sustained fire with a pistol as shown in Chapter VI, USAMU Pistol Marksmanship Guide.

       B. Given the factors of sustained fire control in sequence, DEVELOP a technique of sustained fire based on the known factors and the sequence of rapid fire only as listed in Chapter VI, USAMU Pistol Marksmanship Guide.

       C. Faced with recurring errors in sustained fire, ANALYZE the actions taken in delivery of the sustained fire strings of shots and find the cause of the deficiencies in control and apply positive corrections as shown in Chapter VI, USAMU Pistol Marksmanship Guide.

       D. Given the mission of training a pistol team to win in competition, APPLY the proper sustained fire training methods as described in Chapter VI, USAMU Pistol Marksmanship Guide.

       E. Faced with the task of firing winning pistol scores under adverse conditions, EXPLAIN the techniques used and measures taken to reduce the affects of adverse conditions as shown in Chapter VI, USAMU Pistol Marksmanship Guide.

   III. ADVANCE ASSIGNMENT: Chapter VI, USAMU Pistol Marksmanship Guide.

   IV. INTRODUCTION:

       A. <u>Gain Attention</u>: The pistol shooter must approach the rapid fire stage with the confidence and assurance that comes from having fired an unlimited number of successful strings of sustained fire.

       B. <u>Orient Students</u>:

           1. <u>Lesson Tie-In</u>: We have previously discussed the coordinated employment of the fundamentals and how they apply to slow fire. The same fundamentals apply in timed and rapid fire plus the factors of recovery and rhythm.

           2. <u>Motivation</u>: Can you consistently shoot good timed and rapid fire scores? Have you ever had a chance to win a match and then blown up on Rapid Fire? Timed and rapid fire can be stumbling blocks, especially if attempted in a haphazard manner. However, through the development of proper techniques and careful planning, you can improve and become more consistent in your sustained fire performance.

           3. <u>Scope</u>: The shooter must be familiar with all the factors that contribute to his control of sustained fire stages.

NOTE: SHOW SLIDE #1 - SCOPE AND SUMMARY.

   a. Know the fundamentals and factors that are essential to sustained fire control.

   b. Develop a technique of sustained fire control based on the known factors and sequence of rapid fire only.

   c. Analyze the performance and find the causes of common deficiencies in sustained fire control.

   d. Apply training methods that will enable the pistol shooter to win in competition.

   e. Reduce the effects of adverse conditions.

V. BODY:

 A. <u>First Student Performance Objective</u>: Given the fundamentals and factors that are essential to sustained fire control, EXPLAIN how these fundamentals and additional factors are coordinated in controlling sustained fire with a pistol as shown in Chapter VI, USAMU Pistol Marksmanship Guide.

NOTE: SHOW SLIDE #1a - EMPLOYMENT OF THE FUNDAMENTALS.

  <u>QUESTION</u>: What fundamentals of pistol marksmanship are applicable to sustained fire?

  1. Same fundamentals as in slow fire. The following factors must be employed in coordination with each other:

   a. Minimum arc of movement. A stable stance, natural position of pointing or holding in the center of the aiming area, firm grip that allows natural alignment of sights and breath control for avoiding disturbance of thinking of or breathing during firing, must be utilized during each string of fire.

   b. Sight alignment. The shooter attains and maintains perfect sight alignment by controlling the relationship of the front and rear sights.

   c. Trigger control. Apply positive trigger pressure straight to the rear without disturbing sight alignment.

  2. Factors in addition to the fundamentals that are essential to sustained fire control.

NOTE: SHOW SLIDE #2 - RECOVERY AND RHYTHM.

   a. Recovery from recoil should include a natural realignment of the sights and the arm and pistol should resume a natural point at the center of the aiming area. If sights are not in alignment, grip and head position should be rechecked. If hold recovers to any point other than center of target, position should be rechecked.

   b. Rhythm is the result of proper execution of a planned sequence of action. The development of good rhythm in firing a good score indicates a coordinated control of the employment of the fundamentals.

TRANSITION: When an understanding of how to control the employment of the fundamentals in slow fire is achieved, a technique of maintaining control in sustained fire must be developed.

B. <u>Second Student Performance Objective</u>: Given the factors of sustained fire control in sequence, DEVELOP a technique of sustained fire based on the known factors and the sequence of rapid fire only as listed in Chapter VI, USAMU Pistol Marksmanship Guide.

<u>QUESTION</u>: Should the pistol shooter develop two methods or sustained fire, i.e. timed and rapid fire?

<u>ANSWER</u>: No. The pistol shooter should have only one method of sustained fire. It is based on rapid fire.

<u>NOTE</u>: SHOW SLIDE #3 - AIMING ON EDGE ON TARGET FRAME.

    1. Find aiming area on edge of target frame in line with position of bullseye when target is faced.

<u>NOTE</u>: SHOW SLIDE #4 - MINIMUM ARC OF MOVEMENT AND POINT FOCUS.

    2. Stiffen your shooting arm for solid arm control. Settle into minimum arc of movement.

    3. Establish eye focus on front sight.

    4. Maintain eye focus on front sight throughout string.

<u>NOTE</u>: SHOW SLIDE #5 - POSITIVE TRIGGER PRESSURE AS TARGETS TURN.

    5. Apply positive pressure on trigger as targets turn.

    6. Shift concentration to sight alignment the instant positive trigger pressure has started. Maintain head position.

<u>NOTE</u>: SHOW SLIDE #6 - FOLLOW THROUGH.

    7. Maintain concentration on sight alignment until pistol fires - solid arm control will absorb recoil shock without bending of wrist or elbow.

<u>NATE</u>: SHOW SLIDE #7 - RECOVERY.

    8. Reestablish sight alignment during recovery without focus shift.

    9. Recovery must be natural and uniform. Reestablish hold or minimum arc of movement in center of aiming area and allow proper grip to aid in attaining natural alignment of sights.

<u>QUESTION</u>: Why must recovery be natural and uniform?

<u>ANSWER</u>: There must be no errors in sight alignment or hold in center of aiming area to be corrected.

    10. Reestablish positive trigger pressure during recovery from recoil. Attempt to correct errors in sight alignment and hold after recovery but do not delay prompt reapplication of positive trigger pressure in order to fully accomplish this correction of error.

    11. If the grip you achieved in preparation is correct, firm arm control achieved and head position maintained, the sights will be in near perfect alignment at the end of recovery. However, this ideal situation will not occur constantly.

12. Each shot should have a major effort expended to attain perfect sight alignment. Five perfect sight alignments are necessary for obtaining a perfect score (five x's).

NOTE: SHOW SLIDE #9 - ENLARGED BULLSEYE WITH FIVE TIGHTLY GROUPED X RING SHOT HOLES.

TRANSITION: Before and during the firing of a sustained fire string, errors in performance may be committed. The shooter must familiarize himself with the more common of these errors and learn to analyze and correct all of those committed if progress is to be maintained.

    C. **Third Student Performance Objective**: Faced with recurring errors in sustained fire, ANALYZE the actions taken in delivery of the sustained fire strings of shots and find the cause of the deficiencies in control and apply positive corrections as shown in Chapter VI, USAMU Pistol Marksmanship Guide.

QUESTION: How does a faulty recovery affect performance?

ANSWER: 1. Slow and faulty recovery means unnecessary delay because the hold or sight alignment or both requires correction before each shot.

    2. Understand how lack of follow through affects control.

    3. Calling the shot group accurately requires a visual memory of five sight alignments.

    4. Understand how lack of rhythm indicates failure of coordination in control of employment of fundamentals.

    5. Do not try to correct minor errors in hold while a string of rapid fire is in progress. The resulting delay causes a break in concentration on sight alignment because of anxiety concerning passage of time. The usual outcome is a speed up of employing the fundamentals which in turn jeopardizes control of the later shots in the string.

    6. Understand how lack of a system makes it difficult to repeat a good performance.

    7. Incomplete shot analysis allows errors to remain in the performance of sustained fire.

    8. Lack of positive correction causes errors to continue damaging the performance in sustained fire.

    9. Analyze why you are shooting well.

    10. Overeating during the shooting day will penalize the ability to shoot with maximum control.

    11. Inability to control mental processes during a string of sustained fire indicates a failure to maintain motivation to apply plan of action.

    12. When concentration on sight alignment breaks as target faces, it indicates a lack of continuity in developing and applying plan of action.

TRANSITION: The perfection of sustained fire technique and elimination of errors in performance is dependent upon a training program that provides sound advice, allows time for free practice and assistance in analysis and correction of errors.

D. **Fourth Student Performance Objective**: Given the mission of training a pistol team to win in competition, APPLY the proper sustained fire training methods as described in Chapter VI, USAMU Pistol Marksmanship Guide.

    1. Competition will accelerate your development as a top competitive shooter.

    2. The shooter must have a specific objective for each practice session.

    3. The use of first and second shot drill will accustom the shooter to promptly starting the string as targets face.

    4. Rhythm and sight alignment exercises will improve ability to maintain sight alignment and timing.

    5. Avoid training and shooting alone. A coach or even another shooter can be of great help.

QUESTION: In what ways will dry firing improve performance?

ANSWER: 6. Dry firing practice will improve the shooters coordination, uniform control of employment of the fundamentals, eye focus, analysis and correction of errors, etc.

    7. Recovery must be quick and complete with center hold and good sight alignment so positive trigger pressure can be applied without delay.

TRANSITION: Mastery of a technique of controlled employment of the fundamentals under ideal conditions will of necessity be modified when the wind starts to blow.

E. **Fifth Student Performance Objective**: Faced with the task of firing winning pistol scores under adverse conditions, EXPLAIN the techniques used and measures taken to reduce the affects of adverse conditions as shown in Chapter VI, USAMU Pistol Marksmanship Guide.

QUESTION: Is it wise to practice frequently in the wind?

ANSWER: 1. No. Wind shooting is conducive to jerking the trigger because the shooter attempts to attain a perfect sight picture. Unless the following techniques are observed, control is erratic at best:

    a. Attain a minimum arc of movement, i.e., The smallest that can be expected under the existing conditions. Even a slight wind enlarges the minimum arc of movement to a degree.

    b. Maintain sight alignment in spite of the enlarged holding area on the target.

    c. Apply positive trigger pressure regardless of the fact that the arc of movement is larger. Maintain sight alignment and the bullet will strike the target within your ability to hold. The group may be larger than normal but the shooter has avoided the wild shots generated by faulty trigger action.

    d. Rhythm must be maintained because this indicates smooth coordination of the employment of the fundamentals.

    e. Extensive practice is not advisable under windy conditions.

    2. Compensate for other adverse weather conditions.

    a. Rain. Carry a raincoat to every match.

        b.   Cold. Have warm clothing available.

        c.   Hot. Wear loose, lightweight comfortable clothing.

        d.   Effect of temperature on shot dispersion is reduced when ammunition is not exposed to direct sunlight.

    3.   Compensate for changing light conditions by having a record of the sight setting used under each condition.

        a.   Dim.

        b.   Bright.

    4.   Mental attitude must be one of accepting the existing conditions and work to compensate as much as possible for the enlarged arc of movement by making a special effort to maintain perfect sight alignment.

VI.   CONCLUSION:

    A.   <u>Retain Attention</u>: To shoot winning timed and rapid fire scores, you must know all the factors of shooting control. You must put all of these factors into coordinated action in practice. After a string is fired, analyze your performance so you can apply the necessary corrective action. Your performance will need periodic correction to assure constant improvement.

NOTE: SHOW SLIDE #1 - SCOPE AND SUMMARY.

    B.   <u>Summary</u>:

        1.   Know the fundamentals and factors essential to sustained fire control

        2.   Make use of all known factors of rapid fire technique to develop a technique of sustained fire control.

        3.   Analyze the performance and find the causes of common deficiencies in control.

        4.   Be able to explain how the application of proper training methods will improve performance in sustained fire.

        5.   Explain how to reduce the effects of wind and other adverse conditions during the sustained fire stages.

    C.   <u>Application</u>: Whether you are coaching a shooter or doing the shooting yourself, you will find that the planning and execution of a correct sequence of action is the technique necessary for precise control. Detailed attention to organization, analysis and correction results in being able to successfully repeat a good performance.

    D.   <u>Closing Statement</u>: The many things that contribute to precise control during the firing of the five shot string of timed or rapid fire do not just happen by chance but are carefully planned and executed by the shooter. Successful repetition of a controlled five shot string is possible only by having complete knowledge of how it is done properly.

G. MENTAL DISCIPLINE

AA PMI&CC 307
50 Minutes
March 1975

LESSON OUTLINE

I. LESSON OBJECTIVE: To enable the pistol marksmanship student to control his mental processes while under conditions of competitive stress by developing a mental discipline that will generate self-control, avoid negative thinking and the adverse effects of match pressure.

II. STUDENT PERFORMANCE OBJECTIVES: As a result of this instruction, students must be able to accomplish the following student performance objectives:

A. Given the information on how mental discipline enables the pistol shooter to exercise self-control, EXPLAIN why mental discipline is essential to advanced pistol marksmanship as shown in Chapter VII, USAMU Pistol Marksmanship Guide.

B. Given the factors that aid in developing mental discipline, EXPLAIN how the pistol shooter can achieve a high degree of control over his mental processes while under competitive stress as shown in Chapter VII, USAMU Pistol Marksmanship Guide.

C. Given the emotional and environmental pitfalls the pistol shooter faces in competition, EXPLAIN how negative thinking can lessen the shooter's chance of winning a pistol match, as shown in Chapter VII, USAMU Pistol Marksmanship Guide.

D. Given the causes, effects and control of match pressure, EXPLAIN what causes match pressure, DISCLOSE how it affects the pistol shooter and EMPLOY the control factors of match pressure as stated in Chapter VII, USAMU Pistol Marksmanship Guide.

E. Having been rendered tense by the conditions of match pressure, DISCLOSE the effects of tension and EMPLOY tension relaxing techniques as shown in Chapter VII, USAMU Pistol Marksmanship Guide.

III. ADVANCE ASSIGNMENT: USAMU Pistol Marksmanship Guide, Chapter VII, "Mental Discipline."

IV. INTRODUCTION:

A. _Gain Attention_:

1. The members of this marksmanship class represent a typical cross-section of the personnel in the armed forces and the general adult population. There is a wide variety of mental capacities from infantile to near genius.

B. _Orient Students_:

1. _Lesson Tie-In_: In the preceding instruction the shooter has received information on pistol shooting fundamentals and the technique of employment of the fundamentals.

2. _Motivation_: Mental discipline makes it possible for the shooter to maintain control of employment of the fundamentals. This ability is reflected in the shooter by confidence, positive thinking and the proven ability to repeat the delivery of a successful shot. Full use of mental capacity is the sure way to achieving a more perfect performance and victory in competition.

NOTE: SHOW SLIDE #1-SCOPE AND SUMMARY OF MENTAL DISCIPLINE.

3. _Scope_: You attain mental discipline by having knowledge of:

a. Why mental discipline is essential to advance pistol marksmanship.

b. How to develop mental discipline, confidence and the ability to think positively.

c. Why can't you be a winner?

d. The causes, effects and control of match pressure.

e. How to reduce the effects of tension and attain relaxation under conditions of stress.

NOTE: SHOW CHARTS #2, #3, #4, #5, and #6 IN SUCCESSION-INTERRUPTIONS IN FIRING A STRING OF RAPID FIRE. (GIVE FIRE COMMANDS, "READY ON THE RIGHT, READY ON THE LEFT," ETC.).

V. BODY:

NOTE: SHOW CHART #7 FIRING A SUCCESSFUL STRING OF RAPID FIRE WITH CONTROL OF SIGHT ALIGNMENT.

A. First Student Performance Objective: Given the information on how mental discipline enables the pistol shooter to exercise self-control, EXPLAIN why mental discipline is essential to advanced pistol marksmanship as shown in Chapter VII, USAMU Pistol Marksmanship Guide.

NOTE: SHOW CHART #9 MENTAL DISCIPLINE IS ESSENTIAL.

QUESTION: Why is mental discipline essential to advanced pistol marksmanship?

ANSWER:

1. Mental control has become essential to advanced marksmanship because mastery of the physical skills alone does not provide the consistent, precise performance necessary to compete at the highest level. Emphasis should be placed on how and what to think while endeavoring to control the firing of a shot.

2. Mental discipline provides the thought control the shooter must have to maintain his confidence, positive thinking, and retain the ability to repeat a successful performance. It also aids the ability to channel and sustain one's mental effort. It will help control physical action, avoid over-confidence, pessimism and exposure to conditions that will disrupt mental tranquility.

3. Mental discipline provides the emotional stability necessary to the development of the champion shooter. Confidence in his ability to successfully employ the basic skills of marksmanship produces a dependable performance under various degrees of stress.

4. The self-control attained by the advanced pistol shooter pays off not only in better competitive match scores, but also in combat situations by the improved confidence and skill in using his weapon to hit his target with every round fired.

TRANSITION: It's essential nature requires that the shooter develop mental discipline as part of his shooting skill.

B. Second Student Performance Objective: Given the factors that aid in developing mental discipline, EXPLAIN how the pistol shooter can achieve a high degree of control over his mental processes while under competitive stress as shown in Chapter VII, USAMU Pistol Marksmanship Guide.

NOTE: SHOW CHART #10 DEVELOP MENTAL DISCIPLINE.

QUESTION: How is mental discipline developed?

ANSWER: The continuously repeated, successful execution of a completely planned shot results in the gradual development of mental discipline. The proper degree of mental discipline restricts the thoughts and actions during shooting to an established pattern from which there will be few deviations.

NOTE: PASS OUT ANNEX 1 TO MENTAL DISCIPLINE.

1. Methods of Response to a Shooting Problem.

There are four basic methods of responding to a problem. Two methods are positive and classified as either direct or indirect. Two methods are negative and classified as either retreat and evasion.

a. Positive response to a problem.

(1) The direct, positive approach. This is the self-confident, self-sufficient, direct, positive attack. You realistically face the difficulty or problem, analyze it, identify and eliminate the obstacles to a successful solution. You know what you want to accomplish and you take direct steps to attain it.

(2) The indirect, substitute or compromise approach. There are small, tentative, indirect actions, Sidestepping leads to seeking shortcuts. When the probable solution is tried, there is much fervent hoping that the fates are on your side. You are only hinting and probing instead of establishing definitely what you need to do.

b. Negative Response to a problem.

(1) The negative retreat. The failure to give an honest try toward solution to see what you are capable of accomplishing. Surrendering without a sincere attempt. The flight habit can become chronic. This is the man that cannot accept the responsibility for a mistake or failure. A bad shot produces excuses instead of constructive analysis of performance.

(2) Evading the issue. Evasion is the lack of incentive. Why?, is the approach. Why do I have to do better than anybody else? If the desire to excel is not there, you will never aimlessly achieve the degree of success that crowns the champions' efforts.

NOTE: SHOW SLIDE #11-CONTROL FACTORS, THE BASIS OF A SYSTEM OF SHOOTING A PISTOL.

2. Analyze the problem.

a. Psychologists have discovered that one of the chief reasons for difficulty in the solution of a problem is inability to soundly analyze. Lay out a clearcut plan of action. Where faced with a particular difficulty, make a determined effort to break it down. If it is identified, there is a solution for it because there are shooters on your team that are operating without this specific problem putting a brake on their performance. Talk it over. A cooperative bull session will break it wide open.

b. There is a three-point system of analyzing and solving specific problems. It reduces the whole big problem to many specific small ones. Head four columns on a sheet of paper with the following titles: one, "STEPS IN THE PLANNING" of control for firing on accurate shot; two, "SPECIFIC DIFFICULTIES" in performing each step of the shot plan; three, "SUCCESSFUL SOLUTIONS" to each of the steps that are being performed satisfactorily; four, "DOUBTFUL OR NO WORKABLE SOLUTION" for those difficult steps that aren't working out too well. (Column 1 is based on Chapter IV, "Establish a System".) Follow the complete sequence of firing a shot.

c. The positive approach requires that we have a definite plan. We must support the plan by consistent use of each step of the plan. We must be persistent in the face of difficulty in execution of any step of the plan, and not rest until a solution is reached. Finally, we should be on guard against compromise and negative thinking. The positive approach in overcoming obstacles to mental control can become automatic.

3. Gaining confidence.

Confidence results from repeatedly bringing under control all the factors that create conditions for firing on accurate shot. An accurate shot is one that hits the center of the target within the shooter's ability to hold. People have been telling you for years that you must have confidence to shoot well. Confidence in what? How do you get it? How do you keep it?

a. First and foremost you must have confidence in the fundamentals of advanced pistol marksmanship that you use. You must be convinced that if you control their employment correctly, you will achieve excellent results. The techniques of employment of the fundamentals that you have proven sound and dependable by experience are not going to change suddenly to unreliable factors because of match pressure.

b. Confidence in your ability to execute these proven fundamentals correctly. You have proven your degree of ability to do this in your practice sessions. Go ahead and do it in the big match. To the timid and hesitating, everything is impossible, becuase it seems so.

c. Think big! Think positive! "I will do it," and you will succeed. However, as soon as you admit the slightest possibility of failure, so long as there is an influence in your mind that is preventing you from putting all your energies into your task, your success is questionable.

d. It has been said innumerable times that a pistol shooter must have an open mind, implying that we must have the ability to accept new ideas. What we should also strive for is a mind that is open to positive thoughts and closed to those of a negative character.

4. Ability to channel mental effort.

Channeled mental effort resists the tendency of the mind to drift during the period when intense concentration on sight alignment is essential.

a. Channel your mental effort relentlessly toward controlling the one factor that resists being controlled involuntarily: the relationship of the front and rear sight.

b. Complete Exclusion of Extraneous thoughts for a brief period (three to six seconds) is necessary for controlled delivery of the shot. Concern involving the employment of the involuntarily applied fundamentals reduces the ability to concentrate on sight alignment. All of the other fundamentals are readily adaptable to being executed on the basis of habit or involuntary action. The shooter must also learn to ignore distractions caused by external conditions under which he is firing.

c. Prior Planning of the Sequence of Action is necessary to deliver a controlled shot on the target. This mental involvement gradually enables the shooter to sustain maximum concentration for a longer period. Careful planning of a sequence of events closes the mind to other thoughts. Example: A prior plan is made to apply positive trigger pressure when sights are in alignment and the arc of movement is at the minimum. Uninterrupted positive trigger pressure applied at the right moment becomes almost involuntary.

d. Coordination of Thought and Action is the result of experience obtained through extensive practice and match shooting where the same satisfactory plan of action is followed repeatedly.

TRANSITION: Confidence and the positive approach are essential to success in pistol shooting but the ever present danger of negative thinking makes it necessary to know the characteristics of that inviting trap.

NOTE: SHOW SLIDE #11a ALIBI BOG.

      C. <u>Third Student Performance Objective</u>: Given the emotional and environmental pitfalls the pistol shooter faces in competition, EXPLAIN how negative thinking can lessen the shooter's chance of winning a pistol match, as shown in Chapter VII, USAMU Pistol Marksmanship Guide.

QUESTION: Why is it so difficult to shoot championship scores?

ANSWER:

      1. The fault usually lies in that we open our minds up to thousands of reasons why we cannot shoot good scores. Shooters with a wealth of experience and skill can eliminate their chances of winning by negative thinking.

NOTE: SHOW SLIDE #11-WHY CAN'T I BE A WINNER?

      a. <u>When the weather is bad</u>, it is easy to say "It's raining, snowing, the wind is blowing. My scores are going to be lower." This may be a true assumption. You can follow this vein of thought throughout the match and you probably will shoot only average scores as compared to your competitors. Convince yourself that good scores have been and will be fired under the same bad conditions. Proper application of the fundamentals will produce good results in spite of adverse weather. If your thoughts are directed strongly enough towards planning and executing a controlled performance, you will not have time to worry about the weather.

      b. <u>Distractions on the firing range</u>. In most instances, all it takes to correct an abnormal situation on the range is to have your coach bring the deficiency to the attention of the Chief Range Officer, Executive Officer, or Referee. Don't be distracted from your control of shooting by becoming involved in the dispute.

      c. <u>Lack of incentive to fire winning scores</u>. Have you ever asked yourself, "Why do I have to shoot outstanding scores?" The answer to this question is that you are wasting your time if you are not trying to shoot scores that will win the match. Just hanging on to your place on the team is not enough. You must be motivated to constantly improve your performance or else you should change to a less demanding endeavor. A common excuse for not trying your best is the lack of strong competition. The tendency to perform below your potential becomes a habit. You tolerate a substandard performance without becoming alarmed. You must retain the desire to win <u>and</u> set new records each time you fire in competition. Failure in this area too often will cause a decline into a habit of treating your shooting as a task instead of a challenging adventure.

      d. <u>Inferior shooting equipment and ammunition</u>. How often have you beaten yourself by allowing yourself to think that the cause of your poor performance was that the competitors who beat you had better equipment? How can we keep such thoughts as these from entering our mind? The main factors necessary to shoot championship scores are an accurate gun, good ammunition, an individual with the ability (physical and mental) and the desire to be a champion. <u>Therefore</u> every time you let the thought of inferior equipment enter your mind, Think: "This gun and ammunition will shoot possibles if I control it."

NOTE: SHOW SLIDE #12-COMPETITION, PESSIMISM, ETC.

      e. <u>"The competition is too tough."</u> "Nuts." Critically examine these individuals who look like supermen to you. Analyze a few of them and compare their attributes with yours. They are built just like you, have approximately the same physical ability, hands about the same size, etc. What then, is the determining factor? The potential winner is thinking about applying his plan of action and not about how he is going to beat you. The champion knows that most of the other competitors are defeating themselves with their controlled thoughts. You can be one step ahead of your competitors by directing your mental effort toward your plan of controlling each shot.

f. **Pessimism.** There's a first time for winning in shooting as in everything else. A first time for a national champion to be beaten, and a first time for you to become a national champion. You have never won a major championship? Was it because you didn't have the ability or was it because you over estimated the ability of your strongest competitors and conceded your chance of winning? You have probably won individual matches but that's as far as you have allowed your mental capability to carry you. Now if you really want to win, you can. The best way is to shoot your next pistol tournament as one big match. Let the individual stages and gun aggregates take care of themselves. A good performance on each individual shot is now your aim. Don't let the possibility of winning one little match shatter your composure. **Pessimism** detracts from your ability to concentrate. Anxiety over possible failure undermines the ability to control the shot. Impatience and uncontrolled actions are the results. A negative approach hampers your ability to repeat a uniform, satisfactory performance.

g. **Carelessness.** Do you expect that you will inevitably commit a stupid act in every match you fire in, thereby forfeiting any chance of winning? Carelessness is a state of mind that overwhelms an individual who is aimless and haphazard in his approach to a challenging task. Organization of all the factors having a bearing on the task to be performed will in most instances assure that the action will be successfully executed.

h. **Overconfidence.** You ignore or are not conscious of the development of unfavorable conditions. False assurance can upset the sensitive balance on which your optimum performance depends. Do not relax your determination to work hard even if competition is not keen. Strive to reach a happy medium between overconfidence and pessimism.

i. **Avoid distracting conditions** which you know will upset your mental control. Avoid emotional upset such an anger, worry, giving up under adverse condition or after unsatisfactory shots. Ignore boasts, rumors, misinformation, and snide remarks. Be passive about the other competitors scores. Resist concern over the final results. Dismiss concern over the slight advantage of superior equipment. Avoid adding up individual shots in the buildup to the final score.

QUESTION: Who won the last match in which you participated? If not you, why?

2. To achieve results on a level that will produce winning scores in today's competition, it is necessary to have a coordinated, exacting control of the technique of employment of the fundamentals. Each properly executed sequence of actions that results in a good shot, contributes to the ease with which it can be repeated.

TRANSITION: The man who has never experienced match pressure has never been in a position to win a match.

D. Fourth Student Performance Objective: Given the causes, effects and control of match pressure, EXPLAIN what causes match pressure, DISCLOSE how it affects the pistol shooter and EMPLOY the control factors of match pressure as stated in Chapter VII, USAMU Pistol Marksmanship Guide.

NOTE: SHOW SLIDE #12a MATCH PRESSURE.

QUESTION: What causes match pressure?

ANSWER:

1. **Match pressure is the direct result of the fear of failure** and the loss of self esteem. Are we afraid of winning? No! It is not the actual winning we are afraid of. We are prone to succumb to our fear of performing poorly and having our fellow competitors see our poor performance.

2. **What happens to us physically** when we are subjected to all of these mental gymnastics that result from match pressure? We shake, we drop our magazines, and point our scope on the wrong target. Some of us even shoot in the wrong target. In short, we commit what seems a series of asinine mistakes that normally would never occur.

In addition, you invariably experience a shortness of breath which increases your breathing rate. Your heart beats about twice as fast as necessary. All of this seems to make it impossible for us to hold our pistol reasonably steady, let alone shoot well. To add to our distress, we feel that everyone is witnessing our anxiety and stupidity. Yet, with all the complications, our counterpart the champion appears to be calm and enjoying himself. Let's face it. He is!

   3. <u>There are definite advantages to match pressure</u>. Many of our senses are more acute. For our purposes we see better, and our sense of touch is more exacting (that is why your trigger seems to become heavier in a match; actually it has not changed a bit, but we are more aware of it). Our awareness of the passage of time becomes more vivid. Don't believe it? What about the anxiety you feel just before you shoot the last round of a rapid fire string. All of this sharpened awareness should make us more exacting and consequently better our performance.

   4. <u>How do we control match pressure?</u> First, realize that it can be controlled and actually used to your advantage.

      a. <u>Prior Mental Determination</u>. This is the most helpful factor that is made available to you. By thinking through the correct procedure for firing each shot just before you shoot, you can virtually eliminate distractions in the actual execution. If you fail to do this and approach the shot without a preconceived plan of attack, your results at best will be erratic. You readily appreciate the necessity for concentrating on and aligning your sights. A very effective way to assist in this is to sit down and close your eyes. Imagine the front and rear sights including the blurred target. Try it right now. Most of us find that it is almost impossible to keep them aligned perfectly even in your mind's eye. However, by doing this, you are conditioning your mind to be able to focus the mental concentration where you want it. This technique of mentally aligning the sights is very effective if practiced just before attempting to fire.

      b. <u>Channel Your Thinking</u> to the More Important Fundamentals. You must continually think fundamentals and review them in your mind. Train yourself so that as many as possible of these fundamentals are executed automatically without any tedious effort on your part. When you do this, it leaves you with only the most difficult fundamentals to contend with in the actual firing. This will enable you to place all of your mental and physical effort toward keeping the sights aligned and smoothly releasing the hammer.

      c. <u>Establish a Routine</u>. From routine comes boredom. What is boredom? The lack of excitment. What are we trying to do? Keep from becoming excited. In a more serious vein, however, in establishing a routine, you eliminate the possibility of forgetting some trivial item of preparation that may throw you off balance later if you neglect it.

      d. <u>Work on Each Shot Individually</u>. Each shot must be treated as an individual task. In reality there is no reason to believe that because your first shot was an eight your next one will be poorly controlled. Nor is it logical that if your first three shots were tens that you have a guarantee that those to follow will also be tens. Each shot is only a representation of your immediate present ability to apply the fundamentals correctly or incorrectly. Your ability to do this will vary considerably if you let it. Do not connect the shot you are preparing to fire with the value of those already on the target.

      e. <u>Win the Aggregate, Not just One Match</u>. Why should we become excited or worried when we have cleaned three of the four strings of the 45 caliber timed fire match? Go right ahead and clean the next string. Sure, if you do so, you may win the 45 cal timed fire match. Don't drop a couple of points here just because the possibility of winning one match has arisen. Winning the aggregate is your goal.

      f. <u>Train Yourself to Think Performance Rather Than Score</u>. In employing this technique, an eight or a seven becomes not a shot that subtracts two or three points from your aggregate, but a shot where you allowed yourself to deviate from proper employment of one of the fundamentals. Rest assured that if you control yourself on the firing line, the score will take care of itself.

g. _Who Said "Stay Out of the Scope?"_ If you are shooting a slow fire match, you must move forward to the target and score after ten shots. If it's National Match Course, you must score the slow fire stage before you shoot at 25 yards. Do you think you are going to keep the slow fire score a secret from yourself? Why should a good score scare you? A good score is exactly what you went up to the firing line to accomplish. Of what value is a 98 slow fire if you don't possess the fortitude to continue a good performance? Learn to use the score for the purpose it was intended: A check on your performance and zero. Use you scope as an aid in your analytical procedure, not to score your target.

h. _Relax Your Mind_, right from the time you get up in the morning. Nothing will put you in a greater state of mental agitation than to rush through breakfast and get to the range barely in time to make your relay. Generally when this happens, your slow fire is ruined at about the third red light you have to wait out. Take it easy. Shooting is fun, enjoy it.

i. _Practice Tranquility_. Have you seen the shooter that loses his temper every time he has a bad shot? Who is he mad at? Those individuals who lose their temper are doing nothing more than punishing themselves. They recognize that if they had worked a little harder applying the control factors, the shot would have been better. On the other hand, if we do everything within our power to control the shot and for some reason it isn't a good one, we should have no cause for undue irritation. Although a good shooter must exert all of his mental and physical ability toward shooting a good score, infrequently he will fail to perform with a high degree of skill. When this happens, if he chastises himself severely instead of devoting his mental effort toward analysis of his performance, he will hurt greatly his chances for winning the match. It is not intended that you laugh off or treat lightly a poor performance. You must possess the presence of mind to accept the bitter with the sweet. Preparation, planning, relaxing, delivery of the shot, careful analysis and employing positive corrective measures is the cycle of action you must force yourself to conform to without deviation. You can then be assured that the next shot will be delivered under the most precise control you are capable of exerting at the present moment.

j. _Match Experience_. Without question, competitive experience is one of the ingredients necessary in the development of an accomplished pistol competitor. However, experience alone is of limited value. We must flavor our experience with an accurate and honest evaluation of our performance. Positive corrective measures will improve our ability and eventually our scores. As we compete, we must experience an increasing degree of mental control. Disciplining our thinking is often left out of our training until our physical ability to shoot far outreaches our ability to exercise mental control when the chips are down.

k. _Physical Conditioning_. There is no doubt whatsoever that you can shoot better if you are in good physical condition. Your ability to hold, for example, is no better than the ability of the muscles of your arm and body to do this for you. Your ability to resist the stress and strain of match pressure and anxiety is directly in proportion to the degree of fitness of your physical condition.

l. _Argue with Your Subconscious_. Not only argue with it but win the argument. Even as we read this some of us are hearing that little voice in the back of our minds that keeps saying "Yes, this sort of thing may work for Joe, but I know damn well I'm going to goof up the next time I get close to a winning score." Whose voice is this? Where did all these ideas come from in the first place? Where did this little guy get all his knowledge? Let's be realistic. Our conscious mind puts these ideas into our subconscious. You must believe you can overpower it. It's not easy. He's been saying what he pleased for years and now he isn't going to be routed easily. But don't give in to him and eventually you will find that the subconscious mind is not in conflict with your conscious efforts.

m. _Don't expect immediate results_ the first time you try mental discipline. There is a coordination of employment of the fundamentals to be mastered. There are no hidden secrets. All that we gain is the direct result of hard work. If you find that you can exercise satisfactory control only for a short period of time, work on extending this period by practicing and perfecting your system. Remember that your returns are in proportion to your investments.

TRANSITION: The fear of failure to perform up to your known capability will generate gradually increasing tension.

       E.   <u>Fifth Student Performance Objective</u>: Having been rendered tense by the conditions of match pressure, DISCLOSE the effects of tension and EMPLOY tension relaxing techniques as shown in Chapter VII, USAMU Pistol Marksmanship Guide.

QUESTION: What are the different types of tension?

ANSWER: Normal and pathological tension.

       1.   <u>Types of Tension</u>. Normal tension is a blessing to mankind. Without tension most problems could not be solved; the world's work would not get done and championship scores would not be fired. Normal tension is the prevailing condition of any organism when it is mustering its strength to cope with a difficult situation. All animals, including men, tense in situations which involve the security of themselves and their dependents. But there is a kind of tension that is bad for you: pathological tension. This is an exaggeration of normal tension, and thank heaven fairly rare. The vast majority of people who worry about it have nothing more than normal tension. All they need is a technique for relaxing. We should know what tension really is and a few hints on how to terminate it.

       a.   What happens to you when you attack a challenging problem in shooting and become normally tense? Psychologically you become slightly anxious. This is a highly civilized counterpart of the "fight or flight" reaction of the primitive animal when it perceives danger. This reaction is <u>not</u> anxiety in the pathological sense. Physiologically you undergo certain definite changes. Adrenalin pours into your bloodstream and your liver releases sugar, giving a plentiful supply of energy to your muscles. Your entire nervous system shifts into high gear. It causes your sense of smell, hearing and sight to become sharpened and all your mental faculties to become razor keen. Your stepped up nervous system also causes the large voluntary muscles of your legs, arms and torso to contract, readying for action. The involuntary muscles of your digestive tract cause your digestion to slow down or stop for a while. Your chest and arterial muscles contract so that your breathing becomes shallower and your blood pressure increases. When all these things are happening, you are experiencing normal tension. Most of us experience this kind of tension one or more times a day. When the problem which caused you to be tense has been solved, your tension will subside and you will return to a normal state of relaxation. It may leave slowly but it will leave. Normal tension is self-limiting. It doesn't continue after you need it.

       b.   Pathological tension is not only hard to terminate, your whole body over-reacts, as if the difficulty confronting it were a life-or-death matter. It's the kind of reaction a normal person would have only in a really dangerous situation. In pathological tension, blood pressure, heartbeat and pulse go way up and stay up. Excessive adrenalin may result in jitteriness, flushing and trembling. The digestive actions of the stomach usually stop entirely and will not resume, causing loss of appetite or indigestion. Muscles tense for action but may end by cramping. Rapid, shallow breathing continues to the point of dizziness. The inevitable, and often swift result is a sense of deadening fatigue. Normal tension may make you feel exhausted too, but recovery is quicker. The simple, tension-ending techniques described below, are applicable only for normal tension.

       2.   <u>Tension Reducing Technique</u>:

NOTE: SHOW SLIDE #13 TAKE A BREATHER.

       a.   Take a Breather. Breathe deeply, three times, very slowly. At the end of each exhalation hold your breath as long as possible. When you have finished you should feel noticeably relaxed and much calmer.

Here's what has happened. By forcing yourself to breathe deeply you break the tension of your voluntary breathing muscles causing the involuntary muscles of the lungs, gastro-intestinal tract and heart to relax too. This is the simplest method for relaxing. For some, it can be used to end tension completely. It can be used by others for temporary relief when they do not wish to "let down" completely.

NOTE: SHOW SLIDE #14 LET YOURSELF GO. (RELAX)

        b. Let Go! Sit down and let your head droop forward. In about a minute raise one arm and drop it in your lap as if it were a limp rag. Do the same with the other. Now let your legs go completely limp; now your stomach muscles. Stay in this position for at least ten minutes. This technique, too, is aimed at first relaxing the voluntary muscles. It is especially effective when you've had to maintain normal tension for several hours on end.

NOTE: SHOW SLIDE #15 SHIFT INTO LOW. (AFTER THE SHOOTING DAY)

        c. Shift into low. When you have been overstimulated by highly demanding mental exertion, taper off by becoming involved in a diverting activity. If you like handiwork, pick a kind which interests you but is not too creative. Soap sculpture, finger-painting, woodworking, and gardening all are excellent low-gear activities that will help you to simmer down. This kind of tension-remover is aimed at changing your mental "set." It is helpful for those who have to operate at top capacity such as shooters and are in enforced contact with others all day long. After high stimulation, a part of your mind will continue to be operating at a fast pace. To slow you down when you're in this state of mind, you require something which is engrossing but which demands nothing of you intellectually. The lighter types of television entertainment and simple handicrafts are ideal.

NOTE: SHOW SLIDE #16 TAKE A BREAK (DON'T DRY FIRE BETWEEN RELAYS)

        d. Take a break. This is a "remote control" technique for dealing with normal tension. Simply take a break for ten full minutes every hour. You may find that this allows you to ease out of your working tension more quickly and easily when the day is over. The reason this works; since you have not allowed tension to develop fully, your organism doesn't have so far to go on the road back to normal relaxation.

NOTE: SHOW SLIDE #17 STOP AND THINK. (AFTER THE MATCH IS OVER)

        e. Stop and think. When the tension creating activity of match shooting allows a respite, sit down and calmly review the things in your life that you value highly. Think of the long range purpose of your life, of the people you love, the things you really want. In a few minutes you may notice that you have involuntarily taken a deep breath. This is a sign that tension is dropping away rapidly. When you tense to face a difficult situation, you tend to exaggerate its importance. Judgment and reason can quickly change this mental state when it's time to relax again.

        These techniques are based on the fact that tension can be ended in two distinct ways: through the relaxation of your voluntary and involuntary muscles; and by changing your mental "set." If you achieve either, you set off the other and hasten the process of normal relaxation.

    VI. CONCLUSION:

        A. <u>Retain Attention</u>: Be a hungry shooter. The slashing, no holds barred drive for victory in competition, from the first shot on, destroys the confidence of the lesser competitors.

        B. Summarize:

NOTE: SHOW SLIDE #1 SCOPE AND SUMMARY.

1. Mental discipline is essential for advanced pistol shooters.

2. Develop mental discipline, confidence and the ability to think positively.

3. Negative thinking can prevent a shooter from winning.

4. The shooter must have a knowledge of the causes, effects and control of match pressure.

5. You must know how to reduce the effects of tension and attain relaxation under conditions of stress.

C. <u>Application</u>: The self control attained by the advanced pistol shooter pays off not only in better match scores, but also in combat situations, by the calmness and resolution in using his weapons to hit his target with every round fired.

D. <u>Closing Statement</u>: <u>A chance to be a winner</u> lies in each shooters grasp. Performing at or beyond your potential will catapult you into the lead. Retain the lead by counting on your competitor's inevitable mistakes and gaps in his knowledge of controlling his shooting techniques. You must have confidence that you are capable of a performance exceeding any previous level of personal accomplishment.

<u>Confidence</u> furnishes the alloy to stiffen the will to win and not give up or compromise. Confidence is based on a full grasp of the complete technique of controlling employment of the fundamentals. Confidence combined with knowledge, exacting skills, good physical condition and a seething, consuming determination to win, provides the shooter with an edge from which he can deliver a shattering blow to the composure of the competition. When your competitor realizes his maximum effort is falling short of that necessary to win, the result is no contest.

ANNEX 1 TO MENTAL DISCIPLINE
POSITIVE RESPONSE TO A SHOOTING PROBLEM

STEPS IN THE PLAN	SPECIFIC DIFFICULTIES	SUCCESSFUL SOLUTIONS	NO WORKABLE SOLUTION
1. PREPARATION			
a. PHYSICAL			
(1) PERSONAL WELL BEING			
(2) LIMBER UP EXERCISES			
(3) CHECK FIRING LINE COMMANDS			
(4) CLOTHES & SHOES SUITABLE			
(5) FIRING CONDITIONS NOTICED			
(6) CORRECT FIRING POINT			
(7) SCOPE YOUR TARGET			
(8) CHECK WEAPON			
(9) CHECK AMMUNITION ON HAND			
(10) NATURAL HOLD (STANCE & POSITION)			
(11) NATURAL SIGHT ALIGNMENT (GRIP AND HEAD POSITION)			
(12) BREATH DEEPLY			
b. MENTAL-STIMULATE CONFIDENCE BY THINKING THROUGH SHOOTING METHOD-MENTALLY REVIEW:			
(1) EXTENDING ARM TOWARD TARGET			
(2) BREATH CONTROL			
(3) PICKING UP SIGHT ALIGNMENT			
(4) TRIGGER SLACK & INITIAL PRESSURE			
(5) FINAL BREATH, MINIMUM ARC OF MOVEMENT			
(6) START POSITIVE TRIGGER PRESSURE (COORDINATE WITH ALL OTHER CONTROL FACTORS)			
(7) CONCENTRATE ON SIGHT ALIGNMENT			
(8) APPLYING TRIGGER PRESSURE INVOLUNTARILY			
(9) CONTINUE APPLYING ALL CONTROL FACTORS-FOLLOW THRU			
(10) SURPRISE SHOT			
2. PLAN THE SHOT			
a. STANCE			
(1) STABLE BALANCE			
(2) IMMOBILITY OF BODY PARTS			
(3) HEAD POSITION MAINTAINED			
(4) UNIFORM FROM SHOT TO SHOT			
(5) POSITION OF FEET UNCHANGED			
(6) BODY ERECT			
(7) SHOULDERS LEVEL			
(8) LEGS FIRMLY STRAIGHT			
(9) HIPS LEVEL			
(10) HEAD LEVEL			
(11) NON-SHOOTING ARM & HAND RELAXED			
(12) SHOOTING ARM STRAIGHT AND SOLID-NOBENDING WRIST OR ELBOW			
(13) CENTER OF GRAVITY-LEAN SLIGHTLY FORWARD			

STEPS IN THE PLAN	SPECIFIC DIFFICULTIES	SUCCESSFUL SOLUTIONS	NO WORKABLE SOLUTION
b. NATURAL POSITION ORIENTATION (1) START AT 45 DEGREE ANGLE (2) TURN ONLY HEAD TO TARGET (3) EXTEND ARM TOWARD TARGET CENTER AND SETTLE (4) CLOSE EYES (5) RAISE ARM & SETTLE WITH NATURAL POINT (6) OPEN EYES (7) SHIFT TRAIL FOOT IN DIRECTION OF ERROR IF NOT POINTING AT TARGET CENTER (8) RECHECK			
c. GRIP (1) NATURAL SIGHT ALIGNMENT (2) FIRM TO PREVENT SHIFT (3) UNCHANGING TIGHTNESS (4) INDEPENDENT TRIGGER FINGER (5) UNCHANGING CHARACTER (6) COMFORTABLE (7) RECOIL STRAIGHT TO REAR (8) AVOID FATIGUE OF HAND			
d. BREATH CONTROL (1) SYSTEMATICALLY APPLIED DURING ALL SHOOTING (2) OXYGEN RETENTION BY DEEP BREATHING (3) MINIMIZE MOVEMENT (4) RESPIRATORY PAUSE AFTER EXHALATION (5) COMFORTABLE WITH NO CONSCIOUS NEED TO BREATHE (6) CONCENTRATION ON FRONT SIGHT UNDISTURBED (7) DEEP BREATHING PRIOR TO & DURING FIRE COMMANDS			
e. SIGHT ALIGNMENT (RELATIONSHIP OF FRONT AND REAR SIGHTS) (1) FRONT SIGHT POINT FOCUS (2) REAR SIGHT AWARENESS (3) EXCLUSIVE CONCENTRATION (4) 6 TO 8 SEC. DURATION (5) COORDINATION WITH OTHER CONTROL FACTORS			
f. TRIGGER CONTROL (1) POSITIVE TRIGGER PRESSURE (2) 2 TO 5 SECONDS (3) BASED ON PERFECT SIGHT ALIGNMENT (4) UNDISTURBED SIGHT ALIGNMENT (5) COORDINATE WITH OPTIMUM PERCEPTION OF PERFECT SIGHT ALIGNMENT AND MINIMUM ARC OF MOVEMENT			

STEPS IN THE PLAN	SPECIFIC DIFFICULTIES	SUCCESSFUL SOLUTIONS	NO WORKABLE SOLUTION

3. RELAX BEFORE SHOT
   a. NECK
   b. SHOULDERS
   c. NON-SHOOTING ARM
   d. ABDOMEN
   e. BACK
   f. BUTTOCKS
   g. UPPPER LEGS			

4. DELIVER THE SHOT (SHOT SEQUENCE)
   a. EXTEND ARM & BREATHE
   b. MINIMUM ARC OF MOVEMENT
   c. PICK UP SIGHT ALIGNMENT
   d. TRIGGER SLACK & INITIAL PRESSURE
   e. FINAL BREATH
   f. MAINTAIN SIGHT ALIGNMENT
   g. MAINTAIN MINIMUM ARC OF MOVEMENT
   h. START POSITIVE TRIGGER PRESSURE
   i. CONCENTRATE POINT OF FOCUS ON FRONT SIGHT			

5. ANALYZE THE SHOT
   a. CALL SHOT
   b. CONFIRM THROUGH SCOPE
   c. EVALUATE
      (1) GOOD SHOT-BAD CALL
      (2) BAD SHOT-GOOD CALL
      (3) BAD SHOT-BAD CALL
      (4) GOOD SHOT-GOOD CALL
   d. WHY? GIVE SPECIFIC REASONS FOR ANY BAD SHOTS OR CALLS
      (1)
      (2)
      (3)
      (4)
      (5)
      (6)
      (7)			

6 CORRECTION INCLUDED IN PLAN FOR NEXT SHOT
   a. STANCE
   b. POSITION
   c. GRIP
   d. BREATH CONTROL
   e. SIGHT ALIGNMENT
   f. TRIGGER CONTROL			

H. TECHNIQUE OF FIRE IV
    (INTERNATIONAL SLOW FIRE)

AAPMI&CC 3155
50 Minutes
Mar 1975

LESSON OUTLINE

I. LESSON OBJECTIVE: To enable the pistol marksmanship shooter-instructor student to compete successfully in the International Slow Fire Course of fire, follow the ISU rules, and use the proper weapons and techniques in firing International Slow Fire.

II. STUDENT PERFORMANCE OBJECTIVES: As a result of this instruction students must be able to accomplish the following student performance objectives:

    A. As an International Slow Fire Instructor, EXPLAIN and DEMONSTRATE ISU rules regulating the range, targets, and firearms of International Slow Fire as outlined in the International Shooting Union Rules on 50 Meter Pistol Competition.

    B. As an International Slow Fire competitor, DEMONSTRATE and IMPLEMENT International competition as shown in Chapter XI, USAMU Pistol Marksmanship Guide.

III. ADVANCE ASSIGNMENT: None.

IV. INTRODUCTION:

    A. <u>Gain Attention</u>: Today, International shooting competition between nations can further build our technical and sports loving image. The United States can have the best International shooters in the world by knowing the fundamentals of International Slow Fire, mastering the techniques of controlling them effectively and firing winning scores according the International Shooting Union Rules.

    B. <u>Orient Students</u>:

        1. <u>Lesson Tie-In</u>: This will be the first of four classes on International Pistol Shooting. There are four types of International Pistol events; they are Slow Fire, Center Fire, Rapid Fire and Standard Pistol. In this class we will discuss Free Pistol Shooting or what is known as the International Slow Fire Course of Fire.

        2. <u>Motivation</u>: The International Slow Fire Course of Fire is one of the most difficult and greatest challenges a pistol shooter can face. This type of shooting is especially interesting to those shooters who have become proficient in NRA Slow Fire shooting and are looking for a greater challenge. For the United States, International Pistol Competition is significant for its technical, sporting and political value. During these classes on International Firing, you have an opportunity to find out the level of skill necessary to make the U.S. Team for the World Championships, the Pan American Games, or the next Olympic Games.

        3. <u>Scope</u>: This hour is designed to familiarize you with the International Slow Fire Course, the ISU Rules, some types of hand guns used in the Slow Fire Course, and a discussion of the techniques for effectively competing in this type of program.

V. BODY:

    A. <u>First Student Performance Objective</u>: As an International Slow Fire Instructor, EXPLAIN and DEMONSTRATE ISU rules regulating the range, targets, and firearms of International Slow Fire as outlined in the International Shooting Union Rules on 50 Meter Pistol Competition.

        1. <u>The standard competition course</u> with the Free Pistol is ten (10) sighting shots and 60 competition shots in six (6) series of the (10) shots in a time limit of two and one half (2½) hours. Possible score is 600 points.

2. <u>Competitors are permitted</u> to fire the sighting shots before or between ten (10) shot strings.

3. <u>Shots may be marked</u> in the pits or by the Register Keeper, who is stationed behind the firing point. The Register Keeper may determine the estimated value of each shot with the aid of a telescope. Official scoring will be done by the Official Scorer after firing is completed.

4. <u>A free Pistol or an International Slow Fire Range</u> shall comprise a line of targets, normally so mounted that they can be changed during competition without interrupting the shooting or disturbing the competitors. At a distance of fifty (50) meters from the line of targets and parallel to it, there shall be a line of firing points covered by a roof and each firing point protected from the wind. The distance will be measured from the front limit of the space used by the shooters to the corresponding target. The targets must not be placed more than 75 cms above and not more than 125 cms below the level of the pistol when held in normal shooting position. Each firing point must not be more than two (2) meters to the side of a line 90° to the targets. If the range lies open to strong winds, it is desirable to protect it by lines of trees or walls, so constructed that part of the air is allowed to pass through small openings in the walls. The protection shall not be closer than about 5 meters from the nearest line of fire and so that shadows are not thrown onto the targets.

NOTE: DISPLAY SLOW FIRE TARGET.

5. <u>The International 50 meters Slow Fire Pistol Target</u> is divided into ten (10) circular zones, scoring from ten (10) to one (1). The ten (10) ring has an outside diameter of 50 mms. The changing of targets may be by the pits, or operated mechanically or electrically from each shooting position or from some central place.

6. <u>In Free Pistol competitions the bore</u> must have a caliber of 5.6 mm. We use the caliber 22 LR cartridge which is the equivalent.

7. <u>It is forbidden</u> to have the grip of the pistol extended in such a way that it might gain support from any other part of the arm than the hand. Optical sights are not allowed.

8. <u>The trigger weight</u> is not specified but it must be safe.

9. <u>Any 22 caliber weapon can be used for the Slow Fire Course.</u> Generally a Free Pistol is used.

NOTE: DISPLAY THE HAMMERLI FREE PISTOL AND THE ELECTRIC TRIGGER FREE PISTOL.

TRANSITION: You have an idea of what the course consists of and what the range and the shooting equipment is like. How can we use these facilities and equipment for the best firing results?

B. Second Student Performance Objective: As an International Slow Fire competitor, DEMONSTRATE and IMPLEMENT International Slow Fire techniques to effectively produce winning scores in International competition as shown in Chapter XI, USAMU Pistol Marksmanship Guide.

1. <u>In Free Pistol shooting, the arm holding the pistol must be free</u> of support from the body. You are not allowed to wear a wrist watch or anything else that might give any support to the pistol. The shooter must adopt his position so that it is clearly visible that he does not gain any support from the wall or the bench or any other permanent or temporary structure.

2. *It is necessary to master the peculiarities* of this shooting technique. The basic preparation you need is to know the pistol and how to shoot it.

3. *In knowing the pistol*, first we consider the grips. The hand of the shooter should feel comfortable in the custom made grip. The shape of the grip should let him hold the pistol properly with the least possible effort. The larger the contact surface between the hand and grips the better. It is especially advantageous to have the whole thumb base touching the grips. The hand support should reach as far to the rear as permitted. However, it must not reach behind the wrist.

4. *The Adjustable or hair trigger* fitted in the free type pistol has a trigger pull from 15 to 40 grams (½ to 1½ ozs). Generally the trigger is equipped with a trigger pull adjusting screw. It is the habit of many shooters to adjust the trigger as light as possible. This can produce problems. Due to involuntary movement or accidental triggering, the weapon can fire before the shooter is ready. An unaimed shot in this instance is especially dangerous. A too finely adjusted trigger does not allow the shooter to touch the trigger at all when aiming. You must leave a certain amount of weight on the trigger that enables the shooter to touch and feel the trigger when aiming. This gives more security and confidence to the shooter.

5. *There are three methods for firing the adjustable set or hair trigger.*

a. The shooter puts the trigger finger lightly on the trigger immediately after having raised the arm in shooting position and steadily increases pressure until the shot breaks. Usually a full setting of 1½ oz is used in this method.

b. On aiming, the shooter carefully bends his trigger finger toward the trigger making a slight contact, more or less depending on the front sight's position on the target. If this kind of trigger operation is watched from the side, the trigger finger is constantly moving, bending toward the trigger or straightening out again. The task for the shooter is to choose the best moment for making contact and applying enough trigger pressure to release the shot. This must be done without disturbing the sights.

c. After raising the pistol the shooter makes a somewhat more positive contact of the finger on the trigger only when the sights are perfectly aligned on the aiming point. Up to this moment the trigger finger is kept free without ever touching the trigger. This method should be applied if the trigger weight is set light (½ oz.) and the shot breaks by the slightest touch on the trigger.

In our opinion method a. is the best. Method b. has some danger in that if the pistol is involuntarily jerked or disturbed by an abrupt movement in the first second before releasing the shot it will be a bad shot. Method c. requires exceptional conditioning of the arm muscles because there is a delay in reaction time after attaining correct sight alignment and the firing of the shot and deep concentration is needed in maintaining a steady hold. It is difficult to hold the trigger finger stable for any great length of time. This method sometimes leads to an early nervous fatigue of the shooter.

6. *If the grip is not correctly shaped* to the hand of the shooter, often the finger touches the trigger only after excessive bending. This is uncomfortable. Many shooters mount a small screw or a bent wire on the trigger which can be turned and adjusted to the most favorable position where the trigger finger can reach it. This addition to the trigger should be as short and light in weight as possible. The pressure on the wire or screw comes at an opposing angle to the direction of the pistol and could result in the shooter applying trigger pressure on an angle. If there is some side play in the trigger mechanism, the hair trigger may not function correctly.

7. <u>Determine where the placing of the center of gravity</u> of the pistol is best for you. It should be a little forward of the position of the middle finger, depending on the individual desires of the shooter. Some shooters put additional weight on the forend of the barrel believing that the front sight can be held steadier in the rear sight notch. If the front sight has movement and is unsteady and the shooter is not satisfied with the "feel" of his pistol, he must experiment in order to find the most suitable center of gravity. The goal of this experiment is to find by proper balancing of the pistol in the hand, the condition where the front sight remains as steady as possible.

TRANSITION: When you have found your center of gravity we are ready to determine the technique for firing the weapon effectively.

8. <u>The arm is fully extended and the elbow firmly straight</u>. The weight of the free pistol requires the shooter to build up the tone of the arm muscles. You will find the exercises used in developing the arm for NRA shooting work well for this purpose.

9. <u>To fire one round</u>, the shooter must be able to hold the pistol steady for approximately 12 to 18 seconds. In a series of 10 shots the shooter might raise the arm 25 to 30 times. This requires considerable endurance of the arm and shoulder muscles. To prevent excessive muscle fatigue the shooter has to practice frequently. The fatigue will become apparent mainly in the later portion of the course of fire.

10. <u>To reduce the rapid onset of muscular stress</u>, many shooters bend their body slightly rearward, opposite to the direction of the raised arm. This altered stance helps to compensate for the weight of the extended arm and pistol.

11. <u>The European method</u> of holding the free pistol is with a relaxed hand. This peculiarity of holding the pistol demands special training and self control. In NRA shooting, a loose grip tends to result in wild shots. On the other hand a too tight holding of the pistol causes a trembling movement of the front sight in the rear sight notch. In the course of practice, the shooter must find the correct tension in the muscles of his hand which gives him the most stable holding of the pistol.

12. <u>Usually the free pistol is fired as follows:</u>

    a. The shooter raises his arm from the shooting bench with the loaded and cocked pistol.

    b. The "hair" trigger is set.

    c. During the first 2 to 4 seconds the arm is moving but gradually settling. The front sight stabilizes and comes to rest in the rear notch. The sights then come to rest in the aiming area as a minimum arc of movement.

    d. With decreasing movement of the pistol and the attainment of good sight alignment the finger is laid lightly on the trigger without the application of pressure.

    e. The eye observes the sight alignment and is aware of the steadiness of the hold.

    f. If the sights are perfectly aligned the trigger contact is applied to the point at which the shot breaks.

g. The function of the trigger finger is exacting and careful. The execution of this action is enacted with much greater precision than that generally experienced in NRA type shooting.

h. If for some reason the shot was not released at the optimum moment, the finger is taken carefully away from the trigger and the pistol is lowered to the table in order to relax the arm and the grip.

13. <u>Two and a half hours of concentrated work</u> by the shooter not only tires his arm but the eyes as well. As soon as he notices a diminishing efficiency of his eyesight and a stable focus can no longer be maintained clearly, he must pause for a few minutes. If the arm tires, rest it.

14. <u>Shooting with the free pistol</u> requires highly developed will and great mental concentration for the duration of the shooting. He should keep in mind at all times:

a. The shooter must remember that the set "hair" trigger is likely to release the shot from a careless touching of the trigger, putting the weapons down too hard, and even from a slight tremor when raising the pistol. All movements of the arm have to be flowing and smooth.

b. The shooter must be determined to press positively on the trigger and not let himself be deceived by the minor movements of the sights in the aiming area, as long as the sights are aligned.

c. If the firing is not going well, the shooter must control himself and remain patient. He must make himself raise and lower the pistol until he is satisfied with the sight alignment and the steadiness of the hold before the shot.

d. After a bad shot, the shooter should put the pistol on the table and rest for a few minutes. Take time to analyze the cause of the bad shot. Consider this reason when planning the next attempt to shoot.

e. The shooter must never give up. All the competitors on the firing line are having difficulty with this course of fire. He must remember that one point could make the difference in winning or losing the team event.

VI. CONCLUSION:

A. <u>Retain Attention</u>: Which one of you will be the slow fire champion for the next World Championships? You can do it if you start now with that objective in mind and the will to win.

B. <u>Summary</u>: The range, targets and firearms of International Slow Fire are vastly different from NRA shooting. The range is 50 meters in length, the target center 10 ring is about the size of our NRA X ring and the firearm can be any 22 caliber LR or 5.6 mm as long as the weapon is safe. The fundamentals are the same for free pistol firing as for NRA shooting. The big difference is in trigger control and pacing yourself for 2½ hours of firing.

C. <u>Application</u>: From our discussion today, you are equipped to go back to your home station and start an International Slow Fire Program. You or someone you instruct may one day excel in the International Slow Fire Program and be able to represent the United States in International Pistol Competition.

D. <u>Closing Statement</u>: International Slow Fire Pistol Shooting demands a higher degree of control than any type of competition you have experienced in regular NRA events. Learn to use this higher degree of control in your shooting and win yourself an International Slow Fire Championship and the Distinguished International Shooter's Badge.

<u>NOTE</u>: DISPLAY A DISTINGUISHED INTERNATIONAL SHOOTER'S BADGE.

I. TECHNIQUE OF FIRE V AAPMI&CC 315 C
(INTERNATIONAL CENTER FIRE) Fifty (50) Min
 Mar 1975

LESSON OUTLINE

I. LESSON OBJECTIVE: To enable the pistol marksmanship shooter-instructor student to employ proper techniques in firing International Center Fire Course and thereby attain winning scores in International Competition.

II. STUDENT PERFORMANCE OBJECTIVES: As a result of this instruction, students will be able to accomplish the following Student Performance Objectives:

A. As an International Center Fire competitor, EXPLAIN the rules of International Center Fire and DEMONSTRATE the delivery of effective International Center Fire as outlined in Chapter XIII, USAMU Pistol Marksmanship Guide.

B. As an International Center Fire competitor, IMPLEMENT International Center Fire techniques to effectively produce winning scores in International competition as shown in Chapter XIII, USAMU Pistol Marksmanship Guide.

III. ADVANCE ASSIGNMENT: None

IV. INTRODUCTION:

A. Gain Attention: You are about to witness a demonstration of the firing technique of the International Center Fire Course.

NOTE: PROCEED WITH DEMONSTRATION OF THE INTERNATIONAL CENTER FIRE COURSE TO INCLUDE COMMANDS FOR FIVE (5) ROUNDS OF PRECISION AND FIVE (5) ROUNDS OF THE DUEL PORTIONS OF THE COURSE.

B. Orient Students:

1. Lesson Tie-In: The International Center Fire Course represents one fourth of the International shooting program. Slow fire, rapid fire, center fire and standard pistol competitions make up the International Pistol Program.

2. Motivation: The International Center Fire course challenges such a wide scope of competitors that it has almost universal appeal. Especially those shooters who have been involved in the National Rifle Association (NRA) Three Gun Program. The weapon used is basically the same as the one used in our center fire NRA course.

Those shooters who have competed in International competition can testify to the far reaching International good-will engendered by this extension of personal relationships formed by individual shooters. Even the most hard-bitten, introverted personality thaws to a degree when he performs well in an event in which individuals of other nations are competing. There is an element of warmth in the sharing of feelings experienced by each contestant which promotes better understanding and many times generates lasting friendships.

3. Scope: This hour of instruction is designed to familiarize the student with the International Center Fire course and offer a method for effectively competing in the International Center Fire.

TRANSITION: With the emphasis being placed now on international shooting events and their attendant competitive, social, and political significance, it behooves us as shooters to attain proficiency in these events.

V. BODY:

A. <u>First Student Performance Objective</u>: As an International Center Fire competitor, EXPLAIN the rules of International Center Fire and DEMONSTRATE the delivery of effective International Center Fire as outlined in Chapter XIII USAMU Pistol Marksmanship Guide.

The Center Fire Course was added to the International program to include and accommodate the military shooter and those owners of big bore pistols who felt the need for training with this primarily military sidearm.

<u>NOTE</u>: DISPLAY THE INTERNATIONAL CENTER FIRE PISTOL.

The Center Fire International Course involves the firing of thirty (30) rounds precision slow fire and thirty (30) rounds duel fire (silhouette) for a total of 600 points.

1. The shooter must build a system of shooting Center Fire which will serve as the vehicle which transports him out of his local Army Area, out of the United States and into International Competition in the Center Fire event.

a. You will now see a demonstration in which the shooter is required to accomplish the following:

(1) Five (5) rounds are loaded on command.

(2) Shooter fires at the 50 meter slow fire international target which is positioned at 25 meters. He has eight (8) minutes to accomplish his zero string.

<u>NOTE</u>: DISPLAY A 50 METER SLOW FIRE INTERNATIONAL TARGET.

(3) After repairing targets, the shooter fires five (5) rounds in six (6) minutes at the repaired target. This is done six (6) times for a total of 300 points. This completes the precision stage of fire.

(4) For the duel stage of fire the International Rapid Fire silhouette target is placed 25 yards distant with its edge facing the shooter.

<u>NOTE</u>: DISPLAY AN INTERNATIONAL RAPID FIRE SILHOUETTE TARGET.

(5) The shooter loads five (5) rounds and faces the target with the shooting arm lowered to a 45° degree angle.

(6) The target is edged at the command "THE LINE IS READY" - ten (10) seconds later the target is exposed in the faced position for three (3) seconds, during which the shooter fires. The target edges after three (3) seconds and reappears seven (7) seconds later to be exposed another three (3) seconds. This sequence is repeated until the target face has been exposed to the shooter five (5) times.

<u>NOTE</u>: REMEMBER THAT THE TARGET REMAINS EDGED FOR TEN (10) SECONDS ONLY AFTER THE INITIAL EDGING FOLLOWING THE COMMAND "THE LINE IS READY." ALL SUBSEQUENT EDGINGS OF THE TARGETS ARE SEVEN (7) SECONDS IN DURATION. THE TARGETS ARE ALWAYS EXPOSED FOR A THREE (3) SECOND INTERVAL.

(7) The first five (5) round string is a zeroing string. This procedure is followed in six (6) subsequent five (5) round strings until 30 rounds have been fired for a total of 300 points.

<u>TRANSITION</u>: A demonstration of the International Center Fire Course includes many elements of skill not readily apparent to the average observer.

B. <u>Second Student Performance Objective</u>: As an International Center Fire competitor, IMPLEMENT International Center Fire Techniques to effectively produce winning scores in International competition as shown in Chapter XIII, USAMU Pistol Marksmanship Guide.

    1. The following steps represent an acceptable and effective method of accomplishing a high score in the precision stage of the International Center Fire Course.

        a. Obtain a firm proper grip on the weapon.

<u>NOTE</u>: DESCRIBE AND DEMONSTRATE THE PROPER GRIP.

        b. Obtain an advantageous stance and position keeping in mind that the prime purpose of the stance and position is to give the shooter a natural "point" at the center of the target.

<u>NOTE</u>: DEMONSTRATE BREATH CONTROL.

        d. Align sights and point at center of target.

        e. Attempt to stop all movement while pressing the trigger steadily to the rear.

        f. Concentrate attention upon front sight while attempting to hold as close to the target center as possible.

        g. Continue to press the trigger while holding on the target and exert maximum concentration on the front sight.

        h. Follow through - continue the effort to maintain focus on front sight and maintain minimum arc of movement even after the shot fires.

<u>TRANSITION</u>: This is essentially the precision method of shooting. It encompasses the principles of area type shooting with the addition of the attempt to stop all movement. The size of the 10 ring on the international slow fire target dictates a much more precise method of shooting. You are in effect shooting for the equivalent of X's on the Standard American target. The silhouette target used in duel presents a challenge of a different order.

    2. The duel or rapid fire stage of fire is particularly interesting in that it involves the action of an intermittently reappearing target which tests the reaction time of the shooter as well as his marksmanship skill.

        a. At the command "THE LINE IS READY" the shooter has aligned or positioned himself so as to be able to point at the target without making any lateral movement to acquire the target once the shooter has raised his arm.

<u>NOTE</u>: DEMONSTRATE ASSUMING THE READY POSITION.

        b. A firm grip must be held during the raising of the arm and while aiming. Many shooters maintain this firm grip throughout the approximately 53 seconds it takes to fire one five (5) round string. This method seems tiring at the beginning but the arm soon becomes accustomed to the added strain through training. This method is recommended since it tends to prevent the shooter from varying his grip as he raises the pistol. Variations in the grip will hinder the shooters ability to maintain sight alignment because the variations in grip cause the front sight to be in a different location in the near sight notch each time his arm is raised to fire the next shot.

c. As previously explained, the target will face the shooter after the initial ten (10) seconds delay. The shooter must be totally committed to the task of anticipating the appearance of the target. The acquisition of a "pace" or "rhythm" also is essential here. Through constant observation and an attempt to be aware of the passage of time we are able to build a readiness. We in effect create a reflex response to the appearance of the target. From the moment the target edges or turns away we begin to build this readiness to respond to the suddenly reappearing target. We could compare our action to that of a boiling tea kettle. Let us say that readiness is the steam. The action of the boiling water (the passage of time) is creating an increasing volume of steam (readiness). When the action of the boiling water (time passage) creates enough steam (readiness), the cover on our team kettle pops up (the target appears) and we react by raising the pistol to point at the target.

As the seven (7) seconds of target edge time passes, our readiness to respond to the facing action of the target builds to the point that our whole focus of attention is keyed to the first movement of the target as it begins to move into the faced position.

(1) This total committment to one task is not easily attained. It requires a concentration that must be developed through repeated attempts to accomplish this task of building a systematic readiness to respond as the seven (7) seconds of edge time ticks away.

(2) One of our better duel competitors describes his readiness to respond this way. "As I stand with my shooting arm at the prescribed 45° angle downward, I think of my arm as a tensed spring. One set of muscles wants it to spring into the pointing position but I prevent it from doing so with another set of muscles. When the target appears I release those detaining muscles and <u>allow</u> the set of muscles that will bring the arm up, to have their way."

(3) However we describe it, we must be totally ready for the appearance of the target and respond in a split second!

d. Not only must we raise the pistol as we described, but we must fire at the target. We have taken up any slack in the trigger while still pointed at the ground. The instant the target moves we begin the movement of the gun arm and hand towards it.

e. During the time that the target is edged we look at it. We do not try to define the edge of the target too well. We do not gaze at it with a fixed focus.

f. As the gun moves upward and the muzzle comes through the "hip" area on the target, we pick up our sights, try to align them, begin to press on the trigger and stop the gun hand in the center mass of the target.

(1) Keep in mind that when the muzzle arcs through the hip area we shift the eye from the target to the sights and from that time on through the firing of the shot our task is similar to that encountered in slow fire.

(2) We focus on the sights, align them and keep moving the trigger to the rear until the shot breaks. In this course of fire we must attempt to keep the area of movement small, but we must continue to press positive on the trigger while we align the sights and try to stop all movement. We must fire the shot during the three (3) seconds that the target will face us.

g. There is a tendency to decrease pressure on the trigger or stop its rearward movement when the sights are badly misaligned and the gun arm is moving into the center point on the target. We must train our muscles to continue the pressure on the trigger while aligning sights. The targets will have faced and edged away if we follow the tendency of pulling the trigger only when the sights are in perfect alignment. It must be emphasized that the actions of pulling the trigger, aligning sights and stopping the gun in the center of the target are done in concert with one another. A smooth coordinated effort combining these three factors will cause the hammer to fall within the relatively short three (3) seconds interval during which the target is visible

    h. Most competitors try to breathe normally during the first five (5) seconds of the period of seven (7) seconds of edged time. Approximately two (2) seconds before the target is to face us, we stop breathing and await the appearance of the target.

    (1) After the shot has been fired, we move the arm deliberately downward to the 45° angle and exhale the breath.

    (2) We then breathe normally and begin anew the process of increasing readiness during the time the target is edged. This culminates in total readiness and immediate response to the reappearance of the faced target.

    i. Endurance plays a vital role in shooting effective duel. Early in our training we find that after the second or third shot, the muscles of the forearm begin to burn and ache. This feeling of discomfort must be borne without recourse to relaxation of the grip if possible. However, it may be necessary to relax the grip just slightly between shots during the early stages of training.

    (1) Any exercise which will strengthen the arm is considered advantageous.

    (2) Exercises which more or less duplicate the action of raising and lowering the pistol would be the most desirable.

    (3) Dry firing this event accomplishes the strengthening of the arm plus familiarizing us with the quick alignment of sights and movement of the trigger rearward.

VI. CONCLUSION:

    A. *Retain Attention*: In addition to being one of the most interesting of the several phases of international shooting competition, this event is very demanding. The simplicity of the International Center Fire Course does not seem to square with the relatively low scores produced. WHY?

    It's apparently because we don't put enough effort into mastering the techniques to the point where we make no mistakes. It looks too easy.

    B. *Summary*:

    1. The Center Fire Course is fired with any 32 cal. to 9mm pistol with a trigger pull weight of 1360 grams.

    a. The slow fire or precision portion of the International Center Fire Course is fired at 25 meters.

    (1) Target is the 50 meter Slow Fire International Free Pistol target.

    (2) Five (5) shots for zero in eight (8) minutes.

    (3) Thirty (30) shots fired in five (5) shot strings - five (5) shots in six (6) minutes for a total of 300 points.

        b. The duel or Rapid Fire portion of the International Center Fire Course is fired at 25 meters.

           (1) Target is the 25 meter international rapid fire silhouette target.

           (2) Thirty (30) rounds fired in five (5) round strings for a total of 300 points.

           (3) After the command, "THE LINE IS READY", there is a ten (10) second time period during which the targets are faced away. At the expiration of ten (10) seconds the target faces the shooter for three (3) seconds.

           (4) The competitor raises his arm from the 45° angle (toward the ground) at which he has been holding his pistol arm at the ready position. As he points at the center of the target, he aligns his sights and presses the trigger straight to the rear and causes the hammer to fall.

           (5) He returns his gun arm to the ready (45° down) angle and awaits the appearance of the target after an interval of seven (7) seconds. Remember that the ten (10) second delay is used only after the first edging of the target. Thenceforth the target edges for seven (7) seconds after each exposure.

    2. The essence of precision shooting is:

        To make a concerted effort to stop all movement except that of the trigger finger moving the trigger straight to the rear while aligning the sights.

    3. A main point of duel or rapid fire is:

        Condition oneself to be ready when the target appears. This is done by complete concentration on the task at hand. The shooters attention is on the edge of the target and he is keyed to immediate response to the facing movement of the target.

    C. <u>Application</u>: The principles set out during the foregoing are used by most of the better International Center Fire shooters. There may be some variation in the application of these principles but in essence they apply to all who desire to excel.

    D. <u>Closing Statement</u>: Our goal in presenting this period of instruction on International Center Fire shooting is really an attempt to bring to light the vaulable nature of International competition. It has become obvious that with the daily improvement in communications and transportation media, our lives are increasingly influenced and affected by our neighbors around the world. Ideas concerning our speech, clothing, religion, and sometimes our politics are bending towards one another. No less can be said of our sporting events. If we are to influence our friend and rivals more than they influence us, we have a one sentence task: Win in International Competition. Everyone listens to the winner!

J. TECHNIQUE OF FIRE VI
    (INTERNATIONAL RAPID FIRE)

AAPMI&CC 315R
50 Minutes
Mar 1975

LESSON OUTLINE

    I. LESSON OBJECTIVE: To enable the pistol marksmanship shooter-instructor student to be familiar with the course of fire, employ proper weapons and techniques in firing International Rapid Fire and thereby attain winning scores in International Rapid Fire competition.

    II. STUDENT PERFORMANCE OBJECTIVE: As a result of this instruction, students will be able to accomplish the following Student Performance Objectives:

        A. As an International Rapid Fire Instructor, EXPLAIN and DEMONSTRATE ISU rules regulating the range, targets and firearms used in the International Rapid Fire Course as outlined in Chapter XII, USAMU Pistol Marksmanship Guide.

        B. As an International Rapid Fire competitor, DEMONSTRATE and IMPLEMENT International Rapid Fire procedures for readiness and the techniques of fire to effectively produce winning scores in International competition as shown in Chapter XII, USAMU Pistol Marksmanship Guide.

    III. ADVANCE ASSIGNMENT: None.

    IV. INTRODUCTION:

        A. <u>Gain Attention</u>:

NOTE: ON SIGNAL FROM INSTRUCTOR-DEMONSTRATOR WILL CALL "READY", FIRES ON SERIES OF 4 SECOND RAPID FIRE. (NO COMMANDS, ONLY FIRING)

        "You have just heard and witnessed the firing of the four (4) second stage of International Rapid Fire."

        B. <u>Orient Students</u>:

            1. <u>Lesson Tie-In</u>: The International Rapid Fire Course represents one fourth of the International Pistol shooting program. As you know, the other courses composing the program are, Slow Fire, Center Fire and Standard Pistol.

            2. <u>Motivation</u>: The International Rapid Fire Course fascinates the new shooter with challenges in time and targets that he has never faced before. This type of shooting is especially interesting to those shooters who have become proficient in NRA type shooting and are looking for something more demanding.

        To the special interest in firearms, handguns in particular, that you have displayed by your presence here and being aware of marksmanship's value as an American heritage, we now add the International competitive shooting program. Competitive shooting events between countries have long been recognized as a safe arena for the energies and rivalries between nations.

            3. <u>Scope</u>: This hour of instruction is designed to familiarize you with the International Rapid Fire Course, some types of handguns used in the Rapid Fire Course and a discussion of the techniques for effectively competing in this type of program.

TRANSITION: With emphasis now being placed on International shooting events and their attendant competitive, social and political significance, it behooves us as shooters, to attain proficiency in these events.

V. BODY:

   A. <u>First Student Performance Objectives</u>: As an International Rapid Fire Instructor, EXPLAIN and DEMONSTRATE ISU rules regulating the range, targets and firearms used in the International Rapid Fire Course as outlined in Chapter XII, USAMU Pistol Marksmanship Guide.

   1. The International Rapid Fire Course involves the firing of twenty (20) rounds in eight (8) seconds, 20 rounds in six (6) seconds and twenty (20) rounds in four (4) seconds at the International Rapid Fire silhouette target for a total possible score of 600 points.

   2. The International Rapid Fire range consists of a series of five silhouette targets located twenty-five (25) meters from the firing line. These targets are one (1) meter apart, center to center.

<u>NOTE</u>: DISPLAY INTERNATIONAL RAPID FIRE SILHOUETTE TARGET.

   The ten (10) ring is 10 cms x 15 cms with consecutively decreasing values of scoring rings down to the value of one (1).

   3. The course is fired in the following manner: two (2) strings, five (5) rounds each fired in eight (8) seconds, two (2) strings, five (5) rounds each fired in six (6) seconds and two (2) strings, five (5) rounds each fired in four (4) seconds. This makes a half course of thirty (30) rounds. One shot is fired at each target per string. The entire series is repeated to complete the full course of sixty (60) rounds.

   4. The handgun used is relatively unrestricted so long as it can be fitted into a box 30x15x5 cms. There is a 5% leeway in one direction allowed. Orthopedic grips may be used so long as they do not extend beyond the wrist joint to offer support. The weapon with grips must still fit in the "box." Trigger weight is not specified, but it must be safe, that is, it must not fire automatically. Sights must be metallic. Caliber is restricted to 5.56 mm or 22 cal rimfire. The 22 cal. short is the preferred loading. Weight of the pistol with all attachments and modifications cannot exceed 1260 grams in weight.

<u>NOTE</u>: DISPLAY DOMINO, HAMMERLI AND WALTHER RAPID FIRE PISTOLS.

<u>TRANSITION</u>: This then, gives us an idea of what the course consists of and what equipment is used. Now let us consider the means of employing this equipment to get the desired results.

   B. <u>Second Student Performance Objective</u>: As an International Rapid Fire Competitor, DEMONSTRATE and IMPLEMENT International Rapid Fire procedures for readiness and the techniques of fire to effectively produce winning scores in International competition as shown in Chapter XII, USAMU Pistol Marksmanship Guide.

   Before we compete in any sport or event, we must know the basic fundamentals of that sport or event, and how to employ them. We are fortunate in this case in that the basic fundamentals of pistol marksmanship we have learned in NRA type shooting will also apply in International Rapid Fire. The fundamentals do not change, only the method of employment is modified to meet the needs of the course of fire.

   1. <u>Basic fundamentals:</u>

   a. Stance is the posture or arrangement of parts of the body in relation to each other to provide stability and balance, allowing us to deliver a series of shots without muscular strain. Since our entire performance is so closely related to stance, we must take precautions to insure that our stance is exactly the same each time.

(1) The most important aspect of the stance is the placement of the feet. The feet should be slightly farther apart than in NRA type shooting. This is because while delivering the five shot series, we move the arm across the targets. This will cause a slight and unconscious movement of the torso, rotating in the direction of the movement. Because of this movement, we must have a more stable base than where there is no arm or torso movement.

NOTE: DEMONSTRATE THE PLACEMENT OF THE FEET.

(2) The other major difference in stance is the starting or "Ready" position of the shooting arm. The shooting arm is held at a 45° angle below the horizontal until the targets start to face. We will discuss the movement of the gun to target later in the period.

NOTE: DEMONSTRATE THE STARTING OR "READY" POSITION OF THE SHOOTING ARM.

(3) The position of the head is also important. The head must be held erect so that your vision is out of the center of the eye. If the head is lowered or tilted back, your cone of vision becomes distorted and difficulty is experienced in maintaining sight alignment and in stopping at the proper place on the target. Remember these points: The feet are spread slightly further apart, but not uncomfortably so. The shooting arm is held at a 45° angle until the targets turn, and more attention must be given to the position of the head.

b. Position is the relationship of your body and the stance with the line of targets. In Rapid Fire, we must consider five targets instead of one.

NOTE: SHOW SLIDE #1.

(1) Position, or alignment of the body to the targets is limited to a degree by the shooting rules. These rules provide for a shooting platform at least 1.5 meters wide centered in front of the target bank. A bank of targets is the five targets assigned to a firing point. For all practical purposes, the shooter should limit himself to this area regardless of the actual space provided at the firing point. This is to insure he stations himself so that his body is located at the same place each time in relation to the target bank.

(2) We now are ready to align ourselves with a target preparatory to starting our series of shots.

(a) The method used by many good shooters is to assume a natural stance pointing to the last target, then swing the arm back to the first target, and lower the arm to the "Ready" position. A major advantage of this position is that it allows you to move across with less muscular effort since you are moving toward your natural position. For example, it is easier to relax a spring smoothly than it is to stretch it. A point which must be closely watched is moving the arm to the first target. Since we are holding a slight muscular tension against our natural position, our arm wants to move to the left as we raise it to the firing position.

(b) Another position favored is to align the body with the first target. This gives the advantage of complete control over the arm raising to the shooting position since it is coming to a natural position. A disadvantage is moving against muscles as we move to the subsequent targets in the series, "stretching" the spring, so to speak.

(c) An additional method would be to align the body on the center target. This method has advantages and disadvantages to a greater and lesser degree than the two previous methods described.

NOTE: SHOW SLIDE #2.

    (3) Since we must move the arm in International Rapid Fire, we turn our body away from the target at a greater angle than we would in 3-gun shooting. In American 3-gun NRA type shooting, we normally face away from the targets 40-50 degrees In International Rapid Fire, we are more nearly 90° or at right angles from the target. This limits the arm movement to the least muscular strain. Also, from the 90° position, less tension is created on the back muscles allowing the arm a more favorable angle of swing in front of the body. Regardless of which target the shooter uses as his primary alignment target, his position must meet the requirement of allowing the body to swing on all five targets without undue muscle pull. Since there is going to be a slight amount of torso movement, disregard it. Primarily the arm must move smoothly and evenly.

    c. Breath control is as important in International Rapid Fire as it is in 3-gun NRA shooting. We must still get a sufficient amount of air into the lungs to allow us to accomplish our mission of shooting five good, well aimed shots at five different targets. A recommended method of breath control is to inhale deeply and exhale completely two or three times just before you give the command "Ready." Inhale deeply, give command "Ready," exhale to respiration pause. By this time the targets are turning and at a maximum of eight seconds later you can resume normal breathing. Even though breath control is as important here as 3-gun shooting, it isn't as difficult to master.

    d. <u>Sight Alignment</u>: Sight alignment is just as important in International Rapid Fire as in NRA Slow Fire shooting. Even though the targets are closer and the "10 ring" larger, we still cannot settle for "almost" or "half-way" sight alignment. We must strive for perfect sight alignment each and every time we press the trigger. It makes no difference if it is the first target or the last or a target in between; sight alignment is exactly the same as we are accustomed to.

    (1) In explaining sight alignment in Rapid Fire, we must be conscious of several things. First is the target movement; second is the movement of the gun to the firing position; and third, the sights.

    (a) Since the target movement is our signal to start our gun to target movement, we must see the target move. It is not necessary nor is it desirable to focus on the target while waiting for this movement. By focusing on the target, we must shift our focus to the front sight before firing. This is just something else to think about and accomplish before firing. One of the accepted methods is to fix the focus of the eye on the point in space where the sights will appear.

    (b) This is determined by lifting the arm from the "Ready" position to the "Fire" position, aligning the sights with the aiming area, focus on the front sight. Lower the arm to "Ready" position maintaining the fixed point of focus. Give the command "Ready" and as the targets turn, raise the arm, bring the sights to this point of focus and fire. All this can be done by using what is called "the cone of vision."

NOTE: DISPLAY SLIDE #3.

    (c) When do we pick up our sights and perfect the alignment? Some good shooters pick up the sights as low as the 7 or 8 ring at 6 o'clock. It is felt by most shooters, however, that the best place to pick up the sights is not lower than the nine (9) ring and preferably in the lower part of the 10 ring. By using this latter method, our focus is not shifted so far from the area where our sight alignment will be. After we have moved the arm into the "Fire" position, perfected our sight alignment, we press the trigger and prepare to move to the next target.

(2) The question arises, what about an aiming point? With the focus on the front sight, we will see three white lines on the target. Simply move the sights to the center of these white lines. Since it is natural to point to the center of the mass, we recommend using a center hold in place of a six o'clock hold or some other variation.

e. Trigger control is no more difficult nor any easier in International Rapid Fire than in 3-gun NRA shooting. It makes no difference what type of trigger we have on the gun, light or heavy, smooth or crisp, we must press the trigger smoothly and firmly straight to the rear parallel with the axis of the gun bore. Good NRA trigger control can be transferred to International Rapid Fire with only a speed-up to make it effective.

f. Grip is more critical in Rapid Fire than in the type of shooting we are accustomed to. Our grip must do two things. First it must allow us a natural sight alignment and second it must allow us to control the trigger in such a way that we can fire without disturbing the sight alignment.

We are authorized the use of orthopedic grips so long as they do not extend past the wrist joint. It is permissable to have the back of the hand covered. In fact a perfect fitting wooden glove would be an ideal grip for the rapid fire gun. To determine just what we need in an orthopedic grip, we use the same method as we use for making a custom grip on "3-gun" handgun. Build up, take off, add on, until we finally get a grip that fits us exactly and have it duplicated.

Requirements of the grip by the shooting hand is determined in the same manner we discussed in pistol Fundamentals I. Be sure to grip the pistol with enough tension to cause a tremor, then relax the tension slightly until the tremor stops. This should allow us to exercise independent control of the trigger and keep the sights in natural alignment.

2. <u>Movement of Gun to the First Target</u>: Movement to the first and then subsequent targets is quite different from any shooting we have been accustomed to. Movement of gun to target has already been discussed to a point in both position and sight alignment. To delve deeper into the subject, how do we get this movement started and then, how do we get it stopped? The main point to remember in delivering the first shot: It must be fired in the center of the target with the sights aligned. This can only be done by having a smooth, well coordinated movement from "Ready" to "Fire" position. If any one shot can be given more value, it will have to be the first shot. If this shot is delivered properly, we set the pace for the remainder of the series. If we do something wrong on the first shot, it tends to carry over to the remaining shots. For new and old International Rapid Fire shooters alike, first shot exercises are very beneficial to the training program.

a. Since we know that the "Ready Position" is with the shooting arm lowered 45°, what else must we know and do?

(1) First, the wrist and elbow must be locked, the tricep must be tensed. This aids us in controlling the movement to the first target and aids in reducing the arc of movement of the gun during our five shot series.

(2) Now we come to the problems of actually moving the arm to the firing position. This movement can be fast or slow depending upon the shooter and his state of training.

(a) Let us discuss first the fast movement to the target. We have here the apparent advantage of getting the sights to the center of the first target quickly, allowing us to fire the first shot sooner than we would if we brought the arm up slower. What actually happens here is the arm invariably over-shoots the aiming area and the time is wasted settling back into the center of the 10-ring. This over-shooting and settling negates any advantage gained by the fast movement. This fast movement is not recommended for new shooters, but is a definite advantage for a shooter who is skilled in International Rapid Fire and can overcome the disadvantages of this quick movement.

(b) The movement from "Ready" to "Fire" recommended for beginning shooters and also used by many good International Rapid Fire shooters is the so called "Slow" movement. Bear in mind that when we say "slow" we still mean that the shot must be delivered in no less than 1.25 to 1.50 seconds. In the slow movement the arm starts stopping almost as soon as it starts the movement. This allows us to pick up our sight alignment in the lower part of the ten-ring, stop in the center, squeeze the trigger, almost all in one motion without having to hunt for the sights, settle down, or otherwise waste time delivering the first shot.

3. <u>Movement of the gun to subsequent targets</u> will be difficult for most of the experienced NRA type shooters. This, more than movement to the first target is different than anything we have been accustomed to. As we discussed in movement to the first target and firing, we must do everything correctly to aid us in firing the remainder of the series.

a. As the first shot fires, our recovery should take us to the center of the next target.

b. Stop, align the sights, press the trigger, recover on the next target and so on until we have fired the entire series.

c. No matter what you may hear, no successful shooter goes across the targets, firing without stopping. It is not impossible to fire a good score this way, but it is highly improbable. On each target, you must stop the arm and gun, align the sights and press the trigger, simultaneously.

TRANSITION: There is no substitute for applying the fundamentals. That is the only way to be sure of getting a good shot. In International Rapid Fire, you can not leave anything to chance, nothing just happens of its own accord. It must be thought out, rehearsed and finally accomplished. Now you must employ your stance, position, grip, breath control, sight alignment and trigger control. You must plan exactly what you are going to do and have the intestinal fortitude to carry out your plan of action over all difficulties, actual or otherwise. This brings us to mental discipline.

4. <u>Mental Discipline</u>: is applying all the mental processes toward a chosen goal. Mental discipline is essential in anything we do which requires any degree of concentration. In International Rapid Fire, we need to be more mentally disciplined than in perhaps any other phase of shooting. The effort required to fire a series of shots is perhaps 95% mental effort and 5% physical effort. Therefore we can see why mental discipline plays such a large part in this type of shooting. In your previous class on Mental Discipline in the NRA shooting phase of this course, you learned how to attain mental discipline. You should be able to concentrate on what you are going to do with a nude chorus line dancing the "can-can" accompanied by the Anvil Chorus. When you have achieved this stage of mental discipline, you will then become a good shooter. As in any other field of endeavor, you can practice mental discipline anytime, anywhere. A good way to practice is to set up your chair or if possible your equipment at one end of the range. Dry fire with the competitors. If you can not dry fire, picture in your mind the sequence of events leading up to firing the first shot, set up in your mind how this shot will be fired, then mentally fire it, move to the next target, etc., until the series is completed. Actually, we cannot think about applying each of these fundamentals as they are employed. Rather, we must be trained and disciplined so that when

each mechanical element falls into place, the employment of the fundamental will be automatic. For example, the mechanical movement of the target causes the mental impulse to start the movement of the arm to the aiming area. This mechanical movement of the arm causes the mental awareness of sight alignment. When the mechanical movement of the arm and attainment sight alignment is accomplished, the mental effort of trigger control takes place. The mechanical movement of the trigger finger results in the firing of the weapon. This in turn causes the mechanical movement to the next target and the mechanical/mental process starts all over again. This is nothing more than conditioning the body to react in a given way to a given set of conditions. When we have trained ourselves to the point where we will react properly when the targets turn, we are well on our way to becoming a good shooter.

These mental conditions can be set up only through intense effort and a good training program. When mastering International Rapid Fire, it is desirable to master one phase at a time. When this phase is mastered, move on to the next, still practicing what has been mastered. The reason for this approach is to insure the shooter learns the correct method and maintains his self-confidence.

VI. CONCLUSION:

    A. <u>Retain Attention</u>: The world record International Rapid Fire is 598X600 points.

    B. <u>Summary</u>: During this period of instruction, we have discussed International Rapid Fire:

        1. Handguns and Course of Fire.

        2. Basic Fundamentals as employed in this type of shooting.

        3. Movement of gun to target and movements to subsequent targets.

        4. Mental conditioning.

        5. You will also witness a demonstration of the rapid fire course and then you will fire one half of the International Rapid Fire Course.

    C. <u>Application</u>: Using the correct techniques of fire and following a good training program you will give the international rapid fire shooter the confidence to win in competition.

NOTE: DEMONSTRATOR WILL FIRE ONE STRING EACH OF EIGHT (8) SECONDS, SIX (6) SECONDS AND FOUR (4) SECONDS, USING COMMANDS. AFTER THE DEMONSTRATION, STUDENTS WILL FIRE A HALF COURSE UNDER SUPERVISION OF INSTRUCTORS.

    D. <u>Closing Statement</u>: Competitive shooting is taking on a more important role in the field of International relationships. As interested parties both in shooting skills and in our national destiny, we must carry our flag to the winner's platform.

K. TECHNIQUE OF FIRE VII   AAPMI&CC 315SP
    (INTERNATIONAL STANDARD PISTOL) 50 Minutes
                                     Mar 75
                    LESSON OUTLINE

   I. LESSON OBJECTIVE: To enable the advanced pistol marksman to become familiar with the International Standard Pistol course of fire, weapons used and the limitations placed on their characteristics, and the techniques used in firing the International Standard Pistol.

   II. STUDENT PERFORMANCE OBJECTIVES: As a result of this instruction, the student must be able to accomplish the following student performance objectives:

      A. Given the requirement to fire in competition with the International Standard Pistol, EXPLAIN AND DEMONSTRATE the International Standard Pistol Course as outlined in Chapter XIV, USAMU Pistol Marksmanship Guide.

      B. Given the fundamentals and techniques of fire as described for the International Standard Pistol, EXPLAIN AND APPLY the fundamentals and techniques used while firing the International Standard Pistol as outlined in Chapter XIV, USAMU Pistol Marksmanship Guide.

   III. ADVANCED ASSIGNMENT: None.

   IV. INTRODUCTION:

      A. Gain Attention: The Standard Pistol course is a relatively new course to International style competition. This course was developed to allow persons of all nations to become involved in the international shooting events at relatively low expense. Hopefully, from 1970 on, the International Standard Pistol will become a regular event to be fired in all international competition. Any person who has competitively mastered the .22 caliber pistol becomes a candidate for his or her Country's International Team.

      B. Motivation: To place yourself, or your students on the squad of shooters who are to represent their country in international competition. The Pan American Games, World Championships and the Olympics, should be the goal of each competitor and Coach. This also becomes one of three possible times an individual or firing team member can become Internationally Distinguished. Travel to many places in the world where these events are held should also serve to motivate a person to excel and place themselves on the team which will represent their Nation in one of these events.

      C. Lesson Tie-In: Prior instruction has familiarized you with the commonly understood features of international shooting. You are now ready to move into an interesting field, specificly, the International Standard Pistol.

      D. Scope: This period of instruction will be devoted to describing the course of fire, range facilities, weapons limitations, and explaining fundamentals and techniques used in firing the International Standard Pistol.

   V. BODY:

      A. First Student Performance Objective: Given the requirement to fire in competition with the International Standard Pistol, EXPLAIN AND DEMONSTRATE the International Standard Pistol course as outlined in Chapter XIV, USAMU Pistol Marksmanship Guide.

1. <u>General Provisions</u>:

     a. Range Used: All firing is conducted at a distance of 25 meters from the firing line. The targets are positioned in such a manner as to be in the center of each shooting station. It is the shooters responsibility to insure himself that his target is fastened securely to the frame and in the proper condition with all shot holes from previous firing covered.

     b. Targets Used: All firing is conducted on the International 50 meter Slow Fire Target.

NOTE: SHOW TARGET USED.

Notice the size of the ten ring. In relation to the standard American target, you are in effect shooting for X's. However, learn the distance from each ring to the center of the target. This will help in making any sight corrections during firing.

     c. Weapons and weapons limitations: Weapons used shall be any automatic pistol or revolver capable of firing 5 shots without reloading. .22 caliber long rifle ammunition only shall be used. Each weapon used must conform to the following limitations:

     (1) Sight Radius - Not to exceed 22 cms.

     (2) Barrel length - Not to exceed 152 mms.

     (3) Trigger weight - Must not weigh less than 1000 grams (2.2 Lbs) at moment of discharge when measured horizontally with trigger weights.

     (4) Grips - May not exceed 5 cms when measured at a right angle to the axis of the weapon.

NOTE: SHOW FREE PISTOL STOCKS AND RAPID FIRE STOCKS AND POINT OUT THE STOCKS ON A STANDARD PISTOL FOR COMPARISON.

The rules eliminate such stocks as these. On the Standard Pistol only, a small thumbrest and a slight palm rest may be used. However, the entire weapon with stocks as used must fit inside the regulation measuring box which is provided at each international match.

     2. <u>Course of fire</u>: You are now ready to begin firing. It is the responsibility of the competitor to understand in what manner firing will be conducted.

     a. Distance: All firing is conducted at 25 meters.

     b. Number of shots and time limit: The course of fire involves firing 20 shots of slow fire, time limit, 2½ minutes for each five rounds, scoring after each five rounds. Four strings of five rounds each will constitute the slow fire stage. 20 shots timed fire, time limit, five rounds in twenty seconds, scoring after each five rounds. Four strings of five rounds each constitute the timed fire stage. 20 shots rapid fire, time limit, ten seconds for each five round string, scoring after each five rounds. Four strings of five rounds each constitute the rapid fire stage.

     c. Starting position: The similarity to N.R.A. competition ends here. During the timed and rapid fire stages the shooter must have his shooting arm and weapon lowered to the starting position, which is at a 45 degree angle in relation to the body. The signal to raise the arm to the horizontal or the firing position is the facing action of the targets to the shooter. The arm must be lowered to the 45 degree angle by the command "Ready on the Firing Line" by the range officer.

        d. Sub-summary: A total of 20 shots slow fire, 20 shots timed fire and 20 shots rapid fire make up the entire course. A total of 600 points is the possible score. The present record (1969) is 582. The arm must be lowered to the 45 degree angle by the command "ready on the firing line" for timed and rapid fire. The signal to raise the arm will be the facing action of the targets to the shooter.

TRANSITION: Because of the small size of the scoring rings involved, the shooter must initially ask himself this question. "There must be something I can do to enable myself to fire winning scores in this event. What can I do?" The individual competitor must learn and master the employment of the pistol fundamentals: Position, Stance, Grip, Breath Control, Sight Alignment and Trigger Control.

    B. **Second Student Performance Objective**: Given the fundamentals and techniques of fire as described for the International Standard Pistol, EXPLAIN AND APPLY the fundamentals and techniques used while firing the International Standard Pistol as outlined in Chapter XIV, USAMU Pistol Marksmanship Guide.

        1. **Minimum arc of Movement. (Position-Stance-Grip-Breath Control)**

            a. Stance: Starting with his feet, the shooter must build his body as to be in the most stable position. This insures the shooter that with the correct stance he will reduce sway to a minimum and keep his body muscles from tiring quickly. To do this simply:

                (1) Place the feet approximately shoulder width apart.

                (2) Knees stiff but not locked.

                (3) Weight settled evenly on hips, with hips level.

                (4) Shoulders level.

                (5) Head erect. Eyes looking straight forward.

In conjunction with the correct position, this stance will greatly aid the shooter in natural alignment to the target, reduce strain to the muscles and insure the shooter of proper preparation to fire.

            b. Position: As the shooter moves to the firing line he must first position himself correctly at the firing line. To do this, he must check the pointing of the arm. Close the eyes and rotate the body from the ankles, keeping the head level. Then let the body settle naturally, look along the arm and see if the point is naturally at the center of the target. If he is not, then move the trailing foot in the direction of error. Make any corrections necessary until his natural point to the center of the target is achieved.

            c. Grip: The grip plays an important part in natural sight alignment. The shooter must insure himself that he has the correct and comfortable grip. He can do this by following the following steps:

                (1) Grasp the pistol by the non-shooting hand.

                (2) Form a "V" with the thumb and forefinger of the shooting hand.

                (3) Firmly place the weapon in the "V".

(4) Grasp the stock with the lower 3 fingers of the shooting hand.

(5) Extend the arm and hand towards the target. If natural sight alignment has not been achieved, grasp the weapon by the slide, and by moving the muzzle end, make any corrections necessary to insure the grip is correct and serves the function of creating natural sight alignment. The pressure applied to the grip by the fingers of the shooting hand must remain constant. Applying pressure or decreasing pressure with these fingers will cause the front sight to move from perfect sight alignment and will cause the shooter to lose the concentration needed to keep perfect sight alignment at all times.

d. Breath control: Now that the shooter has attained the proper stance, position and grip, he must insure himself that he has enough oxygen in his blood supply to properly fire one shot or a string of shots without running out of breath. This is accomplished by mastering breath control. Each shooter will establish his own system of breath control. The main objective of breath control is to insure that the shooter has a sufficient supply of air to keep his mind from sensing the need to breathe. Rather, his mind should be concentrating on the fundamentals of pistol shooting and his individual system of delivering that shot or string of shots.

e. Sight alignment: Sight alignment is the most important contribution to firing an accurate shot. In theory, accurate aiming is the ability to maintain the proper relationship between the front and rear sights. It is necessary to be acutely aware of the relationship of the rear sight to the clearly defined front sight. Normal vision is such that the rear sight of the pistol will be as nearly in focus as the front sight. Correct sight alignment must be thoroughly understood and practiced. It appears on the surface as a simple thing - this lining up two objects, front and rear sights. The problem lies in the difficulty in maintaining these two sights in precise alignment while the shooter is maintaining a minimum arc of movement and pressing the trigger to cause the hammer to fall without disturbing sight alignment.

f. Application of Trigger Control: Positive uninterrupted trigger pressure, surprise shot method is primarily the act of completing the firing of the shot once the application of trigger pressure is started. The Shooter is committed to an unchanging rate of pressure, no speed up, no slowdown or stopping. The trigger pressure is of an uninterrupted nature because it is not applied initially unless conditions are settled and near perfect. If the perfect conditions deteriorate, the shooter should not fire, but bench the weapon, relax, replan, and start again. In instances when the pistol is stable and steady, and the periods of minimum arc of movement are of longer duration. It is immaterial whether the release of the trigger is completed a second sooner or a second later. Any time that the shot is fired with minimum arc of movement and the sights are in alignment, it will be a good shot.

TRANSITION: Now that the shooter has done everything in his power to insure the combining of a minimum arc of movement with perfect sight alignment and trigger control, he is ready to begin firing. At this time we shall discuss some of the techniques used by various champion type shooters in firing the standard pistol above and beyond the basic fundamentals of pistol marksmanship.

2. <u>Techniques of Slow and Sustained Fire</u>:

a. Slow Fire: Shooters employ many methods of shooting slow fire. Some shoot only one shot at a time, resting the gun and arm on the bench between shots and then firing another shot after sufficient time has elapsed to plan the next shot. Others shoot one shot and may not lower the arm between shots. They never release the grip on the pistol for the entire five shot string. There are others who after firing one shot, place the weapon on the bench and entirely release the grip prior to shooting another shot. But, all must employ the required techniques.

(1) First, they insure themselves of a minimum arc of movement by attaining a proper stance, position, grip and apply proper breath control. This is accomplished while firing the five sighter shots during the preparation period and by dryfiring occasionally between shots should a problem in zero change or control arise while firing.

(2) Employ a shot sequence while firing slow fire. The most generally accepted sequence is:

    (a) Extend the arm and weapon to the target.

    (b) Apply breath control.

    (c) Settle into the minimum arc of movement with perfect sight alignment.

    (d) Apply a initial pressure to the trigger.

    (e) Shift the point of focus to the front sight.

    (f) Apply positive pressure on the trigger straight to the rear while maintaining perfect sight alignment.

    (g) Follow thru and recover.

    (h) Analyze your performance.

    (i) Apply corrective measures if needed.

(3) Resting between shots gives a shooter ample opportunity to reorganize his mental preparation and clear the vision for the next shot.

Attempt to stop all movement while applying positive trigger pressure. However, due to the physical inability of the shooter to stop all movement, he must apply trigger pressure while some small movement exists. With perfect sight alignment his shot group will be the size of his ability to hold still.

(4) The importance of correct grip while firing slow fire cannot be overemphasized. If the grip on the weapon is released for any reason, the grip must be rechecked prior to firing the next shot. Steadyness of the grip must also be stressed. No additional gripping pressure or no relaxation of pressure can be tolerated, if absolute perfection of sight alignment is to be achieved.

TRANSITION: This then, is the method usually used by the champion shooters in firing their slow fire. Firing a string of sustained fire will be accomplished in somewhat the same way. However there are some additional factors which will have to be taken into consideration.

    2. <u>Sustained Fire</u>: The timed and rapid fire events are shot by most champion type shooters in the same manner as in 3 gun NRA type competition. There are a few factors which should be added to the shooter's sustained fire technique.

    a. Recovery and rhythm: Quick recovery and a firing rhythm must be added to the shot sequence of sustained fire.

    b. Mental Preparation: Mental preparation becomes necessary due to the firing of more than one shot at a time. The shooter must be mentally ready to fire a string of five shots without hesitation in any manner.

c. Starting Position: The starting position is a new addition to the shooters technique. The movement of the shooting arm from the 45 degree starting position to the horizontal firing position must be practiced until it becomes a fluid like, effortless, smooth motion. The method of accomplishing this by most shooters is as follows:

(1) Check position, stance and grip during the commands of "Ready on the right and ready on the left". The breathing must be a little deeper than normal while doing this.

(2) As the command "Ready on the firing line" is given, the shooter must lower the arm and weapon slowly to the starting position. Be careful not to disturb what was achieved during previous preparation, (Grip, sight alignment, etc). At this time the gaze is shifted to the edge of the target as the targets are faced away from the shooter. The shooter must be careful not to stare at the edge of the target but just have the gaze in the general area of the target frame where the 10 ring will appear when the targets are faced to the shooter. Take a last deep breath and let out part of it, cutting the flow of air off in the throat, making sure the lungs are comfortable but enough oxygen is stored to complete the five rounds or the twenty seconds of time required

(3) As the targets appear and the arm is moved to the horizontal, the shooter then goes into the normal techniques of timed and rapid fire. However, the time consumed by the raising of the arm from the starting position to the horizontal usually takes from two to three seconds of the full time allowed. This means that during the ten seconds allowed for rapid fire, the shooter must not hesitate. As soon as the weapon recoils and settles and sight alignment is regained, the next shot must break. The shooter must take care not to panic and rush the string, but there can be no hesitation.

(4) Again, the grip must be stressed here, for any variation of pressure will cause a tendency of misalignment in natural sight alignment causing lost time which could be applied to the other fundamentals.

VI. CONCLUSION:

A. <u>Retain Attention</u>: Remember the additional techniques which are added to regular fundamentals and techniques in shooting slow fire or sustained fire. Items the shooter must pay particular attention are as follows:

B. <u>Summary</u>:

1. Become familiar with the course of fire to include, range commands and the procedure of firing the course.

2. All firing is conducted on the 50 meter International Slow Fire target.

3. Limitations placed on the weapons are:

(a) Sight radius - 22 cms.

(b) Barrel length - 15 cms.

(c) Trigger weight - Not less than 1000 (2.2 Lbs) gms.

(d) Custom grips - May not exceed 5 cms.

4. Follow a sequence to insure that all of the fundamentals of minimum arc of movement, sight alignment and trigger control is accomplished each and every time one shot or a string of sustained fire is fired.

5. Incorporate into your technique the additional factors of firing slow fire and sustained fire that apply to the International Standard Pistol Course.

B. <u>Application</u>: The use of the various factors we have discussed will greatly aid the developing shooter in adding points to his score. By including the techniques of slow and sustained fire into the system, the shooter will gain the confidence so greatly needed to hold still, maintain perfect sight alignment and employ correct trigger control.

C. <u>Closing Statement</u>: The transition from the National Rifle Association Caliber .22 style of shooting to the International Standard Pistol style of shooting is relatively simple because of the similarity in the two courses of fire. However, dedication, desire to excel and perserverence will be needed to lift the shooter from average scores to Championship scores. Learn all the fundamentals needed, incorporate them into your technique, then apply them to each shot fired. You can become a Champion.

L. TECHNIQUE OF FIRE VIII
   (COMBAT PISTOL COURSE)

AAPMI&CC 510
50 Minutes
Apr 1975

LESSON OUTLINE

I. LESSON OBJECTIVE: To enable the pistol marksmanship student to fire the winning score in the Combat Pistol Course by having a complete knowledge of the procedures rules and technique of fire.

II. STUDENT PERFORMANCE OBJECTIVES: As a result of this instruction, the student must be able to accomplish the following student performance objectives:

    A. Given the Combat Pistol Team composition requirements, FORM a Combat Pistol Team for competition as described in FORSCOM/TRADOC Suppl 1 to AR350-6 and Chapter XIV, USAMU Marksmanship Instructors' and Coaches Guide.

    B. Given the Combat Pistol Match procedures and conditions, EXPLAIN the procedures and conditions for the Combat Pistol Match as outlined in FORSCOM/TRADOC Suppl 1 to AR 350-6 and Chapter XIV, USAMU Marksmanship Instructors' and Coaches Guide.

    C. Given the Combat Pistol Match rules for conduct of the match, DESCRIBE the conduct of a Combat Pistol Match as outlined in FORSCOM/TRADOC Suppl 1 to AR 350-6 and Chapter XIV, USAMU Marksmanship Instructors' and Coaches' Guide.

III. ADVANCE ASSIGNMENT: Chapter XIV, Appendix E, USAMU Marksmanship Instructors' and Coaches' Guide.

IV. INTRODUCTION:

    A. <u>Gain Attention</u>: The Combat Pistol Match is the only course of fire authorized for US Army Installation/Division, USAR (ARCOM and GOCOM) and State National Guard Excellence-in-Competition Matches through which the initial four (4) point "leg" for a bronze EIC Badge is awarded.

    B. <u>Orient Students</u>:

        1. <u>Lesson Tie-In</u>: The Combat Pistol Match is a superior pistol marksmanship training vehicle and is highly recommended for service pistol qualification and familiarization from company level through installation/division level.

        2. <u>Motivation</u>: The Combat Pistol Match will be used exclusively in competitions held at the troop unit levels all the way up to installation/division level and is designated as part of the service pistol competition at CONUS Army/major command and the US Army Championships. An Army shooter can start winning at his company competition and by continuing to place high will be able to become a winner of the Combat Pistol Championship at the US Army Championships.

        3. <u>Scope</u>: During this period of instruction the student will cover:

            a. Combat Pistol Team Composition.

            b. Combat Pistol Match Procedure and conditions.

            c. The rules for conducting the Combat Pistol Match.

V. BODY:

<u>NOTE</u>: DEMONSTRATOR WILL MOVE TO FRONT OF CLASS ATTIRED IN THE UNIFORM AND WITH THE EQUIPMENT USUALLY ASSOCIATED WITH FIELD DUTY.

TRANSITION: This is the uniform that the service competitor must wear when shooting the Combat Pistol Course. He may be excused from wearing the steel helmet by the match director. Four personnel similarly equipped, make up the 4 man team.

 A. <u>First Student Performance Objective</u>: Given the Combat Pistol Team composition requirements, FORM a Combat Pistol Team for competition as described in FORSCOM/TRADOC Suppl 1 to AR 350-6 and Chapter XIV, USAMU Marksmanship Instructors' and Coaches Guide.

 1. General:

 a. The Combat Pistol Match fired with the service pistol is designed to increase the proficiency of the individual soldier. The Combat Pistol Match will be used for competition from company through installation/division level in all service pistol competition. This match is a superior pistol marksmanship training vehicle and is highly recommended in training and familiarization. The Combat Pistol Match is the only course of fire authorized for the installation/division/ARCOM/GOCOM/State NG level Excellence-in-Competition Match.

 b. Competition at the company, battalion, and brigade levels may be conducted over the Combat-Pistol Qualification Course as prescribed in paragraph 59-67, FM 23-35, or the Combat-Pistol Match Course as described herein.

 2. Team Composition. Pistol teams entering team matches will be composed of four firing members, a coach, and a team captain and, if desired, two alternates. The coach and team captain may be alternate firing members of the team. At least two of the individual firing members must not have previously fired at that level of competition. All team members will be eligible to compete in the individual portions of the matches.

TRANSITION: The procedures for firing the Combat Pistol Match must test the shooter's skill by simulating some of the firing conditions found in combat.

 B. <u>Second Student Performance Objective</u>: Given the Combat Pistol Match procedures and conditions, EXPLAIN the procedures and conditions for the Combat Pistol Match as outlined in FORSCOM/TRADOC Suppl 1 to AR 350-6 and Chapter XIV, USAMU Marksmanship Instructors' and Coaches' Guide.

 1. Match Conditions:

 a. General. In events not covered by these rules, the NRA Pistol Rules are suggested as a basis for decision.

 b. Clothing and equipment.

 (1) Clothing and equipment normally associated with combat/field duty will be worn by each member of the team through major command competitions. The match director will prescribe a field uniform appropriate to the season and will insure compliance with the following:

 (a) Issue binoculars, not exceeding 7 X 50 power, may be used by the team captain and coach to direct fire during team matches.

 (b) The team captain and coach will wear the equipment required of firers.

 (2) Participation in the US Army Championships will be in duty uniform.

 c. Weapons.

(1) Company through installation/division. The only caliber .45 pistol allowable is that listed under LIN N96741, SB 700-20. Internal of external modification or alteration of parts is not permitted.

(2) Major command US Army Championships. National Match Grade caliber .45 pistol listed under LIN N96878, CTA 60-18, or the same type and caliber of commercially manufactured pistol as described in AR 920-30 is authorized.

d. Ammunition. .45 caliber service grade ammunition will be used through installation/division level. Match grade ammunition is authorized at the FORSCOM Army/major command and US Army Championship Matches.

e. Target. Target, Silhouette: Paper, Standing (FSN 6920-718-5385) which may be trimmed and mounted on a Target, Silhouette, Pasteboard, E, Kneeling (FSN 6920-795-1806) (Fig E-1). When a full silhouette is used, the Target, Silhouette, Pasteboard, Trapezodial (FSN 6920-600-6881) is added to the bottom of the E-type silhouette. Repair centers with "X" ring center may be used. Target, Indoor Silhouette, B-24, NRA approved for 50 feet gallery shooting may be used as a substitute only in the US Army Pistol Championships. Target must be pasted to Standard American target backing for use on standard 25-yard NRA National Match Course Pistol Range (Turning Targets).

f. Range. 25 yards.

g. Timing. When rotating targets are used, time is measured from the time the target starts to move to face the shooter until it starts to move to the edge position.

h. Positions. The STANDING, PRONE, KNEELING, and CROUCH positions used in this course of fire will be as follows:

(1) PRONE: The firer will assume the prone position on command before the command "LOAD" is given and will start firing when the targets face. The shooting hand must not touch the ground or rest on the non-shooting hand that is touching the ground. The firer may use a ground cover that renders no artificial support.

(2) KNEELING: In the kneeling position, either knee may be grounded for support but not both knees simultaneously. Firing with a two-handed support is permitted. The position of the foot used as main support of body weight, i.e., under the buttocks, must be vertical with the body weight resting on heel and toes.

(3) CROUCH: In the crouch position, the firer must face the target with the feet parallel to the target line with the feet placed in relation to each other not exceeding half the length of the firer's foot to the front or rear. Firing with a two-handed support is permitted provided a crouching position low enough to place the nonshooting hand on the knee is used.

(4) STANDING: In the standing position, the firer may either assume the stance permitted by NRA Pistol Rules for participating in National Match type competition or may face the target using the two-handed supported stance.

(5) READY POSITION: The firer is in a STANDING POSITION with the pistol held in one hand with the mechanism cocked and his arm stretched downward so that his arm forms a maximum angle of 45 degrees to the vertical or so that the pistol or his hand touches the bench. The pistol must not be raised until the COMMENCE FIRING signal is heard or the targets start to move to face the firer.
NOTE: The two handed grip is not used in the READY POSITION.

i. Refires.

(1) In the event of a defective cartridge, disabled pistol, or malfunction before a string is completed, the firer will not attempt to clear the malfunction but will fire another five-shot string, except in the PRONE POSITION where he will be allowed 15 seconds per unfired shot, provided he signals the range officer by holding up his hand at the end of the time period. Failure of the firer to notify the range officer of the malfunction, or touching the pistol in an attempt to reduce the malfunction before the range officer inspects the pistol, forfeits the right to refire.

(2) Two refires per table will be authorized due to a defective cartridge, disabled pistol, or malfunction. However, two consecutive refires will not be allowed in a 10-shot string or stage of fire. The firer will not attempt to reduce a failure to fire in a refire string.

j. Scoring.

(1) All shots fired by the firer count and are considered in his score, even if his pistol is not directed toward the target or is accidentally discharged.

(2) All shots fired on the wrong target shall be scored as misses.

(3) If the firer should have excessive hits on his target, he may elect to be scored the value of the lowest hits or refire that stage of the course.

(4) Unfired shots in the refire string shall be scored as misses.

k. Decisions of ties. When two or more firers have the same numerical score, ties will be ranked in the order listed below:

(1) By the greatest number of hits.

(2) By the greatest number of X's (If the X ring is used).

(3) By the greatest number of 10's.

(4) By the highest total score in Table 3.

(5) By the highest score in the 10-second stage of Table 3; if still tied, the highest score in the 15-second stage.

(6) By the highest total score in Table 2; if still tied, the highest last 10-shot string in Table 2.

l. Penalties.

(1) A firer in the READY POSITION who raises his arm too early or does not lower it sufficiently shall first be warned that his position is incorrect by the range officer. The firer may be penalized the value of the string fired should the fault be repeated.

(2) Unsafe handling of the pistol will be grounds for disqualification.

(3) No individual or team member will be allowed to fire unless the equipment prescribed in D, above, is worn in the manner intended. If equipment is lost by the firer during the conduct of the course, he will not be allowed to fire until the equipment is replaced in the proper position without delaying the match or creating a safety hazard. Failure to meet this requirement will result in disqualification of the individual for the match involved.

m. Safety.

(1) Appropriate range safety rules and regulations will be followed.

(2) Loading and reduction of malfunction procedures for the .45 caliber pistol will be as specified in Appendix F.

TRANSITION: When the match procedures are understood and the conditions are met for firing the Combat Pistol Course, the individual shooter must also be familiar with the rules for conducting the match.

C. <u>Third Student Performance Objective</u>: Given the Combat Pistol Match rules for conduct of the match, DESCRIBE the conduct of a Combat Pistol Match as outlined in FORSCOM/TRADOC Suppl 1 to AR 350-6 and Chapter XIV, USAMU Marksmanship Instructors' and Coaches' Guide.

1. Conduct of the Match.

a. General.

(1) The Combat Pistol Match is divided into three tables with 30 rounds fired in each table. Ninety rounds are fired in individual competition with the maximum score of 900 points. One hundred and twenty rounds are fired in 4-man team competition with the maximum score of 1200 points.

(2) Firing may be conducted so that each firer completes each table before starting the next table or completes the entire course while on the line.

(3) Team members shall be squadded so that they will fire as a unit on the line. In the event that all teams cannot fire at the same time, relays will be assigned so that two members of each team will fire together. Team captain/coach may direct the team's firing.

(4) The Team Match will be an abbreviated course, i.e., 10 shots prone position, 10 shots duel stage and 10 shots military rapid fire (with each 5 shots in 10 seconds) for a total of 30 shots.

b. Course of Fire.

(1) Table 1 - Combat Position Firing. This table tests the firer's ability to engage a target with accurate fire from a position he may use in combat: prone, kneeling, or crouch. Thirty shots, 10 shots fired from each position.

(2) Table 2 - Duel. This table tests the firer's ability to rapidly engage a target that is exposed for a limited period of time. Thirty shots total.

(3) Table 3 - Military Rapid Fire. This table tests the firer's ability to rapidly engage a target and place accurate sustained fire on the target under decreasing time limits. Thirty shots total.

c. Firing line commands. Firers will be called to the firing line and be given a 3-minute preparation period. At the conclusion of the preparation period the following commands will be given to control the conduct of firing:

(1) "THE PREPARATION PERIOD HAS ENDED."

(2) "THIS WILL BE THE_____STAGE OF THE COMBAT PISTOL MATCH. TAKE YOUR POSITION ON THE FIRING LINE."

(3) "WITH 5 ROUNDS LOAD."

(4) "IS THE LINE READY?" (Firers must assume the READY POSITION on this command.)

(5) "THE LINE IS READY." (This command will not be given until the firers assume the READY POSITION.)

(6) The targets will be exposed or the signal to "COMMENCE FIRING" will be given in approximately 5 seconds after the command "THE LINE IS READY."

(7) "CEASE FIRING - MAGAZINES OUT - SLIDES BACK - CLEAR ALL WEAPONS." (In Table 1, PRONE POSITION, the firer is commanded, "FIRERS RISE.")

    d. Firing Tables.

    (1) Table E-1 - Combat Firing

POSITION	DESCRIPTION
PRONE	Ten shots fired in two 5-shot strings, reloading second magazine of five rounds without command, 2½ minutes per 10 shots. Firer loads on command while in the prone position and starts firing when the targets face. The firer will remain in the firing position until commanded to rise. CAUTION: The firer must extend the arms far enough to prevent the pistol slide from striking the face during recoil.
KNEELING	Ten shots fired in two 5-shot strings, 20 seconds per string. Firer starts in the READY POSITION and assumes the kneeling position when the targets face. CAUTION: The arms must be extended far enough to prevent the slide from striking the firer in the face during recoil.
CROUCH	Ten shots fired in two 5-shot strings, 20 seconds per string. Firer starts in the READY POSITION and assumes the crouch position when the targets face.

NOTE: THE CONSTRUCTION OF SHOOTING TABLES ON SOME COMPETITIVE RANGES MAY MAKE THE FIRING OF TABLE 1 IMPRACTICAL FROM THE 25-YARD FIRING LINE. TABLE 1 MAY BE FIRED FROM IN FRONT OF THE SHOOTING TABLES AT A DISTANCE OF APPROXIMATELY 22 YARDS.

    (2) Table E-2 - Duel

POSITION	DESCRIPTION
STANDING	Thirty shots fired in six 5-shot strings, 5 seconds per shot. Targets are faced 5 seconds and edged 5 seconds. Firers start in READY POSITION and shoot one shot each time target appears, returning to READY POSITION after each shot.

    (3) Table E-3 - Military Rapid Fire

POSITION	DESCRIPTION
STANDING	Ten shots in two 5-shot strings, 20 seconds per string. Firer starts in the READY POSITION and shoots 5 shots at one target during the time limit.

STANDING   Ten shots in two 5-shot strings, 15 seconds per string. Firer starts in the READY POSITION and shoots 5 shots at one target during the time limit.

STANDING   Ten shots in two 5-shot strings, 10 seconds per string. Firer starts in the READY POSITION and shoots 5 shots at one target during the time limit.

NOTE: HAVE DEMONSTRATOR FIRE FIVE SHOTS FROM EACH POSITION AND FIVE SHOTS DUEL FIRE AND FIVE SHOTS MILITARY RAPID FIRE (10 SECONDS).

    VI. CONCLUSION:

        A. Retain Attention: The US Army Record for the Combat Pistol Course is____ _____.

        B. Summary: During this period we have covered the:

            1. Combat Pistol Team Composition.

            2. Combat Pistol Match Procedures and Conditions.

            3. The rules for conducting the Combat Pistol Match.

        C. Application: With this information, the Pistol marksmanship student will be able to set up, fire in or conduct the Combat Pistol Match.

        D. Closing Statement: This course of fire is your opportunity to get your first four points toward a Distinguished Pistol Shot Badge.

M. COACHING I
    (ATTRIBUTES, AIDS TO COACHING AND DUTIES
    OF A PISTOL TEAM COACH)

AAPMI&CC 308
50 Minutes
Mar 75

LESSON OUTLINE

    I. LESSON OBJECTIVE: To enable the pistol marksmanship student to coach a pistol team by having an understanding of the attributes and duties of a pistol team coach and how to use the various aids to coaching.

    II. STUDENT PERFORMANCE OBJECTIVE: As a result of this instruction the student must be able to accomplish the following student performance objectives:

        A. Given the principal attributes of a qualified pistol team coach, DEFINE and EXPLAIN each of these attributes as shown in Chapter X, USAMU Marksmanship Instructors' and Coaches' Guide.

        B. Given a list of the aids designed to assist the pistol coach, DEFINE and EXPLAIN the use of each of these aids as described in Chapter X, USAMU Marksmanship Instructors' and Coaches Guide.

        C. Given a description of the pistol line coaches' duties, EXPLAIN in detail the pistol coaches' duties as performed before, during and after the pistol match as shown in Chapter X, USAMU Marksmanship Instructors' and Coaches' Guide.

    III. ADVANCED ASSIGNMENT: Chapter X, USAMU Marksmanship Instructors' and Coaches' Guide.

    IV. INTRODUCTION:

        A. <u>Gain Attention</u>: (Have a student ask the question) "Where were you last night, coach?" (Answer) "At home in bed with my wife, sober and asleep. Where else would a good coach be the night before a match?"

        B. <u>Orient Students</u>:

            1. <u>Lesson Tie-In</u>: You have been previously instructed in the fundamentals and techniques of shooting a pistol. We will now discuss the qualities and requirements necessary to properly coach a pistol team.

            2. <u>Motivation</u>: All of you probably will at one time or another find yourselves acting as a line coach of a pistol team during a team match. If you never coach a team, you may coach another shooter.

NOTE: SHOW CHART #1 - SCOPE AND SUMMARY OF PISTOL COACHING

            3. <u>Scope</u>: During this period we will discuss the attributes of a pistol team coach, the various aids to coaching and the coaches' duties before, during and after a pistol match.

NOTE: LIST THE ATTRIBUTES SUGGESTED BY THE STUDENTS ON A CHALKBOARD.

    V. BODY:

        A. <u>First Student Performance Objective</u>: Given the principal attributes of a qualified pistol team coach, DEFINE and EXPLAIN each of these attributes as shown in Chapter X, USAMU Marksmanship Instructors' and Coaches' Guide.

QUESTION: Name an attribute of a pistol team coach.

ANSWER: Temperate, dedicated, self control, patience, compatible, etc.

PERSONAL ATTRIBUTES OF A COACH: A coach's moral character and personal dignity must always serve as a model for those he is training. He must have a profound knowledge of the theory and practice of marksmanship and a serious attitude toward his training responsibilities. He must embrace a love for the sport of pistol shooting, have an abiding respect for each member of his team. He must feel a sincere desire to help his shooters. This positive attitude will in turn instill in his team members a respect for him.

A coach must be strict in his demands upon his shooters and consistent in what he requires. He must insist always on observing discipline and adhering to the day's program. The knowledge that lapses in effort will not go unnoticed, will spur his team members to conscientious work.

There are no perfect pistol coaches. However, there are those who are outstanding because there is something in their makeup that induces excellence of performance from those who shoot under their guidance. Minor shortcomings do not impair the working loyalty of a progressing shooter who has found a coach with merit. To recognize merit in another person is in itself an essential of character. The coach or shooter who scorns all others for even their minor flaws and thinks no one is worth the effort of leading or following, parades his own inferiority before his teammates and competitors.

There are attributes, moral and mental, that the pistol coach must have that will accelerate the shooter's progress and prevent a lapse into habits that may lead to a decline in performance.

1. <u>Temperate</u>: A coach must be temperate in all things. He must have the moral fiber to refrain from loose living. Intestinal fortitude and will power will enable him to deny himself things that are damaging to his character and may compromise the example he is to set. Nothing should preclude the attainment of what he knows to be his prime objective: WIN!

2. <u>Dedicated</u>: Since shooting skill has somewhat less bearing on ability to coach than on being a top shooter, the search for a capable coach should not be directed necessarily toward the top pistol shooter, but toward the dedicated, observant, self-controlled man who possesses a wide knowledge of shooting technique. This does not imply that a Champion Shooter cannot have these desirable qualities. He can, but too often the top shot has time only for improving his own skill rather than in extensively coaching others. The attributes mentioned as desirable in a coach are contagious and will transmit themselves to the shooters.

3. <u>Self Control and Patience</u>: Most successful coaches feel that self control and an infinite amount of patience are the main attributes of a pistol coach. Many times during team matches, under pressure, things could be done and said which would have an adverse effect on the team. The ability of the coach to control himself and the other members of the team will affect greatly the outcome of the team match. For example, irreparable damage can be inflicted upon team morale if the coach loses his temper with a man on his team and calls him uncomplimentary names. From that time forward, there will be a wall between the two. Patience, especially while coaching new shooters will help the shooter to improve more rapidly. For example, it may be necessary for the coach to stand beside a new shooter or an old one, and watch him jerk the trigger on shot after shot. He may prove to the shooter by use of ball and dummy that he is jerking, and then have to bear up under a continuance of a faulty performance. Should he give up? NO! He must keep trying; he must have patience and confidence in the man. Tomorrow may be the day that the developing shooter begins to grasp the meaning of points the coach has continued to stress.

4. **Compatibility**: This is simply a big word for getting along with the shooters on the team. If the coach cannot get along with his shooters, he is worse than useless to the team. The coach must be quick to accept blame for a losing performance and equally quick to give the shooters credit for a winning match.

5. **Inspire Confidence**: The successful coach must be able to inspire confidence in his shooters. First he must have confidence in himself and show his team members that he has complete confidence in them and in their ability, both as individuals and as team shooters. Then, and only then, can he expect them to have confidence in their own abilities and just as important, confidence in him.

6. **Enthusiasm**: The coach must show enthusiasm, not only when the team is winning, but when the team has lost a match. The team must have an enthusiastic desire to win to be able to come back and shoot better in the next team match. The desire to win and unflagging enthusiasm are the things that keep us from giving up in disgust when everything seems to be going wrong.

7. **Observant**: The Pistol Coach will be observant of anything in his shooter's performance that can be improved to make his scores better. This applies not only in practice, when the time can be taken to improve, but also should be observed in matches. He must determine if the training program is achieving the required results. During a match he must be aware of any unusual conditions, noise, or movement that could be distracting to his shooters. A close check of all phases of the match such as scoring, alibis, challenges, and block officer decisions must be made.

8. **Wide Knowledge**: The Pistol Coach must have a wide knowledge of employment of the shooting fundamentals and coaching techniques. In addition he must have specific knowledge of match rules and range procedures. The good line coach does not necessarily have to be a top shooter but he should have sufficient match experience to be familiar with the problems encountered in competition and most of the remedies needed by his shooters. The coach must catalogue in his mind the traits of each shooter. He must know how to handle each member in every shooting situation so that successful results can be obtained.

9. **Exacting Standards of Team Performance**: To excell in pistol competition, the Pistol Coach must set exacting standards for the team to follow. His actions, as a coach, and as a member of the team, must be above reproach at all times. He must require that his team members act in such a way as to reflect credit upon themselves and the US Army. As always, any Army team at a match represents the entire Army. One step out of line will cause some persons to say, "Well, there is the Army making an Ass out of itself again."

The coach must require his team to utilize practice sessions to improve their shooting. This is the time to try new ideas, not just as a chance to burn up ammunition. During the shooting, the coach should require a great amount of effort by each man toward correcting his faults. He must not allow the shooter to accept an average performance. Strive for perfection in shooting, but at the same time retain exacting standards for personal and team conduct.

10. **Open, Progressive Mind**: Because a man has been shooting for 30 years does not necessarily mean that he knows more about shooting or has the best coaching techniques of anybody with less longevity. This business of "It was good enough for the last 30 years, and it's good enough now," is not applicable anywhere today, least of all in pistol shooting and coaching. The good coach should accept, with proper frame of mind, constructive advice from experienced marksmen. He should continually strive to improve himself with better techniques. An open, progressive mind applies also to his approach to the difference in personality and shooting ability in his shooters. No two individuals are alike. The coach who is flexible in his dealings with his shooters will in the long run have the best functioning team.

11. **Avoid Being Over-Eager**: An efficient coach recognizes there are times when his effectiveness is enhanced by being silent and observing the shooter's actions. For example, if a shooter has averaged 98 points for each string of slow fire in the past three months, there is little to be gained by pressing hard for a possible score from that shooter. Know when to let your presence on the firing line be quietly sufficient.

TRANSITION: There are a number of aids to coaching that will enable the pistol coach to accomplish his mission of developing a winning pistol team.

NOTE: SHOW CHART #2 - AIDS TO COACHING.

    B. **Second Student Performance Objective**: Given a list of the aids designed to assist the pistol coach, DEFINE and EXPLAIN the use of each of these aids as described in Chapter X, USAMU Marksmanship Instructors' and Coaches' Guide.

NOTE: SHOW CHART #4 - SHOOTERS WORKSHEET.

    1. **The Shooters Worksheet** is designed for both slow and sustained fire checkout. As the shooter progresses through the factors that control the delivery of an accurate shot, there must be no omission of essentials. This aid is used mainly in the training phases.

NOTE: SHOW CHART #6 - PREPARATION CHECK SHEET.

    2. **The Preparation Check Sheet** is designed to be stapled into the gun box lid for ready reference in conducting complete preparation. Items of readiness are included that must be implemented in the assembly area, firing line and after the command "Load".

NOTE: SHOW CHART #5 - SIGHT ADJUSTMENT CARD.

    3. **The Sight Adjustment Card** is used for a threefold purpose; by knowing the size of the scoring rings and how much one (1) click on the sights will move the strike of the bullet, the shooter can make a hold sight change and move his subsequent shots immediately to center of target; secondly, the shooter can record his sight settings, elevation and windage for all caliber weapons in pistol competition; and last, there is reference on all of the popular makes of adjustable sights as to which direction the strike of the bullet moves when the adjusting screw is turned clockwise. This sheet should be removed from the booklet and stapled in the gun box lid.

NOTE: SHOW CHART #10 - ADJUSTABLE REAR SIGHT.

    4. **In Using the Adjustable Rear Sight**, each shooter should know the capability of each of his adjustable sights, i.e., how far will one click on this sight move the strike of the bullet at 50 yds or 25 yds; which way to turn the windage screw to move the strike to the right or left.

NOTE: SHOW CHART #7 - PISTOL SCOREBOOK.

    5. **The Pistol Score Book** is your personal record of your shooting. Date, time, place, ammunition, weather conditions, sight setting, wind direction, light direction and most important, a shot by shot record, by stages in practice or registered match of every shot fired in the current season. This worksheet booklet contains enough sample pages for all small arms firing school scores.

TRANSITION: The pistol coach must guide and help the shooter before, during and after the match.

C. <u>Third Student Performance Objective</u>: Given a description of the pistol line coaches duties, EXPLAIN in detail the pistol coaches' duties as performed before during and after the pistol match as shown in Chapter X, USAMU Marksmanship Instructors' and Coaches Guide.

NOTE: SHOW CHART #1 - SCOPE & SUMMARY OF PISTOL COACHING.

THE LINE COACH'S DUTIES: The line coach is responsible for the close supervision and the detailed, exhaustive measures necessary to improve the performance of the four shooting members of his team.

1. <u>Before the Match</u>.

    a. <u>Be Available Between Individual Matches</u>: During the firing of the individual matches, the line coach should check the performance of his team members. If any of them are falling below standard, talk to the shooter concerned, closely observe his performance and try to find out the nature of the trouble. The coach should try to steer the shooter's thinking back in the right direction. Remember, the individual matches are used as zeroing matches for the team matches. Any individual matches won are incidental to the major task of winning the team matches. The coach must be available to his men to assist them in every way possible.

    b. <u>Supervise Physical Preparation</u>: After the individual matches and prior to the team matches, the coach should have a team meeting. During this meeting, he should insure that all the team members have the right type and amount of ammunition. Are their weapons clean and in perfect working order? Check everything. (For details of this check-out, refer to the Coach's Worksheet, an Annex to this chapter.)

    c. <u>Assign Relay and Target Numbers</u>: A point to be covered during the team meeting is the assignment of target and relay numbers. Be sure there is no misunderstanding on these numbers. For example, it is demoralizing for a shooter to show up behind target number 40 and find none of his team mates there. By the time the range officer says "Load" the shooter discovers that the team is on target 4. The coach assigns additional duties to members of his team such as scoring the adjacent team in the event this requirement has been announced by the range officer. This duty includes the posting of the score board; carrying flags, chairs, and scopes up to the firing line; and any other duties the coach feels necessary. These matters should have been established during the previous training program and simply confirmed at the team meeting.

    d. <u>Supervise Mental Preparation</u>: From the time the coach begins working with his four shooters, he should be guiding their mental preparation toward winning this team match. Now he must do his best to get their minds off all miscellaneous interests and get them thinking only about this all important team match. (For details of this mental preparation check-out, refer to the Shooter's Worksheet and to the Coach's Worksheet, an Annex to this chapter.)

2. <u>During the Match</u>.

    a. <u>Make Decisions</u>: During the actual firing of a team match, decisions regarding the team must be made by the coach. For example, many shooters are prone to accept a slightly misplaced group rather than move their sights. Their attitude seems to be that the setting on their sights is correct (the zero was perfect during the previous matches), and they will not volunteer to change it. Here the coach must give firm, positive directions to the shooter. Another example is the decision as to whether or not to refire a <u>slow</u> fire string when the shooter's target has excessive hits. Naturally the coach would consult the shooter as to his preference, but the final decision lies with the line coach. He has the responsibility for the team's performance and therefore he has the final authority for decision made on the firing line in the heat of battle.

353

b. <u>Stresses Safety</u>: The coach should be constantly aware of any breaches of safety.

c. <u>Caters</u>: The word may mean different things to different people. What is meant here is that the coach must provide the shooter what he wants and needs, thereby aiding him to achieve his best possible score. As has been mentioned previously the coach should know just how each shooter likes to be coached and what the coach can do for him that is helpful. If this means setting the shooter's gun box in the line and preparing his equipment for shooting, then the coach should be glad to do it. It may also mean for the coach to sit back and watch, and not say a word. The coach must always remember that he is there for the shooter's benefit and not the other way around.

d. <u>Encourages</u>: The coach must encourage the members of his team at all times. When they are shooting well, he must encourage them to shoot better. When they don't shoot so well, he must assist them to mentally prepare themselves to shoot better in the next match. This assistance is in reality, encouragement. The shooter is being shown that the coach believes the team member can do the job required and that he wants to help him to correct any faults that may have lowered his performance. Never say anything to discourage any member of the team. By the same token, he cannot use false praise.

e. <u>Scoring Record</u>: All scoring of record practice and registered matches must be supervised by the coaches. This doesn't mean that every shot fired must be counted for score. The shooters must be given free practice to allow them to try new procedures such as a different position, etc. However, when record scores are fired they must be made a part of the shooter's permanent record. From this record the head coach can ascertain whether the shooter is improving or not, and what weapons or stages he needs more work on. All of the oral and written tests, etc., still won't indicate how well the man can shoot. This can only be determined by the shooter's performance on the firing line.

f. <u>Irregularities</u>: The coach should be a close observer of events and conditions during a pistol match and intercede by his good offices if an argument, protest or an infraction of the match rules concerns an Army team member.

3. <u>After the Match</u>.

<u>Critique the team</u>: There are several things a coach can do after a match. It is not good for him to stand in front of his shooters, rant and rave and tell them (if they didn't win) that they fired like a bunch of ring-tailed baboons. If they did actually shoot that way, the coach does not have to remind them, they already know it. The proper technique at this point should be to:

(1) <u>Discuss and Recommend</u>: For example, the coach and team should go over their performance and suggest ways to improve this shooters rapid fire or that shooter's slow fire when they return to their practice range. If there was a clash of personalities due to the grueling stress of the competition, the ruffled feathers should be smoothed. The opinions and methods of the individual shooter revealed at this time are the building stones of this Coaching Publication and the Pistol Marksmanship Guide.

(2) <u>Evaluation and Observation</u>: Direct coaching of the shooter is only half of the coach's job. True, it might seem to be the most rewarding, but he must plan what to teach him, how to train him, and evaluate his performance to find out if the training is effective. The evaluation of a shooter is achieved primarily from analysis of the shooter by the line coach and from the results of practice and match scores. The line coach will evaluate the shooter's potential ability, since he will

be in the best position to see how the shooter reacts to the problems confronting him when trying to achieve a good score. From his observation he will accurately fill out an Individual Information and Evaluation Sheet on each shooter. This evaluation will show the training needed to make him a better shooter. It can serve as a guide when considering the shooter for future teams.

NOTE: SHOW CHART #11 - INDIVIDUAL INFORMATION AND EVALUATION SHEET.

VI. CONCLUSION:

 A. <u>Retain Attention</u>: The haphazard approach to competitive shooting is self-defeating. Extending the shooting arm to fire a shot and hoping for the best is fruitless. The winner is a man of meticulous detail and savage determination. He leaves nothing to chance.

NOTE: SHOW CHART #1 - SCOPE AND SUMMARY OF PISTOL COACHING.

  B. <u>Summary</u>: The highly qualified pistol marksmanship coach has:

   1. The following attributes:

    a. Temperate

    b. Dedicated

    c. Self control and patience

    d. Compatible

    e. Able to inspire confidence

    f. Enthusiastic

    g. Observant

    h. Wide knowledge of pistol shooting

    i. Exacting standards of team performance

    j. Open mind.

   2. Knows how to use coaching aids:

    a. Shooters worksheet

    b. Preparation check sheet

    c. Sight adjustment card

    d. Adjustable rear sight chart

    e. Pistol score book

   3. Knows the pistol coaches' duties:

    a. Before the pistol match

    b. During the pistol match

    c. After the pistol match

C. <u>Application</u>: The trained pistol coach can transform the wavering beginner into a competent marksman; the expert pistol shooter can be influenced toward the winner's circle; the champion will be sustained in his relentless progress toward national identity.

D. <u>Closing Statement</u>: The nation's destiny is decided during times of conflict by the man with the gun, on the ground, closing with the enemy. The properly trained competitive shooter who knows how to be a winner under pressure is the person who can be relied upon to maintain his control in times of crisis or combat.

ANNEX I To Coaching I - Duties of a Line Coach

A. WORKSHEET FOR PISTOL TEAM COACHES

1. <u>Coach is in charge of team</u>.

2. <u>Each shooter will follow the same complete procedure for each shot or string</u>.

3. <u>Coach has responsibility for assisting shooter</u> between individual matches as well as in team matches.

4. <u>Assemble Team Members on Assembly Line</u> a sufficient time to complete preparation before match time or relay is called. Systematic application of the following factors in pistol line coaching will prevent a haphazard approach to the instruction and thereby properly influence the developing shooter to exert progressively better control over his shooting performance.

    a. Physical Preparation.

        (1) Designate relay and target number.

        (2) Check for clean, proper functioning, and lubricated weapons.

        (3) Check for sufficient proper caliber ammunition.

        (4) Check for proper sight setting and weapon zeroed.

        (5) Blacken sights.

        (6) Use ear plugs and shooting glasses.

        (7) Stop watch for time check.

        (8) Score book and pencil.

        (9) Chair.

        (10) Obtain scorecard.

    b. Mental Preparation:

        (1) Review shot sequence (Slow or rapid fire worksheet)

        (2) Encourage.

        (3) Think and observe.

        (4) "Maintain confidence that a controlled, uniform and exacting performance will produce good results."

        (5) Talk shooter through relaxation.

        (6) Let coach worry about irregularities.

        (7) Review techniques of shot analysis and positive correction of errors.

        (8) Remind shooter to exercise care and safety.

(9) "Damage from mistakes is minimized by continuing to work hard."

(10) "Carefully planning the delivery of each shot will minimize the effect of tension and pressure.

(11) "Concentration on maintaining sight alignment and holding the smallest possible sustained arc of movement while applying positive trigger pressure will result in a surprise shot break that will strike the target within the shooter's ability to hold."

    5. <u>Move to firing line</u> with team member when relay is called.

        a. Coaching equipment:

           (1) Ear plugs

           (2) Pencil and scoresheets

           (3) Scope and stand (or Binoculars)

           (4) If practice, .22, .38, or .45 Cal scoring plugs

           (5) Stop Watch

           (6) Shooting glasses

           (7) If practice, shooter's worksheet

           (8) Guidon (Team Flag) (Registered team matches only)

           (9) Extra pistols and magazines of all calibers available

           (10) Chairs or stools

           (11) Staple gun and staples

    6. <u>During the three (3) minute preparation period</u> have shooter:

        a. Focus scope on proper target. Check target for holes or loose target.

        b. Adjust ear plugs or protectors

        c. Load ammunition into magazines

        d. Assume stance

        e. Practice breath control

        f. Dry fire for natural position that will enable center hold on aiming area without tendency to settle to either side of bullseye.

        g. Dry fire for natural grip that will enable front and rear sights to be alignment without artificial correction by wrist, arm or head movement.

        h. Mentally rehearse the steps involved in delivering a controlled shot.

7. **Plan shot sequence.** (Refer to slow fire or rapid fire worksheet)

   a. During practice or if having difficulty during match, the coach should require the shooter to give his shot plan verbally. (Except for 2650 level master shooters)

   b. Coach and Shooter will converse in low tones (Coach will refrain from talking during actual shooting as this may disturb shooter's concentration.)

8. **Relaxation.** (No unnecessary muscular tension)

   a. Relax all major portions of body not needed to maintain stance and extend the shooting arm toward the target.

   b. The coach will remind the shooter to relax before attempting each shot or string.

9. **Deliver shot as planned.** (No compromise)

   a. Follow through (Continue to employ all control factors until bullet leaves barrel)

   b. Shot should be fired as a surprise. (No reflex action)

   c. Do not hold too long. (If experiencing difficulty in applying any of the Pistol Fundamentals, bench weapon and start over)

10. **Complete shot analysis** after each shot or string.

    a. Call shot (Based on sight alignment, *not* sight picture.)

    b. Scope for hit location (Compare with shot call)

    c. If shot or call is in error, determine cause.

    d. If shooter is unsure or in error on shot call, have him draw a picture of the front and rear sight relationship as it appeared at the time the shot was fired.

    e. If shot and call are good, review the successful method employed, in an effort to duplicate a good performance.

    f. The coach should require the shooter to plot shot calls and hits on appropriate worksheet and enter shot values in scorebook.

11. **Positive corrective measures** to be taken on each shot or string:

    a. Coach and shooter should agree on corrective measures to be taken. (Student understanding is enhanced when coach explains reasons for specific corrective measures.)

    b. Include corrective measures in plan for next shot or string (Prompt application)

12. **Additional Duties:**

    a. Time check after each shot.

    b. Watch shooter or his weapon, not the target. (Do not use shooter's scope while he is firing)

    c. Have team captain or alternate available for scoring between stages of fire.

    d. Have a non-shooting team member of alternate post score on team score board after each stage.

    e. Check and validate team score card with signature if acting as team captain.

    f. Police firing point at completion of firing.

    g. After all relays have completed firing, the coach will conduct a group critique to resolve problems that became apparent during the match or training session. Answer questions and explain techniques. The degree of understanding exercised by the student is in direct proportion to the progress he can expect to bring to improving his shooting skill.

  B. <u>COACHING OF ZEROING, SLOW FIRE, AND THE TIMED AND RAPID FIRE EXERCISE (PISTOL)</u>

NOTE: (REVIEW CHAPTER IV, USAMU PISTOL MARKSMANSHIP GUIDE, "ESTABLISHING A SYSTEM.")

   1. Coach starts supervision of group preparation in the assembly area, before the first relay is called to the firing line.

NOTE: THE COACH WILL INFORM EACH MEMBER OF GROUP TO BE PRESENT ON RANGE THIRTY (30) MINUTES PRIOR TO THE TIME FOR CALLING OF THE FIRST RELAY TO THE FIRING LINE.

    a. The coach will have students check their squadding tickets for relay and target number.

    b. The coach will supervise preparation to fire and have each student check off each completed action during preparation on an appropriate work sheet. Remove the preliminary preparation check list from the shooter's worksheet booklet and staple it on the inside of the gun box lid for easy reference. All assembly area preparation should be complete before the relay is called to the firing line.

    c. Coach will review zeroing procedure.

     (1) Use the shot group method. A minimum of three (3) shots is fired before original sight adjustment, except in team matches of registered competition. Here the sight adjustment is changed when the shooter and coach deem it immediately necessary.

     (2) Make a bold sight change. Do not creep to center of target a click or two at the time in an effort to center the group. Make a bold sight change and accomplish the necessary sight adjustment immediately. Reference will be made to sight adjustment card to determine the total distance your shot group is located away from the center of the target. Divide the distance the shot group must be moved, either in elevation or windage, by the distance one click on your sight moves the strike of the bullet. This figure represents the number of clicks of sight adjustment necessary to center your group.

     (3) Fire a timed fire string to confirm zero.

     (4) Do not "hold off" or use "Kentucky Windage" to establish zero.

     (5) Review the effect on zero, of wind and other adverse conditions.

        (6) Mark the zero sight setting by:

            (a) A lead or colored pencil, marked temporarily on sight adjustment screws for both elevation and windage.

            (b) Remove sight adjustment card from work sheet booklet and staple it on the inside of the gun box lid. Use this card to make a record of normal sight settings after zeroing is completed.

            (c) Recording the conditions under which this zero is obtained in the spaces provided on the sample score book pages. (Last page in worksheet booklet.)

            (d) After zero is confirmed, mark sights (25 yard setting) permanently with finger nail polish or a quick drying paint such as airplane dope.

            (e) Record the 50 yard sight setting by noting on sight adjustment card the number of clicks necessary to center group at 50 yards range. (Elevation and Windage.)

    d. The coach will stress safe handling of weapons.

2. **The coach and first relay shooter move to ready line** when directed by the range officer. At this time all assembly area preparation should be completed.

    a. When called to the firing line by the range officer, set up equipment on assigned firing point but don't handle weapons until directed to do so. When line is clear for handling weapons, the three (3) minute preparation period will start. Sequence of actions taken during three (3) minute preparation period is listed on Slow Fire and Rapid Fire worksheets and on preparation check list stapled in lid of gun box.

    b. Coach will review planning of the delivery of an accurate shot (Sequence of actions in delivering a shot or a string of shots is listed on the Slow Fire or Rapid Fire worksheet.)

3. **When command "load" is given**, the coach and student will strictly adhere to the rules for safe handling of weapons. (The method of loading the weapon will follow the procedure demonstrated during the coaches' briefing class.)

    a. Students will check out the employment of the fundamentals that will enhance attaining a minimum arc of movement: stance, position, grip, and the practice of breath control as previously checked out during the three (3) minute preparation period.

    b. Student will check for proper target number as he extends shooting arm to recheck for natural position which gives him a hold centered in aiming area on the target and effortless sight alignment attained by proper grip and head position.

    c. Relax with pistol at bench rest. Do not change position or allow shift of grip during this period.

    d. Student will continue to mentally review planning of shot. Having the mental processes so engaged will aid in conditioning the mind to control the delivery of the next shot or string. Concentrate on shot sequence. Visualize perfect sight alignment. Mentally reconstruct the smooth trigger control necessary to fire a shot without disturbing sight alignment.

    "YOU - ARE - READY!"

4. <u>When commands are given to "Commence Firing"</u> or the targets are faced as a signal to begin firing, the coach and student will be required to utilize all available time allowed in completing the training exercise. After the required number of shots have been fired, the remaining time will be devoted to review and critique of performance. The coach and student will be allowed time, two (2) minutes between five (5) shot strings of timed and rapid fire where scoring is not in progress, to devote to plotting calls and hits, analysis of performance, correction of errors and planning next shot or string. The one (1) extra minute allowed per shot in each slow fire exercise is sufficient to perform the following functions:

    a. The coach will require the student to give his shot plan verbally before shot or string of shots is fired.

    b. The coach will remind the student to relax before attempting to fire each shot or string of shots.

    c. The coach will not talk or use the students' spotting scope during the delivery of a shot or string. (Observe the shooter or his weapon, not the target.)

    d. The coach will require the student to plot shots calls and hits in the square provided on the Slow Fire or Rapid Fire Worksheet. (Draw a small circle in each square to represent the bullseye. Plot the shot calls with an "X" and hits with a "Dot.")

    e. The coach will require the student to conduct an analysis of each shot or string of shots fired to determine cause of error, if any.

    f. The coach will endeavor to come to an agreement with the student on corrective measures to prevent reoccurrence of error. (Incorporate corrective measures into the plan for the next shot.)

    g. The coach will require student to make a shot value entry on sample scorebook page for each shot fired.

    h. Offer encouragement to student for poor results and compliment him on a good performance in order to stimulate confidence.

<u>NOTE</u>: THE COACH WILL DIRECT THE STUDENT TO FIRE ON A BLANK TARGET (no bullseye) IF LESS THAN 50% HITS ARE RECORD DURING ANY ZEROING STAGE. IF A SCORE OF LESS THAN 75x100 POINTS FOR ANY TEN (10) SHOT STAGE DURING SLOW, TIMED, OR RAPID FIRE EXERCISES IS POSTED BY THE STUDENT, A BLANK TARGET WILL ALSO BE USED FOR THE NEXT TEN (10) SHOT STAGE OF FIRE. SIGHT ALIGNMENT WILL BE STRESSED DURING THESE BLANK TARGET SESSIONS, AS WELL AS REVIEW OF THE OTHER FUNDAMENTALS OF STANCE, POSITION, GRIP, BREATH CONTROL (attaining a minimum arc of movement), AND POSITIVE TRIGGER CONTROL.

N. COACHING II AAPMI&CC 309
(Technique of Coaching) Fifty (50) Min
 Mar 75
LESSON OUTLINE

   I. LESSON OBJECTIVE: To enable the pistol marksmanship student to act as a pistol team coach of military, police or civilian teams by employing the factors of systematic coaching techniques.

   II. STUDENT PERFORMANCE OBJECTIVES: As a result of this instruction students must be able to accomplish the following Student Performance Objectives.

       A. Assigned as a team coach, LIST and EXPLAIN factors in systematic coaching technique as shown in Chapter XI, USAMU Instructors and Coaches Guide.

       B. Faced with the task of coaching a championship caliber pistol team, APPLY the techniques of coaching champions as listed in Chapter XI, USAMU Instructors and Coaches Guide.

       C. Assigned as the Head Coach of a pistol team, LIST and DESCRIBE pistol head coaching responsibilities and techniques as stated in Chapter XI, USAMU Instructors and Coaches Guide.

       D. Given the mission of coaching a team on the firing line, LIST and APPLY the techniques of a line coach as outlined in Chapter XI USAMU Instructors and Coaches Guide.

   III. ADVANCED ASSIGNMENT: Chapter XI USAMU Instructors and Coaches Guide.

   IV. INTRODUCTION:

       A. Gain Attention: The champion pistol shooter is the cutting edge of a competitive effort to penetrate beyond the present frontiers of human accomplishment in marksmanship. The pistol coach helps to fashion this instrument and assist in guiding it toward its objective: A record breaking performance.

       B. Orient Students:

           1. Motivation: The coach can and does influence a pistol team's performance. Whether his influence raises or lowers scores depends on the quality of his technique of coaching.

           2. Lesson Tie-In: In the previous period you were briefed on the attributes of a good coach, aids to coaching and the duties of a pistol line coach.

NOTE: SHOW CHART #1 - SCOPE AND SUMMARY

           3. Scope: During this period we will cover the factors in systematic coaching technique, the pistol champions and the techniques used by the pistol head coach and line coaches that work with each 4 man pistol team.

TRANSITION: Is there a formula, is there a method, a rule of thumb, is there a guide to coaching that can be used to establish a coaching technique? Yes. The factors in systematic coaching technique.

       A. First Student Performance Objective: Assigned as a team coach, list and EXPLAIN factors in systematic coaching technique as shown in Chapter XI, USAMU Instructors and Coaches Guide.

NOTE: SHOW CHART #2 - COACHING FACTORS

       1. The first factor is preparation. This means careful and complete physical and mental preparation. Close attention to detail is required so that nothing is left out. Refer to the pistol coach work sheet.

       2. Plan carefully and in minute detail the exact actions required to deliver controlled shots on the target. This is a joint effort between the shooter and the coach.

       3. Relax your shooter's mind and relieve any unnecessary muscular tension. With some shooters you can tell a joke, with other shooters you can remind them of a past great performance. With a few shooters you just stand and listen to them talk.

       4. Insure that the shooter adheres closely to the plan the two of you have worked out for delivering controlled shots.

       5. As each shot is fired, you hope there will be no flaws in performance to analyze, but you must be prepared to analyze errors. If there are no errors detected and the shot is good, a thoughtful appraisal of the technique used to keep the shot under control is in order.

       6. If an error does creep in, and you analyze the performance and discover the cause; make a positive correction. Suggest to your shooter what he must do to prevent a recurrence of the error. His agreement and an understanding of your suggestion will impart an item of knowledge that he will remember.

TRANSITION: Application of these six factors will result in a systematic coaching technique: Preparation, planning, relaxation, deliver the shot according to plan, analysis of errors and positive corrections when needed. These are the keys to good coaching techniques whether it be in coaching a beginner or a champion.

    B. Second Student Performance Objective: Faced with the task of coaching a championship caliber pistol team, APPLY the techniques of coaching champions as listed in Chapter XI, USAMU Instructors and Coaches Guide.

       1. If you are going to coach a champion, it is necessary that you be able to recognize one. Usually it is a simple matter of reading the score board at a match. The champion's name is on top of the aggregate bulletin. However, as a pistol coach, you have to be able to recognize a champion early in his career of shooting. The potential champion is the shooter who is eager to learn, who wants to compete against the best, the shooter who maintains a calm attitude when the going gets tough, whose scores get better when the competition gets stronger. The shooter who displays these qualities is a potential champion if he receives proper coaching.

       2. Your most important task in coaching a champion is helping him maintain confidence in his ability to control his performance. You have taught him all you can in order to increase his knowledge of shooting. Remind him of techniques he may occasionally forget to employ. Continuing ability to control the employment of the fundamentals builds confidence.

       3. Guide your shooters by suggestion, not orders. We in the military are more apt to give orders than are civilians, but resist that temptation. Use an informal, "Why not try it this way" kind of approach.

       4. Keep the shooter's self-respect at a high level. Do not hesitate to praise a good performance. Maintain a positive attitude. If a shooter's score is not what you expect, let your approach to analysis of his shooting be that of "room for improvement" instead of "you blew it."

5. Realize at all times that a coach can improve any shooter's performance. Even the champion sometimes allows technique errors to creep into his shooting. The coach must recognize the error pattern and help the shooter to correct them. The good coach gets the most out of any shooter. He must know his shooters well enough to know which psychological button to push to improve performance.

TRANSITION: To get the maximum performance out of a pistol team the head coach must plan his training and techniques to develop the potential for winning pistol matches.

    C. <u>Third Student Performance Objective</u>: Assigned as the Head Coach of a pistol team, LIST and DESCRIBE pistol head coaching responsibilities and techniques as stated in Chapter XI, USAMU Instructors and Coaches Guide.

When you hold the title of Head Coach of a pistol team you will find that there are special responsibilities and techniques that apply to training and motivating a larger group of pistol shooters.

    1. <u>Selection of Needed Training Subjects</u>: To determine training requirements, a coach should determine the shooting ability of each of his shooters. This ability varies greatly, not only between the All-Army level and the Post, Camp, or Station level, but, of more interest to the coach, between the individual shooters on the same squad. This is a difficult but very important job. What the coach selects as the needed training subjects may well determine the later success of the team. The Head Coach must take into account the individual needs and preferences of his team members in planning a training program. This does not mean separate training programs for all shooters but it does mean that your training program must be flexible enough to allow the shooters to develop at their own pace. If you schedule extra rapid fire training, you do not have to require every member of the team to shoot rapid fire. You may well have a couple of people who have no problem with rapid fire. Forcing them to shoot rapid fire would be a waste of training time. They may need more slow fire training. The training program should be tailored to the needs of the individual.

    2. <u>Scheduling and Supervising Training</u>: Publication of a training schedule will not suffice to insure progress in training. Careful observation of individual members and correcting faults during the training is one of the primary duties of the head coach. The coach will normally use analysis of the scores fired in various stages in practice to assist in initial selection of team members and in selection of future training subjects. Match scores will become the winning scores if the weak stages are improved and mistakes are eliminated from the shooter's performance.

    3. <u>Limit record shooting</u>: If a shooter feels that his place on the team would be jeopardized by anything less than his best score, he may hesitate to try a new technique during a record match. The technique change may be just what he needs for improvement but he won't try it for fear of losing his place on the team. Allow some time for free practice so your shooters can experiment. Record practice shooting is necessary to keep shooters on their toes, but don't overdo it.

    4. <u>Develop esprit de corps on your team</u>. If you have more than one team, every man should be doing his best to shoot his way onto the first team. There should be a spirit of willing helpfulness on your team. Each man should be out to do what he can to improve the team's performance. Look at the successful pistol teams. Observe the team members. Listen to their conversations and you will detect a cooperative attitude. This is the sort of team spirit you should develop on your squad.

    5. <u>Train for a specific match</u>. It is a good idea to have the training period prior to a given match duplicate the conditions of that match. If you expect to complete a 900 aggregate before coming off the firing line, practice shooting a 900 aggregate without a break. A team's performance would be hurt if they enter a match in which a 2700 aggregate is fired in one day when they have never before fired more than a 900 aggregate in one day.

6. <u>Weapons and Equipment Checks</u>: The coach should periodically check the equipment of each of his shooters. This check serves a quadruple purpose.

    a. Detailed inspection of match pistols serves to assure the proper mechanical functioning of each weapon. Many times a shooter does not bring a weapon in for repair because it malfunctions very seldom. Every time a weapon malfunctions, it should be brought to the attention of the coach and the armorer.

    b. Weapons in constant use tend to lose accuracy over varying periods of use. The handicap of a weapon that shoots groups larger than the shooter's ability to hold will undermine even the performance of a champion. Thus, when a shooter's declining scores indicate a possible loss of accuracy, the weapon should be checked on a static testing device.

    c. The coach should check on the cleanliness of the weapon. Most shooters get pretty lazy about cleaning their weapons, particularly if they are fired every day. Unfortunately, many shooters think that cleaning a dirty, though functioning weapon will somehow cause it to start malfunctioning and to lose accuracy.

    d. The security of government issue weapons is paramount. Regular serial number checks of weapons issued to individual shooters, security storage inspection, travel security supervision insures proper accountability and security of weapons and equipment at all times.

7. <u>Periodic Written and Oral Tests</u>: The coach should conduct periodic written or oral tests in order to check how much the shooters have learned. This means how much of the fundamentals, of the NRA Match Rules, of general match procedure and how well they have mastered techniques. The results of these tests should become part of the shooter's record.

8. <u>Constantly Maintain a Current Evaluation</u> of each shooter regularly assigned to an Army, Police or NRA Club. An evaluation and an estimate of potential should be made concerning outstanding Army shooters in major command level competition. This estimate must include: an analysis of the rate of progress; individual morale and attitudes, on and off the range, the degree of team effort exercised; all of which should be current. A good coach may distinguish between who is shooting for the team and who is merely shooting to hang on for the free ride. (See the Line Coaches' Evaluation Form Chapter XI, USAMU Marksmanship Instructors' and Coaches' Guide, para F.)

9. <u>Refine Doctrine and Raise Performance Standards</u> by constant review of training guides, materials and methods. This research should affirm new ideas and techniques proven to be sound and reliable and weed out unsound doctrine.

10. <u>Improve the Team Potential</u> by conducting periodic, organized, group instruction.

11. <u>Improve the Individual Shooter's Potential</u> by personal and private interview and conducting individual coaching sessions.

12. <u>Supervise the Team Preparation</u> for match participation.

13. <u>Supervise the Coaching Technique</u> of individual line coaches.

14. <u>Assist in Preparation</u> of courses of marksmanship training.

15. <u>Assist in Rehearsal</u> of instructors.

16. <u>Participate</u> in all registered competition. A shooting coach is a person who appreciates the problems faced by the competitive shooter and is not inclined to be arbitrary in his judgment.

17. <u>Influence</u> the morale, attitudes, enthusiasm and stimulate the shooter's will to win by exhibiting individual consideration. Good coaching should promote a feeling of confidence and create an atmosphere of inevitable success. This influence will assure a favorable response if the guidance in personal habits and activities results in measurable improvement in performance. During the shooting season, placing certain limitations of a personal nature on his living habits will help the individual to build and maintain excellence. There must be no use of tobacco, alcohol, coffee or unprescribed use of certain drug preparations that may adversely affect shooting performance. Physical fitness is enhanced by avoiding both late hours and overindulgence in rich or unaccustomed foods. The shooter must be encouraged not to change his normal living routine. Pride and esprit de corps go hand in hand with an attitude that reflects the will to win.

18. <u>Correction and Recommendation</u>: During the match, the head coach should keep notes on any mistakes made by team members or himself. Then he must put into his training program the necessary subject matter to bring about a correction of the errors. He should make recommendations to the team on anything that will cause the group to perform better in future team matches. The coach is the director of the team. If he fails, the team fails; if he is a success, the team will be a success.

TRANSITION: A pistol line coach's work with a four man team presents additional problems and he uses different techniques to overcome them.

    D. <u>Fourth Student Performance Objective</u>: Given the mission of coaching a team on the firing line, LIST and APPLY the techniques of a line coach as outlined in Chapter XI USAMU Instructors and Coaches Guide.

QUESTION: What is the difference in coaching situations faced by the line and head coach?

ANSWER: 1. The biggest difference is that the line coach is on the firing line with his team during competition. He is the man on the spot and has decisions to make in the heat of battle, so to speak. The head coach does all he can in the training program to insure that the members of the team are ready for the team match. Once they reach the line, though, all responsibility passes to the line coach.

    2. The worst enemy of a team is nervousness. The coach must know how to help his shooters cope with the increased nervousness that comes with competition. There is no substitute for experience. The coach must know his team members so well that he knows how to handle each one in any set of circumstances. You may have four men on your team who react in four different ways to match pressure. With one you crack jokes, with another you're dead serious, with the third you listen as he talks, and with the fourth you do all the talking. This is something the coach must learn by trial and error.

    3. The attitude of the coach should be one of cooperation at all times. The coach is there for the shooter's benefit, not the other way around. Four shooters can win a team match without a coach, but no coach in the world can win a team match without shooters. The coach must stand ready, even eager, to do anything in his power to help his shooters.

    4. Another of the coach's duties is deciding who will shoot on which relay. Several factors must be weighted before a decision can be reached. Which pairs shoot well together? You may not want the joker type on the line with Sidney Serious. Which shoots better as anchor man? Which team member performs well regardless of the circumstances? Sometimes the decision boils down to a choice of the lesser of two evils. The coach must decide whether it is more important to have compatible pairs on the line or to have a natural anchor man shoot as lead off man.

5. The line coach should know when not to coach. Don't make unnecessary statements. If Tommy Tenring is shooting the X-ring out of the target in slow fire, don't talk to him about the proper way to shoot slow fire. You may influence him negatively. Avoid over-coaching. Don't disturb the thinking processes of a shooter who is performing well.

VI. **CONCLUSION:**

    A. **Retain Attention:** In a recent Interservice Pistol Match, a prominent Army shooter's first seven shots slow fire in the .45 Cal. Service Pistol team match were ten's. He started to shake when attempting the eighth shot. His coach had the shooter to pause and reminded him of what was required in applying the Pistol Fundamentals to shoot 10's. That shooters last three shots were also 10's. A one hundred point slow fire stage and the timed and rapid fire stages kept under control resulted in the Army Team winning the match. The coach shared the responsibility for victory.

    B. **Application:** Sometime in your shooting career you will be called upon to coach. These methods we have covered here are techniques used by successful coaches. Your team or club can benefit from what you have learned here.

    C. **Summary:** As a head coach you must look at the big picture. You must have a system to your technique of coaching: Prepare, plan, relax, deliver, analyze and correct Lay out a training program for your team that will meet the needs for improving the individual shooter. It must be flexible but not minus essential features. As a line coach your primary concern is getting the most out of each of your four shooters in a team match. You as coach, can help the champion to continue to win. Systematic coaching technique assures that nothing is overlooked that will aid your team in winning the match. Remember: Prepare, plan, relax, deliver the shot, analyze and correct.

    D. **Closing Statement:** If you are satisfied to be an average coach, remember that by being average, your team will place as close to the bottom as to the top.

ANNEX I To Technique of Coaching a Pistol Team

DEMONSTRATION OF PISTOL COACHING TECHNIQUE

The conduct of a demonstration will amplify the factors important to establishing effective pistol coaching technique.

    1. <u>The preparation stages</u> are important insurance that nothing will be overlooked that can help win the match.

        a. Preparation in the assembly area.

<u>NOTE</u>: POINT OUT TO THE CLASS AND COMMENT UPON PREPARATIONS BEING SUPERVISED BY THE LINE COACH. (As listed in the Shooter's Preparation Check Sheet)

<u>"FIRST RELAY - .45 CAL TEAM MATCH TO THE FIRING LINE"</u>

        b. Preparation (3 Min) on Firing Line (Shooters slowfire worksheet)

            (1) Physical)
)
) - (Shooters' worksheet)
)
            (2) Mental)

            (3) Check stance and position for natural hold.

            (4) Check grip for natural sight alignment.

            (5) Take a few deep breaths to increase oxygen level of the bloodstream.

    2. <u>Planning Shot</u>: (Review shot sequence-slow fire worksheet)

<u>FIRE COMMAND - "LOAD"</u>

    3. <u>Relaxation</u>

        a. Relax all muscular systems.

<u>FIRE COMMANDS TO BEGIN SLOW FIRE - "IS THE FIRING LINE READY?" ETC.</u>

    4. <u>Deliver Shots as Planned</u>. (Do not compromise)

    5. <u>Shot Analysis</u>. (Shooter's Remarks)

        a. First Shot: Fired too quickly (Breakdown in follow through - trigger pressure applied abruptly, shot call unsure, sight alignment excellent before trigger pressure was applied.)

        b. Second Shot: No shot call (No clear impression of sight alignment.)

        c. Third Shot: Called good after one unsuccessful try. Benched weapon (No error in sight alignment or hold)

        d. Fourth Shot: Holding too long (Shot broke as arm moved toward two o'clock; gross errors in hold and sight alignment corrected before applying trigger pressure)

    e. Fifth Shot: Shot called good (Describe sight alignment, shot is low ten. Arc of movement was in lower half of aiming area)

  6. <u>Positive Correction</u>: (Coaches Remarks)

    a. First Shot: Probably applied trigger pressure on basis of sight picture, not sight alignment. Unchanged rate of positive trigger pressure will insure surprise break of shot. Once trigger pressure is started, you are committed to continue at the same rate until shot breaks.

    b. Second Shot: Focus allowed to move to target momentarily and shot was fired during this short interval. A prior determination to pinpoint focus and concentrate on front sight until shot breaks will avoid this lapse.

    c. Third Shot: If error is so great as to cause let up in trigger pressure before shot breaks, bench weapon, analyze trouble, correct error, replan delivery of shot and try again. In planning for the next shot, remember the exact sequence in which the factors for controlling a good shot were applied.

    d. Fourth Shot: Holding too long will cause impatience and creates a tendency to speed up trigger pressure rate of application. Anticipation of the shot breaking will cause a somewhat less violent reflex in the form of a straight line arm movement rather than the muscular twitch that usually follows a more abrupt increase in trigger pressure. If error is so great as to cause let up in trigger pressure before shot breaks, bench weapon, analyze trouble, correct error, replan delivery of shot, relax and try again.

    e. Fifth Shot: Normal in all respects. Arc of movement, however small, is present in the ability to hold of all shooters. The law of averages will have some shots break near the edge of your ability to hold. Maintenance of good sight alignment while waiting for shot to break will prevent wild shots.

FIRE COMMAND - "CEASE FIRING"

FIRE COMMAND - "CLEAR WEAPONS"

<u>"SCORE AND REPLACE WITH A 25 YD TARGET"</u>

  7. <u>Overall Critique</u> of slow fire while targets are bring scored.

  8. <u>Physically and Mentally prepare for rapid fire:</u> (Shooters rapid fire worksheet)

    a. Physical (Set sights for 25 yds)

    (1) Check stance and position for natural hold, etc.

"YOU MAY HANDLE YOUR WEAPONS"

    (2) Check grip for natural sight alignment, etc.

    (3) Breath control. Take a few deep breaths, and hold breath at the natural respiratory pause.

    b. Mental (Coaches worksheet)

  9. <u>Plan 5 shot string of rapid fire:</u> (Review shot sequence - rapid fire work sheet)

FIRE COMMAND "LOAD 5 ROUNDS"

    10. <u>Relaxation and Breath Control</u>: (Deep breathing before shooting)

        a. Relax all muscular systems especially those not necessary for maintaining upright posture.

FIRE COMMANDS FOR RAPID FIRE - "IS THE FIRING LINE READY?" ETC.

    11. <u>Deliver 5 Shots as Planned</u>: (Do not interrupt rhythm to make corrections in sight alignment and hold. Grip and position check insures a recovery from recoil that will have few errors that need correction.)

    12. <u>Shot Analysis of 5 Shot String</u>: (Shooter's Remarks)

        a. First Shot: Late. (Looking at target on turn. Focus would not return quickly to clearly defined sight alignment. (Fired shot without focus clearly on front sight.)

        b. Second Shot: Fired too quickly. (Time anxiety caused speed up of trigger pressure; front sight dipped.)

        c. Third Shot: Fired normally. (Settled quickly into minimum arc of movement after recoil, picking up sight alignment and pressing positively on trigger. Shot is called good.)

        d. Fourth Shot: Same.

        e. Fifth Shot: Same.

    13. Positive Correction: (Coach's Remarks)

        a. First Shot: Concentrate point focus on front sight and be aware of proper alignment with rear sight notch. When the targets begin to face, shooter should start application of positive trigger pressure. Do not let your focus leave the front sight and move to the target.

        b. Second Shot: Straight through, positive, unchanging rate of trigger pressure will assure surprise break of shot, no reflex action of muscles to disturb sight alignment because shot is fired before any reaction to the coming recoil takes place.

        c. Third Shot: Fundamentals applied.

        d. Fourth Shot: Same.

        e. Fifth Shot: Same.

The lack of communication between coach and shooter means that many errors are not detected and therefore are not corrected.

O. PHYSICAL FITNESS I   AAPMI&CC 310
   (Physical Conditioning)   One (1) Hour
                                                     Mar 1975
                        LESSON OUTLINE

   I. LESSON OBJECTIVE: To enable pistol marksmanship students to conduct a daily physical conditioning program to condition themselves, mentally and physically, to better withstand the pressures of match shooting conditions.

   II. STUDENT PERFORMANCE OBJECTIVES: As a result of this instruction, the student must be able to accomplish the following student performance objective:

       A. As the coach of a pistol team, EXPLAIN the basis for a good physical condition as outlined in Chapter VIII, USAMU Pistol Marksmanship Guide.

       B. As a pistol team member, ENGAGE in sports and activities that are advantageous to good physical conditioning as outlined in Chapter VIII, USAMU Pistol Marksmanship Guide.

       C. As the coach of a pistol team, PLAN and EXECUTE a program of daily physical conditioning exercises using the pistol team daily dozen exercises as outlined in Chapter VIII, USAMU Pistol Marksmanship Guide.

       D. As a pistol team member, USE a daily program of static tension exercises as outlined in Chapter VII, USAMU Pistol Marksmanship Guide.

   III. ADVANCE ASSIGNMENT: Chapter VIII, USAMU Pistol Marksmanship Guide.

   IV. INTRODUCTION:

       A. *Gain Attention*: If inhabitants of other planets were capable of tuning in on our TV Networks and viewed the commercials for one day they would in all probability classify earth as a planet of sick people. The commercials do us an injustice by exposing us as a race of people with headaches, backaches, earaches, sneezes, wheezes and a variety of other ailments. However, most of us are guilty of neglecting to keep our physical condition up to par. This has been recognized by US Presidential Health Councils seeking to bring to the attention of all Americans the condition of the nation's Health.

       B. *Orient Students*:

           1. *Lesson Tie-In*: You are entering into a challenging endeavor, Pistol Shooting. You will find that good physical condition is an important factor in good shooting performance.

           2. *Motivation*: "Is exercise good for you?" No one would reply negatively to this question and yet so many of us badly in need of such a program keep putting it off because of one excuse or another. For those who become exhausted easily doing exercises such as running, there are strength building exercises which can be accomplished in less than thirty minutes daily.

           3. *Scope*: During this period we will discuss the importance of good physical conditioning to the competitive pistol shooter. We will name various sports activities that are beneficial to shooters. We will show how to perform the pistol shooters daily dozen and point out the value of including static tension exercises in a daily program.

   V. BODY:

A. **First Student Performance Objective**: As the coach of a pistol team, EXPLAIN the basis for a good physical condition as outlined in Chapter VIII, USAMU Pistol Marksmanship Guide.

QUESTION: What degree of physical conditioning should a pistol shooter attain?

ANSWER: Physical training should be progressive, either in repetitions performed or in resistance used. Conditioning should remain short of that sought by athletes. Violent or strenuous athletics which could result in injuries should be avoided. The competitive shooter should possess the following basic physical characteristics:

1. An adequately developed muscular system (especially the muscles of the abdomen, arms and legs.)

2. The ability to relax and to keep from utilizing those muscles which are not required to hold the body in the ready position or to apply pressure on the trigger.

3. Strong breathing muscles so that breathing deeply is an easy function to permit sustaining a supply of oxygen.

4. Quick reactions.

5. A well developed sense of equilibrium.

6. Precision and coordination of bodily actions. Physical conditioning must consist of coordination exercises as well as those of a general nature directed toward strengthening the muscles, toward proper breathing, and toward developing body flexibility and precision of movement.

TRANSITION: Now that we understand the basis for a good physical condition, let's see what we can do to acquire and maintain our bodies in good physical condition.

B. **Second Student Performance Objective**: As a pistol team member, ENGAGE in sports and activities that are advantageous to good physical conditioning as outlined in Chapter VIII, USAMU Pistol Marksmanship Guide.

QUESTION: Is any activity besides formal exercise beneficial?

ANSWER: There are many types of exercises and activities that a shooter can participate in to his advantage.

1. Walking is a very good exercise. Walk briskly if you expect any positive benefit.

2. Golfing offers many benefits for the shooters' conditioning. As with any other physical activity, it must be done regularly if it is to aid shooting performance.

3. Bowling is a good conditioner if done regularly.

4. Swimming exercises almost all the muscles you normally use. However, it is not recommended during match shooting periods but as an off-season training technique.

5. Exercises that strengthen and build the wrist and arm muscles are recommended. Caution should be exercised in the case of weight lifting. It is not necessary to be capable of pressing 100 pounds in order to hold a 39 oz. pistol steady at arms length.

6. Develop the grip by using a sponge rubber ball about 3" in diameter, cut in two halves. Squeeze the ball with the shooting hand whenever you wish or where-ever you may be.

7. The stronger the muscle structure is developed, the surer movement can be coordinated and positions held. Besides general conditioning practices, durable muscular tension exercises of the body, trunk, shoulder and arms make the most sense. Coordination exercises and grip exercises are in order.

Physical training should take place daily for at least 15-30 minutes.

TRANSITION: If you have never followed a program of daily physical conditioning and are interested in starting one that will benefit your shooting performance, we have such a program for you.

C. Third Student Performance Objective: As the coach of a pistol team, PLAN and EXECUTE a program of daily physical conditioning exercises using the pistol team daily dozen exercises as outlined in Chapter VIII, USAMU Pistol Marksmanship Guide.

QUESTION: Is it possible to condition the muscles used only in shooting?

ANSWER: The pistol team daily dozen was especially developed by a former member of the US Army Pistol Team, to condition those muscles used in pistol shooting. To the beginner who does not have a daily exercise program, we recommend starting out slowly by doing only four repetitions, increasing this one each day until twelve repetitions are reached. Demonstrators will perform several of these exercises as the parts of the body that will benefit from each are pointed out.

1. No. 1 - warm up - 4 count exercise with starting positions of hands over-head-feet 12" apart - at count of ONE, bend the waist and knees, reaching between the legs as far back as possible, at the count of TWO, recover. Three and four are a repeat of One and Two. This exercise is a good developer of the legs, particularly the back leg muscles.

2. No II - Body twister - 4 count exercise - starting position - bent forward at the waist, arms extended parallel with ground. At count of ONE, swing right arm so as to touch left toe, keeping shoulders and arms rigid so that twisting motion is from the waist. Another good leg developer plus the abdominal and lower back muscles benefit.

3. No. III - Push-up-4 count exercise done from the leaning rest position - at count of one, bend arms out at elbows and lower the body until the chest touches the floor or ground, count of two, recover, three and four are a repeat. Keep body rigid and straight. An excellent arm and shoulder muscle developer.

4. No. IV - Back bender - 4 count exercise done from the position of hands placed in small of back, feet spread 12" apart. Count of ONE, bend at waist, keeping legs stiff and touch toes with fingertips. Count of TWO, recover, count of THREE, bend backward at waist and count of FOUR, recover. This exercise develops the legs, back and abdomen.

5. No. IX - Side Bender - 4 count exercise done with feet spread and arms extended overhead, palms together. At count of ONE; bend at waist to the right, TWO; recover, THREE: to the left and FOUR: recover. The muscles along the sides are used in this exercise.

TRANSITION: For those, who because of their general physical makeup or who have little time for improving their bodies are unable to enter into a program such as the daily dozen, we have a program which can be done almost anywhere and one that requires little time. It is a sure means of building, conditioning and strengthening muscles.

D. <u>Fourth Student Performance Objective</u>: As a pistol team member, USE a daily program of static tension exercises as outlined in Chapter VIII, USAMU Pistol Marksmanship Guide.

   1. Static tension exercises themselves cannot be regarded as an alternative to other type exercises but rather an addition. The fact that they hasten muscle conditioning and provide added strength will certainly be a factor in reaching good physical condition. They can be done when time or conditions do not permit engaging in daily dozen type exercises. The advantages are that: (1) they can be done in a short period of time, (2) they can be done almost anywhere and at anytime, (3) they do not leave you exhausted, and most important, (4) they increase muscle size, strength and endurance.

   2. The main thing is <u>EFFORT</u>. Each exercise is performed to a count of one thousand - two thousand, and so on to six thousand with maximum pressure or stretching being applied to the full count. Part of the theory is that a muscle builds more rapidly under tension applied vigorously for a short period of time than when put to use over a long period.

   3. There are many types of static tension exercises, some that require simple exercises and others nothing more than effort. The demonstrators will perform several to give you an idea of how to do them.

      a. No. I - Hands held palms together, fingertips in line with chin, elbows raised in line with shoulders. Press hands together with as much effort as possible. A good builder and strengthener of arms, shoulders and chest muscles.

      b. No. II - Both hands at waist level, gripping one another, now with as much effort as possible squeeze hands together. This will develop grip and forearms.

      c. No. III - Arms hanging loosely, slightly bent in front of body palms upward. Suddenly with as much effort as possible contract the bicep and clench fists tightly. Good for arms and hands.

VI. CONCLUSION:

   A. <u>Retain Attention</u>: Good physical condition is something we all want to have but seldom do we give serious thought to acquiring and maintaining ourselves in a physically sound condition. It certainly is necessary to maintenance of good shooting performance but more important it is necessary to good health. Observe those about you who are the top shooters and you will see a person in good physical condition.

   B. <u>Summary</u>:

   1. Having a sound knowledge of the basis for a good physical condition will enable you to select activities which will improve your general health.

   2. Select those sports and activities that are within your capabilities and limitations, but make a habit of engaging in them regularly. Out-of-doors activities are preferable and start into them slowly until your body makes the adjustment.

   3. Be serious about planning a program of daily exercise. Provide a certain period each day to conduct your program. A few repetitions at the beginning of your training is adequate. Increase the number and variety with time.

   4. Static tension exercises can rebuild your muscle strength. Keep in mind that this is just an addition to exercises where muscle tone is derived from exercises requiring motion.

C. <u>Application</u>: If you already have a program which you follow, continue. If not, you should wait until after this match or season. Jumping into a vigorous program in the middle of a shooting season will have certain disadvantages to your shooting performance. When you start, go slowly at first.

D. <u>Closing Statement</u>: Exercise, together with proper diet and moderation of habits, all play an important part in the shooter's success. They are necessary to maintain us in good physical and mental health. An understanding of what is good and what is bad for us should be our approach to improved health. We are never too young nor too old to start understanding more about how our physical condition affects our shooting.

P. PHYSICAL FITNESS II   AAPMI&CC 311
    (Diet)   20 Minutes
                                                              Mar 1975
                              LESSON OUTLINE

   I. LESSON OBJECTIVE: To enable the Pistol Marksmanship Student to know the essential nutrients in a proper diet and identify examples of foods containing each type of nutrient.

  II. STUDENT PERFORMANCE OBJECTIVES: As a result of this instruction students must be able to accomplish the following student performance objectives:

       A. Given the essential nutrients required in an adequate diet, NAME each of the types of nutrients as shown in Chapter IX, USAMU Pistol Marksmanship Guide.

       B. Given the partial list of examples of various foods containing the essential nutrients, IDENTIFY three foods containing of each type of nutrient as indicated in Chapter IX, USAMU Pistol Marksmanship Guide.

 III. ADVANCE ASSIGNMENT: None

  IV. INTRODUCTION:

       A. Gain Attention: The army trains men to be winners. Winners in combat and winners in any competitive activity. If the Pistol Competitor falters, it must not be because he has exhausted his energy.

       B. Orient Students:

           1. Lesson Tie-In: Proper nutrition combined with proper exercise will assure strength and endurance to perform your job well. Your job is to shoot championship scores in every match, not just the first match of the day.

           2. Motivation: The expert marksman must feel well and be energetic to shoot well. No stone can be left unturned in today's level of competition that will provide the all important "edge" that may be the narrow margin of victory.

           3. Scope: Good nutrition is based on a diet that includes all the essential nutrients. These nutrients are found in a wide range of foods.

   V. BODY:

       A. First Student Performance Objective: Given the essential nutrients required in an adequate diet, NAME each of the types of nutrients as shown in Chapter IX, USAMU Pistol Marksmanship Guide.

           1. There are three essential nutrients.

NOTE: SHOW CHART #1 THE ESSENTIAL NUTRIENTS. (Reveal one item at a time)

QUESTION: Name one of the essential nutrients.

ANSWER: The essential nutrients are Protein, Fat, Carbohydrates, and protective ingredients or Vitamins.

TRANSITION: The house builder needs building materials to construct a house.

QUESTION: What basic material is used in building the body?

ANSWER: 2. Proteins: The bodys' building blocks. In the form of Amino Acids, Proteins build and replace the body's tissue of muscle, nerves, tendons and all of the vital organs, etc. For instance, the lining of the gastro-intestinal tract is renewed every three days.

TRANSITION: The protein built body must have fuel to consume as it moves about and also take care of the energy demands of the vital body functions.

QUESTION: How does the body maintain a reserve of energy?

ANSWER: By converting fats to glucose and storing it in the liver.

3. Fats or fatty acids: The long range source of energy. The fat in the diet provides the "full gas tank" reserve for endurance. Part of the fatty acids are converted to easily digested sugar forms that are stored in the liver. The fatty acids that are not held in reserve in the liver provide for part of the present energy needs of the body or they are deposited in and about the muscular system. One of my fat friends once remarked that if he was ever taken prisoner in war and had to walk to Moscow, he had more fuel for the trip than the skinny guys. The complete absence of fat in the diet means slow starvation even if the other nutrients are ample.

TRANSITION: There is another source of energy but a smaller portion of the amount eaten goes into the ready reserve.

QUESTION: What is another source of energy for the body?

ANSWER: 4. Carbohydrates: The "high octane" energy source. Compare fats and carbohydrates to the relative combustion qualities of oil and gasoline respectively. Both will burn but gasoline is ignited easier and burns quicker. Starches and basic sugars are converted into a more readily useable glucose at a relatively faster rate than fats. In this form, they are absorbed by the blood system and supplied to the muscular tissue to furnish needed energy.

QUESTION: Name a starch food that is quickly absorbed into the muscular system and readily converted into energy?

ANSWER: Alcohol. Alcohol can pass through the entire gastro-intestinal tract without being altered in any way and yet be absorbed readily. There is no question about the energetic effect it has on some people.

TRANSITION: Alcohol speeds up the digestive process, is absorbed and used quickly. Vitamins aid the breakdown of nutrients into useable forms.

5. Protective ingredients or vitamins: These substances are not considered to be a food but they are essential to proper nutrition. Vitamins are a means of triggering the chemical reactions that convert the essential nutrients into useable forms in the digestive process. For example, the automobile carburetor converts gasoline from a liquid into a more useable form of a vapor of gasoline molecules mixed with a large proportion of air.

TRANSITION: The pistol shooter can be assured of receiving the important elements of nutrition in his diet if certain representative foods of each type are known to him.

B. Second Student Performance Objective: Given the partial list of examples of various foods containing the essential nutrients, IDENTIFY three foods containing each type of nutrient as indicated in Chapter IX, USAMU Pistol Marksmanship Guide.

1. There are three foods that are the main source of protein in the diet

QUESTION: Name one food that is a rich source of protein:

ANSWER: 2.   Proteins are mainly supplied by eggs, meat and milk. A shooter of 140 pounds should consume 100-120 grams of protein in his daily diet. Approximately one half should come from the animal source foods; eggs, meat and milk. The remaining amount can be obtained from the usual sources of bread, and vegetables.

TRANSITION: The same basic foods provide a source for another essential nutrient.

QUESTION: Name the two main sources of fats.

ANSWER: 3. Fats or fatty acids are obtained from animal fats and vegetable oils. The average shooter consumes too much fat. While in training, a 140 pound man should consume 70 to 90 grams of fat daily. Half of this amount should come from animal sources such as eggs, meat and milk. The remainder should come from vegetable oil sources such as peanuts, corn, soya and olives. Moderation in the use of fats is especially beneficial to the digestive process.

TRANSITION: Other types of energy food provide essential nutrition.

QUESTION: Name the two main sources of carbohydrates.

ANSWER: 4. Carbohydrates in the form of starches and sugar are contained in nearly all foods. If is important to give preference to such starch and sugar sources such as dark bread, fresh fruit, unpolished rice, milk, oatmeal, fresh vegetables and potatoes with the jackets on because of the high content of protective ingredients or vitamins. Stay away from that enormous birthday cake with its gleaming white icing!

NOTE: DISPLAY A CAN OF LIQUID DIET FOOD WITH AN EXPRESSION OF DISGUST. "THIS IS THE FATE OF THOSE WHO WOULD EAT CAKE."

TRANSITION: But take a look at the fine print on this can. This meal of 225 calories contains one fourth (¼) of the minimum adult daily requirement for vitamins and minerals.

QUESTION: What <u>foods</u> contain the vitamins or protective ingredients that the human body requires?

ANSWER: 5. Protective ingredients or vitamins are found in most foods. In addition the human body manufactures some vitamins but usually in amounts too small to meet its needs. Some vitamins cannot be stored in the body and must be replenished daily. Vitamin pills which contain a minimum adult daily requirement of most identifiable vitamins are available at the nearest drug store.

QUESTION: How does a shooter know which vitamins are needed to supplement his daily diet?

ANSWER: Diet supplements such as vitamins should be taken only on the advice of a doctor. Your diet may be furnishing all the vitamins you need. The best source of vitamins are foods in which they are naturally present. Of special importance to the shooter are the following vitamins and sources:

   a. Vitamin A is found in carrots, spinach, sweet potatoes, milk, liver, egg yolk and green and yellow vegetables. This vitamin builds resistance to infection and helps the eyes to function normally in light of varying intensity. This feature is an aid in night vision. Vitamin A prevents and cures pellagra, a disease of the eyes and skin.

   b. Vitamin B1 is found in yeast, most meats, especially beef and pork, whole grain cereals, beans, peas, nuts and green vegetables. A deficiency of this vitamin causes great and persistent fatigue, aching leg muscles and bones and illness of the nervous system. Vitamin B1 prevents and cures beriberi, primarily a disease of the nervous system. Chronic alcoholics sometimes develop symptoms of the disease because alcohol diminishes the appetite and they fail to receive proper nourishment.

        c. Vitamin C: The body does not store this vitamin and it must be replenished daily. It is found in citrus fruits, tomatoes, raw cabbage, strawberries and cantaloupe. Vitamin C promotes a healthy circulatory system, which is important to body energy, develops good sound bones and teeth. This vitamin prevents and cures scurvy a disease known by general listlessness and fatigue, sore, inflamed gums and various other dental disorders.

        d. There are numerous other vitamins that are required in your daily diet. Consult your doctor or research your encyclopedia for a complete list.

VI. CONCLUSION:

    A. <u>Retain Attention</u>: Are you known as the gutless wonder of the National Pistol Championships? Do you start out strong and end up weak and shaky? Nerves like a chicken?

    B. <u>Summary</u>: The lack of forceful energy to carry on under a prolonged condition of stress can be corrected to a degree by proper, balanced nutrition. Various types of food provide this needed nutrition.

        1. Proteins: The body's building blocks. Every muscle, bone and vital organ is composed of proteins.

        2. Fats: The long term energy source. The "full gas tank" reserve is needed for endurance.

        3. Carbohydrates: The "high octane" energy source. Starches and sugar provide for the more immediate energy needs of the body.

        4. Protective ingredients such as vitamins furnish resistance to illness and help promote certain bodily processes involved in digestion and the production of muscular energy and stamina.

    C. <u>Application</u>: A strong, well nourished body will allow sustained physical and mental effort under the stress of competitive shooting.

    D. <u>Closing Statement</u>: The gutless wonder of yesteryear can become the champion this year if he learns to eat the foods that will supply him with the stamina to keep up the fight regardless of the odds against his chances of winning the match.

Q. PHYSICAL FITNESS III AAPMI&CC 311
(The Effects of Alcohol, Coffee, Tobacco and Drugs) 30 Min
  Mar 1975

LESSON OUTLINE

I. LESSON OBJECTIVE: To enable the pistol marksmanship student to be aware of the effect alcohol, coffee, tobacco and drugs has on shooting control.

II. STUDENT PERFORMANCE OBJECTIVES: As a result of this instruction, the student must be able to accomplish the following student performance objectives:

    A. Given the description of the effect of alcohol on the human body, EXPLAIN the detrimental effects of alcohol on shooting control as shown in Chapter X, USAMU Pistol Marksmanship Guide.

    B. Given the information on effects of coffee and tea on the human body, DISCLOSE the adverse effects of coffee and tea on control of shooting as described in Chapter X, USAMU Pistol Marksmanship Guide.

    C. Given the detrimental effects of tobacco on the human body, EXPLAIN the effects of tobacco use on pistol shooting and offer a method of abstaining from the use of tobacco as shown in Chapter X, USAMU Pistol Marksmanship Guide.

    D. Given the detrimental characteristics of various drugs, EXPLAIN how the use of drugs in prescription medicines and certain home remedies are detrimental to shooting as listed in Chapter X, USAMU Pistol Marksmanship Guide.

III. ADVANCE ASSIGNMENT: Chapter X, USAMU Pistol Marksmanship Guide.

IV. INTRODUCTION:

    A. **Gain Attention**: The hangover, the antidote, the nerve calmer, the picker-upper, these are other names for alcohol, coffee, tobacco and drugs respectively. The benefits of good physical condition and proper diet go down the drain when one or all of these habits are indulged in.

    B. **Orient Students**:

        1. **Lesson Tie-In**: You've heard the expression "built-in error" several times during your shooting experience. A few of these built-in errors are caused by intemperate habits.

        2. **Motivation**: The use of stimulants and depressants have varying effects on all of us. As shooters and coaches you should be aware of these effects as they pertain to shooting.

        3. **Scope**: During this period we will duscuss the harmful effects of various stimulants and depressants on the body and how their use is detrimental to pistol shooting.

V. BODY:

NOTE: SHOW CHART #4 SCOPE AND SUMMARY OF THE EFFECTS OF ALCOHOL, ETC.

    A. **First Student Performance Objective**: Given the description of the effect of alcohol on the human body, EXPLAIN the detrimental effects of alcohol on shooting con trol as shown in Chapter X, USAMU Pistol Marksmanship Guide.

NOTE: SHOW CHART #5 EFFECTS OF ALCOHOL

1. Acts as depressant.

2. Dulls the senses.

3. Lessens desire to win.

4. Destroys coordination.

5. Lessens ability to concentrate.

6. Promotes carelessness.

7. Contributes to loss of judgment.

8. Dehydrates the body.

9. Involves the development of permanent damage to health.

10. Established by experience as a habit forming agent.

TRANSITION: Scientific studies have established proof of the damaging effect of alcohol on the human body. The adverse effect of coffee and tea on shooting control needs no scientific proof.

  B. <u>Second Student Performance Objective</u>: Given the information on effects of coffee and tea on the human body, DISCLOSE the adverse effects of coffee and tea on control of shooting as described in Chapter X, USAMU Pistol Marksmanship Guide.

NOTE: SHOW CHART #6 (COFFEE CUP)

QUESTION: Does coffee increase heartbeat enough to be noticeable while shooting?

1. Acts as a stimulant.

2. Stimulates increased heartbeat.

3. Adversely affects nervous system as regards control of muscular reactions.

4. Similar effects of tea.

  a. Caffeine.

  b. Tannic acid.

5. Cola drinks contain caffeine which has a stimulating effect.

TRANSITION: Alcohol and coffee are relatively temporary in the duration of their adverse effect on shooting but the user of tobacco is under the effect of his hard-to-break habit practically 24 hours a day for an indefinite period.

  C. <u>Third Student Performance Objective</u>: Given the detrimental effects of tobacco on the human body, EXPLAIN the effects of tobacco use on pistol shooting and offer a method of abstaining from the use of tobacco as shown in Chapter X, USAMU Pistol Marksmanship Guide.

NOTE: SHOW CHART #7 (CIGAR SMOKER)

QUESTION: How long after the last cigarette does the shooter continue to feel the effects?

ANSWER: Approximately three hours.

    1. Nervous system disturbances are continuous and damaging to shooting results when the shooter smokes cigarettes a minimum of one every three hours.

    2. Nicotine is a deadly poison.

    3. Lung damage is inevitable after protracted use of tobacco.

    4. Circulatory system effect is great enough to cause the heart to overwork.

    5. Increased heartbeat is noticeable during shooting.

    6. Loss of appetite and weight.

    7. Expense.

    8. Danger of other serious illness or death.

    9. Habit forming properties established beyond doubt.

    10. You can quit smoking.

TRANSITION: Many shooters that are aware of their ill effects and as a result avoid tobacco, alcohol and coffee innocently fall prey to the devastating effect of certain drugs found in commonly used medicines and pills.

    D. *Fourth Student Performance Objective*: Given the detrimental characteristics of various drugs, explain how the use of drugs in prescription medicines and certain home remedies are detrimental to shooting as listed in Chapter X, USAMU Pistol Marksmanship Guide.

NOTE: SHOW CHART #8 (MEDICINE CHEST GUN BOX)

QUESTION: If a shooter has a bad cold, should he accept treatment before shooting?

ANSWER: No, if the shooter is too ill to shoot, he must be excused from firing. If he takes drugs to eliminate symptoms of his sickness, he may not shoot well.

    1. Depressants slow down the bodily processes in the following manner:

        a. Slow reflexes are detrimental to fine coordination.

        b. Alertness is affected by the slowed down mental state.

        c. Lessens the desire to win.

        d. Loss of ability to concentrate intensely for the accustomed length of time.

        e. Promotes carelessness in adhering strictly to techniques of control.

        f. Usually found in sleeping pills and other barbituate preparations.

2. Stimulants abnormally accelerate certain bodily processes and give a false sense of well being.

    a. A feeling of nervousness and a slight trembling are the usual effects.

    b. Increased anxiety is experienced concerning results of the match.

    c. Involuntary movement of the extremities (trembling movement of hands).

    d. Increases heartbeat to a noticeable level.

    e. Some are as habit forming as are alcohol and tobacco.

    f. Usually found in keep awake pills, benzedrine inhalers, appetite curbers, etc.

3. Drugs in daily use.

    a. Sedatives and depressants.

    b. Analgesics (pain relief).

    c. Antihistamines (relief of colds and associated illness).

    d. APC tablets (Relief of colds and associated illness).

    e. Decongestants (relief of nasal and sinus congestion, etc.)

VI. CONCLUSION.

    A. <u>Retain Attention</u>: How many shooters in this class have stopped smoking since the 1st of January?

NOTE: SHOW CHART #4a SUMMARY OF EFFECTS OF ALCOHOL, COFFEE, TOBACCO AND DRUGS.

    B. <u>Summarize</u>:

        1. Alcohol.

        2. Coffee and tea.

        3. Tobacco.

        4. The use of drugs.

    C. <u>Application</u>:

Try giving up alcohol and tobacco, stop prescribing remedies for your aches and pains. Desist completely or try to reduce consumption of coffee drastically for several months. Decaffeinated coffee will help. You will feel better and shoot better.

    D. <u>Closing Statement</u>: Remember that to be a fine pistol shooter you must train diligently. Indulge in nothing that will cancel out the beneficial results of adhering to your training program. The "EDGE" you have thereby obtained will give you a solid chance at attaining top honors in competition.

R. COMPETITIVE REGULATIONS I AAPMI&CC 701
   (PISTOL MATCH RULES)   One (1) Hour
                               Mar 1975

LESSON OUTLINE

I. LESSON OBJECTIVE: To enable pistol instructor-shooters and coaches to use their knowledge of NRA pistol match rules to prevent being penalized or disqualified while participating in competition.

II. STUDENT PERFORMANCE OBJECTIVES: As a result of this instruction, the pistol coach/instructor student should be able to accomplish the following student performance objectives:

   A. Given the task of training instructor/coaches for participation in pistol matches, EXPLAIN the many types of competition for pistol as defined in Section 1, NRA Pistol Rule Book.

   B. Given the task of training instructor/coaches for participation in pistol matches, EXPLAIN NRA pistol competition regulations as defined in Section 9, NRA Pistol Rule Book.

   C. Given the task of training instructor/coaches for participation in pistol matches, EXPLAIN how to make challenges and protests when necessary as defined in Section 16, NRA Pistol Rule Book.

   D. Given the task of training instructor/coaches for participation in pistol matches, USE the proper match weapons, equipment and ammunition in a pistol match as defined in Section 3, NRA Pistol Rule Book.

   E. Given the task of training instructor/coaches for participation in pistol matches, EXPLAIN the scoring of pistol match targets as defined in Section 14, NRA Pistol Rule Book.

   F. Given the task of training instructor/coaches for participation in pistol matches, EXPLAIN the competitor's duties and responsibilities as defined in Section 18, NRA Pistol Rule Book.

   G. Given the task of training instructor/coaches for participation in pistol matches, EXPLAIN the team officers duties and position during a pistol team match as defined in Section 12 NRA Pistol Rule Book.

   H. Given the task of training instructor/coaches for participation in pistol matches, EXPLAIN the eligibility to competitors in a pistol match as defined in Section 2, NRA Pistol Rule Book.

III. ADVANCE ASSIGNMENT: NRA Pistol Match Rule Book - current edition.

IV. INTRODUCTION:

   A. <u>Gain Attention</u>: The NRA referee's decision is final except in the National Matches.

   B. <u>Orient Students</u>:

      1. <u>Lesson Tie-In</u>: In order to compete in a Pistol Match, It is essential that all the competitors have a complete knowledge of the rules of pistol competition.

2. **Motivation**: Knowledge of NRA pistol match rules will help prevent any injustice being inflicted on you or your team; also it will help prevent mistakes for which the team could be disqualified.

3. **Scope**: During this period we will discuss the different types of competition, the competition regulations with which the shooter frequently comes into contact, challenges and protests, types of weapons, how to score, competitors duties, team officers duties and eligibility of competitors.

V. BODY:

A. <u>First Student Performance Objective</u>: Given the task of training instructor/coaches for participation in pistol matches, EXPLAIN the many types of competition for pistol as defined in Section 1, NRA Pistol Rule Book. (Section 1 - Types of Competition)

QUESTION: I fire in an NRA registered league. Can we enter a league team from this league in a registered tournament using shooters from several clubs?

ANSWER: The type of teams which are eligible to enter a tournament is determined by the provisions of the tournament program. While a tournament sponsor may take entries from a league team, it is seldom done. To be eligible the tournament program must indicate that league teams may enter. (1.75)

QUESTION: Can I make entry in an aggregate match after the firing has started in a tournament but before I have fired in any match in the aggregate concerned?

ANSWER: Yes. Your entry can be accepted for an aggregate match if you have not started firing in any of the matches making up that aggregate. (1.13)

QUESTION: What is the meaning of "stage of a match" in connection with competitive courses of fire?

ANSWER: "In a match fired at more than one range or class of firing each range or class of firing is referred to as a "stage of the match." (1.16)

    1. (1.2) Open Match.

    2. (1.2.1) National Trophy Matches (See requirements for earning a Distinguished Pistol Shot Badge - AR 672-5-1 and AR 350-6)

    3. (1.2.2) National Matches.

    4. (1.3) Restricted Match.

    5. (1.4) Classified Match.

    6. (1.5) Invitational Match.

    7. (1.6) NRA Competition.

    8. (1.7) League Competition.

    9. (1.8) Squadded Individual Match.

    10. (1.11) Squadded Team Match.

    11. (1.13) Aggregate Matches.

    12. (1.13.1) Tournament.

B. _Second Student Performance Objective_: Given the task of training instructor/coaches for participation in pistol matches, EXPLAIN NRA pistol competition regulations as defined in Section 9, NRA Pistol Rule Book. (Section 9 - Competition Regulations)

QUESTION: Is a pistol considered "loaded" when the loaded clip is in the gun but the slide is locked open, or a revolver when the cylinder is loaded but swung out of the frame.

ANSWER: Yes. To be unloaded there can be no ammunition in the gun. A clip may be loaded but must not be placed in the gun until after the Range Officer's command "With five rounds load." (9.3)

QUESTION: I have a pistol with interchangeable barrels. When firing the National Match Course can I use the long barrel for slow fire and change to the short barrel for the timed and rapid fire stages?

ANSWER: Yes. Rule 9.6 covers the regulation for changing pistols and provides in part, "The exchange of barrels, portable weights, etc., shall not be restricted." Therefore, the changing of pistol barrels would not be considered changing the gun. (9.6)

QUESTION: Can a competitor use one gun for slow fire and another for timed and rapid fire in multiple stage pistol competition?

ANSWER: No. Pistol Rule 9.6 provides in part, "No competitor will change his pistol during the firing of any match (except aggregate matches), unless it has become disabled and has been so designated by the Chief Range Officer." Therefore, he cannot change guns during the firing of a single match. (9.6)

1. (9.1) Pistols unloaded.
2. (9.2) Actions open.
3. (9.3) Pistols loaded.
4. (9.4) Cease firing.
5. (9.5) Not ready.
6. (9.6) Changing pistols.
7. (9.9) Defective cartridge.
8. (9.10) Disabled pistol.
9. (9.11) Malfunction.
10. (9.14) Weighing triggers.
11. (9.16) Competitors position.
12. (9.17) Individual coaching.
13. (9.22) Interruption of fire.
14. (9.23) Failure to function in slow fire.
15. (9.24) Failure to function in timed and rapid fire.
16. (9.25) Interference with target.

17. (9.29) Score and classification falsification.

18. (9.32) Disorderly conduct.

19. (9.33) Refusal to obey.

20. (9.34) Evasion of rules.

21. (9.35) Disqualification.

22. (9.36) Suspension.

C. <u>Third Student Performance Objective</u>: Given the task of training instructor/coaches for participation in pistol matches, EXPLAIN how to make challenges and protests when necessary as defined in Section 16, NRA Pistol Rule Book. (Section 16 - Challenges and Protests)

<u>QUESTION</u>: What incidents may be challenged by a competitor?

<u>ANSWER</u>: When a competitor feels that a shot fired by himself or by another competitor has been improperly evaluated or scored, he may challenge the scoring. Such challenge must be made immediately upon announcement of the score. No challenge will be accepted after a target has been pasted. Any other injustice must be handled by a protest, see Rule 16.2 (16.1)

<u>QUESTION</u>: If I am not satisfied with the scoring of my target, can I protest?

<u>ANSWER</u>: No. Rule 16.2 provides, "A competitor may formally protest any injustice which he feels has been done him <u>except the evaluation of a target</u>, which may be challenged as outlined in Rule 16.1." Under Rule 16.1, it provides that in NRA competitions to which the NRA assigns a Referee or Supervisor, the decision of the Referee or Supervisor will be final on scoring except in the National Championships. (16.2)

<u>QUESTION</u>: What is the formal procedure for making a protest?

<u>ANSWER</u>: It must be initiated immediately upon the occurence of the protested incident. It will first state the complaint verbally to the Chief Range Officer or Chief Statistical Officer as the case may be. If not satisfied with his decision, file with the Official Referee or Supervisor a formal protest stating all the facts in the case. Such protest must be filed within 12 hours of the occurence of the protested incident. If not satisfied with the decision of the Referee or Supervisor, he will file with the NRA Executive Committee a written appeal stating all facts. This appeal must be filed with the Referee or Supervisor within 12 hours after his decision has been made known to the competitor. The Referee of Supervisor will add to the appeal a complete statement of facts and forward it to the NRA office within 12 hours of the time it is filed with him. The information forwarded to the NRA should be complete as it will be all the Protest Committee will have to make their ruling. (16.3)

1. (16.1) Challenges.

2. (16.2) Protests.

3. (16.3) How to protest.

4. (16.4) Team Matches.

D. <u>Fourth Student Performance Objective</u>: Given the task of training instructor/coaches for participation in pistol matches, USE the proper match weapons, equipment and ammunition in a pistol match as defined in Section 3, NRA Pistol Rule Book. (Section 3 - Equipment and Ammunition)

QUESTION: The pistol rules provide limitations to the distance between sights. How are these distances measured?

ANSWER: The distance between sights is measured from the rear face of the front sight to the rear face of the rear sight. (3.1)

QUESTION: Is it legal to wrap the grip of my .45 semi-automatic pistol with tape when the match rules require arms as defined in Rule 3.1?

ANSWER: No. Pistol Rule 3.1 covering the Service pistol provides that all standard safety features of the weapon must operate properly. Usually taping the grip will prevent the grip safety from functioning. Also, the Service Pistol must be used "as issued (3.1)

QUESTION: Can the .44 caliber revolver be used when firing a .45 caliber pistol match?

ANSWER: No. Pistol Rule 3.2 provides for any .45 caliber semi-automatic pistol, and Rule 3.4 provides for .45 caliber revolvers. A .44 caliber is not a .45 caliber. Therefore, it cannot be used in a match restricted to the .45 caliber. (3.2)

1. (3.1) Service Pistol
2. (3.2) Pistol cal .45 (Wad Cutter)
3. (3.3) Service Revolver cal .45
4. (3.4) Any Center Fire Pistol or Revolver
5. (3.5) .22 Caliber Pistol or Revolver
6. (3.8) Spotting Scopes
7. (3.9) Shooting Kits
8. (3.9.1) Deflecting screens
9. (3.17) Ammunition

E. **Fifth Student Performance Objective**: Given the task of training instructor/coaches for participation in pistol matches, EXPLAIN the scoring of pistol match targets as defined in Section 14, NRA Pistol Rule Book. (Section 14 - Scoring)

QUESTION: If a pistol competitor accidentally fires a shot while loading, before the command "Is the Line Ready" is given, may he reload the fired round before firing that string?

ANSWER: No. All shots fired by the competitor after he has taken his position at the firing point will be counted in his score, even if the pistol is accidentally discharged (14.6)

QUESTION: In a pistol timed fire match there were 11 hits on my target and 10 hits on the targets on each side of me. The officials ruled that when the correct number of hits are on all other targets, it is proof that I fired 11 shots on my target and I was disqualified for that match for firing excessive shots. Is this ruling correct?

ANSWER: No. This rule provides that when there are excessive hits on a target and none of the shots can be identified by the type of bullet hole as having been fired by some other competitor, or as having been fired in a previous string, or if all hits are not of equal value, the competitor may accept the score equal to the required hits of lowest value or he may fire a complete new score. (14.10)

QUESTION: How are ties ranked in pistol team matches?

ANSWER: By considering the team score as though it were a single score fired by an individual. The same precedent applies as that indicated in Rules 15.3, to 15.5 inclusive. (15.7)

1. (14.1) When to Score
2. (14.2) Where to Score
3. (14.3) How to Score
4. (14.4) Misses
5. (14.5) Early or Late Shots
6. (14.6) All Shots Count
7. (14.7) Hits on Wrong Target
8. (14.8) Ricochets
9. (14.9) Visible Hits and Close Groups
10. (14.10) Excessive Hits
11. (14.14) Scorer's Duties
12. (14.15) Score Cards
13. (14.15) Score Cards
14. (18.14) Signatures
15. (10.5) & (18.3.1) Competitors Will Score
16. Decisions of Ties (Section 15)
    a. (15.3) Single stage ties
    b. (15.4) Multiple stage ties
    c. (15.5) Aggregate match ties
    d. (15.7) Team Match ties
    e. (15.10) Unbreakable ties
    f. (9.12) Perfect Score Ties

F. Sixth Student Performance Objective: Given the task of training instructor/coaches for participation in pistol matches, EXPLAIN the competitor's duties and responsibilities as defined in Section 18, NRA Pistol Rule Book. (Section 18 - Competitors duties and responsibilities)

QUESTION: A competitor in a pistol tournament made a mistake in his squadding and fired his rapid fire match during a timed fire match. Because no one was squadded on this firing point for that relay the error was not discovered by the tournament officials before the competitor fired. He fired a good score and wanted it used as his rapid fire score. Could this be done?

ANSWER: No. Pistol rules provide that rapid fire is in strings of 5 shots at 10 seconds per string. Timed fire is in strings of 5 shots at 20 seconds per string. Therefore, the timed fire score could not be used for the rapid fire match. (18.6)

QUESTION: If at the start of a match, one of the competitors finds his gun needs a slight adjustment for proper operation, must the entire line be held up until the necessary adjustments or quick repair has been done?

ANSWER: Rule 18.7 provides, "Competitors must report to their firing point immediately when the relay is called by the range officer. The proper gun and ammunition for that particular match must be ready and in safe firing condition. Time will not be allowed for gun repairs, sight blacking, sight adjustments or search for missing equipment after a relay has been called to the firing line." Also Rule 10.3 provides, "No competitor may delay the start of a match through tardiness in reporting or undue delay in preparing to fire. In all cases all competitors will be allowed 3 minutes to take their places at their firing points and prepare to fire after the firing point has been cleared by the preceding competitor." (18.7)

QUESTION: On my entry card, I marked "across the board." After the firing, I found I was not entered in one aggregate. Isn't it the Statistical Officer's responsibility to make the entry and advise me if there is not enough money to cover the entry fee?

ANSWER: No. This rule provides it is the duty of the competitor to make his own entries on the forms and in the manner prescribed for that tournament. Errors due to illegibility or improper filling out of forms are solely the competitor's responsibility. (18.5)

1. (18.1) Discipline
2. (18.2) Knowledge of program
3. (18.3) Eligibility
4. (18.3.1) Competitors will score
5. (18.4) Classification
6. (18.5) Individual Entries
7. (18.6) Squadding Tickets
8. (18.7) Reporting at Firing Point
9. (18.8) Timing
10. (18.9) Loading
11. (18.10) Cease Firing
12. (18.11) Checking Score Card and Signing
13. (18.12) Clearing the Firing Point
14. (18.13) Checking Bulletin Board
15. (18.14) Score Cards Must be Signed
16. (18.15) Responsibility

G. <u>Seventh Student Performance Objective</u>: Given the task of training instructor/coaches for participation in pistol matches, EXPLAIN the team officers duties and position during a pistol team match as defined in Section 12, NRA Pistol Rule Book. (Section 12 - Team Officers Duties and Position)

QUESTION: One of the members of our team was injured and unable to continue firing in a team match. Under such circumstances, can another shooter be substituted for the injured team member?

ANSWER: No. Rule 12.3 provides in part, "No alternates may be used unless alternates are provided for in the match conditions and have been named on the entry form. If alternates are permitted in the match conditions and are to be used, they must be named on the entry form at the time the card is filed. A shooter may not be listed as an alternate on more than one team in any match. Competitors listed as principals on one team cannot be named as alternates on another team entered in the same match. An alternate must be substituted for the regular team member he is replacing before that regular member has fired any record shots in the team match concerned." (12.3)

QUESTION: What are the duties of a team coach?

ANSWER: A team coach is permitted to coach in all team matches within the team only. The coach may assist team members by calling shots, checking time, checking scoring, ordering sight changes, etc., but he must so control his voice and actions as not to disturb other competitors. He will not physically assist in loading or in making sight corrections. (12.5)

QUESTION: Is it absolutely necessary to have a team captain as well as a team coach in a pistol team match?

ANSWER: Yes, except in the case of two (2) man teams (12.1)

       1.   (12.1)   Team Captain

       2.   (12.2)   Team Coach

       3.   (12.3)   Team Entries

       4.   (12.4)   Team Captain and Coach Position

       5.   (12.5)   Team Coach in Team Matches

    H. **Eighth Student Performance Objective:** Given the task of training instructor/coaches for participation in pistol matches, EXPLAIN the eligibility to competitors in a pistol match as defined in Section 2, NRA Pistol Rule Book. (Section 2 - Eligibility of competitors)

QUESTION: I belong to the U.S. Army Reserves. I attended a tournament where the competitors were divided into 2 categories - Service and Civilian. The program provided that Reserves would fire in the Service category. My Reserve unit did not send me to the tournament nor pay my expenses or entry fee. It was all my own idea and my personal expense. Therefore, I entered as a Civilian. Is this correct?

ANSWER: No. Rule 2.7 explains that Reserves are, "Officers and enlisted men of any Reserve components of the Armed Forces, exclusive of the Army National Guard and the Air National Guard of the United States, not on extended active duty." Any officer or enlisted man who is a Reserve as outlined in this rule must enter all tournaments as a Reserve competitor. If the Reserve competitors are included with the Service, they must fire in that category. If the tournament has no categories the competitor should still show his Reserve connection to be used in the event a new National Record is established for Reserves. It does not matter if expenses or entry fees are paid by the Reserve or not or if the competitor is in uniform or not. (2.7)

QUESTION: In registered competition, if a competitor belongs to several shooting clubs, fires in a team match as a member of one club, must he wait 6 months before he can fire as a member of a team representing another club?

ANSWER: No. Rule 2.10 provides, "No competitor may fire on more than one team in any one match." Therefore, if there is more than one team match for a tournament, he can represent a different club, in which he holds membership, for each team match. (2.10)

QUESTION: What are the requirements an individual must fulfill to be able to fire for a particular club in a team match?

ANSWER: "Members of such teams must (a) have been active, fully paid members of the club for a period of at least 10 days immediately prior to the date of the competition, (b) the club must be affiliated with the NRA and in good standing." (2.11)

    1.   (2.1)   Members of NRA

    2.   (2.2)   Civilian

    3.   (2.3)   Junior

    4.   (2.4)   Police

    5.   (2.5)   National Guard

    6.   (2.6)   Regular Service

    7.   (2.7)   Reserve

    8.   (2.8)   College

    9.   (2.9)   School

    10.   (2.10)   Team Representative

    11.   (2.13)   Regular, N.G. or Reserve Teams

    12.   (2.20)   Residence

VI.   CONCLUSION:

    A.   <u>Retain Attention</u>:   If a pistol competitor is one second late arriving at the firing line, he may lose as many as three hundred points from his aggregate score.

    B.   <u>Summarize</u>:   The pistol shooter coach or instructor must be familiar with all the NRA rules of pistol shooting. We have covered those rules that are most applicable to the shooter.

    1.   Types of Competition

    2.   Competition Regulations

    3.   Challenges and Protests

    4.   Match Weapons, Equipment and Ammunition

    5.   Scoring and Ties

    6.   Competitor's duties and Responsibilities

    7.   Team Officer's Duties and Positions

    8.   Eligibility of Competitors

C. **Application**: NRA Pistol Match Rules will be used in all the different shooting phases of this course, in all NRA registered matches and most of the service sponsored pistol matches.

D. **Closing Statement**: A competitor who has full knowledge of the pistol match rules has relieved his mind of anxiety during and after the shooting phases as to whether any of these numerous essentials have been overlooked. This will allow him to concentrate on controlling his shooting to the fullest extent of his ability.

ANNEX I TO COMPETITIVE REGULATIONS I

STUDENT QUESTIONAIRE

PISTOL MATCH RULES

1. Section 1 - Types of Competition

QUESTION: I fire in an NRA registered league. Can we enter a league team from this league in a registered tournament using shooters from several clubs?

ANSWER:

QUESTION: Can I make entry in an aggregate match after the firing has started in a tournament but before I have fired in any match in the aggregate concerned?

ANSWER:

QUESTION: What is the meaning of "stage of a match" in connection with competitive courses of fire?

ANSWER:

2. Section 9 - Competition Regulations

QUESTION: Is a pistol considered "loaded" when the loaded clip is in the gun but the slide is locked open, or a revolver when the cylinder is loaded but swung out of the frame?

ANSWER:

QUESTION: I have a pistol with interchangeable barrels. When firing the National Match Course can I use the long barrel for slow fire and change to the short barrel for the timed and rapid fire stages?

ANSWER:

QUESTION: Can a competitor use one gun for slow fire and another for timed and rapid fire in multiple stage pistol competition?

ANSWER:

3. Section 16 - Challenges and Protests

QUESTION: What incidents may be challenged by a competitor?

ANSWER:

QUESTION: If I am not satisfied with the scoring of my target, can I protest?

ANSWER:

QUESTION: What is the formal procedure for making a protest?

ANSWER:

4. Section 3 - Equipment and Ammunition

QUESTION: The pistol rules provide limitations to the distance between sights. How are these distances measured?

ANSWER:

QUESTION: Is it legal to wrap the grip of my .45 semi-automatic pistol with tape when the match rules require arms as defined in Rule 3.1?

ANSWER:

QUESTION: Can the .44 caliber revolver be used when firing a .45 caliber pistol match?

ANSWER:

5. Section 14 - Scoring

QUESTION: If a pistol competitor accidentally fires a shot while loading, before the command "Is the Line Ready" is given, may he reload the fired round before firing that string?

ANSWER:

QUESTION: In a pistol timed fire match there were 11 hits on my target and 10 hits on the targets on each side of me. The officials ruled that when the correct number of hits are on all other targets it is proof that I fired 11 shots on my target and I was disqualified for that match for firing excessive shots. Is this ruling correct?

ANSWER:

QUESTION: How are ties ranked in pistol team matches?

ANSWER:

6. Section 18 - Competitors Duties and Responsibilities

QUESTION: A competitor in a pistol tournament made a mistake in his squadding and fired his rapid fire match during a timed fire match. Because no one was squadded on this firing point for that relay, the error was not discovered by the tournament officials before the competitor fired. He fired a good score and wanted it used as his rapid fire score. Could this be done?

ANSWER:

QUESTION: If at the start of a match one of the competitors finds his gun needs a slight adjustment for proper operation, must the entire line be held up until the necessary adjustments or quick repair has been done?

ANSWER:

QUESTION: On my entry card I marked "across the board". After the firing, I found I was not entered in one aggregate. Isn't it the Statistical Officer's responsibility to make the entry and advise me if there is not enough money to cover the entry fee?

ANSWER:

7. Section 12 - Team Officers, Duties and Position

QUESTION: One of the members of our team was injured and unable to continue firing in a team match. Under such circumstances, can another shooter be substituted for the injured team member?

ANSWER:

QUESTION: What are the duties of a team coach?

ANSWER:

QUESTION: Is it absolutely necessary to have a team captain as well as a team coach in a pistol team match?

ANSWER:

8. Section 2 - Eligibility of Competitors

QUESTION: I belong to the U.S. Army Reserves. I attended a tournament where the competitors were divided into 2 categories - Service and Civilian. The program provided that Reserves would fire in the Service category. My Reserve unit did not send me to the tournament nor pay my expenses or entry fee. It was all my own idea and my personal expense. Therefore, I entered as a Civilian. Is this correct?

ANSWER:

QUESTION: In registered competition, if a competitor belongs to several shooting clubs, fires in a team match as a member of one club, must he wait 6 months before he can fire as a member of a team representing another club?

ANSWER:

QUESTION: What are the requirements an individual must fulfill to be able to fire for a particular club in a team match?

ANSWER:

S. COMPETITIVE REGULATIONS II   AAPMI&CC 702
    (Pistol Range Procedure and Safety)   One (1) Hour
                                                            Mar 1965

                                  LESSON PLAN

   I. LESSON OBJECTIVE: To enable the pistol marksmanship instructor-shooters and coaches to use their knowledge of pistol match range procedure and safety rules to avoid penalty, disqualification or injury to competitors while participating in pistol competition.

   II. STUDENT PERFORMANCE OBJECTIVES: As a result of this instruction, the pistol coach/instructor student should be able to accomplish the following student performance objectives:

        A. Given the mission of training pistol coach/instructor students to participate in pistol matches, EXPLAIN the provisions of pistol match range procedure as outlined in Section 10, NRA Pistol Rule Book.

        B. Given the mission of training pistol coach/instructor students to participate in pistol matches, EXPLAIN pistol range safety rules as outlined in Chapter XXV, USAMU Marksmanship Instructors' and Coaches' Guide.

   III. ADVANCE ASSIGNMENT: Chapter XXV, USAMU Marksmanship Instructors' and Coaches Guide.

   IV. INTRODUCTION:

        A. <u>Gain Attention</u>: It is the duty of each competitor to sincerely cooperate with tournament officials in the effort to conduct a safe, efficient tournament. Competitors are expected to promptly call the attention of proper officials to any infraction of rules of safety or good sportsmanship.

        B. <u>Orient Students</u>:

            1. <u>Lesson Tie-In</u>: In order to compete in a pistol match, it is necessary that each competitor have a working knowledge of pistol range procedure and range safety rules.

            2. <u>Motivation</u>: The safety of competitors, range personnel and spectators requires a high degree of self-discipline, constant attention to range control commands, the careful handling of firearms, exercising caution in moving about the range area, acting and speaking with consideration toward other competitors.

            3. <u>Scope</u>: During this period of instruction, there will be a discussion of:

                a. Pistol range procedure

                b. Pistol range safety rules

   V. BODY:

        A. <u>First Student Performance Objective</u>: Given the mission of training pistol coach/instructor students to participate in pistol matches, EXPLAIN the provisions of pistol match range procedure as outlined in Section 10, NRA Pistol Rule Book. RANGE CONTROL (Rule 10)

1. (10.1) Discipline - The safety of competitors, range personnel and spectators requires continuous attention by all to the careful handling of firearms and caution in moving about the range. Self-discipline is necessary on the part of all. Where such self-discipline is lacking is the duty of range officials to enforce discipline and the duty of competitors to assist in such enforcement.

2. (10.2) Loud Language: Loud or abusive language will not be permitted. Competitors, scorers, and range officers will limit their conversation directly behind the firing line to official business.

3. (10.3) Delaying a Match: No competitor may delay the start of a match through tardiness in reporting or undue delay in preparing to fire.

4. (10.3.1) Preparation Period: In all cases, competitors will be allowed three minutes to take their places at their firing point and prepare to fire after the firing point has been cleared by the preceding competitor.

5. (10.4) Policing Range: It is the duty of competitors to police the firing points after the completion of each string. The range officer will supervise such policing and see that the firing points are kept clean.

6. (10.5) Competitors Will Score: Competitors will act as scorers when requested to do so by the Executive Officer or Chief Range Officer, except that no competitor will score his own target.

7. (10.6) Repeating Commands: A range officer will repeat the Chief Range Officer's commands only when those commands cannot be clearly heard by competitors under his supervision.

8. (10.7) Firing Line Commands: When ready to start firing a match the range officer (usually the Chief Range Officer) commands, "RELAY NO. 1, MATCH NO. _____ (or naming the match) ON THE FIRING LINE." Each competitor in that relay then immediately takes his assigned place at his firing point and prepares to fire but does not load. If this is a new relay, the range officer will announce, "THE PREPARATION PERIOD STARTS NOW." At the end of three (3) minutes, the range officer states, "THE PREPARATION PERIOD HAS ENDED," then proceeds with the match.

    a. The range officer having made sure that the range is clear (in timed and rapid fire, targets must also be turned out of firing position) then commands, "WITH 5 ROUNDS, LOAD."

    b. The range officer then asks, "IS THE LINE READY?" Any competitor who is not ready or whose target is not in order will immediately raise his arm and call, "Not ready on target...." The range officer will immediately state, "The line is not ready," and the range officer will immediately investigate the difficulty and assist in correcting it. When the difficulty has been corrected, the range officer calls, "THE LINE IS READY."

    c. When the range officer asks, IS THE LINE READY?" and the line is ready, he then calls, "THE LINE IS READY."

    d. The range officer then commands, "READY ON THE RIGHT. READY ON THE LEFT." Competitors may point their guns toward the target after the command, "Ready on the right." The range officer will then command, "READY ON THE FIRING LINE." The targets will be exposed or the signal to commence firing will be given in approximately three seconds.

e. The range officer then commands, "COMMENCE FIRING" which means to start firing without delay as timing of the string is started with this command. "Commence Firing" may be signalled verbally, by a short sharp blast on a whistle or by moving the targets into view.

f. "CEASE FIRING" is the command given by the Range Officer at the end of time limit for each string or at any other time he wishes all firing to cease. Firing must cease immediately. Even if a competitor is about to let off a carefully aimed shot, he must hold his fire and open the action of his gun. Failure to immediately obey this command is one of the worst infractions of range discipline as it may result in the wounding or death of some man, woman or child who has wandered into the line of fire somewhere on the range or behind the targets. On this command, cylinders shall be opened or slides locked back and all guns placed on the shooting stand and not handled until the next command of the Range Officer. "Cease firing may be signalled verbally, by a short sharp blast on a whistle or by moving the targets out of view.

g. After the command to "cease firing" is given at the end of a string, a subsequent command is given, "CEASE FIRING - UNLOAD - CYLINDERS OPEN - SLIDES BACK - GUNS ON THE TABLE." On this command all assistant range officers and scorers check their competitors to make sure each one obeys the command before signalling the Range Officer that their portion of the firing line is clear.

h. When all assistant range officers and scorers have given the "clear" signal the range officer commands "SCORE TARGETS AND PASTE" (OR CHANGE")

i. Other commands used less frequently are:

(1) "POLICE FIRING POINTS" means pick up fired cartridge cases, empty cartridge cartons and "tidy-up" the firing line.

(2) "AS YOU WERE" means disregard the command just given. For example, if the commands were given "Ready on the right" followed by "As you were", it would mean someone was not ready.

(3) "CARRY-ON" means proceed with whatever was being done before some interruption occured.

B. <u>Second Student Performance Objective</u>: Given the mission of training pistol coach/instructor students to participate in pistol matches, EXPLAIN pistol range safety rules as outlined in Chapter XXV, USAMU Marksmanship Instructors' and Coaches' Guide.

The coach and the shooter must exercise utmost care in all phases of handling weapons and ammunition. Cleaning weapon, dry fire practice, transporting guns and ammunition to and from quarters or on trips to matches seems to be more dangerous than when the equipment is being used in a match or organized practice. The average person is less apt to commit a dangerous act when in a crowd where many weapons are present. This, no doubt, is why you seldom hear of anyone being wounded or killed at pistol or rifle matches. The constant supervision and caution in handling firearms is well served. A coach or range official will not hesitate to correct a shooter when observing even the slightest infraction of range safety.

The following items are listed to enable the shooter to avoid jeopardizing the safety of others or himself:

1. Clear the pistol every time it is picked up for any purpose; never trust your memory. Consider every pistol as loaded until you have proved it otherwise.

2. Always unload and clear the pistol if it is to be left where someone else may handle it. A cleared pistol has the slide back, magazine out and the chamber is inspected for unextracted round.

3. Always point the pistol up when snapping it after examination. Keep the hammer fully down when the pistol is not loaded.

4. Never place the finger within the trigger guard until you intend to fire.

5. Never point the pistol at anyone you do not intend to shoot, nor in a direction where an accidental discharge may do harm. On the range, do not "dry fire or snap" for practice while standing back of the firing line. "Dry fire or snap shoot" only with a weapon that has been cleared and point weapon only at a backstop that has been approved by range officials.

6. Before loading the pistol, draw back the slide and look through the bore to see that is free from obstruction. When loading pistol, insert loaded magazine until fully seated, place left thumb on hammer, release slide stop with left index finger while gripping stock firmly with right hand.

7. On the range, do not insert a loaded magazine until the time for firing.

8. Never turn around at the firing point while you hold a loaded pistol in your hand, because by so doing, you may point it at the man firing alongside of you.

9. On the range, do not load the pistol (with a cartridge in the chamber) until immediate use is anticipated. If there is any delay, lock the pistol and only unlock it while extending the arm to fire. Do not lower the hammer on a loaded cartridge; the pistol is much safer cocked and locked.

10. In reducing a jam first remove the magazine.

11. To remove a cartridge not fired, first remove the magazine, and then extract the cartridge from the chamber by drawing back the slide.

12. Safety devices should be frequently tested. A safety device is a dangerous device if it does not work when expected.

13. Don't mix alcohol with gun powder.

14. Make sure range is clear before practice firing if you are not participating in organized firing or when firing alone. Fire only on command of range officer if organized practice is being conducted.

15. Do not handle weapon when any person is forward of firing line.

16. No government issue ammunition to leave the firing range. Return all excess ammunition to supply personnel.

17. Do not fire over and above barrier or backstop.

18. Protect weapons from theft, thereby preventing lost weapons from falling into the hands of irresponsible persons. Do not leave weapons unattended either on the range or in a car. Lock all shooting equipment in trunk of car when transporting. Equipment exposed in an unlocked car is a temptation needlessly advertised.

19. Register personal weapons on post, camp or station and with your local police department if local laws require it.

VI. CONCLUSION:

   A. <u>Retain Attention</u>: Loud or abusive language on a pistol range during match firing is ample cause to be removed from the range area, especially if a warning has been given previously by one of the range officials.

   B. <u>Summarize</u>: The pistol shooter or coach must have a working knowledge of pistol range procedure and have an acute consciousness of every element of safe handling of firearms while present on a pistol range during firing.

   C. <u>Application</u>: The pistol range procedure will be very similiar on all pistol ranges and safety rules are identical on every range. The pistol shooter must comply fully or run the risk of penalty, disqualification or injury.

   D. <u>Closing Statement</u>: The pistol shooter can devote his full mental effort to employing his shooting skill if he acts in accordance with established range procedure and does not have to worry about carelessness in his or the other competitors' handling of firearms.

T. COMPETITIVE REGULATIONS III   AAPMI&CC 701A
    (National Trophy Pistol Match Rules and   One (1) Hour
    Earning a Distinguished Pistol Shot Badge)   May 75

LESSON OUTLINE

I. LESSON OBJECTIVE: To enable the pistol marksmanship instructor-shooters and coaches to use their knowledge of national trophy match rules to avoid any penalty, disqualification or injury to competitors while participating in a national trophy individual or team pistol match and earn the award of the Distinguished Pistol Shot Badge.

II. STUDENT PERFORMANCE OBJECTIVES: As a result of this instruction, the pistol coach/instructor student should be able to accomplish the following student performance objectives:

   A. Given the mission of training pistol coach/instructor students to participate in national trophy pistol matches, EXPLAIN the provisions of National Trophy Pistol Match rules applicable to pistol as outlined in AR 920-30.

   B. Given the provisions for earning a Distinguished Pistol Shot Badge, EXPLAIN the requirements for earning a Distinguished Pistol Shot Badge as shown in Chapter XXVI, USAMU Marksmanship Instructors' and Coaches Guide and AR 350-6.

III. ADVANCE ASSIGNMENT: AR 920-30 and Chapter XXVI, USAMU Marksmanship Instructors and Coaches Guide and AR 350-6.

IV. INTRODUCTION:

   A. Gain Attention: The most prized award in pistol shooting competition is the Distinguished Pistol Shot Badge.

   B. Orient Students:

      1. Lesson Tie-In: In order to compete in the National Trophy Pistol Match, it is necessary that the service pistol competitor have a working knowledge of the provisions of AR 920-30 as applicable to national trophy competition.

      2. Motivation: By successfully competing in the National Trophy Pistol Match, points can be earned toward the award of the Distinguished Pistol Shot Badge.

      3. Scope: During this period of instruction, there will be coverage of:

         a. National Trophy Pistol Match Rules.

         b. Requirements for earning a Distinguished Pistol Shot Badge.

V. BODY:

   A. First Student Performance Objective: Given the mission of training pistol coach/instructor students to participate in national trophy pistol matches, EXPLAIN the provisions of national trophy match rules applicable to pistol as outlined in AR 920-30.

      1. National Trophy Matches:

         a. The National Trophy Matches are conducted by the National Match Director for the National Board for the Promotion of Rifle Practice in accordance with the provisions of AR 920-30.

b. These matches consist of the following events:

   (1) The National Trophy Individual Match, both rifle and pistol.

   (2) The National Trophy Team Match, both rifle and pistol.

2. <u>Mandatory Participation</u>: (See Para 12, AR 920-30) Military competitors, whose transportation or other travel expenses are paid, either partially or wholly from Government funds (appropriated) or quasipublic funds (exchange, recreation, welfare and morale) and/or who uses Government issued equipment will be required to:

   a. Attend the appropriate Small Arms Firing School.

   b. Enter and complete firing in the appropriate National Trophy Individual Match, unless eliminated under the provisions of paragraph 20.

   c. Enter and complete firing in the appropriate National Trophy Team Matches as eligible under this regulation.

   d. Fire a service rifle authorized by paragraph 25 at ranges of 200 to 600 yards inclusive in the NRA as well as National Trophy Rifle Matches, except as may be specifically exempted in the current National Match program. Military personnel will fire the type service rifle as directed by their parent Service. The firing of other weapons at greater ranges or in international-type matches is permissible.

3. <u>Rules and Regulations</u>: (See Para 18 AR 920-30)

   a. Except as specifically set forth herein the National Trophy Matches will be conducted under the rules and regulations as prescribed in the latest edition of Official Rules of the NRA for the conduct of high power rifle and pistol competitions.

   b. The Executive Officer, National Matches, may alter the rules or conditions of the National Matches only if necessary for the successful completion of the matches. In the event such alteration becomes necessary, the Executive Officer, National Matches, will immediately submit a written report of details to the President of the National Board for the Promotion of Rifle Practice.

4. <u>Competitor's Responsibilities</u>: (See Para 21, AR 920-30)

   a. It is the competitor's responsibility to know and abide by the provisions of:

   (1) AR 920-30.

   (2) The Official National Match Program.

   (3) NRA rules as applicable.

   (4) Appropriate National Match Executive Officer's Bulletins.

   b. In team matches, the team captain is charged with this responsibility for his team.

5. <u>Entries</u>: (See Para 19, AR 920-30)

   a. Entries of individuals and teams in the National Trophy Matches will be made for each match as provided hereinafter. No entries will be accepted after the closing time for the submission of entries for each match.

b. Competitors are urged to make advance entry in the Small Arms Firing School and individual matches by mail to the Statistics Director, National Matches, Camp Perry, OH 43452. The following entry blanks for individual matches may be obtained from the Statistics Director:

(1) DA Form 1342 (Entry and Score Card for NBPRP Individual Service Pistol Match).

(2) DA Form 1344 (Entry and Score Card for NBPRP Individual Service Rifle Match).

c. Mail entries will not be accepted for the team matches. Entry in team matches must be made at the National Match site by the team captain on official entry blanks provided by the Statistics Director as follows:

(1) DA Form 1343 (Entry and Score Card for Pistol Team Match) for National Trophy Pistol Team Match.

(2) DA Form 1345 (Entry and Score Card for Rifle Team Match) for National Trophy Rifle Team Match.

d. Entries for individual matches will close not later than twelve noon 10 days prior to the day on which the match is scheduled to be fired. (Consult current National Match Program for exact date.)

e. Entries for team matches will close not later than twelve noon two days prior to the day on which the match is scheduled to be fired.

f. Entries must be complete and accurate. Incomplete or inaccurate entries may be refused or cause disqualification.

g. Entries will be accepted to range capacity on a first-come, first-served basis.

6. Entry Fee: (See Para 6, AR 920-30) When approved by the President, NBPRP, an entry fee will be charged for any National Trophy Match. No entry fee for the Small Arms Firing Schools will be charged except to defray necessary expenses for ammunition, if required.

7. Individual Eligibility: (See Paras 10 and 11, AR 920-30)

a. The National Trophy Match competitions are open to all United States citizens, male and female, who are at least 16 years of age, and to all members of the United States Armed Forces, both active duty and Reserve Components.

b. Certain competitive events and certain awards are restricted to special categories of shooters as a matter of grouping contestants in the best interest of fair competition.

c. The Small Arms Firing School is open to individuals, both military and civilian, 16 years of age or older, attending but not necessarily competing in the National Matches.

8. Competitor Category: (See para 13, AR 920-30)

a. In National Trophy Matches, competitors will register and participate in only one of the following categories: Regular Service, National Guard, Reserve, ROTC, civilian, police (police category is for pistol matches only).

b. The following eligibility requirements will govern in determining proper categorical match participation:

(1) Regular Service Category. Regular Service, National Guard, and Reserve personnel on extended active duty (90 days or more) with the regular Service will participate in this category only.

(2) Reserve category. Members of any Reserve branch of the US Armed Forces (Naval and Marine Corps Reserve includes Fleet and Fleet Marine Corps Reserve), except Army and Air National Guard will participate in this category unless qualified as regular Service under (1) above, or, if a bona fide member, an individual may select and participate only in police or ROTC category. If qualified, individual may elect to fire as a civilian within the criteria specified in (6) (c) below.

(3) National Guard category. Members of the Army or Air National Guard will participate in this category unless qualified as regular Service under (1) above, or, if a bona fide member, an individual may select and participate only in police or ROTC category. If qualified, may elect to fire as a civilian within the criteria specified in (6) (c) below.

(4) Police category (Pistol Trophy Matches only). Full-time, paid members of any police organization will participate in the police category or, if a bona fide member, an individual may select and participate only in National Guard or Reserve category. A policeman if qualified as a civilian within the criteria specified in (6) (c) below, may also participate in the civilian category if he attends as one of two NBPRP-authorized police members of a State Civilian Pistol Team provided he is not a member of a police organization that enters a team in the National Trophy Pistol Team Match. In the National Trophy Individual Pistol Match all policemen, except those who enter as National Guard or Reserve, will participate in the police category.

(5) ROTC category. Members of any high school or college ROTC unit (Army, Navy, or Air Force) and service academies of the Armed Forces will participate in this category or, if a bona fide member, an individual may select and participate in the National Guard, Reserve, police, or civilian category.

(6) Civilian category.

(a) Individuals who are not in any of the above-listed categories will fire as civilians.

(b) Retired service personnel will fire as civilians or, if a bona fide member, an individual may select and participate only in police category.

(c) Individuals of any Reserve or National Guard component who, during the present calendar year, have not competed as Reservists, National Guardsmen, or members of a Service component team, and have not been provided Service support for competition, in the form of weapons, ammunition, payment of travel or other expenses, wholly or in part, may fire as civilians. The provision of weapons and ammunition for a specific competition, i.e., National Matches or NBPRP Regional Leg Matches, when such is available to both military and civilian competitors, is not considered Service support or a bar for subsequent competition in the civilian category.

(d) Reserve and National Guard personnel who elect to compete as "civilians" may not revert to their Service category to compete in open competition within any one calendar year, unless called to active Federal service.

9. __Ammunition:__   (See para 27, AR 920-30)

    a.   In the National Trophy Matches service ammunition will be issued by range personnel at the firing line. Competitors are required to fire this ammunition and none other. A competitor will be disqualified if any other ammunition of the same caliber as issued is found about the person while in position at the firing line. Exception: Until Match Grade 5.56 ammunition is available to match sponsor, competitors using M16 rifle or equivalent, shall use safe ammunition provided by themselves.

    b.   Under certain circumstances, when approved by the President, NBPRP, it may be necessary to charge competitors for the ammunition expended.

10. __Weapons:__   (See para 25, AR 920-30)

    a.   US Rifle, Caliber .30 M1, as issued by the US Army, or modified with a 7.62 barrel, having not less than a 4½ pound trigger pull, with standard-type stock and standard-type leather or web sling. Sling cuffs and sling pads are not permitted. External alterations to the stock will not be allowed. The application of synthetic coatings, which includes those containing powdered metal, to the interior of the stock to improve bedding is authorized provided the coatings does not interfere with function or operation of safety features. The front and rear sights must be of US Army design but may vary in dimensions of rear sight aperture and front sight blade. The internal parts of the rifle may be specially fitted and include alterations which will improve the functioning and accuracy of the arm, provided such alterations in no way interfere with the proper functioning of the safety devices as manufactured. Personnel concerned are responsible that weapons are inspected to insure adequate safety, and otherwise meet the requirements of this paragraph.

    b.   US Rifle, Caliber 7.62 mm, M14 as issued by the US Army, having not less than a 4½ pound trigger pull, with standard-type stock and standard-type leather or web sling. Sling cuffs and sling pads are not permitted. The rifle must be so adjusted as to be incapable of automatic fire without removing the stock and changing parts. In all courses and in all positions the 20-round box magazine will be attached. The hinged butt plate will be used only in the folded position. The gas system must be fully operational. External alterations to the stock will not be allowed. The application of synthetic coatings which includes those containing powdered metal to the interior of the stock to improve bedding is authorized provided the coating does not interfere with the function or operation of safety features. The front and rear sights must be of US army design, but may vary in dimensions of rear sight aperture and front sight blade. The internal parts of the rifle may be especially fitted and include alterations which will improve the functioning and accuracy of the arm. provided such alterations in no way interfere with the proper functioning of the safety devices as manufactured. Personnel concerned are responsible that weapons are inspected to insure adequate safety, and otherwise meet the requirements of this paragraph.

    c.   US Pistol, Caliber .45 M1911 or M1911A1 or the same type and caliber of commercially manufactured pistol. The pistol must be equipped with issue or similar factory-type standard stocks, i.e., without thumb rests. Trigger pull must be not less than 4 pounds. This pistol must be equipped with open sights. The front sight must be nonadjustable. The pistol may be equipped with an adjustable rear sight with open U or rectangular notch, the distance between sights measuring not more than 7 inches from the apex of the front sight to the rear face of the rear sight. The forestrap of the receiver grip may be checkered. The mainspring housing may be either the flat or arched type, checkered or uncheckered. Trigger shoes may be used. Trigger stops, internal or external, are acceptable. Otherwise, external alterations or additions to the arm will not be allowed. The internal parts of the pistol may be specially fitted and include alterations which will improve the functioning and accuracy of the arm, provided such alterations in no way interfere with the proper functioning of the safety devices as manufactured. All standard safety features of the weapon must operate properly. Personnel concerned are responsible that weapons are inspected to insure adequate safety, and otherwise meet the requirements of this paragraph.

d. US Rifle, Caliber 5.56 mm M16 Series, as issued by the US Armed Forces, or commercial equivalent, without bipod, grenade launcher, or other attachments, having not less than 4½ pound trigger pull, with standard-type stock, pistol grip, handguard, and leather or web sling. Sling cuffs and sling pads are not permitted. The rifle must be so modified as to be incapable of automatic fire without replacing or altering parts. In all courses of fire and in all positions the 20-round or 30-round box magazine will be attached. The gas system must be fully operational. External alterations to the stock maintaining contour similar to the standard stock and butt plate is permitted. The application of synthetic coatings to the interior of the stock, pistol grip, and handguard to improve bedding is authorized provided the coating does not interfere with the function or operation of safety features. The front sight shall be of open metallic post or blade type mounted on the standard front sight assembly. The rear sight shall be metallic, not containing a lens or system of lenses, mounted on the carrying handle, so mounted that the carrying handle may be used as intended. Distance between front and rear sights shall be not less than 19 or more than 22 inches, measured from the apex of the front sight to the rear face of the rear sight. The internal parts of the rifle may be especially fitted and include alterations which will improve the functioning and accuracy of the arm, provided such alterations in no way interfere with the proper functioning of the safety devices as manufactured.

11. <u>Use of weapon by more than one competitor</u>: (See para 28, AR 920-30) Two or more competitors may fire the same rifle or pistol in any NBPRP Excellence-in-Competition Match if tournament squadding permits.

12. <u>Firing positions authorized</u>: (See para 35, AR 920-30) In the National Trophy Matches no special consideration will be given to physically handicapped shooters. NRA rules will apply except as shown below.

a. Standing. Standing - erect on both feet, no other portion of the body touching the ground or any supporting surface. The sling must be attached to the rifle but may not be used for support. The rifle will be supported by both hands and one shoulder only. The elbow of the forward arm may be placed against the body or rested on the hip.

b. Sitting. The sitting position will be any of those authorized under NRA rules.

c. Kneeling. The kneeling position will be any of those authorized under NRA rules; however, use of the kneeling roll or pad is not authorized.

d. Ground cloths or ground pads.

(1) Ground cloths or ground pads normally will not be authorized for use during the National Trophy Matches.

(2) Exceptions to this rule may be by the Executive Officer National Matches if, in his opinion, the ground conditions resulting from inclement weather result in undue disadvantage to the competitor.

(3) Ground cloths or ground pads, when authorized for use, must not be constructed or used in a manner that provides artificial support.

e. With the M14 or M16 series, the magazine may touch the person of the shooter, but may not touch the ground when in firing position.

13. <u>Loading and reloading</u>. (See para 40, AR 920-30)

a. Rifle Caliber .30 M1. In sustained fire stages of National Trophy Matches, individual or team, the rifle will be loaded initially with a clip containing two rounds. After the two rounds are fired, the rifle is reloaded with a clip containing eight rounds from the belt or the ground at the option of the firer.

b. Rifle, 7.62mm, M14 and rifle 5.56mm, M16 series. In sustained fire stages of National Trophy Matches, individual or team, the rifle will be loaded initially with a magazine containing two rounds. After the two rounds are fired, the rifle is reloaded with a magazine containing eight rounds from the pouch or the ground at the option of the firer.

14. <u>Field glasses and telescopes</u>: (See para 41, AR 920-30) Competitors may use field glasses or telescopes on the firing points unless otherwise prescribed in the conditions of the match.

15. <u>Competitor must verify score</u>: (See para 42, AR 920-30) The competitor acknowledges the correctness of all data on the card when he signs it. Should a competitor or team captain sign an incorrect card or leave the firing line without signing the score card, no challenge or protest will be allowed. If the competitor or team captain desires to protest, he will write the word "Protested" on the score card above his signature.

16. <u>Competitors present punctually</u>: (See para 34, AR 920-30) Competitors will be present at the firing points punctually at the time stated on their squadding tickets. It is the competitor's responsibility to appear at the firing point to which he has been assigned, prepared to fire, when his relay in any match is called to the firing line. Any competitor who fails to report on the proper relay or firing point will forfeit the right to fire in that match unless the competitor presents satisfactory evidence to the National Match Executive Officer that he is late through fault of the National Match Staff. In team matches, the first pair only need be present at the time set for firing to begin.

17. <u>Station of competitors</u>: (See para 32, AR 920-30) Each competitor will remain on or in rear of the assembly line in rear of his firing point until called by the range officer to take his position at the ready line or firing point. No one except the officials of the matches, members of the National Board for the Promotion of Rifle Practice, team officials, the competitors on the ready line and on the firing points, scorers and others on duty will be permitted in front of the assembly line without special permission of the Range Director.

18. <u>Elimination of teams or individuals</u>: (See para 20, AR 920-30) The Executive Officer, National Matches, may, at his and by such standards as he may prescribe eliminate teams and/or individuals of the lowest standing at any time after the first stage of a team or individual trophy match is completed.

19. <u>Teams authorized</u>: (See para 15, AR 920-30) The following teams are authorized in the National Trophy Team events:

a. Active Service Teams -- One each representing the major services each US Army Region, each Naval District, each Marine Corps competition area, each Air Force area, each Coast Guard area and the Defense Atomic Support area.

b. National Guard Teams -- One each representing each State, the Commonwealth of Puerto Rico and the District of Columbia. Such teams may be composed of mixed Army National Guard, Air National Guard personnel.

c. Reserve Teams -- Not to exceed two each representing the Army Reserve, the Naval Reserve, the Air Force Reserve, the Marine Corps Reserve, the Coast Guard Reserve, and one each representing the Army Reserve within any one major oversea command; the Naval Reserve within any one Naval District; the Air Force Reserve within any one Continental Air Command Reserve Region; the Marine Corps Reserve within any one Marine Corps Reserve District; and the Coast Guard Reserve within any one Coast Guard area.

d. ROTC Teams -- One each representing the Reserve Officer's Training Corps (Army, Navy or Air Force) from each of the respective Army Area commands. Naval Districts and Air Force ROTC Liaison Areas within the Continental United States, ROTC students from units in Alaska and Hawaii are authorized to compete as members of the Sixth US Army Area ROTC Rifle Team. ROTC students from Puerto Rico are authorized to compete as members of the First US Army Area ROTC Rifle Team.

e. Service Academy Teams.

f. Civilian Teams:

(1) One each representing each of the States, the Commonwealth of Puerto Rico and the District of Columbia.

(2) Civilian shooting clubs organized under the rules of the NBPRP and in good standing on the rolls of the DCM as of 30 June 1968, or enrolled with the DCM subsequent to that date, may enter one or more teams. However, entry of more than one team will be subject to available range facilities as determined by the Match Executive Officer.

g. Police Teams in Pistol Team Match -- One or more teams representing each regularly organized Federal, State, County or Municipal law enforcement agency in the United States. Subject to available range facilities.

20. <u>Team Membership Requirements</u>: (See para 16, AR 920-30)

a. At least one of the firing members of each team will be an individual who has never before fired as a member of any team which has competed in that particular event. Firing members who meet the requirements of this paragraph may be counted against the requirements of paragraphs c and d below.

b. No one who has fired on any team which has placed in the top 15% of competing teams in the same event in any two out of three National Trophy Team Matches immediately proceeding may be a firing member or alternate.

c. At least three of the firing members of rifle teams and two of the firing members of the Pistol teams selected to represent the Armed Forces (Army, Navy, Air Force, Marine Corps and Coast Guard), active Reserve or National Guard, will be individuals who have never before fired on any team which has placed in the top 15% of teams competing in the same event in past respective rifle or pistol National Trophy Team Matches.

d. Teams representing ROTC, Service Academies, civilian or police organizations, will be required to have at least two firing members on rifle teams and one firing member on pistol teams who have never before fired on any team which has placed in the top 15% of teams competing in the same event in past respective rifle or pistol National Trophy Team Matches.

e. Participation as a shooting member of a Service Academy team or ROTC team will not be considered as previous participation within the above eligibility requirements provided the team did not place in the top 15% of winning teams.

f. No individual may be a member of any of the authorized teams unless he has been a bona-fide member of that particular group or organization from which the team is selected for at least 30 days prior to the opening date of the National Matches except that: Members of State civilian teams will be confined to bona-fide residents of the state which the team represents, who have lived in the state for at least 30 days prior to the opening date of the National Matches and are otherwise qualified.

21. <u>Alternates</u>: (See para 48, AR 920-30) Not later than the time designated for the closing of entries, each team captain will submit to the Statistics Director at his office, on score cards furnished by the Statistics Director for the purpose, a legible list of members and alternates. A team captain and/or team coach may also serve as firing member or alternate, provided he is so listed on the entry form and is otherwise eligible. Substitution for a firing member may be made at any time before he fires his first shot in a match, provided the individual has been listed on the score card as an alternate. Once a team member has fired a shot, substitution may be made for him only in case of incapacitating injury and upon approval by the National Match Executive Officer.

22. <u>Team Certification</u>: (See para 17, AR 920-30) Team captains will present to the Statistics Director lists of their respective team members to include designation of team officials and other members of the team, the correct first name, middle initial, last name, service number and grade of members as applicable, their individual home addresses, official military mailing addresses if applicable, and proper certification as to the eligibility of the members of their team under these rules and regulations at the time entry in the match is made. Civilian teams will be recognized as representing a particular state only if attested to by the State Adjutant General or President, Vice President or Secretary of the respective State Rifle or Pistol Association, as accredited by the National Rifle Association or under conditions established by the DCM. Individual club presidents will attest to the teams representing their organizations. Teams representing law enforcement agencies and service teams of the Armed Forces of the United States will be certified as provided under appropriate regulations governing each such agency or service. ROTC and Service Academy teams will be certified by appropriate military commanders.

23. <u>Coaching</u>: (See para 49, AR 920-30)

    a. Individual Matches: No coaching is permitted.

    b. Team Matches: Coaching is permitted in all team matches within the team only; any member of a team (captain, coach, firing member or alternate) may function as coach provided such member fulfills the eligibility requirements set forth below:

    (1) A team captain or coach must be at least 16 years of age on his last birthday prior to the beginning date of the National Matches and must otherwise meet the eligibility rules of the National Matches;

    (2) Officers, warrant officer, or enlisted personnel of the active military services may act as coaches for teams of their respective services and for civilian, ROTC and Service Academy teams;

    (3) Teams representing the National Guard and teams representing other components of the Armed Forces Reserves may be coached only by officers, warrant officer, or enlisted personnel of the particular reserve component concerned;

    (4) ROTC and service academy teams will be coached only by members of the Armed Forces;

    (5) A civilian rifle or pistol team may request the National Match Executive Officer to assign a coach, and upon application, the National Match Executive Officer will assign a coach from the officers and enlisted men or resident civilians.

    (6) A coach once assigned will not be changed except in an emergency or other valid reason and upon approval or the National Match Executive Officer

24. **Station of Team Officials:** (See paras 50 and 51, AR 920-30)

    a. In team matches a team coach may take position on the firing line between the competitors of the pair firing in slow fire stages and immediately in rear of the firing member in rapid fire stages.

    b. In pistol team matches one coach will be allowed for each target assigned to the team.

    c. The coach cannot shift position nor shift the position of the competitors of the pair firing for the purpose of forming a windshield for the firer.

    d. The coach must confine himself to the normal position of a coach and his activities to those normally expected of a coach.

    e. The coach may assist team members by calling shots, checking time checking scoring, ordering sight changes, etc., but must control his or her voice and actions so as not to disturb other competitors. Natural voice commands will be unaided by electrical, mechanical, or other means.

    f. The coach will not physically assist in loading or making sight corrections.

    g. In team matches the team captain and one assistant may be seated in front of the assembly line, but not in advance of a line established three paces in rear of the line of scorers except as provided herein. A team captain may coach but only if he actually occupies the coaching position on the line.

25. **Safety Precautions:** (See para 22, AR 920-30)

    a. No arms will be loaded except at the firing point and under direction and command of a range officer.

    b. During and after loading, the pistol will be kept at raised (or ready) pistol or on the bench, muzzle down range until unloaded, except when aimed at the target for authorized practice or firing.

    c. Pistols will have their magazines withdrawn and the slides pulled back except when the competitor is at the firing point in the act of loading, firing, or ready to fire.

    d. During and after loading, the muzzle of the rifle will be kept pointed in the direction of the target until unloaded.

    e. Rifles will be carried at all times with actions open except when the competitor is at the firing point in the act of firing or ready to fire.

16. **Challenges:** (See para 54, AR 920-30) The procedure outlined in NRA rules for scoring challenges will be modified to the extent indicated below.

    a. No change for rifle matches.

    b. In pistol matches the competitor, when dissatisfied with the score announced by the scorer and the line officer, may upon payment of the challenge fee have the disputed score settled by the scoring referee whose decision is final.

    c. The official referee is appointed by the National Match Director.

    d. The challenge fee in these matches will be $1. All challenge money will be delivered to Executive Officer, National Matches, for payment to the National Match Fund.

27. <u>Protest</u>:   (See para 55, AR 920-30)

    a.   A competitor may formally protest:

        (1)   Any injustice it is felt has been exercised. (This excludes evaluation of a target which may be challenged as outlined elsewhere.)

        (2)   The conditions under which another competitor has been permitted to fire.

        (3)   The equipment which another competitor has been permitted to use.

    b.   A protest must be initiated immediately upon the occurence of the protested incident. Failure to comply with the following procedure will automatically void the protest:

        (1)   State the complaint verbally to the Chief Range Officer or the Statistics Director, as appropriate. If not satisfied with that decision then-----

        (2)   State the complaint orally to the official referee. If not satisfied with his decision, then within 24 hours-----

        (3)   File a written protest stating all the facts with the official referee. If not satisfied with that decision, then within 5 hours-----

        (4)   File with the NBPRP protest committee a protest in writing stating all the facts in the case. The decision of the protest committee will be final.

    c.   The NBPRP Protest Committee will consist of not less than three members. The members of the committee shall be appointed by the President, NBPRP.

    d.   Challenges and Protests in team matches must be made by the team captain.

28. <u>Penalties</u>:   (See para 57, AR 920-30)

    a.   Any person interfering with a competitor on the firing line or annoying the competitor in any way will be warned to desist. If upon said warning, the offense is repeated, such person will be ordered off the range at once.

    b.   Any competitor or team will be disqualified from competing further in the matches and may be denied any prize won during the current matches when found guilty by the National Match Executive Officer of any of these offenses listed below:

        (1)   Firing under a name other than that under which entered;

        (2)   Firing twice for the same prize;

        (3)   Falsifying scores or being an accessory thereto;

        (4)   Offering a bribe of any kind to any official or other person;

        (5)   Evading any of the conditions prescribed for the conduct of any match;

        (6)   Refusing to obey any instructions of the National Match Director, Range Director or Range Officer;

(7) Being guilty of disorderly conduct;

(8) Violating range safety regulations;

(9) Being guilty of any conduct considered by the National Match Executive Officer to be discreditable.

TRANSITION: Pistol competitors who fire high enough scores in National Trophy pistol competitions can earn a Distinguished Pistol Shot Badge.

 B. <u>Second Student Performance Objective</u>: Given the provisions for earning a Distinguished Pistol Shot Badge, EXPLAIN the requirements for earning a Distinguished Pistol Shot Badge as shown in Chapter XXV, USAMU Marksmanship Instructors' and Coaches' Guide and AR 350-6.

 1. Distinguished badges are awarded by all the major services in the US military establishment, as a symbol of the highest achievement in the field of competitive marksmanship.

 2. In order to enhance and ensure the continued prestige of these badges and to provide uniform conditions for their award to servicemen, it is agreed that:

 a. Award of Excellence-in-Competition badges and credit toward Distinguished badges shall be made on the basis of individual unassisted performance in recognized individual matches.

 b. Award of the appropriate Distinguished badge shall be made when an individual has earned a minimum of 30 credit points for excellence in competition in such recognized matches.

 c. Credit points shall be awarded to the highest scoring 10% of all non-distinguished participants completing the match ranked in order of merit. Fractions of .5 and over will be resolved to the next higher whole number. Smaller fractions will be dropped.

 d. Credit points shall be awarded to winning personnel as determined in paragraph c above on the following basis:

To the highest scoring 1/6	10 points
To the next highest scoring 1/3	8 points
To the remaining personnel authorized credit points	6 points

 e. Each individual shall be authorized to compete for credit points in not more than four recognized matches with the Service Rifle or Pistol each calendar year.

 f. Matches recognized for award of credit points and authorized participation are as follows:

COMPETITIONS	ACTIVE ARMY	USAR ARNG	CADETS USMA
US Army Installation/Division, USAR (ARCOM and GOCOM) (4 points only)	X	X	X
Major Command Championships	X	X	X
US Army Championships	X	X	X
Interservice Championships	X	X	X
National Trophy Individual Matches	X	X	X

COMPETITIONS	ACTIVE ARMY	USAR ARNG	CADETS USMA
NRA Regional Championship Matches) or) Command Matches of other Services)	*	X	X
National Guard Championships	X	X	X
National Guard State Matches (4 pts only)		X	

   g. Courses of fire, weapons and ammunition for matches in which credit awards are authorized shall be in accordance with those prescribed in the current edition of Rules and Regulations for the National Matches AR 920-30.

   h. Non-distinguished personnel who hold "leg" credits for Distinguished award, as of 1 January 1963, shall be credited with 10 points for each such leg, not to exceed a total of 20 points.

   i. Badges will be provided eligible Army personnel by the CG, FORSCOM, in accordance with the provisions of AR 672-5-1.

   j. Members of the other Services and civilian competitors who qualify for credit points toward Distinguished award in major command matches will be reported by the CG, FORSCOM, as follows:

    (1) Members of the other services direct to the Service concerned.

    (2) Civilians, to include ROTC students and retired military personnel, to the Director of Civilian Marksmanship, Headquarters, Department of the Army, Forrestal Bldg, Washington, DC 20314.

NOTE: *AUTHORIZED ONLY WHEN CIRCUMSTANCES PRECLUDE PARTICIPATION IN A MAJOR COMMAND CHAMPIONSHIP.

 VI. CONCLUSION:

  A. <u>Retain Attention</u>: Leg Match!!! I used to get excited about earning three (3) legs to get my Distinguished Badge. Now I have to worry about winning thirty (30) points. Point day!!!!

  B. <u>Summarize</u>: The pistol shooter and his coach must know the conditions prescribed in AR 920-30 for firing in the National Trophy Pistol Match and the requirements for earning a distinguished pistol shot badge as shown in AR 350-6.

  C. <u>Application</u>: Knowledge of the provisions for firing the National Trophy Pistol Match is necessary if the pistol shooter would avoid penalties and disqualifications while firing the National Trophy Matches. Every shot fired must count toward the award of a Distinguished Pistol Shot Badge.

  D. <u>Closing Statement</u>: You can receive that letter someday from the Department of the Army stating: SGT Joe Roe, serial no. 7-11 is hereby awarded the Distinguished Pistol Shot Badge.

V. REVIEW OF PISTOL FUNDAMENTALS

AAPMI&CC 600
(50) Min
Mar 75

LESSON OUTLINE

I. LESSON OBJECTIVE: To review and summarize the employment of basic fundamentals of pistol marksmanship which will enable coaches and instructor-shooters to have maximum control of their shooting abilities.

II. STUDENT PERFORMANCE OBJECTIVES: As a result of this review, the student must be able to accomplish the following student performance objectives:

　　A. GIVEN THE FACTORS of how proper minimum arc of movement is obtained, EXPLAIN how to attain the minimum arc of movement as shown in Chapter I, USAMU Pistol Marksmanship Guide.

　　B. REVIEWING INSTRUCTION given on proper sight alignment, EXPLAIN the factors necessary to employ and maintain proper sight alignment as described in Chapter II, USAMU Pistol Marksmanship Guide.

　　C. GIVEN A DESCRIPTION of the controlled action of applying positive trigger control, EXPLAIN how positive trigger control is applied to obtain championship scores as shown in Chapter III, USAMU Pistol Marksmanship Guide.

III. ADVANCED ASSIGNMENT: Chapters I, II, and III, USAMU Pistol Marksmanship Guide.

IV. INTRODUCTION.

　　A. *Gain Attention*: The basic fundamentals are so important to attaining good scores that it is necessary to review them.

　　B. *Orient Students*:

　　　　1. Lesson Tie-In: Previous classes have furnished detailed information on the fundamentals and techniques required by the pistol marksman. Fundamentals are the foundation of pistol shooting. This is a period of instruction to make sure each of you understand the fundamentals of pistol marksmanship.

　　　　2. Motivation: All of you can become champions. Before you do you must learn the fundamentals of marksmanship and how to employ them. A good shooter knows from experience that all the fundamentals must be coordinated with each other if he is to become a champion.

NOTE: SHOW SLIDE #1 - SCOPE AND SUMMARY.

　　　　3. *Scope*: During this period you can employ the fundamentals of advanced pistol marksmanship by knowing:

　　　　　　a. First, how to attain the minimum arc of movement.

　　　　　　b. Second, the factors necessary to employ and maintain proper sight alignment.

　　　　　　c. Third, the controlled action of applying positive trigger control.

TRANSITION: By Question: How would you explain to a beginner if he asked, "What is minimum arc of movement"? The degree of body and arm movement when holding on the target

V. BODY

   A. <u>First Student Performance Objective</u>: GIVEN THE FACTORS of how proper minimum arc of movement is obtained, EXPLAIN how to attain the minimum arc of movement as shown in Chapter I, USAMU Pistol Marksmanship Guide.

<u>NOTE</u>: SHOW SLIDE #2 - REQUIREMENTS OF STANCE.

   1. Requirements of Stance:

   The shooter must try to fire a shot when conditions are the most favorable and he has settled into his minimum arc of movement, therefore he must devote attention to selecting a stance that will afford the least body and head movement and provide balance.

<u>NOTE</u>: In pistol marksmanship, the words stance and position have been used to describe the shooters body in relationship to the target. For the purpose of this instruction, stance is defined as the arrangement or posture of the body only.

   Body configuration (height, weight and muscle development) permits no all-purpose stance but whatever the variation, the individual must follow certain basic rules which allow him to meet the requirements.

<u>QUESTION</u>: What is one of the requirements of the pistol shooters stance?

<u>ANSWER</u>: a. Greatest possible degree of balance and stability without muscle strain. b. Greatest possible degree of immobility, ie. no independent movement of any part of the body. c. Correct head position for the best use of eyes and preservation of equilibrium.

<u>TRANSITION</u>: To arrange the body into a proper stance, the shooter should be acquainted with certain important features of the human body.

   2. Characteristics of the Human Body:

   a. Passive apparatus: There are 200 or more bones in the human body. The bones, ligaments and joints position the various parts of the body. The basis of the skeleton is the spinal column which has great durability. The shoulders are fastened by muscle and ligaments to the spinal column. In the standing position it is undesirable to hold the weight of the pistol at arms length by muscle support alone. The shooter should try to transfer most of this weight to the bones and ligaments.

   b. Active apparatus: The active portion contains the muscles, nervous system and vestibular system.

   (1) Muscles: There are over 600 muscles in the body. The use of these muscles should be just enough to hold the body erect in balance and motionless with the least strain.

   (2) Nervous System: Controls the flexing of muscles due to various stimuli.

   (3) Vestibular apparatus (Equilibrium): The stimulus that controls the body balance comes from the vestibular apparatus.

   (4) Neck tendons and skin in the area of the neck also effect balance.

(5) The eyes perform an important function in maintaining the balance by also transmitting impulses for restoring balance any time the eyeball axis is not parallel with the horizon.

TRANSITION: These are some of the factors that control the body. The main goal is stability and immobility.

NOTE: DEMONSTRATOR TAKE A POSITION IN FRONT OF CLASSROOM.

NOTE: SHOW SLIDE #3 - STEPS OF ASSUMING PROPER STANCE. (DEMONSTRATOR IS TALKED THRU THE STEPS OF GETTING INTO THE STANCE)

3. Stance:

a. Position of feet: width, angle, distribution of weight and support area.

b. Knees - firm straight but not locked.

c. Hips - forward to compensate weight of weapon away from body.

d. Abdomen - relaxed.

e. Non-shooting arm - no tension, place hand in pocket.

f. Shooting arm - solid arm control.

g. Weight distribution: A slight shift of weight forward to the toes to reduce the action of the balance correction mechanism. This action is caused by the alternate tensing and relaxing of various back muscles, leg muscles and abdomen.

TRANSITION: It is insufficient merely to assume a stable stance. The shooter must be able to place his body so the shooting arm is pointing naturally toward the target center

4. Position:

a. Position is defined as the relationship of the shooter's body and the target. Before each shot or string it is necessary to check this relationship to the target.

NOTE: SHOW SLIDE #4 - POSITION. (DEMONSTRATOR IS TALKED THROUGH THE METHOD OF ORIENTATION)

b. Method of Orientation:

(1) Face to left of target 40-50 degrees

(2) Turn head to face directly to target

(3) Raise arm to target and close eyes-swing body and arm from side to side as a unit from the ankles, come to rest with natural point at what is assumed to be the target center.

(4) Open eyes--check alignment of arm to target

(5) Move rear foot in direction of error

(6) Recheck

NOTE: DEMONSTRATOR EXCUSED.

    Devote serious attention to perfecting stance and position. Make an intelligent approach to any changes in position or stance by careful analysis. A greatest weakness of getting the proper stance and position is the frequent unwarranted changes caused by the shooter's constant efforts to find a perfect stance and position.

TRANSITION: Now that his body has a natural position in relation to the target, the shooter must maintain control of the pistol and sight alignment by proper grip.

    5. Grip: The proper grip provides maximum control in maintaining sight alignment while applying pressure to the trigger.

QUESTION: What is the most important feature of attaining the grip?

ANSWER:
    a. Uniformity

    b. Requirements of a good grip

        (1) Natural sight alignment

        (2) Firm grip to prevent shifting of weapon

        (3) No change in tightness

        (4) Independent movement of index finger

        (5) Grip must be comfortable

        (6) Force of recoil straight to rear

TRANSITION: Your ability to consistently place the strike of the bullet in the center of the target depends on to a great extent how well you grip the pistol.

NOTE: SHOW SLIDES #5, 6, 7 GRIP.

    c. Method of getting proper grip.

        (1) Hold pistol by barrel with non-shooting hand.

        (2) Spread index finger and thumb to form a "V".

        (3) Bend wrist down.

        (4) Seat pistol in "V" of hand.

        (5) Fit pistol firmly into gripping space.

        (6) Place index finger above trigger guard.

        (7) Grasp stock with lower three fingers.

        (8) Thumb placed in high side of stock.

        (9) Place index finger on trigger.

        (10) Tighten grip to maximum with tremor.

        (11) Relax grip tightly until tremor disappears.

        d.   Check grip to see if it meets requirements.

TRANSITION: Stance, position and grip are fundamentals that most shooters are aware of. The importance of breath control is fundamental and is often overlooked by many shooters.

      6.   <u>Breath Control</u>: The correct method of breathing or holding the breath is essential part of the shooters system of control. It is often overlooked as it is an involuntary reflex action. The object of breath control is to hold sufficient amounts of oxygen in the lungs with comfort during firing so there is no conscious need to breathe. Thinking about having to breathe undermines concentration on sight alignment. Also, breathing is accompanied by a rhythmic movement of the chest, shoulders and stomach and therefore it enlarges the arc of movement and interferes with firing an accurate shot.

<u>NOTE</u>:   SHOW SLIDE #8 - BREATH CONTROL

      When breathing, the chest alternately rises and falls and has periods of rest. This period of rest is the end of the respiratory cycles (exhalation). It is necessary that the shooter hold his breath only after partial exhalation. This method causes less diaphram muscle strain because near this point there is a natural pause in breathing.

        a.   Deep breathing prior to fire commands decreases temporarily the carbon monoxide level of the blood and thereby lengthens the period of comfort while not breathing.

        b.   Breathing coordinated with fire commands aids the shooter to be systematic about deep, regular inhalation immediately before firing.

TRANSITION: The stable foundation provided by the careful selection of a stance, position grip and the means of breath control will permit the shooter to approach the stress of competition firing with confidence. They will be able to apply the other fundamentals of sight alignment and trigger control with greater precision. The proper habits formed during extended practice and match shooting gradually transform the attainment of a minimum arc of movement into a series of involuntary acts.

      B.   <u>Second Student Performance Objective</u>: REVIEWING INSTRUCTION given on proper sight alignment, EXPLAIN the factors necessary to employ and maintain proper sight alignment as described in Chapter II, USAMU Pistol Marksmanship Guide.

<u>QUESTION</u>: What is the definition of perfect sight alignment?

<u>ANSWER</u>: Perfect sight alignment is the proper relationship of the front sight to the rear sight as viewed by the shooter's eye.

      1.   We must consider three factors for proper understanding of sight alignment. First, we must have something at which to aim. This is the target that we use in competition. The target is a stationary object and obviously the shooter does not have to be concerned about the stability of its position.

<u>NOTE</u>:   SHOW SLIDE #1 - A STANDARD AMERICAN FIFTY-YARD PISTOL TARGET.

      2.   The second factor that we must consider is the rear sight.

<u>NOTE</u>:   SHOW SLIDE #2 - REAR SIGHT

      It is necessary to control the rear sight because it contains the notch wherein the front sight is to be centered. No great mental effort is needed here because the rear sight is located almost directly over the pivot point of the wrist and is relatively stable.

3. The third factor that must be considered is the front sight.

NOTE: SHOW SLIDE #3 - FRONT SIGHT

It is centered in the rear sight notch with an equal amount of light on either side of the front sight. The top of the front sight is level with the top of the rear sight.

4. If you focus your vision on the front sight, this big black bullseye becomes a fuzzy, grey color to you. You would then be looking at a perfect sight alignment.

5. As you have been told before, it is physically impossible to get a clear picture of the target and the sights at the same time. Therefore you must find a point of focus. The point of focus must be on the one point that is most difficult to control.

QUESTION: What do you want to look at? The target? The rear sights? Or the front sight?

ANSWER: 6. The most important factor, the front sight. It is by far the most difficult to maintain in a precise manner in relationship to the other two factors. This element alone must be our point of focus. The target must remain out of focus if the shooter is to attain perfect sight alignment. The relationship of the front sight to the rear sight is the primary consideration. To obtain a precise relationship, focus on the front sight and be aware of the rear sight notch relationship.

NOTE: SHOW SLIDE #4 - AN OUT OF FOCUS, FUZZY BULLSEYE, AND A FRONT AND REAR SIGHT SUPERIMPOSED AT A 6 O'CLOCK HOLD.

TRANSITION: This is what you are looking for to enable yourself to attain higher scores, and some day become a champion. But this is only one step. After you have the proper sight alignment, you want the shot to hit the target in the area in which you are moving and holding. Your trigger finger, in applying pressure to fire the pistol, might cause the sight alignment to be disturbed. One one-hundredth (1/100) of an inch error in sight alignment causes three (3) inches or error at fifty (50) yds. Therefore sight alignment and proper trigger control go hand in hand.

C. **Third Student Performance Objective:** Given the description of the controlled action of applying positive trigger control, EXPLAIN how positive trigger control is applied to obtain championship scores as shown in Chapter III, USAMU Pistol Marksmanship Guide.

1. Factors providing for correct trigger control.

a. The pistol is fired by applying positive pressure to the trigger You are in control of a mechanical object and the degree of control applied is measured in most part by the distance away from X ring that the bullet strikes the target.

b. Trigger control is a series of controlled actions. The action of applying positive pressure is necessary to overcome certain conditioned and unconditioned reflexes. Trigger control requires the independent action of the index finger. There is no other movement or further tightening of the hand required to fire the weapon. The index finger bisects the trigger and applies straight-to-the-rear pressure. There is no lateral movement allowed when pressure is applied to the trigger. Otherwise, there will be deviations of sight alignment on a horizontal plane.

QUESTION: Where is the index finger placed so pressure is applied straight to the rear?

NOTE: SHOW SLIDE #5 - CORRECT PLACEMENT OF TRIGGER FINGER.

ANSWER: The placement is such that trigger pressure causes the trigger to travel straight to rear. This control factor aids in maintaining perfect sight alignment.

QUESTION: How does the shooter know that trigger pressure is being exerted straight to the rear?

ANSWER: By dry firing, the shooter can detect any disturbance of sight alignment as the hammer falls.

TRANSITION: Mistakes in applying pressure to trigger can be eliminated if you have a definite plan but it is first necessary to understand unconditioned and conditioned reflexes.

    2. Nerve Processes

QUESTION: What is a reflex action?

ANSWER: a. A reflex action is the active apparatus or muscular parts of the body responding to a stimulus.

        b. These reflex actions are divided into two classes; conditioned and unconditioned.

NOTE: HAVE SOME SHARP NOISE MADE TO REAR OF STUDENTS SUCH AS FIRING A BLANK ROUND.

QUESTION: What happens when you hear an unexpected noise?

ANSWER: You may jump or flinch.

        c. This is an example of an unconditioned reflex. They are instinctive, are not trained to respond but can be influenced.

        d. Conditioned reflexes are reflexes trained to react in a desired manner.

QUESTION: Who has played baseball?

NOTE: WHEN A STUDENT INDICATES THAT HE HAS PLAYED BASEBALL - PITCH A BALL TO HIM. FOLLOW UP WITH:

QUESTION: Did the instructor ask or tell you to catch the ball?

ANSWER: (Student) No.

QUESTION: What type of reflex is this?

ANSWER: Conditioned.

        e. Describe how a child learns to catch a ball. Further reinforce the idea of conditioned reflex by describing how to eat food.

        f. These are examples of a conditioned reflex they must develop, they are subject to the individual's environment and are temporary unless reinforced by training.

        g. The shooter must be able to control to a great degree the unconditioned reflexes, develop and reinforce the conditioned reflexes by constant repetition of carrying out a plan for application of trigger pressure.

3. Sequence of uninterrupted positive trigger control (area shooting).

    a. Slack

    b. Initial Pressure

    c. Positive Pressure - pressure is increased when minimum arc of movement and perfect sight alignment are accomplished. It is increased until the weapon fires. The shooter is committed to continue pressure as long as conditions remain ideal. This method is called the area method of trigger control. If conditions should deteriorate, stop the process and start over.

4. Sequence of indefinite interrupted trigger pressure (Point shooting).

    a. Point shooting trigger control is based on a perfect sight picture. The hold must be motionless, which is of extremely short duration. Then the pressure is applied to the trigger. If the hold or arc of movement enlarges, the trigger pressure is not further increased. Thus, the arc of movement becomes the cue for applying trigger pressure, not the combination of perfect sight alignment and minimum arc of movement. The shooter is setting up the conditions for an uncontrollable reflex action as he presses the trigger quickly before normal arc of movement resumes.

NOTE: SHOW SLIDE #6 - COMMON ERRORS IN TRIGGER CONTROL.

5. Recognition of errors in trigger control.

    a. Jerking (Abrupt application of trigger pressure)

        (1) Straining of muscles in hand arm and shoulder

        (2) Abrupt tightening of hand on stock

        (3) Failure to press trigger straight to rear.

    b. Holding too long. (Trying for a perfect sight picture)

    c. Mechanical adjustments of trigger (eliminate back lash, too heavy or too light triggers and creeping triggers)

TRANSITION: Trigger control is very important in the delivery of a good shot, but it must be used in coordination with all the other fundamentals.

VI. CONCLUSION:

    A. <u>Retain Attention</u>: Ask a top shooter "How can I improve my shooting performance?" His advice to you will be--"start with the fundamentals."

    B. <u>Summary</u>:

        1. The smallest arc of movement must be attained by choosing of a stance that provides balance with the least body movement. The shooter then selects the position that naturally aligns his body to the target, employs a uniform grip on the weapon that gives him natural sight alignment and controls his breathing before and during firing

        2. Proper sight alignment is the relationship between the front and the rear sights. Remember for every .01 inch of error in alignment there will be three (3) inches of error at fifty (50) yds. The front sight must be your point of focus. Concentrate on it and be aware of the relationship with the rear sight notch.

3. The best method for overcoming unwanted reflexes in trigger control is to have a definite, positive plan of action. When this plan is executed, the student should not fail to recognize any errors and take steps to correct them.

C. Application: There is no substitute for proper application of the pistol fundamentals. Before you can attain good scores in pistol marksmanship, you must master all of the fundamentals. You will prove to yourself and to your competition that these fundamentals can be mastered and you can become a champion.

NOTE: DISPLAY THE USAMU PISTOL MARKSMANSHIP GUIDE BY HOLDING IT UP IN VIEW.

D. Closing Statement: This is your shooting bible. The first 71 pages are devoted exclusively to fundamentals. Throughout the rest of the book constant references are made back to the fundamentals.

W. PANEL DISCUSSION   AAPMI&CC 703
                                                              One (1) Hour
                                                              Mar 1975

LESSON OUTLINE

I. LESSON OBJECTIVE: To enable Pistol Marksmanship student to ask questions pertaining to subjects taught during the Army Area Pistol Marksmanship Instructors' and Coaches' Clinic.

II. STUDENT PERFORMANCE OBJECTIVES: As a result of this instruction students must be able to accomplish the following student performance objectives:

   A. As panel moderator, INTRODUCE guest speaker(s), instructors, and name the classes presented by them during the course of instruction as outlined in USAMU Pistol Marksmanship 40 Hour Program of Instruction for instructors and coaches.

   B. As panel moderator, CONDUCT a question and answer session between students and panel, as outlined in USAMU Pistol Marksmanship 40 Hour Program Instruction for Instructors and Coaches.

III. ADVANCE ASSIGNMENT: None

IV. INTRODUCTION:

   A. <u>Gain Attention</u>: I am sure there are many of you that have questions that you would like to ask about pistol marksmanship.

   B. <u>Orient Students</u>:

      1. <u>Lesson Tie-In</u>: In all previous periods of the Pistol Marksmanship Instructors' and Coaches' Course we have been presenting instruction to you. During this hour we would like to hear from you. If you have questions pertaining to any subject taught during the course, direct your question to the instructor of that phase.

      2. <u>Motivation</u>: If there is doubt in your mind about any of the material taught during the course, the panel members will assist you in getting a better understanding of pistol marksmanship.

      3. <u>Scope</u>: During this hour we hope to clear up any weak or vague points that you may have about pistol marksmanship principles taught during the course.

V. BODY:

   A. <u>First Student Performance Objective</u>: As panel moderator, INTRODUCE guest speaker(s), instructors, and name the classes presented by them during the course of instruction as outlined in USAMU Pistol Marksmanship 40 Hour Program of Instruction for Instructors and Coaches.

      1. Introduce guest speaker(s).

      2. Introduce the Pistol Marksmanship Instructors and name the classes presented by them.

      3. Questions asked should be directed to the principle instructor who presented the subject. You may ask for additional comments from other panel members.

4. Please ask questions only about subjects taught during the present course of instruction. If you have questions on other Marksmanship material, we will discuss these queries after the conclusion of the scheduled panel discussion or refer you to persons knowledgeable in that matter.

B. <u>Second Student Performance Objective</u>: As panel moderator, CONDUCT a question and answer session between students and panel, as outlined in USAMU Pistol Marksmanship 40 Hour Program Instruction for Instructors and Coaches.

1. When you have a question, please raise your hand so that an assistant instructor can reach you with a microphone so your question may be heard by everyone in the class.

2. Proceed with panel discussion.

VI. CONCLUSION:

A. <u>Retain Attention</u>: I hope that your questions have been answered satisfactorily.

B. <u>Summary</u>: We state with assurance that the material we have presented to you will aid you in becoming a better pistol competitor and encourage you to use the information received here to train others.

C. <u>Closing Statement</u>: On behalf of the Commander, USAMU and the members of the panel, we hope that you have a successful shooting year. Thank you.

X. PISTOL MARKSMANSHIP EXAMINATION   AAPMI&CC 313
                                                          One Hour
                     LESSON OUTLINE   Mar 1975

   I. LESSON OBJECTIVE: To enable the pistol marksmanship student to successfully complete the Pistol Marksmanship Written Examination and be familiar with the purposes and techniques of writing marksmanship examinations.

   II. STUDENT PERFORMANCE OBJECTIVE: As a result of this instruction the student must be able to accomplish the following student performance objectives:

       A. As an examination writer, STATE the purpose of testing the pistol marksmanship student and EXPLAIN the characteristics that affect the quality of an examination as outlined in Chapter XXII, USAMU Pistol Marksmanship Instructors' and Coaches' Guide.

       B. As the examination supervisor, EXPLAIN test subdivisions, STATE how to obtain assistance if needed during the Pistol examination, EXAMINE test material and make necessary corrections of test as announced by the principal instructor as outlined in Chapter XXII, USAMU Pistol Marksmanship Instructors' and Coaches' Guide.

       C. As the examination supervisor, EXPLAIN student procedure at completion of test: STATE time allowed to complete test, and ANNOUNCE starting time as outlined in Chapter XXII, USAMU Pistol Marksmanship Instructors' and Coaches' Guide.

   III. ADVANCE ASSIGNMENT: None.

   IV. INTRODUCTION:

       A. Gain Attention: Everyone here has learned a little or a lot more about pistol marksmanship. If there is any information about marksmanship that was not explained clearly for you during class, please mention it during the panel discussion which follows.

       B. Orient Students:

           1. Lesson Tie-In: This class is the evaluation period of your marksmanship training. Have you learned enough in this school to improve your performance on the firing line? Can you coach and teach someone to shoot better scores?

           2. Motivation: The test that you are about to take will tell us how effective the instruction was, and how much you know about Pistol Marksmanship. A passing score is _____ or above, and earns an AMU diploma. By achieving a passing grade, you will have the personal satisfaction of knowing that you have a better understanding of those techniques so necessary for successful shooting, coaching, and teaching of pistol marksmanship.

           3. Scope: This test will contain questions about every subject taught during the class to include the three hours you received on Methods of Instruction. At this time the assistant instructors will pass out a test booklet to each student. Do not open this booklet until told to do so by the principal instructor.

   V. BODY:

       A. First Student Performance Objective: As an examination writer, STATE the purpose of testing the pistol marksmanship student and EXPLAIN the characteristics that affect the quality of an examination as outlined in Chapter XXII, USAMU Pistol Marksmanship Instructors' and Coaches' Guide.

The use of tests or examinations to evaluate student performance is a necessary step in the teaching process. Instructors must use tests to determine overall training progress and they must also use tests to check on the effectiveness of instruction. It cannot be assumed that men have learned until the examination stage of instruction has revealed a desirable standard of achievement.

1. PURPOSES OF TESTS OR EXAMINATION.

   a. Tests aid in improving instruction by-

   (1) Providing a basis for evaluation of student performance. Testing enables the instructor to determine which students have attained the minimum standard of performance and which have not. In many cases it is desirable to indicate the extent to which students exceed or fall below the standards required. Students learn different amounts; the grade recorded for each student should be an accurate index of what he has learned. Unless a sound testing program is employed, it is impossible to determine the relative achievement of students.

   (2) Discovering gaps in learning. Properly constructed tests reveal gaps and misunderstandings in student learning. If frequent tests are given, such weaknesses can be discovered and instructors can correct them by reteaching their material.

   (3) Emphasizing main points. A test is actually a valuable teaching device in that students tend to remember longer and more vividly those points which are covered in an examination. Tests encourage students, as well as instructors, to review the materials that have been presented and to organize various phases of instruction into a meaningful set of skills, techniques, and knowledge.

   (4) Evaluating instructional methods. Tests measure not only student performance but also instructor performance. By studying the results of test, instructors can determine the effectiveness of their various methods and techniques of teaching. Specific comparison of subject coverage with student comprehension can be made.

   b. Tests provide an incentive for learning. Students learn more rapidly when made to feel responsible for learning. For example, they are more likely to pay close attention to a training film if they know a test will be given when the picture is over. Generally, instructors who frequently give tests will find that their students will be more alert and learn more. There is a danger, however, in overemphasizing tests and test scores as a basic motivation for learning. Student interest in test scores is a superficial one which can easily lead to efforts to "hit the test" rather than learn the subject matter for its value in the future. Students who study primarily to pass tests may forget what they learn much faster than those who are interested in learning because of the real value to be derived. The instructor should give rigid tests and give them frequently, but they should be designed to require the student to apply what he has been taught.

   c. Tests furnish a basis for selection of subject matter for further training. The results of training tests furnish valuable information upon which to measure student performance. The test results become a basis for determining which subjects should be emphasized in subsequent training and whether the student should advance to a higher training level.

2. CHARACTERISTICS OF AN EXAMINATION. There are six important factors which affect the quality of an examination. These factors, while not considered to be separate and distinct, are defined and discussed separately in order to develop a clear understanding of the characteristics of an examination.

a. *The test must be valid.*

(1) The test must measure what it is supposed to measure; this is its most important characteristics. A test designed to measure what students have learned in a specific training program should measure achievement in that training program and nothing else.

(2) The instructor should, whenever possible, invite the opinion of other competent persons as to the validity of his tests. The test results obtained should be compared with other measures of student achievement. A variety of tests and other evaluating devices must be used in obtaining a valid measure of achievement.

b. *The test must be reliable.*

(1) A test is said to be reliable when it measures accurately and consistently. If the test measures in exactly the same manner each time it is administered, and if the factors that affect the test scores affect them to the same extent every time the test is given, the test is said to be highly reliable. This characteristic of a test is especially important when tests are used to compare the proficiency of several classes.

(2) There are several factors which affect the reliability of a test. In general, the reliability of a test can be raised by increasing its length. The more responses required of students, the more reliable is the measurement of their achievement. Test items should be designed to make it difficult to guess the correct answer. Also, the way in which a test is administered, and the conditions under which it is given should be consistent.

(3) Other characteristics of the test, such as validity and objectivity, also contribute to its reliability.

c. *The test must be objective.* A test is objective when instructor opinion, bias, or individual judgment is not a major factor in scoring it. Objectivity is a relative term. Some tests, such as written examinations which are machine graded, are highly objective; others, such as essay examinations, written exercises, and observation techniques, are less objective. Sometimes observation is the only effective way of determining proficiency. In such cases the instructor must strive to make his observations as objective as possible.

d. *The test should discriminate.* The test should be constructed in such a manner that it will detect or measure small differences in achievement or attainment. This is essential if the test is to be used for ranking students on the basis of individual achievement or for assigning marks. It is not an important consideration if the test is used to measure the level of the entire class or as an instructional quiz where the primary purpose is instruction rather than measurement. As is true with validity, reliability, and objectivity, the discriminating power of a test is increased by concentrating on and improving each individual test item. After the test has been administered, an item analysis can be made which will show the relative difficulty of each item and, of greater importance, the extent to which each discriminates between good and poor students. Often, as with reliability, it is necessary to increase the length of the test to get clear-cut discrimination. Three things will be true of a test that has discrimination:

(1) There will be a wide range of scores when the test is administered to the students who have actually achieved amounts that are significantly different.

(2) The test will include items at all levels of difficulty. Some items will be relatively difficult and will be answered correctly only by the best students; others will be relatively easy and will be answered correctly by most students.

(3) Each item contained in the test will possess discrimination. If <u>all</u> students answer an item correctly, it is probably lacking in this respect.

e. <u>The text must be comprehensive</u>. It must sample liberally all phases of instruction which are covered by the test. It is neither necessary nor practical to test every point that is taught in a course; but a sufficient number of points should be included to provide a valid measure of student achievement in the complete course.

f. <u>The test must be readily administered and scored</u>. It must be so devised that a minimum amount of student time will be consumed in answering each item. The test items must also be constructed so that they can be scored quickly and efficiently.

B. <u>Second Student Performance Objective</u>: As the examination supervisor, EXPLAIN test subdivisions, STATE how to obtain assistance if needed during the pistol examination, EXAMINE test material and make necessary corrections of test as announced by the principal instructor as outlined in Chapter XXII, USAMU Pistol Marksmanship Instructors' and Coaches' Guide.

1. This test has three subdivisions.

   a. <u>Multiple Choice</u>: Multiple choice questions have only one correct answer. Where there is doubt, mark the most correct answer.

   b. <u>True-False</u>: A true answer is true only if it is 100% correct.

   c. <u>Fill in the Blank</u>: Fill in the blank with word or words which are a proven fact or have been included within a statement made by the instructor(s) during the course of instruction.

2. If in doubt, assistance may be obtained from an assistant instructor by raising your hand.

3. If there are any typographical errors or changes to be made in the wording of the test questions, the principal instructor will point out each item.

C. <u>Third Student Performance Objective</u>: As the examination supervisor, EXPLAIN student procedure at completion of test: STATE time allowed to complete tests and ANNOUNCE starting time as outlined in Chapter XXII, USAMU Pistol Marksmanship Instructors' and Coaches' Guide.

1. At the completion of the test, move to the rear of the classroom, turn your test answer sheet in to_____. Answer to the test will be posted at_____.

2. The last page of the booklet is the answer sheet that will be turned in. You may keep the test booklet if you desire. At the top of the answer sheet, fill in the following.

   a. Name

   b. Competitor No.

   c. Unit or address

   d. Post or City

3. You have 50 minutes to complete the test. <u>Begin</u>.

4. The time is_____.

VI. <u>CONCLUSION</u>: N/A (Students will leave the classroom upon completion of examination.

Y. GRAUDATION

AAPMI&CC 704
One Half (½) Hour
Mar 1975

LESSON OUTLINE

I. LESSON OBJECTIVE: To enable Army Area Pistol Marksmanship Instructors' and Coaches' Clinic students who attained a passing grade in the pistol coaching clinic to have their accomplishment recognized and to be awarded an appropriate certificate at a graduation ceremony.

II. STUDENT PERFORMANCE OBJECTIVE: As a result of this instruction, the students must be able to accomplish the following student performance objective:

A. As an instructor in the Army Area Pistol Marksmanship Instructors' and Coaches' Clinic, RECOGNIZE the students' accomplishments by presenting the honor graduate certificate and certificates of completion to all students who attained a passing grade in the pistol Marksmanship clinic as outlined in the Army Area Pistol Marksmanship Instructors' and Coaches' Clinic 40 Hour Program of Instruction.

III. ADVANCE ASSIGNMENT: None

IV. INTRODUCTION:

A. <u>Gain Attention</u>: This period concludes the 40 hour clinic in pistol marksmanship instruction and coaching. It is our hope that you have increased your knowledge of pistol team instruction and coaching techniques.

B. <u>Motivation</u>: Unless you retain this knowledge and pass it on to your new shooters, this clinic will be of limited value to the US Army Marksmanship Training Program.

C. <u>Scope</u>: During this period, you will receive your certificates of accomplishment and be addressed by our guest speaker.

V. BODY:

A. <u>First Student Performance Objective</u>: As an instructor in the Army Area Pistol Marksmanship Instructors' and Coaches' Clinic, RECOGNIZE the students' accomplishments by presenting the honor graduate certificate and certificates of completion to all student who attained a passing grade in the pistol Marksmanship clinic, as outlined in the Army Area Pistol Marksmanship Instructors' and Coaches' Clinic 40 Hour Program of Instruction.

1. Announce seating arrangement for the commanding officer, his staff, marksmanship students and guest speaker.

2. Orient students on procedure to be followed in the graduation ceremony

3. Coordinate with post commander or representative as to the starting time of the graduation ceremony and time needed for an opening or closing statement.

4. Have students seated five minutes prior to the graduation exercise.

NOTE: UPON ARRIVAL OF THE COMMANDING OFFICER, ANNOUNCE HIS ARRIVAL AND TAKE SEATS UPON HIS COMMAND.

5. Introduce guest speaker.

6. Following the guest speaker's remarks, present certificates as rehearsed.

NOTE: HONOR STUDENT IS INTRODUCED AT THIS TIME.

       7. Have guest speaker present honor graduate certificate.

NOTE: AN APPROPRIATE TROPHY MAY BE AWARDED, IF AVAILABLE.

       8. Rise as commanding officer and guest speaker leave the classroom.

       9. Resume command of group. Ask for questions.

VI. CONCLUSION:

    A. <u>Retain Attention</u>: Any time you feel that the Army Marksmanship Program can be of assistance to you in your marksmanship effort, don't hesitate to write the USAMU a letter or contact your local Installation Marksmanship Detachment.

NOTE: ADDRESS: COMMANDER,
                THE UNITED STATES ARMY MARKSMANSHIP UNIT
                FORT BENNING, GEORGIA 31905

    B. <u>Summary</u>: None

    C. <u>Application</u>: When you go back to your respective organizations and clubs apply what you were taught during this pistol marksmanship clinic and your team will develop the potential to move into the winner's circle.

    D. <u>Closing Statement</u>: If this clinic is to be of any value to your unit in the Army, or your pistol club, you will be the ones to make it so. By properly using the material that you were taught during this 40 hour block of instruction in your local marksmanship program, you will be raising marksmanship standards throughout the Nation and the Army.

CHAPTER XVIII
LESSON PLANS FOR SERVICE RIFLE MARKSMANSHIP COACHES" COURSE

A. ORIENTATION   AA/RMCC 124
                                                   1 Hour
                    LESSON OUTLINE   Mar 1975

I. LESSON OBJECTIVE: To enable Army Area Rifle Marksmanship Coaching Clinic students to: explain the purpose and scope of the clinic; gain a general knowledge of International Shooting, the marksmanship ladder, and the requirements for earning the Distinguished Rifleman Badge.

II. STUDENT PERFORMANCE OBJECTIVE: As a result of this instruction the students must be able to accomplish the following student performance objectives:

   A. Given the mission to conduct an Army Area Rifle Marksmanship Instructor/ Coaching Clinic (AARMCC), INTRODUCE the instructor personnel and EXPLAIN the purpose and scope of the clinic, as outlined in AARMCC 40 hour program of instruction and Chapter 1, USAMU Service Rifle Marksmanship Guide.

   B. As leaders to competitive marksmanship, EXPLAIN the International Program for competitive shooting outlined in the International Rifle Marksmanship Guide.

   C. As part of the mission given to conduct an AARMCC, EXPLAIN the competitive marksmanship ladder as outlined in FORSCOM/TRADOC Supplement 1 to AR 350-6.

   D. As part of the mission given to conduct an AARMCC, STATE the requirements for earning the Distinguished Rifleman Badge as outlined in AR 350-6 and FORSCOM/TRADOC Supplement 1 to AR 350-6.

III. ADVANCE ASSIGNMENT: None.

IV. INTRODUCTION:

   A. <u>Gain Attention</u>: Past record-breaking scores that were fired in competition are attributed largely to outstanding coaching. For any team to produce its maximum score, it is necessary for them to have a well trained coach. This clinic is devoted to making each of you a better and more informed coach.

   B. <u>Orient Students</u>:

      1. <u>Lesson Tie-In</u>: This orientation is the first period of a 40 hour course of instruction in the Rifle Marksmanship Instructor/Coaching Clinic. The course includes the basic principles and advanced techniques of Rifle Marksmanship Instruction and Coaching.

      2. <u>Motivation</u>: You, as coaches, will be expected to pass on the information received from this clinic to your present and future team members, both shooters and prospective coaches alike.

      3. <u>Scope</u>: During this hour, I will introduce the instructor personnel. I will discuss with you the purpose of this clinic, the International Shooting Program, the marksmanship ladder, the Distinguished Rifleman Badge, and present to you an outline of the instruction.

V. BODY:

A. <u>First Student Performance Objective</u>: Given the mission to conduct an Army Area Rifle Marksmanship Instructor/Coaching Clinic (AARMCC), INTRODUCE the instructor personnel and EXPLAIN the purpose and scope of the clinic, as outlined in AARMCC 40 hour program of instruction and Chapter 1, USAMU Service Rifle Marksmanship Guide.

    1. Introduce the instructor personnel.

    2. The purpose of this clinic is to better qualify each of you as individual rifle coaches, team coaches, and marksmanship instructors.

    3. Discuss the 40 hour Program of Instruction, and the pertinent training schedule.

<u>TRANSITION</u>: Now that you have been introduced to the instructors and the Program of Instruction, let us discuss other types of shooting.

B. <u>Second Student Performance Objective</u>: As leaders in competitive marksmanship, EXPLAIN International Shooting as outlined in the International Rifle Marksmanship Guide.

    1. Explain the International Program and how shooters are selected.

<u>TRANSITION</u>: No matter what type of shooting you are involved with you must have a starting point, so let's discuss the marksmanship ladder.

C. <u>Third Student Performance Objective</u>: Given the mission to conduct an AARMCC, EXPLAIN the marksmanship ladder as outlined in AR 350-6 and FORSCOM/TRADOC Supplement 1 to AR 350-6.

    1. The marksmanship ladder is a continuous climbing process starting at the company or lowest level of shooting, and climbing to the top rung at All Army with the opportunity of firing as a member of a U.S. Team in international competition.

<u>NOTE</u>: SHOW SLIDE #1 (MARKSMANSHIP LADDER).

    2. Discuss all levels of competitive marksmanship.

        a. Company.

        b. Battalion.

        c. Division.

        d. Major command or CONUS Army.

        e. All-Army and Interservice.

        f. National.

        g. International.

<u>NOTE</u>: TURN OFF SLIDE #1.

<u>TRANSITION</u>: As progress is made up the marksmanship ladder, the shooter's interest toward earning the Distinguished Badge increases.

D. <u>Fourth Student Performance Objective</u>: Given the mission to conduct an AARMCC, DEFINE the requirements for earning the Distinguished Rifle Badge as outlined in AR 350-6 and FORSCOM/TRADOC Supplement 1 to AR 350-6.

1. The Distinguished Rifleman Badge is the highest achievement in service rifle competition.

NOTE: SHOW SLIDE #2 (DISTINGUISHED BADGE).

2. Discuss the levels of competition in which the individual may compete to earn credits for the distinguished badge.

NOTE: TURN OFF SLIDE #2.

a. Define who is eligible to compete for the badge.

3. Discuss the point system used to determine credits for the distinguished badge.

NOTE: SHOW SLIDE #3 (POINT SYSTEM OF CREDITS FOR DISTINGUISHED DESIGNATION).

a. Define who gives the authority for issuing the badge.

NOTE: TURN OFF SLIDE #3.

b. Define who issues the badge.

VI. CONCLUSION:

A. <u>Retain Attention</u>: By paying particular attention to the instruction presented in this clinic, you can aid your team members to have a higher objective to reach as they progress through the various levels of competition.

B. <u>Summary</u>:

1. Review purpose of clinic.

2. Review International Rifle Shooting.

3. Review marksmanship ladder.

4. Review how Distinguished Badge is obtained.

C. <u>Application</u>: The information you have received in this hour will aid you in introducing a program of instruction for your present and future rifle shooters.

D. <u>Closing Statement</u>: This clinic is a fast moving 40 hour course in which you must earnestly apply yourself if you expect to master the material presented.

B. SQUAD SELECTION, ORGANIZATION AND TRAINING   AA/RMCC 125
1 Hr 30 Min
Mar 1975

LESSON OUTLINE

I. LESSON OBJECTIVE: To enable the Army Area Rifle Coaching Clinic students to establish methods for squad selection, squad organization, use of the evaluation file, training of the individual and teams, and equipment used by the shooter and coach.

II. STUDENT PERFORMANCE OBJECTIVES: As a result of this instruction, students must be able to accomplish the following student performance objectives:

   A. Given the mission to organize a rifle team to compete at any level of competition, STATE the policies and procedures of the US Army Marksmanship Training Program, as outlined in FORSCOM/TRADOC Suppl 1 to AR 350-6.

   B. Assigned as rifle squad coach, SELECT a rifle squad to compete in company level competition, as outlined in Chapter 1, USAMU Service Rifle Marksmanship Guide.

   C. Assigned the mission of coaching a rifle team at a higher command level, SELECT the most desirable shooters and coaches from the lower level rifle squads and form a rifle team to compete at the higher command levels of competition as outlined in the USAMU Service Rifle Marksmanship Guide.

   D. Assigned the mission of coaching a rifle team, STATE the minimum equipment requirements to promote maximum performance from a rifle squad, and DESCRIBE the equipment that adds to shooter comfort and convenience, as outlined in the USAMU Service Rifle Marksmanship Guide.

   E. Assigned the mission of coaching a rifle team, PROGRAM the individual and team training necessary to improve the performance of a rifle team in competitive shooting as outlined in the USAMU Service Rifle Marksmanship Guide.

   F. Assigned the mission of coaching a rifle team, EVALUATE, as the training program progresses, the shooters' capabilities, attributes, and equipment in order to assemble the strongest team for representation in team competition, as outlined in the USAMU Service Rifle Marksmanship Guide.

III. ADVANCE ASSIGNMENT: None.

IV. INTRODUCTION:

   A. _Gain Attention_: The rifle team coach is the leader of a rifle team.

   B. _Orient Students_:

      1. _Motivation_: To be an efficient coach, you must be able to select, organize and train your squad into an efficient unit capable of producing winning team scores.

      2. _Lesson Tie-In_: As coaches of the various unit rifle squads in this Army area, many of you may or may not have had experience and previous instruction on team selection, organization and the training of individual and team shooters.

      3. _Scope_: For a rifle team to successfully compete at any level of competition, the rifle coach must be able to:

         a. State the policies and procedures of the US Army Rifle Marksmanship Training Program.

b. Select the most desirable candidates and organize a rifle team at the company level of competition.

c. Select the most desirable candidates and organize a rifle team to compete at a higher command level of competition.

d. Specify the equipment requirements necessary to promote maximum performance from a rifle squad in competition.

e. Program the individual and team training necessary to improve the performance of a rifle team in competition.

f. Evaluate the rifle team members so as to assure the best possible team representation in team match competition.

A. <u>First Student Performance Objective</u>: Given the mission to organize a rifle team to compete at any level of competition, STATE the policies and procedures of the US Army Rifle Marksmanship Training Program as outlined in FORSCOM/TRADOC Suppl 1 to AR 350-6.

1. FORSCOM/TRADOC Suppl 1 to AR 350-6 prescribes the objectives, responsibilities, policies and procedures applicable to the selection and training of individuals and teams for Army-wide competition from company level units through International type competition.

a. Promote training interest and raise standards of proficiency in use of individual service weapons.

b. Establish a broad base of highly qualified marksmanship instructors.

c. Improve firing techniques.

d. Maintain the skill of shooters in the US Army as national leaders in the marksmanship field.

e. Assist the United States in attaining leadership in International marksmanship competition.

2. This regulation requires that one half of the firing members of National Match teams at post, division, major command and the US Army Championships be individuals who have not previously fired in a team match at that current level.

<u>TRANSITION</u>: The first problem that a rifle coach will face in organizing a team is the selection of personnel.

B. <u>Second Student Performance Objective</u>: Assigned as rifle squad coach, SELECT a rifle squad to compete in company level competition, as outlined in Chapter 1, USAMU Service Rifle Marksmanship Guide.

1. The initial selection of squad members at the lower levels of marksmanship participation is generally accomplished as follows:

a. The person who is responsible for organizing the squad will generally ask for experienced or interested volunteers.

b. In the absence of volunteers, personnel firing the highest scores during the annual qualification firing should be considered for selection.

c. The selected group should be interviewed to determine their attitude and desires, and if possible given the instruction and practice to determine their potential.

TRANSITION: Once you are able to organize a rifle team at the lowest level of competition your next logical assignment will be to select qualified shooters for a rifle squad to compete at the next level of competition.

C. <u>Third Student Performance Objective</u>: Assigned the mission of coaching a rifle team at a higher command level, SELECT the most desirable shooters and coaches from the lower level rifle squads and form a rifle team to compete at the higher command levels of competition, as outlined in the USAMU Service Rifle Marksmanship Guide.

1. The task of organizing a squad for the next level of competition is somewhat simpler because random selection is avoided. It is accomplished generally as follows:

a. Consider the opinions and recommendations of the lower level coaches prior to making your squad selection.

b. Review the match bulletins to determine the top competitors and to supplement recommendations as a basis for selection.

c. Personally observe the individual on the firing line during practice and matches as an aid in evaluating his ability.

TRANSITION: Since you are responsible for the selection of individuals that you think will make the best shooters, it is necessary for you to know the attributes of a good shooter.

2. The characteristics or qualifications that are important for the individual firer to possess are:

<u>NOTE</u>: SHOW CHART #1 (ATTRIBUTES OF A SHOOTER).

a. Interest in marksmanship.

b. Previous experience.

c. Eligibility (as outlined in FORSCOM/TRADOC Suppl 1 to AR 350-6).

d. Availability.

e. Physically qualified.

f. Cooperative, honest, ambitious, and reliable.

g. Sportsmanship.

TRANSITION: The attributes listed on this chart are not all of the characteristics necessary to becoming a good shooter. The shooter also needs conditioning and practice in applying the fundamentals of shooting.

D. <u>Fourth Student Performance Objective</u>: Assigned the mission of coaching a rifle team, STATE the minimum equipment requirements to promote maximum performance from a rifle squad, and DESCRIBE the equipment that adds to shooter comfort and convenience, as outlined in the USAMU Service Rifle Marksmanship Guide.

1. To successfully participate in competitive marksmanship, the firer must have certain items of equipment. The items listed below are considered to be the absolute minimum requirements.

NOTE: SHOW CHART #2 (INDIVIDUAL EQUIPMENT)

    a. Rifle and sling with rifle case

    b. Two magazines

    c. Shooting jacket

    d. Shooting glove

    e. Pistol belt and ammo pouch

    f. Shooting glasses

    g. Carbide lamp

    h. Spotting scope

    i. Scorebook

    j. Cleaning equipment

    k. Stool

    l. Shooting box

    m. Ear plugs

    n. Rifle fork

QUESTION: What are some additional items of equipment that will help to promote better scores and provide for shooter comfort and convenience?

ANSWER: 2. Some additional items of equipment that will help to provide shooter comfort and higher scores are:

    a. Shooting mats

    b. Wet weather gear

    c. Sweatbands

    d. Different color glasses

    e. Firmgrip

    f. Ear defenders

3. Those items that are considered necessary equipment for the coach are:

NOTE: SHOW CHART #3 (COACHING EQUIPMENT)

    a. Team spotting scope

    b. Stop watch

    c. Clipboard with pencil

    d. Stool

    e. Plotting sheets

    f. Allen wrench, screwdriver, combination tool, and flash suppressor lock nut pliers

    g. Lens tissue

    h. Ear plugs

    4. Department of the Army has published a table of allowances, CTA 50-900/906/913/915/970 and TDA for Marksmanship Units that prescribes the allowance of equipment normally authorized for rifle teams representing various levels of command.

TRANSITION: With the squad selected and adequately equipped, you now can proceed to train them.

    E. <u>Fifth Student Performance Objective</u>: Assigned the mission of coaching a rifle team, PROGRAM the individual and team training necessary to improve the performance of a rifle team in competitive shooting, as outlined in the USAMU Service Rifle Marksmanship Guide.

    1. After selection of individuals for the squad, at least two weeks, time permitting, should be devoted to the instructional training phase. This instructional training should include:

      a. Physical conditioning

      b. Mental conditioning

      c. Rules and regulations

      d. Safety

      e. Fundamentals of marksmanship

NOTE: SHOW CHART #4 (FUNDAMENTALS OF MARKSMANSHIP)

        (1) Aiming

        (2) Positions

        (3) Trigger control

        (4) Rapid fire

        (5) Sight adjustment

        (6) Effects of weather

        (7) Zeroing

        (8) Use of scorebook

        (9) Use of scope

      f. Detection and correction of errors

        g.   Dry firing

        h.   Range firing

      2.   The primary goal for the rifle squad coach is to win the team match at any level of competition. Therefore, at the conclusion of the instructional training phase, and based upon the progress of each individual, team groups should be organized. A concentrated effort must be placed on team shooting. However, time must be reserved for individual shooting, since a greater portion of all rifle competition is of an individual nature.

      3.   Having formed the team(s) in accordance with each individual's ability and adhered to organizational requirements, a training program which has proven successful in previous experience should include the following:

NOTE: SHOW CHART #5 (TRAINING PROGRAM)

        a.   Physical Conditioning, Limbering Up Exercise, and Organized Athletics. These exercises should be performed daily during practice and discontinued two days prior to and during competitive firing.

        b.   Individual Shooting: Monday and Thursday of each week should be devoted to individual firing. During this phase of training, those deficiencies noted by the coaches on team days, can and should be corrected.

        c.   Team Shooting: Tuesday and Wednesday should be devoted to team practice. Fridays should be devoted to team record firing. The shoulder to shoulder team record competition on Friday will aid in the mental conditioning of the team shooter.

        d.   Match Shooting: All available competitive matches should be entered. There is no substitute for this type of training which offers the true competitive spirit of marksmanship. Here the individual shooter is afforded the opportunity to build and develop the mental conditioning and discipline so necessary to developing winning teams.

      4.   When a weakness is noted, a review of the techniques and fundamentals pertinent to the area of poor performance is recommended.

TRANSITION: The head coach is responsible for choosing the individuals who will fire in team matches. To accomplish this after the selection and training of the firers, it is usually necessary to rely on the observations of the team coaches. An efficient method of recording the evaluation of individuals on each team is by use of the Evaluation File.

    F.   <u>Sixth Student Performance Objective</u>: Assigned the mission of coaching a rifle team, EVALUATE as the training program progresses, the shooters' capabilities, attributes, and equipment in order to assemble the strongest team for unit or US Army representation in team competition, as outlined in the USAMU Service Rifle Marksmanship Guide.

      1.   The Evaluation File provides the team coach with a systematic method of recording data concerning each firer. This file, if properly used and maintained, will assist in the planning of the training program.

        a.   Cover sheet (Instructions)

NOTE: SHOW CHART #6 (PERSONAL INFORMATION)

        b.   Personal Information

        (1)   Determine the quality of a shooter's mental attitude by answering these questions.

      (a) Is he easily perturbed?

      (b) Does he give up easily?

      (c) Is he easily discouraged by unfavorable conditions?

      (d) Is he susceptable to rumors?

      (e) Is he disturbed by scores fired by other competitors?

      (f) Does he worry about equipment?

      (g) Does he lack the will to win?

    (2) Personal behavior indicates other traits of character which measure the capabilities and shortcomings of a potential team member. The team member should be:

      (a) Cooperative

      (b) Ambitious

      (c) Aggressive

      (d) Honest

      (e) Reliable

      (f) Neat in appearance

    (3) Remarks: Describe the shooters abilities, characteristics and potential.

    (4) Personal Information Sheet. This information sheet is considered confidential and should be only for the use of the coach and squad officials. Each coach will make his remarks in such a manner as to reflect his considered opinion and what action he has taken to orient the shooter concerning his particular problem if a problem is noted. Extreme care is indicated here, and it is suggested that this portion of the file be filled in after the shooter has been under observation for at least a week. Because of the subjective nature of these comments, they should be used only for corrective action by the coach who maintains the file. Critical remarks are worthless unless they are followed up with appropriate counseling. If improvement is noted, these remarks should be destroyed. This sheet should be destroyed when the shooter changes coaches.

  c. National match shooters' graph and daily log section.

NOTE: SHOW CHART #7 (SHOOTERS" GRAPH)

    (1) Shooter's Graph: The shooter's graph is used for the entry of all practice, match, and average scores, whether they are fired as a member of a team or as an individual. All other factors being equal, this is the information which will provide the basis for the final selection of team members. The following color code is used to record and qualify scores:

      (a) Red - record practice

      (b) Blue - match

      (c) Black - average

(d) Solid Color - team shooting

(e) Diagonally striped - individual shooting

NOTE: SHOW CHART #8 (DAILY LOG)

    (2) Daily Log: The daily log is used to record individual strengths and weaknesses of the firers. Comments must be of a factual nature and used to qualify each score on the Shooter's Graph. Examples are: effects of weather, coaching or shooting errors, equipment failure, etc. These comments present a picture of the shooter's progress and are particularly useful in programing instruction and remedial training.

VI. CONCLUSION:

 A. <u>Retain Attention</u>: Winning scores were fired in the National Team Matches by the All Army Team in eleven of the last thirteen years. Next year will be another victory year for the U.S. Army Rifle Team.

 B. <u>Summary</u>:

NOTE: SHOW CHART #9 (SUMMARY)

  1. It takes a well organized team to produce record breaking scores. To build an Army Area rifle team capable of winning the All Army matches requires a great amount of hard work. Starting at the lowest level of competition, each head coach is responsible for the selection and training of his team members.

   a. Initial selection: Attributes of a shooter

   b. Subsequent Selection: As we proceed up the marksmanship ladder through the different levels of competition, the best shooters and coaches are chosen for the succeeding level of competition where the training program continues.

  2. Individual Training.

   a. Fundamentals of Marksmanship

   b. Physical Conditioning

   c. Care of the Weapon

   d. Dry firing

   e. Mental conditioning

   f. Range firing

   g. Match firing

  3. Team Training: Winning the team match is the overall goal. Therefore, the majority of the training program is devoted to team shooting.

  4. Evaluation of your Shooters: The evaluation file will help you to choose the shooters that will make up your team in the big team match.

 C. <u>Application</u>: Use these methods of selecting and training your shooters whenever you are responsible for organizing a Rifle Team.

 D. <u>Closing Statement</u>: By effectively organizing a Rifle Team at every level of competition, you will be promoting the rifle marksmanship program throughout the Army.

C. SELECTION OF A RIFLE AND CARE AND CLEANING   AA/RMCC 126
                                                        One (1) Hour
                                                        Mar 1975

LESSON OUTLINE

I. LESSON OBJECTIVE: To enable the Army Area Rifle Coaching Clinic students to select and assemble the best weapon available for competitive shooting; whether the M-14 or M-16 rifle. Also explain the care and cleaning procedure of the M-14 and M-16 rifle used for competitive firing.

II. STUDENT PERFORMANCE OBJECTIVE: As a result of this instruction, Army Area Rifle Marksmanship Coaching Clinic students must be able to accomplish the following student performance objectives:

   A. Assigned the duty as an instructor on the competitive M-14 rifle, DESCRIBE the selection of parts and the assembly techniques as outlined in Chapter 1, USAMU Service Rifle Marksmanship Guide.

   B. Assigned the duty as an instructor on the competitive M-14 rifle, EXPLAIN the cleaning procedures used on the M-14 as outlined in Chapter 1, USAMU Service Rifle Marksmanship Guide.

   C. Assigned the duty of instructing on the competitive M-16 rifle, DESCRIBE the method of selecting component parts and the assembly techniques as outlined in Chapter 1, USAMU Service Rifle Marksmanship Guide.

   D. Given the mission of instructing the competitive M-16 rifle, EXPLAIN the cleaning procedures used as outlined in Chapter 1, USAMU Service Rifle Marksmanship Guide.

   E. Given the mission of instructing on the competitive M-14 and M-16 rifles, DESCRIBE the special care that is afforded these weapons as outlined in Chapter 1, USAMU Service Rifle Marksmanship Guide.

III. ADVANCE ASSIGNMENT: None.

IV. INTRODUCTION:

   A. <u>Gain Attention</u>: (Walk to center of stage and pause) This is one of the best National Match M-14 Rifles available today. This rifle was accurized in the USAMU Ordnance Shop at Fort Benning, Georgia.

   B. <u>Orient Students</u>:

      1. <u>Lesson Tie-In</u>: During the class on team selection, you were taught the requirements for equipping a rifle team. The most important piece of equipment is your weapon.

      2. <u>Motivation</u>: If you are given the responsibility of organizing and training a rifle team, the only way you can measure the shooter's ability or progress is by his shot groups or scores on the rifle range. The only way that consistently dependable shot groups and scores can be obtained is with an accurate rifle.

      3. <u>Scope</u>: During this period of instruction I will explain and demonstrate the selection and assembly of the M-14 and M-16 rifles. Also included will be the care and cleaning procedures used on M-14 and M-16 rifles used in competition.

V. BODY:

    A. <u>First Student Performance Objective</u>: Assigned the duty as an instructor on the competitive M-14 rifle, DESCRIBE the selection of parts and the assembly techniques as outlined in Chapter 1, USAMU Service Rifle Marksmanship Guide.

        1. The selection of a barrel and receiver is most important.

            a. Air gauge or visually inspect the bore.

            b. Breechbore gauge the throat of the bore (304 tolerance).

            c. Check headspace (1.631 to 1.636) and don't interchange the bolts without first gauging with headspace gauge.

            d. Inspect the crown for burrs. If a burr is found the muzzle must be crowned and faced.

<u>TRANSITION</u>: Once you have selected a barrel and receiver you can proceed to selecting and fitting the remaining parts.

        2. The component parts of the rifle must be hand fitted to the barrel and receiver.

            a. Knurl or peen the barrel and replace the operating rod guide.

            b. Attach the front band to gas cylinder.

            c. Peen the gas cylinder barrel splines.

            d. Tighten the gas cylinder lock to 6:00 with the barrel.

            e. Gas cylinder plug tightened to 120 inch pounds.

            f. Ream the flash suppressor with #7 taper reamer.

            g. Peen the flash suppressor barrel splines.

            h. Place and align the flash suppressor onto the barrel.

            i. Cut and fit the handguard, allowing clearance between it and the stock and between it and the front of the receiver.

            j. The operating rod should ride close to the receiver without touching, and must strike the center of the gas piston.

            k. Attach the connector. It should be bent, holding the rod in place.

            l. Insert operating rod spring guide and spring. Guide should be replaced with a round, burr free, tapered type. Lubricate outside of operating rod and spring guide.

            m. Firing mechanism should have slack but no creep. It should be clean breaking at about 4 3/4 pounds of pressure.

            n. Stock should not be oil soaked or cracked. There are two methods of eliminating free play between the stock and receiver.

                    (1) Glass bedding - Rout and glass the stock for a custom fit between stock and receiver.

                    (2) Shimming - Using manila folder material, shim the top rails horseshoe and recoil shoulders; also the trigger housing.

          3. Assemble the three main groups.

               a. Place the receiver into the stock.

               b. Place the firing mechanism into the stock. Pressure should be required to lock the trigger guard in position.

          4. After the rifle has been assembled, give it a visual and functional check.

               a. Tension on foreend of stock 8 to 12 pounds.

               b. Front sight tight.

               c. Rear sight tight and operating properly.

               d. Piston free.

               e. All safety features operational.

TRANSITION: Having seen how much time and effort goes into assembling a match condition weapon, you can understand the need for proper cleaning.

     B. Second Student Performance Objective: Assigned the duty as an instructor on the competitive M-14 rifle, EXPLAIN the cleaning procedures used on the M-14 as outlined in Chapter 1, USAMU Service Rifle Marksmanship Guide.

          1. Cleaning of the rifle is a must in order to retain its accuracy. There are several items of equipment that should be obtained in order to properly clean the weapon.

               a. Military issue materials.

NOTE: POINT OUT EACH ITEM AS IT IS DISCUSSED.

                    (1) Steel cleaning rod.

                    (2) Chamber brush (30 cal).

                    (3) Lubricating oil.

                    (4) Bore cleaner.

                    (5) Patches.

                    (6) Bore brush.

                    (7) Rifle grease.

                    (8) Rag.

                    (9) 45 Cal. cleaning rod (steel).

                    (10) 45 Cal. bore brush.

    b. Purchased items.

      (1) Plastic coated cleaning rod.

      (2) Toothbrush.

      (3) Shaving brush.

      (4) Dri-slide.

  2. The rifle should be inspected and cleaned daily. The M-14 should not be disassembled except when in need of repair.

    a. Swab bore four or five times with a brass brush and bore cleaner

    b. Swab the chamber with chamber brush and bore cleaner.

    c. Swab flash suppressor with a 45 caliber bore brush and bore cleaner.

    d. Clean the sights and remove all external dirt.

    e. Close the bolt, invert the rifle and place one or two drops of dri-slide on the piston. Activate the operating rod several times with the muzzle of the weapon up until the piston moves freely.

    f. Dry the suppressor with clean patches.

    g. Remove excessive lubricant if necessary.

    h. Swab bore with clean patches.

    i. Swab chamber with clean patches.

    j. Clean face of bolt and interior of receiver with brush and rag.

    k. Relubricate if necessary.

      (1) Lip of the receiver.

      (2) Locking lugs.

      (3) Rod guide grooves.

      (4) Bolt camming lug.

      (5) Bolt camming recess.

      (6) Guide groove of connector.

TRANSITION: The same effort must be put into the accurizing and cleaning of the M-16 rifle as went into the M-14 competitive rifle.

  C. <u>Third Student Performance Objective</u>: Assigned the duty of instructing on the competitive M-16 rifle, DESCRIBE the method of selecting component parts and the assembly techniques as outlined in Chapter 1, USAMU Service Rifle Marksmanship Guide.

    1. The first step is again the selection of a barrel and receiver.

      a. Air gauge or use a barrel erosion gauge.

     b. Check muzzle and crown for burrs. The muzzle can be crowned and faced if needed.

     c. Check gas system for tightness.

     d. Check flash suppressor for tightness.

     e. Check handguards for cracks or looseness.

     f. Check upper receiver for large pits or cracks.

     g. Rear sights should be tight and operational.

<u>TRANSITION</u>: Once you have selected the upper receiver and barrel you can proceed to the selection of the lower receiver and stock.

    2. Select the lower receiver and stock.

     a. Check take down pins for tightness.

     b. Check fit between stock and lower receiver.

     c. Check that trigger and sear pin holes are not oversize.

     d. Check buffer and spring for freedom of movement inside lower receiver extension.

<u>TRANSITION</u>: After selecting the lower receiver and stock, then you can assemble the weapon.

    3. Assembly of the upper and lower receivers.

     a. Seat the receiver pivot pin. The upper and lower receivers should fit tight together with no slope.

     b. Seat the take down pin, this should pull the receivers together making them one piece.

     c. The trigger should have a clean break at about 4 3/4 pounds of pressure.

<u>TRANSITION</u>: After the M-16 rifle has been assembled, a function check and visual inspection should be made.

    4. Function check and visual inspection.

     a. Pull the bolt to the rear and release it; pull the trigger and it should have no creep and weigh about 4 3/4 pounds.

     b. Front sight tight.

     c. Rear sight tight and operational.

     d. All safety features operational.

<u>TRANSITION</u>: Having learned how to select and assemble the M-16, you need to know how to clean the rifle.

   D. <u>Fourth Student Performance Objective</u>: Given the mission of instructing on the competitive M-16 rifle, EXPLAIN the cleaning procedures used as outlined in Chapter 1, USAMU Service Rifle Marksmanship Guide.

1. Cleaning the M-16 rifle is a must for accurate shooting.

    a. Military issue of cleaning materials.

        (1) Steel cleaning rod - 22 cal.

        (2) LSA.

        (3) Borebrush - 22 cal.

        (4) Chamber brush - M16A1.

        (5) Patches.

        (6) Pipe cleaners.

        (7) Bore cleaner.

        (8) Rag.

    b. Purchased materials.

        (1) Plastic cleaning rod - 22 cal.

        (2) Carbon solvent.

        (3) Toothbrush.

        (4) Shaving brush.

2. The M-16 should be inspected and cleaned daily. The bore should be cleaned after 30 to 40 rounds, because of lead buildup in the bore.

    a. Swab the bore 4 or 5 times with bore cleaner and brass brush.

    b. Swab the chamber with chamber brush and bore cleaner.

    c. Disassemble bolt and remove all carbon deposit.

    d. Clean gas extension and tube with pipe cleaner.

    e. Clean inside of lower receiver with brush and rag.

    f. Swab bore with clean patches.

    g. Swab chamber with clean patches.

    h. Assemble bolt with one drop of LSA on piston rings and gas extension.

    i. Relubricate.

        (1) Locking lugs on barrel.

        (2) Guide groove in bolt.

        (3) Hammer and trigger.

TRANSITION: Having seen how much time goes in accurizing and cleaning the M-14 and M-16 rifles, you must also know how to care for them.

    E.   <u>Fifth Student Performance Objective</u>: Given the mission of instructing on the competitive M-14 and M-16 rifles, DESCRIBE the special care that is afforded these weapons as outlined in Chapter 1, USAMU Service Rifle Marksmanship Guide.

        1.   Taking care when storing or transporting these rifles will extend the life of the accurized weapon.

            a.   Transporting of rifle.

                (1)   Enclosed in rifle case.

                (2)   Place in secure "sights up" position.

                (3)   Bolt forward, hammer released.

            b.   Storing the rifles.

                (1)   Hung by lower sling swivel, bolt forward, hammer released and flash suppressor free of contact.

                (2)   Bore should have light coat of oil.

            c.   On the range.

                (1)   Protect from weather with case or plastic cover.

                (2)   Don't jar or drop weapon.

                (3)   Keep weapon sights up on a rifle fork.

NOTE: SHOW PLASTIC CASE AND RIFLE FORK.

   VI. CONCLUSION:

    A.   <u>Retain Attention</u>: To properly equip your team with highly accurate weapons, certain steps should be taken.

    B.   <u>Summary</u>:

        1.   Select the best weapons available. Careful selection and correct assembly will result in a more accurate weapon.

        2.   All the techniques of assembling rifles that have been demonstrated here can be used by you, if you have the know how and the material.

        3.   Care and cleaning is directly related to the accuracy of the rifle.

    C.   <u>Application</u>: Select and accurize your weapon carefully, have your shooters clean them on a daily basis and the result will be better shot groups and scores.

    D.   <u>Closing Statement</u>: Accurate weapons are a must for winning any team match.

D. EFFECTS OF THE WEATHER AND USE OF THE
SCOREBOOK AND TELESCOPE

AA/RMCC 127
1 Hr
March 75

LESSON OUTLINE

I. LESSON OBJECTIVE: To enable students of the Army Area Rifle Clinic to determine what effects wind, light, and temperature have on the strike of the bullet, how to compensate for these effects, and record this information for future reference.

II. STUDENT PERFORMANCE OBJECTIVE: As a result of this instruction, students must be able to accomplish the following student performance objectives:

A. Before firing, DETERMINE the direction of the wind, ESTABLISH its value, COMPUTE its velocity and APPLY the correction to the rear sight to compensate for the strike of the bullet as outlined in Chapter 1, Section IV, USAMU Service Rifle Marksmanship Guide.

B. During the preparation period, properly POSITION the spotting scope and ADJUST the focus so that any change in direction or velocity of the wind may be detected by observing the mirage as outlined in Chapter 1, Section IV, USAMU Service Rifle Marksmanship Guide.

C. While aiming, DETERMINE the direction of the light and what effect it may have on the bullseye as outlined in Chapter 1, Section IV, USAMU Service Rifle Marksmanship Guide.

D. Before firing, DETERMINE what effect temperature may have on the shooter and the strike of the bullet as outlined in Chapter 1, Section IV, USAMU Service Rifle Marksmanship Guide.

E. As a shooter, IDENTIFY those portions of the scoresheet that are filled out before firing, during firing, and after firing for slow and rapid fire as outlined in Chapter 1, Section IV, USAMU Service Rifle Marksmanship Guide.

III. ADVANCE ASSIGNMENT: None

IV. INTRODUCTION:

A. <u>Gain Attention</u>: Winning or losing a rifle match may depend on your first shot. This shot will measure your ability to use the knowledge that is available to all of us.

B. <u>Orient Students</u>:

1. <u>Lesson Tie-In</u>: You have received instruction on aiming, positions, and trigger control. This is all you would need if you were shooting in a vacuum. But you will be shooting on a range where the weather will effect you, and the strike of your bullet on the target. Therefore, you must know what the effects are and how to compensate for them.

2. <u>Motivation</u>: You may be the best shot on the range as long as the weather is good. But what if the weather is bad? By learning the effects of the weather, the use of the spotting scope, and scorebook; and by putting this information to work for you. You will improve your shooting performance.

3. <u>Scope</u>: During this period of instruction, I will explain and demonstrate the methods of determining the number of clicks of windage and elevation that are needed to compensate for the effects of wind, light, and temperature and how to properly record this information in your scorebook for future reference.

V. BODY:

    A. <u>First Student Performance Objective</u>: Before firing, DETERMINE the direction of the wind, ESTABLISH its value, COMPUTE its velocity and APPLY the correction to the rear sight to compensate for the strike of the bullet as outlined in Chapter 1, Section IV, USAMU Service Rifle Marksmanship Guide.

QUESTION: What effect will the wind have on the bullet?

<u>ANSWER</u>:   1. The wind will effect the bullet laterally, and has a greater effect than any other element of the weather.

    2. A shooter can determine the wind direction by watching various indicators on the range.

QUESTION: What are these indicators?

<u>ANSWER</u>:
    a. Range flags

    b. Grass and trees

    c. Sense of feel

    3. After determining the direction of the wind, you must be able to classify its value.

QUESTION: How would you classify the value of the wind?

<u>ANSWER</u>:   a. By use of the clock system.

NOTE: SHOW CHART #1 (WIND VALUE CHART).

    b. With the shooter at the center of the clock and the target at twelve, the wind is assigned three values:

    (1) Full value

    (2) Half value

    (3) No value

QUESTION: If the wind is blowing from six and twelve o'clock, what effect will it have on the bullet?

<u>ANSWER</u>:   4. Six and twelve o'clock are no value winds.

    a. Laterally, the wind has no affect on the bullet.

    b. At long range, no value wind may cause slight vertical displacement.

    5. After deciding the value of the wind, its velocity must be computed, this may be done in different ways.

NOTE: COVER CHART #1 (WIND VALUE CHART)

    a. Estimation: Under 5 mph, smoke will drift; 5 to 10 mph, leaves will be in constant motion; 10 to 15 mph, dust will blow about and trees sway.

        b. Using the range flag, estimate the angle from the bottom edge of the flag to the flagpole, divide this angle by the constant figure four, the answer will be velocity of the wind in miles per hour.

NOTE: WORK EXAMPLE ON CHALKBOARD

        6. The number of clicks of windage needed to compensate for a full value wind are determined by the use of the wind formula, (velocity times range in hundreds of yards divided by the constant figure ten equals the number of clicks of windage).

NOTE: WORK EXAMPLE ON CHALKBOARD

QUESTION: The wind is blowing from 11 o'clock at a speed of 15 mph, how many clicks of windage are needed to compensate for its affect, and which direction the rear sight should be moved if you are firing from the 600 yard line?

ANSWER: 7. Eleven o'clock is a half value wind, divide the answer by two. (4½ clicks) Use next higher number, move rear sight left.

TRANSITION: Another method of estimating the wind, and the best method of spotting small changes in it is by reading the mirage through the spotting scope.

        B. _Second Student Performance Objective_: During the preparation period, properly POSITION the spotting scope and ADJUST the focus so that any change in direction or velocity of the wind may be detected by observing the mirage as outlined in Chapter 1, Section IV, USAMU Service Rifle Marksmanship Guide.

QUESTION: What is mirage as we know it in shooting?

ANSWER: 1. Mirage is the reflection of light through layers of air of different temperature and density.

TRANSITION: Before you can effectively read the mirage you must position the spotting scope and properly adjust the focus.

NOTE: DEMONSTRATOR WILL SHOW HOW TO POSITION AND FOCUS THE SPOTTING SCOPE IN THE PRONE POSITION.

        2. To position the spotting scope you:

        a. Assume a position on the firing line.

        b. Place the scope beside you so you may look through it without straining or shifting your position.

        3. The scope must be focused at a point at mid-range since that is where the wind will have the most effect on the bullet.

        4. There are two methods to adjust the focus of the scope to read the mirage.

        a. Focus the scope clearly on the target, then turn the eyepiece one quarter turn counter clockwise.

        b. Focus the scope clearly on an object midway between the firing line and the target, then move the scope so you are looking at the target.

5. Wind direction can be checked by rotating the scope until you see the mirage come to a boil, the axis of the scope is then directly in line with the wind direction.

TRANSITION: Mirage moves with the wind, with practice you will be able to estimate the velocity and spot changes in wind direction.

NOTE: SHOW CHART #2 (MIRAGE CHARTS)

      6. Mirage is divided into three classifications:

         a. Boiling, which shows no wind, or wind from six or twelve o'clock.

         b. Medium, the mirage seems wavy in a 0-7 mph wind.

         c. Fast, the mirage seems to flatten out at 8-12 mph.

      7. Over 12 mph, any wind is hard to estimate by reading mirage alone. Therefore, you must include one or more of the other methods in estimating the velocity of the wind when it is over 12 mph.

NOTE: COVER CHART #2 (MIRAGE CHART)

TRANSITION: The wind and its affects is not the only weather element you must be aware of Many shooters are affected by changing light conditions.

    C. **Third Student Performance Objective**: While aiming, DETERMINE the direction of the light and what effect it may have on the bullseye as outlined in Chapter 1, Section IV, USAMU Service Rifle Marksmanship Guide.

      1. A change in light does not affect all shooters the same, the general tendency is to shoot high in dull light and low in bright light.

NOTE: SHOW CHART #3 (EFFECTS OF LIGHT)

      2. The charts will show how different lights can appear to change the bullseye and therefore change the shooters' point of aim on the target.

         a. A perfect sight picture. (describe)

         b. Sight picture seen on normal days. (describe)

         c. Sight picture seen on dull day with low humidity.

         d. Sight picture seen with bright light directly on the face of the target with low humidity.

         e. Sight picture seen with bright light and high humidity.

         f. Targets appearance with very bright light coming from the side.

      3. The general rule is "light up - sights up, light down - sights down". However, as stated before, this does not happen to all shooters and each of you will have to decide how the changing light affects you.

      4. The faster the wind velocity and heavier the mirage, the further the bullseye is displaced.

      5. The bullseye may look as if it is moving around on the target. This is caused by light refraction.

TRANSITION: Light may not affect all shooters in the same way but a change in temperature does. Actually, it changes the muzzle velocity of the ammunition.

       D.   <u>Fourth Student Performance Objective</u>: Before firing, DETERMINE what effect temperature may have on the shooter and the strike of the bullet as outlined in Chapter 1, Section IV, USAMU Service Rifle Marksmanship Guide.

         1.   A $20°$ change in temperature will change the muzzle velocity about fifty feet per second and move the strike of the bullet about six inches at six hundred yards, when using match ammunition.

NOTE: SHOW CHART #4 (TEMPERATURE CHANGE EFFECTS)

            a.   At $50°$ the muzzle velocity is 2500 feet per second.

            b.   At $70°$ the muzzle velocity is 2550 feet per second.

            c.   At $90°$ the muzzle velocity is 2600 feet per second and the bullet will strike the target about one foot higher at six hundred yards than it did with a $50°$ temperature.

TRANSITION: The only way you can keep up with effects of the weather is to keep an accurate record of how it affects you.

       E.   <u>Fifth Student Performance Objective</u>: As a shooter IDENTIFY those portions of the scoresheet that are filled out before firing, during firing, and after firing slow and rapid fire as outlined in Chapter 1, Section IV, USAMU Service Rifle Marksmanship Guide.

NOTE: HAND OUT SCORESHEETS.

NOTE: STUDENTS FILL OUT SCORESHEET AS PI EXPLAINS.

         1.   The scorebook is used to record every shot fired by the shooter. It is also used to record the weather conditions and their effects on the strike of the bullet and the shooter. If used properly, it will provide the necessary information for initial sight settings at each range. It provides a basis for analyzing the performance of the shooter and his rifle, and is a valuable aid in making bold and accurate sight changes.

         2.   The following procedure should be used for filling out and maintaining the scorebook in <u>slow fire</u>.

            a.   Before firing, the date, hour, rifle number, ammunition type, temperature, target number, place, light (word description and direction), wind (word description and direction), sight picture to be used, windage zero, elevation used, and any other appropriate remarks to aid the shooter are entered in the space provided.

            b.   During firing, a strict sequence must be followed.

               (1)   If a wind is blowing, the value must be determined and set on the sights.

               (2)   Entries in the scorebook should be made while the target is in the pit.

                  (a)   After firing the first round the windage used should be entered in the appropriate space in the scorebook and the call plotted.

(b) When the target is marked, the call is compared to the hit, any needed sight changes are made, and the second shot fired.

(c) After the second shot is fired the call is plotted along with any sight changes made on the previous shot, and then the location of the first shot plotted.

c. Upon completion of firing, all remaining entries in the scorebook are filled out. The results should be analyzed and studied very carefully.

3. The following procedure should be used for fillout and maintaining the scorebook in rapid fire.

a. Before firing, the shooter records the same information as he did for slow fire.

b. On the firing line the sequence is different than slow fire.

(1) He makes his final windage correction shortly before the targets appear and applies this to the sights. While firing he should mentally note any shots called out of the group.

(2) Immediately after firing, the shooter plots his calls. He does this by noting any erratic shots on the plotting bullseye.

(3) When the target is marked he should plot all visible hits with an "X" and compare his calls to his hits.

c. Upon completion of firing, all remaining entries in the scorebook are filled out. The results should be analyzed and studied very carefully.

4. The most competent rifleman would not be able to consistently hit the center of the target if he were unable to analyze his performance, or if he had no record of his performance or of the conditions that affect his firing.

VI. CONCLUSION:

A. Retain Attention: To shoot winning scores you must understand the effects of the weather and use them to your advantage. Let's watch an experienced shooter fire his first shot at six hundred yards.

B. Summary:

NOTE: DEMONSTRATOR WITH SHOOTING EQUIPMENT ACTS OUT INSTRUCTOR'S SUMMARY.

1. While on the ready line he determines the direction of the wind by noticing:

a. Range flags

b. Grass

c. Trees

d. Smoke

2. He decides its value by using the clock system, it is half value, coming from five o'clock.

        3. He computes its velocity by dividing the angle of the range flag from the flagpole by four, it is 15 mph.

        4. Using the formula, range times velocity divided by ten, he needs nine clicks of wind for a full value wind at this range, however, since this is a half value wind he divides it in half and rounds it off to five clicks.

        5. He chooses a level place on the firing line and assumes a good position.

        a. He positions his scope so he can look through it without moving his position.

        b. He focuses the scope clearly on the target, then backs it off a quarter turn, counter clockwise, so he can read the mirage at midrange.

        c. He checks the mirage and verifies the wind velocity.

        d. He sets his windage and elevation zero. The temperature was 55° when he zeroed and if it is 70° today and he is using the same ammunition, he comes down one click.

        e. He loads.

        f. He makes a last check of the mirage. It has not changed and the wind is steady. He moves the windage knob five clicks right.

        g. The light is normal and will not affect him.

        h. He places the rifle in his shoulder, goes into position and:

        (1) Breathes

        (2) Relaxes

        (3) Aims

        (4) Takes up slack

        (5) Applies proper trigger control

NOTE: WHEN TRIGGER SNAPS ASSISTANT PLACES SPOTTER IN CENTER OF "X" RING.

        i. And his first shot is a "X".

        j. He then records his call and his initial windage setting in his scorebook and prepares for his second shot.

        C. <u>Application</u>: Study how the weather affects you, and use this knowledge when you go to the range to shoot. Don't wait until you shoot a nine to move your sights

        D. <u>Closing Statement</u>: Winning scores are fired by shooters who use their knowledge of how to react to existing conditions. Make your first shot an "X".

E. MENTAL AND PHYSICAL CONDITIONING   AA/RMCC 128
                                                            30 Minutes
                        LESSON OUTLINE   Mar 1975

I. LESSON OBJECTIVE: To enable Army Area Rifle Marksmanship Coaching Clinic students to recognize the mental and physical requirements necessary to a competitive shooter.

II. STUDENT PERFORMANCE OBJECTIVE: As a result of this instruction, students must be able to accomplish the following student performance objectives:

   A. Given the mission of training a rifle team, EXPLAIN mental conditioning and what must be done to control the mental process while under competitive stress as outlined in Chapter 1, Section IV, USAMU Service Rifle Marksmanship Guide.

   B. Given the mission of training a rifle team, EXPLAIN the necessity of maintaining a good physical condition, as outlined in Chapter 1, Section IV, USAMU Service Rifle Marksmanship Guide.

III. ADVANCE ASSIGNMENT: None.

IV. INTRODUCTION:

   A. <u>Gain Attention</u>: The objective of mental and physical conditioning in the marksmanship program is to condition the shooter to withstand the pressure of match conditions.

   B. <u>Orient Students</u>:

      1. <u>Lesson Tie-In</u>: This class is presented to you as coaches and potential coaches to show you the methods recommended by the US Army Marksmanship Unit for training your shooters both mentally and physically.

      2. <u>Motivation</u>: You, as coaches, are responsible for the training of your teams. If a shooter becomes physically exhausted or loses mental control of himself in a match, the responsibility lies on your shoulders.

      3. <u>Scope</u>: During this period of instruction I will explain mental conditioning, what can be done to control it, and the necessity of maintaining a good physical condition.

V. BODY:

   A. <u>First Student Performance Objective</u>: Given the mission of training a rifle team, EXPLAIN mental conditioning and what must be done to control the mental processes while under competitive stress, as outlined in Chapter 1, Section IV, USAMU Service Rifle Marksmanship Guide.

      1. The primary emotion felt by most shooters is fear of failure and anxiety over the match results, the degree depending on the individual. This results in mental and physical reactions which are natural and involuntary. The most detrimental of these effects are:

         a. Rapid pulse.

         b. Rapid breathing.

         c. Excessive muscular tension.

460

d. Impairment of the mental control of bodily action.

TRANSITION: How can the shooter suppress the emotions which cause the mind to lose some of its control over the body?

  2. Breathing slowly and deeply. In this way, both the rapid breathing and to the extent the excessive muscular tension will be reduced.

  3. Prior to and during a match, avoid distracting thoughts that create anxiety over the match results. Ignore rumor and stay away from the scoreboard.

  4. Try to doze between relays to eliminate all conscious thoughts (CAUTION - If you do this, make sure that you have someone to wake you in time for your next relay).

  5. Try to think about something that will help you to relax.

  6. Establish a system to prepare equipment prior to the match and between relays.

TRANSITION: Having knowledge of the mental emotions and how they may be suppressed is not a sufficient basis for insuring a firm, stable body for a shooting base. You must also have a well developed muscle and lung system.

  B. *Second Student Performance Objective*: Given the mission of training a rifle team, EXPLAIN the necessity of maintaining a good physical condition as outlined in Chapter 1, Section IV, USAMU Service Rifle Marksmanship Guide.

  1. The goal of any shooter is to be a winner and in order to do this, good health and physical endurance is a must. The following characteristics are desirable for a competitive rifleman.

   a. Control of an adequately developed muscle system.

   b. Endurance to fire over long periods of time without lowering of score.

   c. Highly efficient heart and lung system.

   d. Good reflexes and coordination.

  2. Physical training exercises: These exercises should be discontinued at least two (2) days prior to participating in a record match. Indulge in no strenuous activity during the match day.

   a. Isometric exercise will improve muscle tone and to some degree, strength.

   b. Weight lifting and calisthenics to some extent, as warm up exercises only.

   c. Exercises that are helpful to develop strength and coordination are outlined on page 19, USAMU Service Rifle Marksmanship Guide.

   d. Running, fast walking, swimming and cycling helps the shooter have a higher developed heart and lung system.

  3. Shooting Practice:

        a. Actual practice of the positions, whether live or dry firing, helps keep the muscles toned to perform well and easily in these positions.

      4. Sports:

        a. Tennis, volleyball, softball, handball and bowling help develop mobility, reflexes and precision.

VI. CONCLUSION:

    A. <u>Retain Attention</u>: When the expert shooter has learned to control his mind and body, he will be a valuable asset to your team.

    B. <u>Summary</u>:

      1. Mental Conditioning.

        a. The primary emotion felt by the shooter in competition is fear of failure.

        b. This condition results in anxiety that can be suppressed in various ways.

      2. Physical Conditioning.

        a. Good physical endurance is a must for the rifle shooter.

        b. Participate in exercise and activities that will build the necessary muscle tone, endurance and coordination.

    C. <u>Application</u>: In handling your shooters you will find that each shooter has an individual method of exercising mental control and developing a physical condition that will give him muscular tone, endurance and coordination.

    D. <u>Closing Statement</u>: When the shooter has achieved complete confidence in himself and his equipment, the one who has the best chance to win is the competitor whose mental state involves an overwhelming desire to win, and who refuses to accept defeat.

F. RIFLE RANGE SAFETY　　　　　　　　　　　　　　　　　　　AA/RMCC 129
　　　　　　　　　　　　　　　　　　　　　　　　　　　　　　Thirty (30) Min
　　　　　　　　　　　　　　LESSON OUTLINE　　　　　　　　　Mar 75

I. LESSON OBJECTIVE: To enable Army Area Rifle Marksmanship Coaching Clinic Students to observe proper range safety procedures.

II. STUDENT PERFORMANCE OBJECTIVE: As a result of this instruction the students must be able to accomplish the following student performance objectives:

　　A. Assuming duties behind the firing line, in the assembly area and on the ready line, APPLY and ENFORCE good safety habits as outlined in Chapters 1 & 2, USAMU Service Rifle Marksmanship Guide.

　　B. Assuming firing positions on the firing line, APPLY and ENFORCE good safety habits as outlined in Chapters 1 & 2, USAMU Service Rifle Marksmanship Guide.

　　C. Assuming duties of a range official on a rifle range, APPLY and ENFORCE good safety habits in general as outlined in Chapters 1 & 2, USAMU Service Rifle Marksmanship Guide.

III. ADVANCE ASSIGNMENT: None.

IV. INTRODUCTION:

　　A. <u>Gain Attention</u>: Range safety is a point that can never be over emphasized or elaborated upon too much.

　　B. <u>Orient Students</u>:

　　　　1. <u>Lesson Tie-In</u>: While completing your preparations to fire and when on the firing line, you must handle your rifle safely. Anytime a person has a weapon in his hand, range safety must be rigidly enforced.

　　　　2. <u>Motivation</u>: Rifle ranges must continue to be the safest places where weapons are handled.

　　　　3. <u>Scope</u>: This period is devoted to overall range safety. We will cover good safety habits behind the firing line, on the firing line, and safety in range operation.

<u>TRANSITION</u>: First of all let's cover safety behind the firing line.

V. BODY:

　　A. <u>First Student Performance Objective</u>: Assuming duties behind the firing line, in the assembly area and on the ready line, APPLY and ENFORCE good safety habits as outlined in Chapter 1 & 2 USAMU Service Rifle Marksmanship Guide.

　　　　1. Rifle will be cleared at all times. The belt will be open, magazine removed, and the safety locked.

　　　　2. When handling rifles, keep the muzzle pointed up in the air.

　　　　3. There will be no aiming, dry firing, or position work behind the firing line.

4. Do not run on the range.

5. Draw and handle ammunition according to ground rules.

6. When called to the ready line, inspect the bore to insure that it is clear.

7. Use shooting glasses and ear plugs.

    B. <u>Second Student Performance Objective</u>: Assuming firing positions on the firing line, APPLY and ENFORCE good safety habits as outlined in Chapter 1 & 2 USAMU Service Rifle Marksmanship Guide.

1. Inspect ammunition for cleanliness, serviceability, and proper caliber.

2. Keep rifle clear and pointed down range until the range has been declared safe for firing.

3. Load only on the command of the range officer.

4. After firing, remain in position until the rifle is cleared and permission to leave the firing line has been granted. In many matches, shooters must remain in position until the firing line has been cleared.

5. After firing, each shooter, while still in position, must open the bolt, remove the magazine, inspect the chamber, lock the safety.

6. No one will go forward of the firing line before the firing line is cleared by the range officer.

7. In addition, Range or Match standing operating procedures will be followed.

    C. <u>Third Student Performance Objective</u>: Assuming duties of a range official on a rifle range, APPLY and ENFORCE good safety habits in general as outlined in Chapter 1 & 2 USAMU Service Rifle Marksmanship Guide.

1. Any person who observes an unsafe condition on or in front of the firing line <u>will</u> give the command "Cease Fire". Also, he <u>will</u> correct any unsafe condition observed behind the firing line.

2. Before use, dummy rounds must be inspected to insure that no live rounds are present. Dummy rounds should be stored separately from live rounds.

3. All rifles should be inspected before conducting dry fire exercises.

4. Before firing on any range, the range officer must insure that Range Regulations are observed which pertain to range fans, range guards, range flags, and range clearance.

VI. CONCLUSION:

    A. <u>Retain Attention</u>: There is much more involved in range safety than merely keeping the bolt of the weapon to the rear.

<u>Summary</u>: Range Safety is practiced before you go onto the firing line while you are on the firing line, and again at the rear of the firing line after shooting. Range Safety is practiced from the moment you arrive at the range, to the moment you leave.

C. <u>Application</u>: Keep Range Safety uppermost in your mind to avoid the inevitable result of careless handling of firearms.

D. <u>Closing Statement</u>: Accidents result from violations of "common sense" safety rules. Any violation can result in the disqualification of the shooter or his team. Safety consciousness can only be developed if all squad officials and experienced shooters set the example and <u>insist</u> that everyone adheres to these same rules.

G. COMBAT RIFLE MATCH PRINCIPLES   AA/RMCC 131
                                                       Two (2) Hours
                                                       Mar 75
                        LESSON OUTLINE

   I. LESSON OBJECTIVE: To enable Army Area Rifle Marksmanship Coaching Clinic students to fire the Combat Rifle Match both individual and team.

   II. STUDENT PERFORMANCE OBJECTIVE: As a result of this instruction students must be able to accomplish the following student performance objectives:

      A. Assigned the duties of a Combat Rifle Match team leader, GIVE the composition of the Combat Rifle Team, the courses of fire (individual and team), and the match conditions, as outlined in Chapter 4, USAMU Service Rifle Marksmanship Guide, and the FORSCOM/TRADOC Supplement 1 to AR 350-6.

      B. Assigned the duties of a combat rifle match team leader, SELECT a team as outlined in Chapter 4, Section 3, USAMU Service Rifle Marksmanship Guide.

      C. Shown the Combat Rifle Match targets, DESCRIBE the types of targets used for each stage of the match and DISCLOSE the disking procedures, as outlined in Chapter 4, The USAMU Service Rifle Marksmanship Guide.

      D. Assigned the duties of a combat rifle match team leader, EXPLAIN the duties of the team leader and assistant, number of rounds of ammunition issued (individual and team) and describe the fire plans, as outlined in Chapter 4, The USAMU Service Rifle Marksmanship Guide.

      E. Assigned the duties of a combat rifle match umpire, EXPLAIN duties of the umpire, as outlined in Chapter 4, The USAMU Service Rifle Marksmanship Guide.

      F. Given the mission of firing the Combat Rifle Match, DESCRIBE the conduct of the match (individual and team), and EXPLAIN the scoring system used, as stated in Chapter 4, The USAMU Service Rifle Marksmanship Guide.

   III. ADVANCE ASSIGNMENT: None

   IV. INTRODUCTION:

      A. Gain Attention:

NOTE: SKIT - SEE ANNEX # I

      B. Orient Students:

         1. Lesson Tie-In: In the previous classes, you were taught the fundamentals of rifle marksmanship. This instruction will reveal how these fundamentals are put to use in a combat situation.

         2. Motivation: The purpose of the training in the Combat Rifle Match is to instill confidence in the individual soldier to a degree that he will engage an armed enemy instinctively and without hesitation. This match is fired at division level and lower.

         3. Scope: During this period, we will explain the composition of the Combat Rifle team, its course of fire (individual and team), match conditions, target, disking procedure, duties of the team leader and assistant, ammunition, fire plans, duties of the umpires, conduct of the match and scoring system used.

V. BODY:

　　A. **First Student Performance Objective**: Assigned the duties of a Combat Rifle Match team leader, GIVE the composition of the Combat Rifle Team, the courses of fire (individual and team), and the match conditions, as outlined in Chapter 4, USAMU Service Rifle Marksmanship Guide, and the FORSCOM/TRADOC Supplement 1 to AR 350-6.

　　　　1. The Combat Rifle Team consists of 6 firers, 2 alternates, 1 team leader and 1 assistant team leader. The team is divided into 2 three-man fire teams. The team leader and the assistant team leader are responsible for one fire team each. Both are on the firing line during the match, similar to Infantry Trophy Team Match.

　　　　2. The Combat Rifle Match (team) fires the same course of fire (ind) except there are no sighters at the 400 yard line and the team of firing at a bank of 8 targets. In stages 1 and 3, only the six middle targets are raised and in stages 2 and 4 all 8 targets are raised.

　　　　3. The Combat Rifle Match (individual) is composed of four stages of fire - Slow Fire, Rapid Fire, Quick Fire, and Fire and Movement. The first stage of the Individual Match is 2 sighters and 10 rounds for record, slow fire, prone position from the 400 yard line, 1 minute per round, pull and mark targets after each round. The second stage is 10 rounds rapid fire, in the prone position from the 300 yard line. The third stage is 10 rounds quick fire, firing at a snap target from the prone position. The fourth stage is an assault firing exercise requiring movement from the 450 yard line.

TRANSITION: The combat nature of the training places a special requirement on the type of equipment that all individual and team members must wear or have. This includes the team leader and assistant. Each competitor will wear a full combat field uniform (as prescribed below):

NOTE: ASSISTANT INSTRUCTORS MOVE FORWARD (FATIGUE UNIFORM W/FIELD GEAR AND WEAPON)

　　　　4. In addition to carrying an M16 Rifle, personnel must wear a steel helmet with liner and chin strap must be fastened. Equipment must include pistol belt with suspenders, full canteen with cup and carrier, first-aid packet and pouch, magazine pouch with two magazines and is issued a bayonet.

NOTE: AI'S DISMISSED AFTER EACH ITEM OF EQUIPMENT IS POINTED OUT.

　　　　5. The team leader and assistant must also carry an M16 rifle unloaded and locked. A pair of binoculars not to exceed 7 X 50 power may be used to direct fire during team matches.

　　　　6. No individual or team member will be allowed to fire without having the prescribed equipment, worn or carried in the manner intended for it. During a match, if equipment is lost by the firer, he will not be allowed to continue firing until the equipment is replaced in its proper position.

TRANSITION: The composition of the Combat Rifle Team must meet the demands of closely coordinated team work. Let's see how we go about selecting a team.

　　B. **Second Student Performance Objective**: Assigned the duties of a combat rifle match team leader, SELECT a team as outlined in Chapter 4, Section 3, USAMU Service Rifle Marksmanship Guide.

　　　　1. In the selection of squad members, it is important to utilize the best talent available. Follow the provisions of FORSCOM/TRADOC Supplement 1 to AR 350-6, Appendix C, paragraph 3, Team Composition and Eligibility. To select a Combat Rifle Team you would follow the same guide lines as explained in National Trophy period of Instruction, Squad Selection, Organization, and Training (attributes of a shooter).

2. The Combat Rifle Team leader and assistant leader must be selected with extreme care. They must:

   a. Have leadership qualities.

   b. Be able to dope wind.

   c. Be able to move fast, think fast, and be alert.

   d. Be able to observe accurately the wake or trace of the bullet with binoculars.

   e. Have a voice which is loud enough to be heard by all firers while firing.

3. The Combat Rifle Team Match shooters should be selected from among the best qualified shooters available. The most desirable characteristics are:

   a. Ability to shoot well; both rapid fire and at 400 yards.

   b. Ability to think fast and react instantly.

   c. Ability to cooperate and act as a team member.

TRANSITION: With the team members selected, let's see what type of targets we will be firing at.

   C. <u>Third Student Performance Objective</u>: Shown the Combat Rifle Match targets DESCRIBE the types of targets used for each stage of the match and DISCLOSE the disking procedures, as outlined in Chapter 4, the USAMU Service Rifle Marksmanship Guide.

NOTE: SHOW "D" PRONE SILHOUETTE TARGET.

   1. In stages 1, 2, and 4, the standard "D" prone silhouette target is used. The silhouette is 19" high and 26" wide with a 4, 3, and 2 ring.

NOTE: COVER "D" SILHOUETTE AND SHOW "F" SILHOUETTE.

   2. In stage 3, the "F" prone silhouette target is used. The "F" silhouette target is attached to a pole so that it may be raised, and lowered by the pit personnel. The "F" silhouette is 19" high and 26" wide.

NOTE: COVER "F" SILHOUETTE AND "D" TARGET.

   3. As for the disking procedure, it is somewhat different from the National Match because the "F" target has 5 scoring rings: V, 5, 4, 3, and 2.

NOTE: AI'S DISK TARGET AS PRINCIPAL INSTRUCTOR EXPLAINS.

   a. To signal a "V", the white disk is raised vertically up across the right side of the target until horizontal with the silhouette, moved across the face of the target, then lowered vertically down the left side of the target.

   b. To signal a "5", the white disk is raised vertically up the center of the target, held over the center of the silhouette, then lowered vertically.

   c. To signal a "4", red disks raised vertically up the center of the target, held over the center of the silhouette, then lowered vertically.

      d. To signal a "3", the red disk is raised vertically up the RIGHT side of the target to the upper right hand corner, held momentarily then lowered vertically.

      e. To signal a "2", the red disk is raised vertically up the LEFT side of the target to the upper left hand corner, held momentarily, then lowered vertically.

      f. To signal a "miss", the red flag is waved once across the face of the target from RIGHT to LEFT.

      g. When disking the "F" prone silhouette target, the white disk is raised vertically for each hit on the target (Stage 3 Combat Rifle Match). Misses are not indicated.

NOTE: AI'S COVER "D" TARGET.

TRANSITION: You have been given the composition of the Combat Rifle Match Team, the type of targets used and how the targets are disked to indicate score. Coverage is needed of the duties of the team leader and his assistant, how much ammunition is issued for the individual and team matches, and a discussion of the rifle plans that may be used for the team match.

    D. <u>Fourth Student Performance Objective</u>: Assigned the duties of a combat rifle match team leader, EXPLAIN the duties of the team leader and assistant, number of rounds of ammunition issued (individual and team) and DESCRIBE the fire plans, as outlined in Chapter 4, The USAMU Service Rifle Marksmanship Guide.

      1. During team matches, the team leader and assistant team leader will distribute the team's ammunition, repeat range commands, give sight changes, assist with dropped equipment, and if a firer's weapon is disabled, the team leader will let that firer use his weapon. Also, the team leader gives the fire plan to be used by the team in Stages 2 and 4 of the team matches.

TRANSITION: You must know how much ammunition is issued during the individual match and team match.

      2. a. Individual - 52 rounds.

      b. Team - 300 rounds plus 60 rounds for stage 2 and 4 which may be fired at discretion of the team leaders.

      c. Only issue type service-grade ammunition may be used.

TRANSITION: Knowing how much ammunition a team is issued will give you a better understanding of how fire plan is devised.

      3. In stages 1 and 3 of the team match, the six man team is firing at 6 targets and stages 2 and 4, a block of 8 targets. So, in order to receive the bonus points for stages 2 and 4, the team must have a minimum amount of hits on <u>all</u> targets.

      a. Stage 2. You must have a minimum of 35 points on each target to receive the maximum bonus. For example, the bonus score is determined by a formula: targets squared times 2 which would be 8 X 8=64 X 2=128 points maximum bonus for stage 2 and 4.

NOTE: SHOW FIRE PLAN CHART AS PRINCIPAL INSTRUCTOR EXPLAINS STAGES 2 & 4.

      b. Stage 4. You must have a minimum of 75 points per target to receive a bonus.

NOTE: SEE APPENDIX #2 (STAGE #4 FIRE PLAN)

This is just one of many fire plans that may be used for Stage 4.

NOTE: COVER FIRE PLAN CHART.

TRANSITION: One important individual that you have in the Combat Rifle Match that you do not have in other competitive type combat firing is an umpire.

    E. <u>Fifth Student Performance Objective</u>: Assigned the duties of a combat rifle match umpire, EXPLAIN duties of the umpire, as outlined in Chapter 4, The USAMU Service Rifle Marksmanship Guide.

    1. During the Combat Rifle Match, one umpire is assigned to each firer both individual and team - and it is the umpire's duty to see that:

      a. All safety rules and regulations are complied with.

      b. Check each competitor to see that he has the prescribed equipment and that it is worn or carried in the manner intended.

      c. Issue ammunition in the individual matches.

TRANSITION: If you will observe the firer to your front, we will now fire the individual Combat Rifle Match using blank ammunition and reduced targets.

    F. <u>Sixth Student Performance Objective</u>: Given the mission of firing the Combat Rifle Match, DESCRIBE the conduct of the match (individual and team), and EXPLAIN the scoring system used, as stated in Chapter 4, The USAMU Service Rifle Marksmanship Guide.

NOTE: AI'S MOVE TO THE ASSEMBLY LINE-ONE FIRER AND ONE UMPIRE.

NOTE: UMPIRE WILL CALL OUT THE VALUE OF ALL HITS AND RECORD THEIR VALUE ON THE INDIVIDUAL SCORE CARD FOR EACH STAGE OF THE MATCH.

    1. <u>Stage 1</u>. RANGE COMMAND: "RELAY 1 TO THE FIRING LINE FOR STAGE 1 OF THE COMBAT RIFLE MATCH. YOUR 1 MINUTE PREPARATION PERIOD STARTS NOW." The targets are raised at this time, the umpire issues 12 rounds of ammunition. The shooter loads all 12 rounds in one magazine and checks out his position. At the completion of the preparation period, the targets are lowered and the Range Officer gives a <u>RANGE COMMAND</u>: YOUR PREPARATION PERIOD HAS ENDED. WITH A MAGAZINE OF 12 ROUNDS LOCK AND LOAD. YOU WILL HAVE 12 MINUTES TO FIRE 12 ROUNDS SLOW FIRE FROM THE PRONE POSITION. IS THE LINE READY? THE LINE IS READY. YOU MAY COMMENCE FIRING WHEN THE TARGETS APPEAR. The first stage is 2 sighters and 10 rounds for record slow fire, prone position. Targets pulled and marked after each round. All positions used in the Combat Rifle Match are the unsupported positions. Use of the hasty sling is optional. At the completion of 12 minutes, the targets are pulled, then the range officer gives a <u>RANGE COMMAND</u>: "CEASE FIRE, CLEAR ALL

WEAPONS. UMPIRES CLEAR WEAPONS, BOLTS OPEN, SAFETY ON. IS THE LINE CLEAR? THE LINE IS CLEAR. Total possible score for stage 1 is 50 points. The shooters remain in position.

2. <u>Stage 2</u>. <u>RANGE COMMAND</u>: "RELAY 1 PREPARE FOR STAGE 2 OF THE COMBAT RIFLE MATCH. YOUR 1 MINUTE PREPARATION PERIOD STARTS NOW". The targets are not raised during this preparation period. Again the umpire issues 10 rounds of ammunition and the shooter loads 5 rounds each in 2 magazines. When the targets appear, the shooter will rise, and move to the 300 yard line. Prior to any movement, shooters must have their stomach and one elbow touching the ground. (Prone Ready Position). At the 300 yard line, the shooter will assume the prone position, unlock and fire 5 rounds, unload, insert another magazine and fire 5 more rounds. At the completion of the preparation period, the range officer commands: RANGE COMMAND: "YOUR PREPARATION PERIOD HAS ENDED. WITH A MAGAZINE OF 5 ROUNDS, LOCK AND LOAD. THIS EXERCISE IS 10 ROUNDS RAPID FIRE IN 75 SECONDS FROM THE PRONE POSITION AT THE 300 YARD LINE. IS THE LINE READY? THE LINE IS READY. YOUR WEAPON WILL BE LOCKED DURING MOVEMENT. YOU MAY RISE AND MOVE FORWARD WHEN YOUR TARGET APPEARS. WATCH YOUR TARGETS." At the end of 75 seconds, the range officer commands: <u>RANGE COMMAND</u>: "CEASE FIRE. CLEAR ALL WEAPONS. IS THE LINE CLEAR? THE LINE IS CLEAR". When the targets are ready for scoring, they are raised and disked starting with the highest value first, or they may be given by telephone for stages 2, 3, and 4. If they are given by phone, then the targets will have spotters placed in all hits and raised for 30 seconds, for the shooter to see. In stages 1, 2, and 3 or the individual matches, only the 10 hits of the highest value will be counted on each target. In stage 5 only, the 20 hits of highest value will be counted. If an individual fires less than the prescribed number of rounds and there should be more hits on his target than shots fired, he will only be scored the number of shots of highest value equal to the number of shots fired. The total possible score for stage 2 is 50 points. When the targets are spotted and ready for scoring the range officer commands: RANGE COMMAND: "STAND BY, YOUR TARGETS ARE COMING UP FOR SCORING". At the completion of scoring, the range officer commands RANGE COMMAND: "IS SCORING COMPLETED? RELAY 1 PREPARE FOR STAGE THREE OF THE COMBAT RIFLE MATCH. YOUR 1 MINUTE PREPARATION PERIOD STARTS NOW.

3. <u>Stage 3</u>. This stage is quick fire from the 300 yard line. Five 6-second exposures of 2 rounds per exposure, firing at the "F" prone silhouette attached to a pole, lowered and raised by the pit personnel. The targets are raised for 6 seconds and lowered 6 seconds. The umpire issues 10 rounds of ammunition and the shooter loads 10 rounds in one magazine. The targets will not be raised during the preparation period. At the completion of the preparation period, the range officer commands <u>RANGE COMMAND</u>: "YOUR PREPARATION PERIOD HAS ENDED. WITH A MAGAZINE OF 10 ROUNDS, LOCK AND LOAD. TARGETS WILL BE EXPOSED FIVE TIMES FOR SIX SECONDS ON EACH EXPOSURE. UNLOCK, WATCH YOUR TARGET. After the fifth exposure, the range officer commands: <u>RANGE COMMAND</u>: "CEASE FIRE, CLEAR ALL WEAPONS. IS THE LINE CLEAR? THE LINE IS CLEAR. STAND BY FOR YOUR SCORE". Spotters are placed in each hit in the target, then they are raised for 30 seconds to show the hits to the shooter. Total possible score is 50 points. When the pits are ready to score, the range officer is notified. RANGE COMMAND: "STAND BY, YOUR TARGETS ARE COMING UP FOR SCORING." At the completion of the 30 second exposure, the hits are phoned back to the line. At completion of scoring, the range officer commands: <u>RANGE COMMAND</u>: "IS SCORING COMPLETED? RELAY 1 MOVE TO THE 450 YARD LINE FOR STAGE 4 OF THE COMBAT RIFLE MATCH."

4. <u>Stage 4</u>. This will be an assault firing exercise requiring movement from the 450 yard line to the 100 yard line. Twenty rounds (2 magazines of 10 rounds each) will be fired in this exercise: 2 rounds from prone position at the 400 yard line, 8 rounds from the sitting or squatting position at 300 yards, and 5 rounds each from the kneeling position at 200 yards, and the standing position at 100 yards. The shooter will have 30 seconds to rise and move to the 400 yard line, assume the prone position, unlock his weapon and fire 2 rounds. When his target disappears, he locks his weapon, assumes the <u>prone ready position</u>, and prepares to rise and move to the 300 yard line, assume a sitting or squatting position, unlock and fire 8 rounds. When the target disappears, he

will lock his weapon, load his second magazine of 10 rounds, assume the <u>prone ready position</u>, and prepare to move to the 200 yard line. When the targets appear, the shooter will have 65 seconds to rise and move to the 200 yard line, assume the kneeling position, unlock and fire 5 rounds. Then the target disappears, he will lock his weapon, assume the <u>prone ready position</u>, and prepare to rise and move to the 100 yard line. When the target appears, the shooter will have 65 seconds to rise and move to the 100 yard line, assume the standing position and fire 5 rounds. When the target disappears, the shooter will lock his weapon, remove the magazine, and clear his weapon. The range officer moves the relay to the starting line by commanding: <u>RANGE COMMAND</u>: "RELAY 1 TO THE STARTING LINE FOR STAGE 4 OF THE COMBAT RIFLE MATCH. YOUR 1 MINUTE PREPARATION PERIOD STARTS NOW." At this time the shooter(s) move to the 450 yard starting line and assume the prone ready position. The umpire issues 20 rounds of ammunition and the shooter loads two magazines of 10 rounds each. The targets are not raised during the preparation period. When the exercise starts, the umpire must see that all movement is made with the rifle locked and with the muzzle pointed downrange. Any firer not moving out at any particular stage as the targets appear will be disqualified. At the completion of the preparation period, the range officer commands: <u>RANGE COMMAND</u>: "YOUR PREPARATION PERIOD HAS ENDED. WITH ONE MAGAZINE LOCK AND LOAD. WEAPONS WILL BE LOCKED DURING MOVEMENT. IS THE LINE READY? THE LINE IS READY. WATCH YOUR TARGET." At the completion of the fourth phase (100 yard line) firing, the range officer commands: RANGE COMMAND: "CEASE FIRE CLEAR ALL WEAPONS. IS THE LINE CLEAR? (Weapons checked by umpire). THE LINE IS CLEAR. STAND BY FOR YOUR SCORE." The firer must fire 2 rounds at the 400, 8 rounds at the 300, 5 rounds at the 200, and 5 rounds at the 100 yard line. If not, he will only receive credit for the rounds fired. The total possible score for Stage 4 is 100 points. As soon as the targets are spotted and ready for scoring, the range officer commands: <u>RANGE COMMAND</u>: "STAND BY, YOUR TARGETS ARE COMING UP FOR SCORING." The targets are run up and disked or the score is phoned to the line. After scoring is completed, the range officer commands: RANGE COMMAND: "IS SCORING COMPLETED? SLING YOUR RIFLE AND MOVE TO THE REAR OF THE RANGE. The senior umpire will fall the relay in and march the shooters and umpires to the rear of the range.

<u>TRANSITION</u>: After seeing the Combat Rifle Match fired, let's look at the score card as filled out by the umpire.

NOTE: AI'S PASS OUT INDIVIDUAL AND TEAM SCORECARD TO EACH STUDENT.

      5. Individual Score Card. (Explain each stage, number of hits, and total points for course. Two-hundred and fifty points possible score.)

      6. Team Score Card. (Explain each stage, how it differs from individual score card. Total possible score 2056 points.)

      7. <u>MISFIRES, STOPPAGES AND ALIBIS</u>.

        a. NO REFIRES WILL BE AUTHORIZED FOR STOPPAGES DUE TO FAULTY AMMUNITION, RIFLES, OR COMPETITOR ERRORS.

        b. REFIRES <u>(RANGE ALIBIS)</u> MAY BE AUTHORIZED IN CASE OF FAULTY TARGET OPERATION OR IMPROPER RANGE MANAGEMENT IF:

        (1) IN THE JUDGEMENT OF THE CHIEF RANGE OFFICER, THE FIRER WAS PENALIZED BY THE FAILURE AND THE FIRER CONCERNED PROTESTS THE FAULTY PROCEDURE PRIOR TO EXPOSURE OF THE TARGET FOR SCORING.

        (2) WHEN AN ALIBI HAS BEEN ALLOWED, THE PROTESTED TARGET AND/OR SCORE WILL NOT BE SHOWN OR INDICATED TO THE FIRER.

    VI. CONCLUSION:

A. <u>Retain Attention</u>: In the future, the Combat Rifle Match will be the principal marksmanship training vehicle below CONUS Army level. The Combat Rifle Match is a superior combat training match for team or squads.

B. <u>Summary</u>: During this period, we have explained the composition of the Combat Rifle team, its course of fire (individual and team), match conditions, targets, disking procedure, duties of the team leader and assistant, ammunition, fire plans, duties of the umpires, conduct of the match and scoring system used.

C. <u>Application</u>: After hearing the explanation and seeing the demonstration, you will be better prepared when you actually fire the combat rifle individual and team matches.

D. <u>Closing Statement</u>: The Combat Rifle Match was fired first at Camp Perry in 1967 and we plan to see it fired at all command levels to CONUS Army matches. Correct use of the described techniques will enable you to see your team in the winner's circle at the next US Army Championships.

ANNEX #1 AA/RMCC 131
Two (2) Hours
THE COMBAT RIFLE MATCH Mar 75

SKIT

<u>NOTE</u>: WHEN THE PRINCIPAL INSTRUCTOR SAYS "COMBAT", ONE OF THE DEMONSTRATORS WILL CHARGE OUT FROM BEHIND THE BLEACHERS DRESSED IN A MIXED UNIFORM WITH SNEAKERS, OLD FATIGUES AND AN OLD HAT. HE WILL BE ARMED WITH A RIFLE AND AN ENTRENCHING SHOVEL. WHEN HE REACHES THE CENTER, HE WILL ATTEMPT TO TAKE UP A PRONE POSITION, SHOUTING, "Is this the place where they are holding the Combat Rifle Match?" HE STARTS TO DIG HOLES WITH THE SHOVEL FOR HIS ELBOWS. AT THIS TIME, ANOTHER DEMONSTRATOR WILL COME OUT, SHOUTING, "You can't shoot this match equipped like that." THE FIRST DEMONSTRATOR ANSWERS BACK WITH, "Why not?" "We did it this way back in 1948." THE SECOND DEMONSTRATOR THEN TELLS HIM TO GET UP AND MOVE BACK TO THE BLEACHERS AND LISTEN TO THE CLASS, ON COMBAT RIFLE MATCH AND HE WILL LEARN HOW IT IS SUPPOSED TO BE SHOT UNDER THIS YEAR'S RULES.

ANNEX #2

AARMCC 131
Two (2) Hours
Mar 75

THE COMBAT RIFLE MATCH

COMBAT RIFLE MATCH PRINCIPLES

Stage #4 - Team Fire Plan

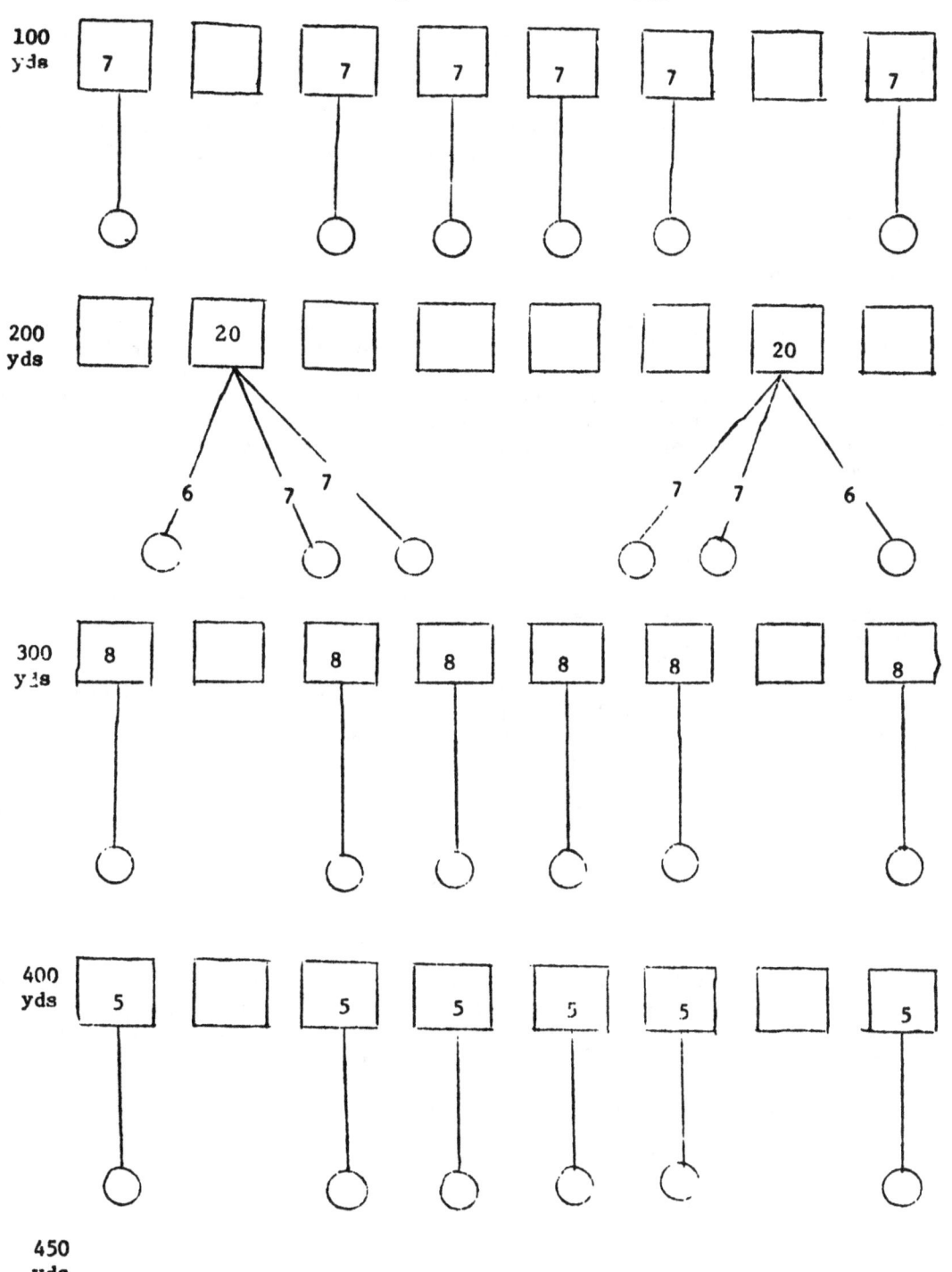

H. PRECISION COMBAT RIFLE MATCH   AA/RMCC 130.1
    (Practical Exercise)   Four (4) Hours
                                                           Mar 75
                         LESSON OUTLINE

   I. LESSON OBJECTIVE. To enable the Army Area Rifle Marksmanship Coaches Clinic students to shoot and coach the Precision Combat Rifle Match.

   II. STUDENT PERFORMANCE OBJECTIVE: As a result of this instruction, the students must be able to accomplish the following student performance objectives:

       A. Given the assignment as Precision Combat Match team coach, PERFORM correctly all the duties of a shooter, team coach, and target operator as outlined in Chapter 2, Section V, the USAMU Service Rifle Marksmanship Guide, and Appendix D, FORSCOM/TRADOC Supplement 1 to AR 350-6.

   III. ADVANCE ASSIGNMENT: None.

   IV. INTRODUCTION:

       A. <u>Gain Attention</u>: Two AI's will come forward with 2 M16A1 rifles and ammunition. There will be one "E" type silhouette set up at 25 meters and one "F" type set up at 100 meters. One AI will fire one magazine of 20 rounds on full automatic at the 25 meter target from the standing position. The other AI will fire 10 rounds from the standing position at the 100 meter target, slow fire. The targets will be brought forward to the class area and displayed to show how much more effective aimed fire is over the automatic mode.

       B. <u>Orient Students</u>:

           1. <u>Lesson Tie-In</u>: This period of instruction will be practical exercises in which the student coach will be able to apply all the material presented in the course that applies to shooting techniques and team coaching techniques of the Precision Combat Rifle Match.

           2. <u>Motivation</u>: Almost anyone can talk about how good a shooter or coach they are, but it is only the "professionals" who can actually do it. To be a good shooter or coach, you must know what the P.C.R.M. is, how to properly coach it and how to apply the rifle fundamentals when you are doing the job.

           3. <u>Scope</u>: Now you will receive instruction on how the Precision Combat Rifle Match is conducted. You will learn how to properly coach a team shooting this course. You will do a little target pulling, shooting and coaching so you will better understand just how the Precision Combat Rifle Course is fired.

   V. BODY:

       A. <u>First Student Performance Objective</u>: Given the assignment as Precision Combat Match team coach, PERFORM correctly all the duties of a shooter, team coach, and target operator as outlined in Chapter 2, Section V, the USAMU Service Rifle Marksmanship Guide, and Appendix D, FORSCOM/TRADOC Supplement 1 to AR 350-6.

           1. The student body will be organized into 6 man groups, setting up a system of rotating functions so that each student performs duties in the target pits or the duties of the coach while 2 members of the group are shooting.

           2. Conduct of Match:

a. General. The Precision Combat Rifle Match is fired in four stages: sustained fire 100 yards, rapid fire 200 yards, rapid fire 300 yards, and slow fire 400 yards. The maximum possible score for individual competition is 250 points for a team 1500 points.

b. Targets. The current Rifle Target "D" FSN 6920-922-7450 with replacement center FSN 6920-922-7541.

c. Stages of fire.

(1) Stage 1. Fired from the 100 yard line. Two sighters (to be fired during the 3-minute preparation period) and 10 rounds sustained fire, standing position, 2-minute time limit, possible score 50 points.

(2) Stage 2. Fired from the 200-yard line. Ten rounds rapid fire from the standing to the sitting position, 50-second time limit. Possible score 50 points.

(3) Stage 3. To be fired from the 300-yard line. Ten rounds rapid fire from the standing to the prone position, 60-second time limit. Possible score 50 points.

(4) Stage 4. To be fired from the 400-yard line. Two sighters and 20 rounds slow fire, prone position, 22-minute time limit. Possible score 100 points

d. Team matches.

(1) Course of fire. Same as for individual match per firing member except the 100 through 300-yard stages will be fired individually and by relays, with assistance of the coach as in National Trophy Rifle Team Matches. The fourth stage, 400 yards, will be fired in pairs.

(2) Two sighter rounds will be fired by each firing member of the team at the 100-yard line only. No sighter rounds are authorized for the remainder of the match.

e. Scoring. Targets will be scored in the same manner as the National Match Course; i.e., slow fire stages, targets will be pulled, marked, and scored after each shot; rapid fire at the completion of a 10-round string.

VI. CONCLUSION:

A. <u>Retain Attention</u>: There is a great challenge to the team coach and his six shooters in being able to fire the winning team score in the Precision Combat Rifle Match.

B. <u>Summary</u>:

1. Critique class--point out errors in coaching duties and shooting techniques.

a. 100 yard sustained fire-standing position.

b. 200 yard rapid fire-sitting position.

c. 300 yard rapid fire-prone position.

d. 400 yard slow fire-prone position.

2. Critique of target pit operation and scoring.

C. <u>Application</u>: The exercises performed in this class will be beneficial to you when you form your combat rifle team for training.

D. <u>Closing Statement</u>: How well your team does in competition will depend on how well you train your shooters. You, as a team coach, are responsible for the over all training and organization of your team.

I. SHOOTING TECHNIQUES   AA/RMCC 224
                                                            10 Hours
                                                            Mar 75
                          LESSON OUTLINE

   I. LESSON OBJECTIVE: To enable Army Area Rifle Marksmanship Coaching students to: apply proper aiming techniques: assume stable positions with the use of the sling; apply correct trigger control; employ rapid fire cadence and techniques; and obtain a zero for each stage of the National Match Course with the M-14 Service Rifle.

   II. STUDENT PERFORMANCE OBJECTIVES: As a result of this instruction students must be able to accomplish the following student performance objectives:

      A. While aiming, OBTAIN the proper eye relief as outlined in Chapter 1, Section IV, USAMU Service Rifle Marksmanship Guide.

      B. While aiming, USE the correct sight alignment as outlined in FM 23-71 and Chapter 1, Section IV, USAMU Service Rifle Marksmanship Guide.

      C. While shooting, IDENTIFY and USE the correct sight picture as outlined in Chapter 1, Section IV, USAMU Service Rifle Marksmanship Guide.

      D. When aiming and firing, USE the proper breath control as stated in Chapter 1, Section IV, USAMU Service Rifle Marksmanship Guide.

      E. When teaching aiming, USE the "Rifle Rest and Aiming Box" exercise as stated in Chapter 1, Section IV, USAMU Service Rifle Marksmanship Guide.

      F. Given an M-14 or M-16 rifle w/sling, ADJUST the proper loop sling as outlined in Chapter 1, Section IV, USAMU Service Rifle Guide.

      G. While firing from the 600 yard line with the M-14 or the 400 yard line with the M-16, ASSUME the prone slow fire position and APPLY correct trigger control as outlined in Chapter 1, Section IV, USAMU Service Rifle Marksmanship Guide.

      H. When firing from the 300 yard line with the M-14 or M-16 rifle, ASSUME the prone rapid fire position, APPLY correct trigger control, and DEVELOP a rapid fire cadence as outlined in Chapter 1, Section IV, USAMU Service Rifle Marksmanship Guide.

      I. While on the 200 yard line with the M-14 or M-16 rifle, ASSUME the sitting position, APPLY correct trigger control, and DEVELOP a rapid fire cadence through firing as outlined in Chapter 1, Section IV, USAMU Service Rifle Marksmanship Guide.

      J. While on the 200 yard line with the M-14 or the 100 with the M-16, ASSUME the proper standing position and APPLY correct trigger control as outlined in Chapter 1, Section IV, USAMU Service Rifle Marksmanship Guide.

      K. Given the job of zeroing your M-14 or M-16 rifle, DEMONSTRATE and EXPLAIN the positions, ranges, and course of fire used as outlined in Chapter 1, Section IV, USAMU Service Rifle Marksmanship Guide.

      L. Given the method of zeroing the individual weapons in your squad, DEMONSTRATE, and EXPLAIN the accepted procedure for obtaining a zero for a rifle as stated in Chapter 1, Section IV, USAMU Service Rifle Marksmanship Guide.

   III. ADVANCE ASSIGNMENT: None.

   IV. INTRODUCTION:

NOTE: SHOW "B" TARGET WITH DUPLICATE OF GROUP FIRED BY THE US ARMY RIFLE TEAM AT 1,000 YARDS.

  A. <u>Gain Attention</u>: This is a duplicate of the shot group fired at 1,000 yards, by members of the US Army Rifle Team that holds the four-man team record. The record is 400 w/67 "V" out of a possible of 400 w/80 "V".

  B. <u>Orient Students</u>:

    1. <u>Motivation</u>: By using correct aiming process combined with the other fundamentals of shooting, your teams will be able to shoot groups like this.

NOTE: REMOVE "B" TARGET.

    2. <u>Lesson Tie-In</u>: You have been taught how to select and equip a rifle squad, now you must know how to train it in the fundamentals of rifle marksmanship.

    3. <u>Scope</u>: During this fourteen hour block of instruction, we will discuss the phases of aiming, how to assume stable positions with the use of the sling, how to apply correct trigger control, how to employ rapid fire cadence and techniques, and how to obtain a zero for each stage of the National Match Course with the M-14 and M-16 Service Rifles.

 V. BODY:

  A. <u>First Student Performance Objective</u>: While aiming, OBTAIN the proper eye relief as outlined in Chapter 1, Section IV, USAMU Service Rifle Marksmanship Guide.

    1. Aiming is the first fundamental taught to the shooter.

QUESTION: Why should aiming be taught first?

ANSWER: Not only is it one of the most important fundamentals, it also provides a means whereby the shooter can check the effectiveness of his position and trigger control in later phases of training and should be practiced throughout the teaching of the other fundamentals.

    2. Aiming is divided into five phases as follows.

NOTE: SHOW CHART #1 (PHASES OF AIMING)

      a. Eye Relief

      b. Sight alignment

      c. Sight picture

      d. Breathing and aiming process

      e. Aiming exercises

    3. In order to see what is required during the process of aiming the shooter must have the proper eye relief.

QUESTION: What is eye relief?

ANSWER: a. Eye relief is the position of the eye with respect to the rear sight.

      b. Proper eye relief, subject to minor variations, is approximately three inches.

QUESTION: Why is it important to have proper eye relief?

ANSWER: c. Any variation in the position of the eye with respect to the rear sight will cause a variation in the image received by the eye. The best method of obtaining the same eye relief for each shot is by the use of the "spot weld", which will be covered in the period of instruction on positions.

    4. To clarify the use of the eye in the aiming process, you must understand that the eye is capable of instanteneous focus from one object to another at a different distance. But, it cannot be focused at two objects at different distances at the same time.

NOTE: HAVE STUDENTS DO EYE EXERCISE. FOCUS ON A SPOT ON THE CHALKBOARD, THEN CHANGE THE FOCUS TO THE THUMB EXTENDED AT ARM'S LENGTH. POINT OUT THE RAPID CHANGE OF FOCUS AND THE FACT THAT THEY CANNOT CLEARLY SEE THE SPOT ON THE CHALKBOARD WHILE THE FOCUS IS ON THE THUMB.

    5. To achieve an undistorted image while aiming, you must look straight out of the aiming eye and not out of the top or corner of the eye.

QUESTION: Why is the straight forward position of the eye the best position for aiming?

ANSWER: You have different sets of muscles called orbital muscles, of equal strength that control the position of the eyeball in the socket. If you move the eye up or down or sideways, the set of muscles are trying to return the eye to the straight forward position In a short while, the muscles become fatigued and the vision becomes blurred. An example of a person whose orbital muscles are not of equal strength, is the person who is cross-eyed.

NOTE: COVER CHART #1.

    B. <u>Second Student Performance Objective</u>: While aiming, USE the correct sight alignment as outlined in FM 23-71 and Chapter 1, Section IV, USAMU Service Rifle Marksmanship Guide.

TRANSITION: The purpose for obtaining the proper eye relief is to allow you to obtain the correct sight alignment.

QUESTION: What is sight alignment?

ANSWER: 1. Sight alignment is the relationship between the front and rear sight with respect to the eye.

QUESTION: When using as aperture rear and a post front sight, what is the correct sight alignment?

ANSWER: 2. When using an aperture rear and a post front sight, the correct sight alignment is as follows: The top of the front sight is centered in the rear aperture both vertically and horizontally.

NOTE: SHOW CHART #2 (SIGHT ALIGNMENT)

    3. It has been found that this is the most natural method of aligning the sights. The eye will instinctively accomplish this task with little training. This method also causes the least amount of inconsistency from shot to shot.

    4. Sight alignment is the most important element in aiming.

QUESTION: Why is sight alignment the most important element in aiming?

     a. An error in sight alignment increases as the range increases, while an error in sight picture remains constant.

NOTE: WORK EXAMPLE ON CHALKBOARD.

     b. The sight radius of the M-14 rifle is 26.7 inches. If you have a 1/32 inch error in sight alignment at 300 yards, this error will be multiplied 400 times, giving you an error of 12.5 inches when the bullet reaches the target. A six (6) inch error in sight picture will still be a six (6) inch error when the bullet reaches the target because it is a parallel error rather than an angular error. An error at the same range with the M-16 will be much greater because of the shorter sight radius. At 300 yards with the M-16 a 1/32 error in sight alignment would be a 16.8" error on the target.

TRANSITION: We have the proper eye relief and correct sight alignment. We need one more element to complete the process of aiming. This element is the sight picture.

NOTE: COVER CHART #2.

    C. <u>Third Student Performance Objective</u>: While shooting, IDENTIFY and USE the correct sight picture as outlined in Chapter 1, Section IV USAMU Service Rifle Marksmanship Guide.

QUESTION: What is sight picture?

ANSWER: 1. Sight picture is the relationship of the bullseye to the front and rear sight, as seen by the eye. It differs from sight alignment only with respect to adding the bullseye to the front sight.

NOTE: SHOW CHART #3 (6 O'CLOCK HOLD).

     2. The sight picture used by most shooters is known as the 6 o'clock hold. This is the sight picture that should be taught to new shooters.

QUESTION: Why should the 6 o'clock hold sight picture be taught to a new shooter?

ANSWER: It offers distinctive aiming point and is more readily understood and applied.

     3. All experienced shooters do not use the same sight picture. But, whatever sight picture is used, it must be uniform from shot to shot in order to obtain accuracy. Some of the other sight pictures are:

NOTE: COVER SHART #3.

NOTE: SHOW CHART #4 (NAVY HOLD, LINE OF WHITE, AND FRAME HOLD).

     a. The Navy Hold or Point of Aim, is used primarily for off hand shooting.

     b. The Line of White for slow fire at long ranges.

     c. The Frame Hold is sometimes used in slow fire at 600 yards and 1,000 yards with the M-14 during periods of reduced visibility. Only through experience can you determine which sight picture is correct for you.

TRANSITION: To obtain the correct sight picture, you must know how to control your breath while aiming.

    D. <u>Fourth Student Performance Objective</u>: When aiming and firing USE the proper breath control as outlined in Chapter 1, Section IV, USAMU Service Rifle Marksmanship Guide.

1. The control of the breath while aiming is important.

QUESTION: What will happen if you breathe while aiming?

ANSWER: If you breathe while aiming, the rise and fall of the chest will cause the rifle to move vertically and you will be unable to obtain the correct sight picture.

2. Sight alignment is accomplished during the breathing, but to complete the process of aiming, you must be able to hold your breath.

QUESTION: When should you hold your breath?

ANSWER: a. To properly hold your breath, you must inhale, then exhale normally, stop at the moment of natural respiratory pause, and hold the breath at this time. If you do not have the correct sight picture, your position must be changed to conform.

b. During rapid fire, it is difficult to change the position while firing. Therefore, the breath must sometimes be held at some point other than the point of natural respiratory pause.

c. The breath should not be held for over ten seconds.

QUESTION: Why should the breath be held no longer than ten seconds?

ANSWER: d. If the breath is held for longer than ten seconds, it will cause dimming of the vision and increased muscular tension, because of the lack of oxygen. If the round is not fired, you should take your rifle down, relax, and take three or four breaths before resuming the aiming and breathing process.

3. As previously mentioned, the eye plays an important part in the process of aiming. While exhaling to move the front sight up to the bullseye, the focus should be repeatedly shifted from the front sight to the bullseye until you have the correct sight picture. When the sight picture has been obtained, the focus should be shifted to the front sight and should remain there until the round has been fired.

QUESTION: Why should the focus be on the front sight when the round is fired?

ANSWER: The final focus must be on the front sight to: "call the shot" accurately and detect errors in the sight picture and sight alignment.

E. **Fifth Student Performance Objective:** When teaching aiming, USE the Rifle Rest and Aiming Box exercise as stated in Chapter 1, Section IV, USAMU Service Rifle Marksmanship Guide.

TRANSITION: Once the aiming process is learned, practical work in the aiming exercise is most valuable.

1. There are three aiming exercises recommended in FM 23-71. They are as follows:

a. The M-15 Sighting Device. Used to teach the proper way to obtain the correct sight alignment and sight picture.

b. The Aiming Bar. Used to teach sight alignment and placement of the aiming point for correct sight picture.

c. The Rifle Rest and Aiming Box. Used to teach sight alignment, sight picture and the importance of a steady hold.

QUESTION: How can you determine which of the exercises to start with in training your shooters?

ANSWER: 2. You should determine the level of experience among the members of your squad.

    a. If you were teaching basic trainees, you would have to start with the M-15 Sighting Device.

    b. Since you will be teaching more experienced competitive shooters we will concern ourselves with the Rifle Rest and Aiming Box Exercise.

    3. This aiming box exercise has been modified by the US Army Marksmanship Unit for competitive shooters by increasing the range to 600 and 400 yards, depending on the weapon used.

QUESTION: What **adva**ntage does the long range exercise have over the old 50 foot exercise?

    a. By increasing the range, you add realism and it magnifies the aiming errors.

    b. This exercise eliminates the effects of the weather, rifle and ammunition dispersion, trigger control and position difficulties.

    4. The aiming exercise is accomplished in the following manner:

    a. Organize your shooters into four-man teams; with two men on the firing line and two men at the target.

    b. At the firing line, one man will act as the shooter with his rifle firmly fixed in the rifle rest. In the prone position, with his chin supported by the left hand, he will aim at the target, being very careful not to disturb the position of the rifle.

    c. The second man on the firing line will act as a signaler. By using arm signals, he will relay commands from the shooter in order to move the aiming disk to the desired point.

    d. One man at the target will observe the signaler through a telescope. He in turn, will relay the commands to the man with the disk.

    e. The disk holder will then move the 20 inch aiming disk in the face of the target in accordance with the commands received from the man on the telescope

    f. The target frame should be faced with blank paper. When the signal, "Mark" is received at the target, the disk holder should hold the disk in its last position and mark this position by inserting a pencil through the center of the disk The dot on the face of the target should be numbered.

    g. The shooter should make two 10 shot groups. For one group the disk should be started from various positions on the target. For the other it should be started from the top.

QUESTION: Why make the second group with the disk starting from the top of the target only?

ANSWER: This will prove to the shooter that he will get better groups by approaching the bullseye from the same direction for each shot.

    h. The shot groups should be traced on onion skin paper in order to facilitate the critique by the coach. A satisfactory group should be no more than three (3) inches in diameter.

      5. Change over and repeat the process until all four members of the team have made two groups.

TRANSITION: Correct eye relief; sight alignment; and sight picture; with proper breath control, will eliminate errors in aiming. These fundamentals must be thoroughly understood before positions and trigger control are taught.

NOTE: GIVE CLASS A TEN MINUTE BREAK.

TRANSITION: When the coach is convinced that the shooter understands eye relief sight alignment, sight picture and breath control, then he is ready to be taught positions and trigger control. I'll present to you the methods used by the All-Army Rifle Team to build stable positions, good rapid fire techniques and correct trigger control. I hope to accomplish this by explanation and demonstration of the slow fire prone position which will be followed by practical work in this position, the prone rapid fire position, followed by practical work, the sitting position with practical work. The explanation and demonstration of positions will pertain to right handed shooters. Left handed shooters will have to reverse the procedure.

    F. <u>Sixth Student Performance Objective</u>: Given a M-14 or M-16 rifle with sling ADJUST the proper loop sling as outlined in Chapter 1, Section IV, USAMU Service Rifle Marksmanship Guide.

QUESTION: What is the purpose of the sling?

ANSWER: 1. The sling helps steady the weapon.

NOTE: DEMONSTRATE ADJUSTMENT OF LOOP SLING.

    2. Place the butt of the rifle on the right hip and cradle the rifle in the crook of the right arm.

QUESTION: What is the purpose of holding the weapon in this manner?

ANSWER: 3. This leaves both hands free to adjust the sling.

     a. Disengage from, form a loop around the sling swivel, and engage the hooks in the sling holes.

     b. The loop to be used is formed by that portion of the long strap between the D-ring and the lower keeper.

     c. Twist the loop a half turn to the left, and insert the left arm through the loop, well up on the arm.

     d. Tighten the sling by pulling down on the outside strap and then placing the top keeper against the frog.

TRANSITION: With the sling properly adjusted, the position is ready to be assumed.

    G. <u>Seventh Student Performance Objective</u>: While firing from the 600 yard line with the M-14 or the 400 yard line with the M-16, ASSUME the prone slow fire position and APPLY correct trigger control as outlined in Chapter 1, Section IV, USAMU Service Rifle Marksmanship Guide.

QUESTION: What is the first thing the shooter should consider before assuming a prone position?

ANSWER: 1. The shooter should select the best ground surface upon which to build his position.

       a. Level ground with relation to target.

       b. Free of holes and depressions.

NOTE: DEMONSTRATOR COME FORWARD AND ASSUME THE PRONE SLOW FIRE POSITION.

    2. The prone position checklist is used to determine if a shooter's position is balanced and stable.

       a. Sights vertical.

       b. Left hand forward to sling swivel.

       c. Rifle supported by heel of left hand.

       d. Left elbow under rifle.

       e. Position and tension of sling.

       f. Right elbow position.

       g. Shoulders approximately level.

       h. Rifle butt close to neck in hollow of shoulder.

       i. Torso and legs relaxed.

       j. Grip of right hand.

       k. Daylight between trigger finger and stock.

       l. Spotweld.

    3. A good prone slowfire position has three elements:

       a. Bone support.

       b. Muscular relaxation.

       c. Natural point of aim.

NOTE: DISMISS DEMONSTRATORS.

QUESTION: If, when you look through your sights, you see that you are aiming to the right of the target and you muscle the front sight over to the bullseye, what may happen when you fire this shot?

NASWER: 4. We have a tendency to relax as the rifle is fired, causing the rifle to move to it's natural point of aim, in this case to the right, causing a wild shot.

NOTE: DEMONSTRATE THIS USING FRONT SIGHT SIMULATOR.

TRANSITION: Since it is important to have a natural point of aim on the bullseye, it is necessary to know how to adjust the natural point of aim.

    5. Check and adjust natural point of aim.

       a. Adjust horizontally.

        b.   Adjust vertically.

        c.   Breath control.

NOTE: DEMONSTRATE ADJUSTING VERTICALLY WITH FRONT SIGHT SIMULATOR.

TRANSITION: With the shooter's position adjusted so that he is aiming naturally at the target, his final step is to fire the shot. This brings to us the most important marksmanship fundamental, trigger control.

    6.   Trigger control is defined as the independent action of the forefinger on the trigger, with a uniformly increasing pressure straight to the rear until the rifle is fired.

    7.   Trigger control procedure:

        a.   Attain sight alignment.

        b.   Take up slack.

        c.   Exhale to move the front sight up.

NOTE: DEMONSTRATE EXHALING TO MOVE THE FRONT SIGHT UP TO BULLSEYE WITH FRONT SIGHT SIMULATOR.

        d.   Concentrate on front sight.

        e.   Apply pressure straight to rear.

    8.   Anticipation is a natural reaction.

        a.   Anticipation that results in misalignment of the sights will cause a wild shot.

        b.   If you see movement the instant the rifle fires, it will be a bad shot.

        c.   Anticipation can be overcome by concentration on the front sight.

NOTE: EXPLAIN CONDUCT OF PRACTICAL WORK AND MOVE STUDENTS TO PRACTICAL WORK AREA.

TRANSITION: In addition to the prone slowfire position as fired from 600 and 400 yard line, the prone position is also fired from the 300 yard line. The 300 yard stage is a rapid fire exercise.

    H.   **Eight Student Performance Objective:**   When firing from the 300 yard line with the M-14 or the M-16, ASSUME the prone rapid fire position, APPLY correct trigger control and DEVELOP a rapid fire cadence as outlined in Chapter 1, Section IV, USAMU Service Rifle Marksmanship Guide.

NOTE: DEMONSTRATOR FORWARD AND ASSUME PRONE RAPID FIRE POSITION.

    1.   The 300 yard rapid fire stage makes a tight position necessary for rapid recovery following a shot.

        a.   Sling tension.

        b.   Body behind weapon.

        c.   Grip with right hand.

        d.   Pressure on spotweld.

  2.   Marking position, rising, and retaking position.

        a.   Mark left elbow.

        b.   Rise from knees to feet.

        c.   Take short step forward.

        d.   Retake premarked position.

  3.   Trigger control for prone rapid is identical to prone slowfire with the exception that is applied slightly faster.

NOTE: DISMISS DEMONSTRATOR.

TRANSITION: Now that we have seen the basics of a prone rapid position, let's watch as a shooter is called to the firing line for a three minute preparation period and prone rapid fire string using blank ammunition.

NOTE: POST AND INTRODUCE DEMONSTRATOR.

NOTE: GIVE RANGE COMMANDS AND EXPLAIN DEMONSTRATION OF PREPARATION PERIOD AND A PRONE RAPID FIRE STRING.

  4.   Actions taken during 60 second string.

NOTE: SHOW CHART #5 (CADENCE CHART)

        a.   Take position in 8 seconds.

        b.   4 second cadence.

        c.   10 second reload.

        d.   2 second buffer.

  5.   Breath, sight control for rapid fire.

NOTE: DEMONSTRATE RAPID FIRE BREATH CONTROL WITH FRONT SIGHT SIMULATOR.

NOTE: DISMISS DEMONSTRATOR.

NOTE: EXPLAIN CONDUCT OF PRACTICAL WORK TO INCLUDE ONE SHOT, RELOAD & 10 SHOT DRILLS AND MOVE STUDENTS TO PRACTICAL WORK AREA.

TRANSITION: This completes our coverage of the prone position, the next stage to be demonstrated will be the sitting position.

  I.   <u>Ninth Student Performance Objective</u>: While on the 200 yard line with the M-14 or the M-16 rifle, ASSUME the sitting position, APPLY correct trigger control, and DEVELOP a rapid fire cadence through firing as outlined in Chapter 1, Section IV, USAMU Service Rifle Marksmanship Guide.

NOTE: DEMONSTRATOR FORWARD AND ASSUME SITTING RAPID FIRE POSITION.

  1.   Introduce sitting position.

  2.   The three sitting positions.

           a.   Open leg.

           b.   Crossed leg.

           c.   Crossed ankle.

TRANSITION: Since all three sitting positions are similar with the exception of the placement of the legs, we will use the crossed ankle to point out the sitting position check list.

      3.   The sitting position check list.

           a.   Sights vertical.

           b.   Position of left hand.

           c.   Left elbow under weapon.

           d.   Sling tension.

           e.   Shoulders level.

           f.   Contact between arms and legs.

           g.   Butt high in shoulder close to neck.

           h.   Grip of right hand.

           i.   Daylight between trigger finger and stock.

           j.   Spotweld.

TRANSITION: After building a position, adjust onto the target.

      4.   Adjusting natural point of aim.

           a.   Adjust horizontally.

           b.   Adjust vertically.

      5.   Application of trigger control.

TRANSITION: Since the sitting position is fired rapid fire, it is necessary to build a position, rise, and then retake the position.

      6.   Marking, rising, and retaking position.

NOTE: DEMONSTRATE BOTH METHODS OF RISING AND RETAKING POSITIONS.

           a.   The cross ankle stand and squat method.

           b.   The one arm support method.

TRANSITION: Let's observe a shooter as he is called to the firing line to fire a string using blank ammunition.

NOTE: GIVE RANGE COMMANDS AND EXPLAIN DEMONSTRATION FOR PREPARATION PERIOD AND A SITTING RAPID FIRE STRING.

7. Actions taken during a 50 second string.

NOTE: SHOW CHART #5 (CADENCE CHART)

    a. Take position in 8 seconds.

    b. 3 second cadence.

    c. 10 seconds reload.

    d. 2 second buffer.

NOTE: DISMISS DEMONSTRATOR.

NOTE: EXPLAIN CONDUCT OF PRACTICAL WORK TO INCLUDE FIRING ONE SHOT, RELOADING, AND DRILLS IN FIRING TEN (10) SHOTS. MOVE STUDENTS TO PRACTICAL WORK AREA.

TRANSITION: This completes the sitting position, the next stage is the standing position.

  J. <u>Tenth Student Performance Objective</u>: While on the 200 yard line with the M-14 or the 100 yard line with the M-16, ASSUME the proper standing position and APPLY correct trigger control as outlined in Chapter 1, Section IV, USAMU Service Rifle Marksmanship Guide.

NOTE: DEMONSTRATOR FORWARD AND ASSUME STANDING POSITION.

   1. Introduce the standing positions.

   2. The 3 standing positions.

    a. Military standing.

    b. NRA standing.

    c. International standing.

   3. Assuming the standing position from sitting.

    a. Keep feet in place.

    b. Place rifle high in shoulder.

    c. Place check on stock.

    d. Lower rifle onto target.

    e. Relax torso down to hips.

TRANSITION: You have seen how to assume position. We will now cover the standing position check points.

   4. Standing positions check point.

    a. Feet level.

    b. Knees not locked.

    c. Legs supporting equal weight.

    d. Right arm position.

    e. Butt high in shoulder.

    f. Spotweld or stockweld with neck relaxed.

    g. Trigger finger clear of stock.

TRANSITION: After assuming a correct position, the shooter then adjusts his natural point of aim.

   5. Adjusting natural point of aim:

    a. Check location of aim.

    b. Adjust horizontally.

    c. Adjust vertically.

NOTE: DEMONSTRATOR DISMISSED.

NOTE: DEMONSTRATE WOBBLE AREA WITH FRONT SIGHT SIMULATOR.

   6. Trigger control in the standing position is of utmost importance because of the wobble area.

NOTE: DEMONSTRATE TRIGGER CONTROL WITH FRONT SIGHT SIMULATOR.

    a. Take up slack with a heavy initial pressure.

    b. Apply trigger pressure when sight picture is correct.

    c. When sight moves away from the bullseye, don't release that pressure that you have applied.

    d. Whenever sight returns to bullseye, apply pressure until rifle is fired.

    e. By concentrating hard on sight picture and some practice you can develop your trigger control so that is is a natural reaction for the rifle to fire when the sight is properly aligned on the bullseye.

TRANSITION: This concludes the standing position. I will now explain how the practical work will be conducted.

NOTE: EXPLAIN CONDUCT OF PRACTICAL WORK AND MOVE STUDENTS TO THE PRACTICAL WORK AREA.

NOTE: AT COMPLETION OF PRACTICAL WORK, MOVE STUDENTS BACK TO CLASS AREA.

   7. Sub-Summary.

    a. The three elements of a good position are bone support, muscular relaxation, and natural point of aim on an aiming point.

    b. The fundamental that is considered the most important is trigger control; the act of firing the rifle without disturbing the aim.

    c. To become a consistently good shooter, you must master positions and trigger control; and have good rapid fire cadence and techniques in order to obtain a zero for your rifle.

NOTE: GIVE CLASS A BREAK BEFORE STARTING ZEROING.

TRANSITION: In order for you to consistently put your first shot in the black, you must know your zero. During this period of instruction, I will explain the method used by USAMU to zero for each stage of the National Match Course. You will then move to the range and perform each exercise as it has been demonstrated.

    K.   Eleventh Student Performance Objective: Given the job of zeroing your M-14 or M-16 rifle, DEMONSTRATE and EXPLAIN the positions, ranges, and course of fire used as outlined in Chapter 1, Section IV, USAMU Service Rifle Marksmanship Guide.

QUESTION: What is a zero?

ANSWER: 1.   A zero is the sight setting, both windage and elevation, that will allow you to hit the center of the bullseye at a given range in a no-wind condition.

    2.   It has been found that the best way to zero a rifle is to use the position and cadence at the range that we intend to fire the rifle. However, the course of fire for zeroing is different than that fired in the National Match Course.

QUESTION: What are the differences and what are the reasons for the differences?

ANSWER: 3.   We must obtain a fast accurate initial zero which can be the "key" with which we can make a normal change for different positions and/or ranges.

    a.   This leaves us with but one alternative; to shoot sitting first.

    b.   Any zeroing exercise must utilize the least amount of time and ammunition and still be accurate. Keeping this in mind, we find that it is best to fire single rounds of small groups of rounds initially, to center our groups on the bullseye.

    c.   Any zeroing exercise should give an indication of a probable N.M.C. score. Therefore, we must fire full strings, or at least half strings, if the time does not permit a full string.

    d.   The course of fire for zeroing the M-14 rifle should be:

    (1)   Sitting 200 yards, three (3) rounds slow fire, pull and mark after each shot.

    (2)   Standing to sitting 200 yards, two (2) three (3) round shot groups rapid fire.

    (3)   Standing to sitting 200 yards, one (1) ten (10) round rapid fire string fired in fifty seconds.

    (4)   Standing 200 yards, twelve (12) rounds in twelve (12) minutes, pull and mark after each shot.

    (5)   Standing to prone 300 yards rapid fire, same as the 200 yard rapid fire, except there is no three (3) rounds slow fire.

    (6)   Prone slow fire 600 yards, twenty-two (22) rounds in twenty-two (22) minutes.

    (7)   Zeroing the M-16 should be the same in both slow and rapid fire; only it is done at the 1, 2, 3, and 400 yard lines.

L. **Twelfth Student Performance Objective**: Given the method of zeroing the individual weapons in your squad, DEMONSTRATE and EXPLAIN the accepted procedure for obtaining a zero for a rifle as stated in Chapter 1, Section IV, USAMU Service Rifle Marksmanship Guide.

    1. The initial firing will be on the 200 yard line with the M-14 and the 100 yard line with the M-16. It will be three (3) rounds slow fire from the sitting position. During this time, the coach will move the front sight on the M-14 rifle to insure that the no-wind zero is less than four (4) clicks from the mechanical zero.

QUESTION: You are the coach and your shooter is shooting to the left of the bullseye. What direction would you move the M-14 front sight to move the shot to the center of the bullseye?

ANSWER: a. Move the front sight in the direction of the bullet hole, left. This movement of the front sight can be accomplished with the M-14 rifle only. The M-16's front sight is movable for elevation only.

    b. Move the rear sight in the direction you wish to move the bullet; right.

    2. The second stage of zeroing is two (2), three (3) round groups. During this stage, the groups will be centered in the bullseye by moving the rear sight only.

    3. After completion of the regular full-time rapid fire exercise, the coach must establish a correct zero and be sure their shooters record their zero in their scorebook.

    4. In both the slow fire stages you, as the coach, may instruct the shooter to change his no-wind zero whenever you deem necessary, and again you must insure that he records this zero in his scorebook.

VI. CONCLUSION:

    A. **Retain Attention**: Ask yourself: "Can I zero my rifle?" "Can I help my shooters zero their rifles?" Your answer will be yes if you follow the techniques of zeroing covered in this period of instruction.

    B. **Summary**: What are some of these zeroing techniques?

    1. We zero in the position, cadence, and range in which we intend to fire the rifle.

    2. Never have the windage zero more than four (4) clicks from mechanical zero.

    3. Always record the correct zero in your scorebook.

    4. Always start zeroing at the most stable position at the shortest range (sitting).

    5. Obtain proper eye relief.

    6. Correct sight alignment.

    7. Correct sight picture.

    8. Proper breath control.

9. When teaching aiming, use rifle rest and aiming box.

10. Adjust proper loop sling.

11. Apply correct trigger control in proper prone position.

12. Develop a rapid fire cadence.

13. Assume proper sitting position.

14. Assume proper standing position.

    C. <u>Application</u>: You will have the opportunity to apply the zeroing techniques during the next 5½ hours. You will have the opportunity to actually zero your own rifle.

    D. <u>Closing Statement</u>: A match winner is a shooter who knows his rifle zero at every range, for all positions, and with the cadence in which he fires.

J. NATIONAL TROPHY TEAM COACHING TECHNIQUES   AA/RMCC 132
                                                  2 Hours
                    LESSON PLAN   March 1975

   I. LESSON OBJECTIVE: To enable students of the Army Area Rifle Clinic to explain the coaching techniques employed, the arrangement of the shooters and coaching equipment used for each of the stages of the National Match Team Course, and to explain the purpose, preparation, and use of the coach's plotting sheet.

   II. STUDENT PERFORMANCE OBJECTIVE: As a result of this instruction, students must be able to accomplish the following student performance objectives.

      A. Given mission to train a rifle team, STATE where information on the training and coaching of a rifle team can be found, as outlined in Chapter 1, USAMU Service Rifle Marksmanship Guide.

      B. Assigned the duties of coaching, EXPLAIN techniques employed by the team coach and DESCRIBE how the shooters and the coaching equipment are arranged at each of the stages of the National Match Team Course, as outlined in Chapter 1, USAMU Service Rifle Marksmanship Guide.

      C. Assigned as an instructor, EXPLAIN the purpose, preparation, and use of the coach's plotting sheet, as outlined in Chapter 1, USAMU Service Rifle Marksmanship Guide.

      D. Assigned as an instructor EXPLAIN how the National Trophy Team Coaching Technique may be presented to a small group of students with the use of improvised training aids, as outlined in Chapter 1, USAMU Service Rifle Marksmanship Guide.

   III. ADVANCE ASSIGNMENT: None

   IV. INTRODUCTION:

      A. Gain Attention:

NOTE: FROM THE SIDELINES, A COACH AND A SHOOTER APPEAR AND GO THROUGH A 10 ROUND RAPID FIRE SITTING STRING. IMPROPER AND EXAGGERATED COACHING TECHNIQUES ARE USED. A SHOT GROUP IS MISPLACED AND SCATTERED ON "A" TARGET. SEE APPENDIX I.

NOTE: PI'S ONE AND TWO WALK FROM BEHIND BLEACHERS TO THEIR INDIVIDUAL PODIUMS. PI ONE STARTS INSTRUCTION.

   I am certain that if this shooter had a good coach, his group would have been much tighter and probably in the center of the bulls-eye.

      B. Orient Students:

         1. Motivation: What is the mission of the Coach? The mission of the coach is to select, train, and organize a rifle team to produce a winning score. This mission can only be accomplished when you as a coach have a complete understanding of the fundamentals of shooting, the coaching techniques, and the use of the coaching equipment. The proficiency of a coach is measured by his ability to detect and correct errors. This is a continuous process. Many errors found during the first stage of a match, if corrected, will probably be the difference between winning or losing. As a coach, you should be guided by a rule of thumb. Do everything that can possibly be done for the shooter with the exception of loading or reloading the rifle, aligning the sights, or squeezing the trigger.

2. <u>Lesson Tie-In</u>: All subjects or classes during the school are given for the purpose of refreshing your memory or toning your muscles on the fundamentals of shooting. In this class we will explain and demonstrate how the coach induces the shooter to apply those fundamentals, and how he is able to produce winning scores by employing the recommended team coaching techniques.

3. <u>Scope</u>: To be able to select, train, and organize a winning team, you must know:

    a. Where to obtain information on training of a rifle team.

    b. The importance of coach-shooter relationship.

    c. The technique required while coaching a rifle team.

    d. The proper way to use the coaching equipment.

    e. The importance of the Coach's Plotting Sheet.

<u>TRANSITION</u>: So if you are planning to become a coach, there is only one place where all this information or knowledge is compiled - "The Coach's Guide".

V. BODY:

    A. <u>First Student Performance Objective</u>: Given mission to train a rifle team, STATE where information on the training and coaching of a rifle team can be found, as outlined in Chapter 1, USAMU Service Rifle Marksmanship Guide.

    1. United States Army Marksmanship Unit Service Rifle Marksmanship Guide gives a complete and recommended program for the organization and training of a team.

    a. The information in this guide is brought up to date every year.

    b. The guide is written with the instructor and coach in mind.

    c. The individual shooter can also benefit, especially if he reads the Fundamentals of Marksmanship section.

<u>TRANSITION</u>: Now that you know where to get the information to <u>train</u> a rifle team, lets move into the portion of this class where you learn the techniques of <u>coaching</u> a rifle team. To facilitate bringing the remaining of this class across to you, we will present a demonstration on how to coach a team through the National Match Team Course. You can follow this demonstration by using our Coach's Checklist and Examination, which is also a part of the book.

<u>NOTE</u>: AI'S PASS OUT COACH'S CHECKLIST AND EXAMINATION.

    B. <u>Second Student Performance Objective</u>: Assigned the duties of coaching, EXPLAIN the techniques employed by the team coach and DESCRIBE how the shooters and the coaching equipment are arranged at each of the stages of the National Match Team Course, as outlined in Chapter 1, USAMU Service Rifle Marksmanship Guide.

    1. The coach's responsibilities start before he reports to the range. He must:

    a. Receive instructions from the head coach:

    (1) Team member's names (Evaluation File).

                (2) Target Assignment (Squadding Ticket).

            b. Analyze previous Coach's Plotting Sheets.

            c. Check coaching equipment for completeness and serviceability.

                (1) Team spotting scope.

                (2) Stop watch.

                (3) Clipboard and pencil.

                (4) Coach's Plotting Sheets.

                (5) Stool.

                (6) Lens Tissue.

                (7) Ear Plugs.

                (8) Allen Wrench, screwdriver and combination tool.

TRANSITION: After this preparation, the coach is ready to tackle his coaching assignment on the range. The coach should be the first one to arrive at the range and should request that all shooters arrive at least 20 minutes before the scheduled start of the match or practice.

NOTE: EXIT PI-I, ENTER PI-II.

NOTE: DEMONSTRATION TEAM OF SIX SHOOTERS, TEAM CAPTAIN AND COACH MOVE TO ASSEMBLY LINE. COACH HAS MICROPHONE AND FOLLOWS "COACHING CHECKLIST" WITH TEAM. AS THE COACH COMPLETES EACH STAGE, INSTRUCTOR WILL STOP DEMONSTRATORS AND RECAP THE MAIN POINTS OF THE DEMONSTRATION. SEE APPENDIX 2.

NOTE: COACH FOLLOWS STEPS IN "2" (BELOW).

        2. Coaching actually starts at the assembly line where the coach must:

            a. Check that team is present and ready.

            b. Fill in appropriate entries in the day's Coach's Plotting Sheets

            c. Minimize mental anxiety and human error by talking to and checking the team on:

                (1) Course of fire

                (2) Firing order

                (3) Equipment serviceability (rifle, sights, etc.)

                (4) Scorebook analysis

                (5) Weather conditions

                (6) Correction of individual errors

TRANSITION: Many matches have been won at this line by the application of a good coach - shooter relationship. This relationship does not start or stop at any stage of the course of fire but is a continuous process from shot to shot, day to day, on and off the range. You will be able to observe throughout this demonstration the importance of the coach - shooter relationship and how the coach's decisions are influenced by coach - shooter relationship.

NOTE: THE FIRST STAGE WILL BE OFFHAND. THROUGHOUT THIS CLASS, DEMONSTRATORS WILL FOLLOW RANGE COMMANDS.

NOTE: EXIT PI-II, ENTER PI-I.

RANGE COMMAND: "TEAM CAPTAINS AND COACHES, YOU MAY MOVE YOUR EQUIPMENT TO THE FIRING LINE YOUR FIRST PAIR TO THE READY LINE. THIS WILL BE THE 200 YARD SLOW FIRE STAGE OF THE NATIONAL TROPHY TEAM MATCH, TEN SHOTS PER TEAM MEMBER FROM THE STANDING POSITION. TEAMS WILL FIRE IN PAIRS AND SHOULD SPLIT THE STAKE. THE MAN ON THE RIGHT SHOULD FIRE FIRST. YOU ARE ALLOWED ONE SHOT OUT OF SEQUENCE. TOTAL TIME FOR THIS STAGE, 66 MINUTES."

RANGE COMMAND: "COACHES, MOVE YOUR FIRST PAIR TO THE FIRING LINE. YOUR THREE MINUTE PREPARATION PERIOD HAS STARTED."

NOTE: EXIT PI-I, ENTER PI-II.

NOTE: COACH FOLLOWS STEPS IN "a" (BELOW).

3. The 200 yard slow fire (offhand) stage is the first stage to be fired.

    a. During the <u>preparation period</u>, the coach must:

        (1) Help shooters find a suitable location on the firing line.

        (2) Confirm the sight setting to be used.

        (3) Check that sights are blackened and set properly.

        (4) Remind shooter of his target number.

        (5) Have shooters dry-fire, and check natural point of aim.

        (6) Check that ammunition is protected from the weather.

        (7) Position the scope properly.

        (8) Check that target is suitable.

        (9) Focus scope on target.

TRANSITION: But before we continue with the domenstration, I must explain the many things the coach had to take into consideration before he got this far.

    b. The firing order for this stage can be of critical importance.

        (1) It is usually best to pair two men who get along well and have somewhat the same characteristics.

        (2) The first pair usually are fast firing and know their zeros. The best shooter is usually on the right and fires first.

        (3) The last pair is usually strong and not liable to pressure under the stress of time or score.

c. The arrangement of the shooters and their equipment is somewhat flexible.

(1) As the range command mentioned, the shooters (pairs) should split the stakes.

(2) The individual scorebook should be maintained throughout all the stages.

(3) Shooting stools should be on the line and used for the shooters convenience. (Either to sit on, or to lay the rifle and scorebook on).

d. The coaching equipment must be arranged and organized to take advantage of the available time.

(1) A coaching scope which for this stage can be placed anywhere as long as it does not interfere with the shooters.

(2) A clipboard with a Coach's Plotting Sheet, (to be explained later).

(3) A stop watch, visible to the coach.

(4) A stool with lens tissue, allen wrenchs, screwdriver, combination tool, etc.

e. The position of the team captain is the same for all stages, just to the rear of the line of scorers, and he is responsible for:

(1) Checking the scoring.

(2) Initiating all challenges.

(3) Observing other teams for violations of rules.

(4) Assisting the coach in:

(a) Keeping time during the slow fire stages.

(b) Verifying the target exposure time during the rapid fire stages.

(c) Counting the number of "X's" as they are fired in rapid fire.

NOTE: DEMONSTRATORS WILL FIRE BLANKS AT METALLIC CENTERED "A" AND "B" TARGETS. ONE AND THREE INCH MAGNETIC SPOTTERS WILL BE USED TO PLOT GROUPS OR SHOTS.

NOTE: EXIT PI-II, ENTER PI-I.

RANGE COMMAND: "YOUR PREPARATION PERIOD HAS ENDED. LOCK AND LOAD. IS THE LINE READY? THE LINE IS READY. READY ON THE RIGHT? READY ON THE LEFT? READY ON THE FIRING LINE? YOU MAY COMMENCE FIRING AND THE TIME WILL START WHEN YOUR TARGETS APPEAR. TARGETS UP."

NOTE: COACH DOES ALL STEPS IN "f" (BELOW). NUMBER ONE FIRER SHOOTS AN "X" AT 1200 AND CALLS A "10" AT 1200. NUMBER 2 SHOOTER FIRES A "10" AT 9:00 AND CALLS A "X" AT 3:00. COACH GIVES #2 SHOOTER A CHANGE OF TWO CLICKS RIGHT. NUMBER 1 FIRER STANDS UP AND COACH CAUTIONS, "WATCH YOUR FRONT SIGHT, YOUR FIRST SHOT WAS OFF CALL IN ELEVATION." NUMBER 1 FIRER SHOOTS A "10" AT 5:00 AND CALLS A "X" AT 3:00. BOTH COACH AND FIRER DECIDE TO COME UP 1 CLICK. NUMBER TWO SHOOTER STANDS AND AFTER 10 OR 12 SECONDS OF UNSTEADY,

WOBBLY HOLDING, THE COACH DIRECTS HIM TO "SIT DOWN, UNLOAD, RELAX, AND DRY-FIRE."
NUMBER 2 MAN STANDS AGAIN AND FIRES A "X" AT 1200 AND CALLS A "X" AT 1200.

NOTE: EXIT PI-I, ENTER PI-II.

    f. <u>During firing</u> the coach must:

    (1) Start the stop watch and record each relay's starting time on the Coach's Plotting sheet (66 minutes total time).

    (2) Insure firers have set sights.

    (3) Wait until shooters are physically and mentally ready.

    (4) Call of target number before each shooter fires.

    (5) Have shooters call each shot.

    (6) Have shooter maintain the scorebook correctly.

    (7) Give correct sight changes when necessary.

    (8) Maintain Coach's Plotting Sheet properly.

    (9) Check on time periodically.

    (10) Insure scores are recorded correctly and have team captain challenge when necessary.

TRANSITION: Changing of pairs should be completed quickly and orderly. As a pair is about to complete their string, the coach or captain signals the next pair to proceed to the ready line. Observe the demonstrators and let us assume that the next shot to be fired is the last shot for the pair.

NOTE: COACH FOLLOWS STEPS OUTLINED IN "g" BELOW. PAIRS CHANGE OVER.

    g. During the change over of the firing pairs, the coach must:

    (1) Insure that the weapons are cleared.

    (2) Prepare the firing line for the next relay.

    (3) Confirm elevation and windage settings used.

    (4) Analyze performance (detection and correction of errors) and confirm or establish zeros.

NOTE: DEMONSTRATORS MOVE BACK TO ASSEMBLY AREA.

NOTE: EXIT PI-II, ENTER PI-I

TRANSITION: Most of the coaching techniques observed so far will apply to the other three stages of the National Match Team Course with slight variations. The second and third stages are the rapid fire stages and they are a true test of a firer's grouping ability and a coach's ability to employ the coaching techniques.

    4. The 200 yard rapid fire, from standing to sitting or kneeling, is the first rapid fire stage.

a. The coach directs firers with commands and signals.

   (1) He gives sight changes in elevation and windage, direction first, then the number of clicks needed.

   (2) He gives <u>favors</u> in elevation and windage.

NOTE: DEMONSTRATE FAVORS WITH WOBBLE STICK AND TARGET

   (a) "Favor right"

   (b) "Favor left"

   (c) "Hold closer"

   (d) "Take white"

   (3) Any rounds saved are the responsibility of the coach so he must be able to let the shooter know that he must shoot a faster cadence.

   (a) "Speed it up" (Shoot faster, but keep them in the black)

   (b) "Shoot 'em" (Two nines are better than a ten and a miss.)

   (4) The coach calls the first rounds out of both magazines "Good" if they are well in the 10 ring. The same is true for the first round fired after a favor.

   (5) Through training and advancement of the coach - shooter relationship, favors and commands may be expanded.

b. The coach may again fire his shooters in any sequence he desires

   (1) The first shooter should be experienced and <u>know</u> his zero.

   (2) The last shooter should be able to perform under pressure.

NOTE: EXIT PI-I, ENTER PI-II.

RANGE COMMAND: "RELAY 1, TO THE FIRING LINE FOR STAGE TWO, NATIONAL TROPHY TEAM MATCH. YOUR THREE MINUTE PREPARATION PERIOD HAS STARTED. RELAY 2 TO THE READY LINE."

NOTE: EXIT PI-II, ENTER PI-I.

NOTE: DEMONSTRATORS FOLLOW COMMANDS, COACH DOES STEPS OUTLINED IN "C" BELOW.

c. During the rapid fire <u>preparation period</u>, the coach must:

   (1) Help the shooter find a suitable location on the firing line.

   (2) Confirm sight setting to be used.

   (3) Check that sights are set and blackened properly.

   (4) Remind shooter of his target number.

   (5) Help shooter dry-fire and check natural point of aim.

(6) Check that ammunition is clean and serviceable and magazines properly loaded.

(7) Focus the scope on the target and check to see that the target is suitable.

(a) Place the scope close to the firer without interfering with his position. Close contact is a must if the coach is going to give his shooter commands.

(b) Position the scope low and directly over the barrel of the rifle. If, in poor light, you are unable to see the bullet holes, this position will allow you to watch the bullet turbulence which indicates the path of the bullet.

(8) Use a 24 power eye piece and assume a kneeling position behind the scope in order to be readily available to the shooter.

NOTE: EXIT PI-I, ENTER PI-II.

RANGE COMMAND: "YOUR PREPARATION PERIOD HAS ENDED. FIRERS RISE. THIS IS THE SECOND STAGE OF THE NATIONAL TROPHY TEAM MATCH, 10 SHOTS RAPID FIRE, STANDING TO KNEELING OR SITTING, 50 SECONDS TIME LIMIT. WITH A MAGAZINE OF 2 ROUNDS, LOCK AND LOAD. IS THE LINE READY? THE LINE IS READY. READY ON THE RIGHT? READY ON THE LEFT? READY ON THE FIRING LINE."

NOTE: COACH DOES ALL STEPS IN "d" (BELOW). COACH GIVES SHOOTER "ONE RIGHT". FIRST SHOT IS A CLOSE TEN AT 9 O'CLOCK, NOTHING IS SAID. SECOND SHOT IS ANOTHER TEN AT NINE, THEN COACH GIVES "RIGHT TWO." THIRD SHOT IS GOOD AND COACH GIVES "GOOD". THE REST OF THE SHOTS ARE GOOD AND NOTHING IS SAID BY THE COACH.

RANGE COMMAND: "CEASE FIRE, LOCK AND CLEAR ALL WEAPONS. THE TIME WAS CORRECT."

NOTE: EXIT PI-II, ENTER PI-I.

TRANSITION: You noticed that after the command, "firers rise", the coach did many things, some of which were not too apparent.

d. After the command, "firers rise," the coach:

(1) Moved to the side of the shooter.

(2) On the command, "Load", he insured that the safety was engaged, magazine was latched, and a round was chambered.

(3) Insured safety was disengaged on the command, "Ready on the firing line."

(4) Gave wind adjustment.

(5) Started the stop watch when the targets appeared.

(6) Reminded the shooter of his target number.

(7) Gave appropriate favors.

(8) Moved from behind the scope after the first 2 rounds and insured the weapon was reloaded - then gave sight changes correctly.

(9) Reminded the firer of his target number after reloading.

                              (10) Checked shooters time periodically during firing with a
stop watch.

                         (11) Insured that the team captain:

                              (a) Noted the target exposure time.

                              (b) Counted the X's the shooter fired.

NOTE: EXIT PI-I, ENTER PI-II.

RANGE COMMAND: "IS THE FIRING LINE CLEAR ON THE RIGHT? IS THE FIRING LINE CLEAR ON THE LEFT? THE FIRING LINE IS CLEAR. MOVE BACK TO THE LINE OF SCORERS TO RECEIVE YOUR SCORES - SECOND RELAY TAKE YOUR PLACE ON THE FIRING LINE. SCORE ALL TARGETS."

NOTE: EXIT PI-II, ENTER PI-I.

NOTE: COACH AND SHOOTER DO STEPS IN "e" BELOW. NUMBER 2 SHOOTER TAKES POSITION.

                    e. *After firing*, the coach:

                         (1) Insured the weapon was cleared (unless the firer has an
alibi).

                         (2) Had the firer call bad shots?

                         (3) Prepared the firing line for the next relay.

                         (4) Completed the entries in the Coach's Plotting Sheet,
specifically, the plotting of the group.

                         (5) Made sure the shooter filled out his scorebook correctly.

                         (6) Analyzed the performance and confirmed or established a
new zero.

                         (7) Insured the score was recorded correctly. If necessary,
he would have been ready to challenge. (Team Captain handles this).

NOTE: DEMONSTRATORS REGROUP IN THE ASSEMBLY AREA.

TRANSITION: The third stage of the National Match Team Course is the 300 yard, rapid fire, from standing to prone.

          5. The 300 yard prone rapid fire is the "problem" of many shooters.

NOTE: EXIT PI-I, ENTER PI-II.

RANGE COMMAND: "RELAY 1, FOR THE THIRD STAGE OF THE NATIONAL TROPHY TEAM MATCH, TAKE YOUR PLACE ON THE FIRING LINE. YOUR THREE MINUTE PREPARATION PERIOD HAS STARTED. RELAY 2, TO THE READY LINE."

NOTE: COACH FOLLOWS STEPS OUTLINED IN "c" OF 4 (SITTING - PREPARATION PERIOD)

                    a. As you can see, the coach's duties in the preparation period
for rapid fire prone is the same as it was for sitting.

RANGE COMMAND: "YOUR PREPARATION PERIOD HAS ENDED. FIRERS RISE. THIS IS THE THIRD STAGE OF THE NATIONAL TROPHY TEAM MATCH, 10 SHOTS RAPID FIRE, STANDING TO PRONE, 60

SECONDS TIME LIMIT. LOCK AND LOAD. IS THE LINE READY? THE LINE IS READY. READY ON THE RIGHT? READY ON THE LEFT? READY ON THE FIRING LINE."

NOTE: COACH FOLLOWS STEPS OUTLINED IN "d" OF 4 (SITTING AFTER COMMAND "FIRERS RISE"). FIRST SHOT IS JUST OFF THE "X" RING AT 7:30. COACH CALLS "GOOD". THE REST OF THE FIRST MAGAZINE IS GOOD, NOTHING IS SAID. SHOOTER RELOADS AND FIRES 3RD ROUND AND IS ON THE "X" LINE AT 8:00. COACH CALLS "GOOD". EIGHTH ROUND IS A LOW 10 AT 6:00. COACH SHOUTS "HOLD CLOSER." NINTH ROUND IS IN "X" RING AND COACH SHOUTS "GOOD". TENTH ROUND IS OKAY.

RANGE COMMAND: "CEASE FIRE, LOCK AND CLEAR ALL WEAPONS. THE TIME WAS CORRECT. IS THE FIRING LINE CLEAR ON THE RIGHT? IS THE FIRING LINE CLEAR ON THE LEFT? THE FIRING LINE IS CLEAR. MOVE BACK TO THE LINE OF SCORERS TO RECEIVE YOUR SCORES. RELAY TWO, TAKE YOUR PLACE ON THE FIRING LINE." TARGETS ARE SCORED.

NOTE: COACH AND SHOOTER FOLLOW STEPS OUTLINED IN "e" of 4 (SITTING* AFTER FIRING).

NOTE: EXIT PI-II, ENTER PI-I.

NOTE: INSTRUCTOR SUMS UP DEMONSTRATION MENTIONING ONLY THOSE THINGS WHICH WERE DIFFERENT FROM THE 200 SITTING STAGE.

    b.   While firing, the coach gave two commands, "good" and "hold closer".

    c.   After firing, the coach and shooter determined that a zero change was not needed.

NOTE: EXIT PI-I, ENTER PI-II.

TRANSITION: The fourth and last stage of the National Match Team Course is the prone slow fire from 600 yards.

    d.   600 yard prone.

NOTE: EXIT PI-II, ENTER PI-I.

RANGE COMMAND: "COACHES MOVE YOUR EQUIPMENT TO THE FIRING LINE AND EXAMINE YOUR TARGETS. THIS WILL BE THE 600 YARD SLOW FIRE STAGE OF THE NATIONAL TROPHY TEAM MATCH, 20 SHOTS PER TEAM MEMBER FROM THE PRONE POSITION. THE TEAM WILL FIRE IN PAIRS SPLITTING THE STAKE. THE MAN ON THE RIGHT SHOULD FIRE FIRST. TOTAL TIME FOR THIS STAGE IS 126 MINUTES. COACHES, MOVE YOUR FIRST PAIR TO THE FIRING LINE. YOUR THREE MINUTE PREPARATION PERIOD HAS STARTED."

NOTE: COACH FOLLOW STEPS OUTLINED IN "a" of 3 (STANDING-PREPARATION PERIOD).

NOTE: EXIT PI-I, ENTER PI-II.

TRANSITION: As you can see, this preparation period is almost identical to the preparation period in the offhand demonstration.

    a.   600 yard preparation period.

    (1)   The scope is positioned as close to the firers without interfering with their position.

    (2)   The 24 power scope is focused at 300 yards to read the mirage at mid-range.

    b.   The 600 yard stage is difficult by virtue of the range at which fired. Of all the stages it requires the utmost in mirage reading ability by the coach.

(1) The scanning technique of scoping should be used to protect the eye from fatigue.

(2) The individual scopes are used by the firers to plot the shots in order to be able to keep up with their elevation.

(3) The coach must keep abreast of wind changes and sight changes given to the firers.

(a) To do this he must maintain and use the Coach's Plotting Sheet.

(b) When pair firing, the coach should start with the synchronized coaching technique but be ready to change to individual coaching if any pecularities are developed by any rifle during the strings.

(4) Favors at long ranges should be kept to a minimum, but in the case of a fast shifting wind where it becomes fatiging and distracting to click, favors may be given.

NOTE: EXIT PI-II, ENTER PI-I.

RANGE COMMAND: "YOUR PREPARATION PERIOD HAS ENDED. LOCK AND LOAD. IS THE LINE READY? THE LINE IS READY. READY ON THE LEFT. READY ON THE RIGHT. READY ON THE FIRING LINE. YOU MAY COMMENCE FIRING AND THE TIME WILL START WHEN YOUR TARGETS APPEAR."

NOTE: COACH FOLLOWS STEPS OUTLINED IN "f" of 3 (STANDING DURING FIRING). COACH GIVES SHOOTERS "THREE RIGHT" FOR THE INITIAL WIND. FIRST FIRER CALLS A 10 AT 3:00 AND HITS A 10 AT 6:00. THE COACH COMMANDS "ONE RIGHT." SECOND FIRER SHOOTS AND CALLS IT GOOD. HIT IS A 10 AT 9:00. COACH GIVE #2 MAN "ONE RIGHT BY YOURSELF." FIRST MAN CONSULTS WITH COACH AND COMES UP ONE CLICK, SHOOTS A GOOD SHOT AND TARGET SHOWS AN "X" AT 1200. SECOND FIRER PREPARES TO SHOOT AND COACH HALTS HIM AND GIVES "BOTH RIFLES LEFT 4 CLICKS." SECOND FIRER SHOOTS A GOOD SHOT AND IT IS A CENTER "X".

NOTE: EXIT PI-I, ENTER PI-II.

TRANSITION: After a pair finished their string, the coach critiques their performance (Detection and correction of errors).

NOTE: COACH FOLLOWS STEPS OUTLINED IN "g" of 3 (STANDING, AFTER FIRING) AND CRITIQUES THEIR DEMONSTRATED PERFORMANCE).

NOTE: EXIT PI-II, ENTER PI-I.

NOTE: INSTRUCTOR DISMISSES DEMONSTRATORS AND RECAPS DEMONSTRATION, POINTING OUT ONLY THOSE THINGS WHICH WERE DIFFERENT FROM THE 200 YARD SLOW FIRE STAGE.

TRANSITION: In order for a coach to observe the progress he or his shooters are making, he must record everything that pertains to the team firing. To facilitate his record keeping, we have designed and are giving you a Coach's Plotting Sheet.

NOTE: ASSISTANTS PASS OUT ONE COACH'S PLOTTING SHEET TO EACH STUDENT.

C. <u>Third Student Performance Objective</u>: Assigned as an instructor EXPLAIN the purpose, preparation, and use of the Coach's Plotting Sheet as outlined in Chapter 1, USAMU Service Rifle Marksmanship Guide.

1. The Coach's Plotting Sheet is to the coach, what the scorebook is to the individual shooter.

a. It is a daily record of:

(1) The firer's performance

(2) The Coach's performance

(3) The rifle and ammunition performance

(4) The effect the weather has on the firers.

b. The maintenance of this sheet is self explanatory and it is filled out just like an individual score sheet. A filled out sample of the plotting sheet is on page 91 of the USAMU Service Rifle Marksmanship Guide.

2. The Coach's Plotting Sheet is designed to accomodate two shooters for a day of firing. This facilitates:

a. Pair shooting recording during the slow fire stages.

b. Comparing pairs as pairs and not as individuals.

c. Filing on a permanent basis in the individual evaluation file (page 99 of the USAMU Service Rifle Marksmanship Guide.)

TRANSITION: During the class on methods of instruction, you were shown how to make improvised training aids. Here are a few methods that can be used to teach this class.

D. <u>Fourth Student Performance Objective</u>: ASSIGNED as an instructor, EXPLAIN how the National Trophy Team Coaching Techniques may be presented to a small group of students with the use of improvised training aids as outlined in Chapter 1, USAMU Service Rifle Marksmanship Guide.

1. National Trophy Team Coaching Techniques may be taught to a small group with the use of the following aids.

a. Utilizing the individual spotting scope in place of the 100 mm team scope.

b. A watch with a second hand instead of a stop watch.

c. By firing at reduced ranges, such as a smallbore range using small bore targets.

VI. CONCLUSION:

A. <u>Retain Attention</u>: By now, you are probably asking yourself a question. "Can I coach?" Your answers will undoubtably vary, but more important is the answer to this question: "Do you want to learn?"

B. <u>Summary</u>: The USAMU Service Rifle Marksmanship Guide gives us a well organized program for the coaching of a team. In it is:

1. The techniques of coaching.

2. The proper use of the team equipment.

3. The Coach's Check List and Examination.

4. The Coach's Plotting Sheet.

C. <u>Application</u>: For many of you, the opportunity to apply what you have learned today will come very soon - for others, perhaps never. But even if you never have an opportunity to coach, this class will be of benefit to you because all members of a team must know and understand each other if they are to <u>perform</u> like a team.

D. <u>Closing Statement</u>: Coaching a winning team at any level of competition is quite an accomplishment. I hope to see you in the winners circle as a winning coach in the not too distant future.

K. NATIONAL TROPHY TEAM COACHING TECHNIQUES   AA/RMCC 225
(Practical Exercise)   Four (4) Hours
      Apr 1975

LESSON OUTLINE

I. LESSON OBJECTIVE: To enable Army Area Rifle Marksmanship Coaching Clinic students to perform the duties of a rifle coach.

II. STUDENT PERFORMANCE OBJECTIVES: As a result of this instruction, the students must be able to accomplish the following student performance objectives:

    A. Given the assignment as a National Trophy Rifle Team Coach, PERFORM correctly all the duties of the team coach, as outlined in Chapter 1, Section V, The USAMU Service Rifle Marksmanship Guide.

III. ADVANCE ASSIGNMENT: None

IV. INTRODUCTION:

    A. *Gain Attention*: Do you consider yourself a good rifle coach? During this period of practical exercise on National Trophy Team coaching, you will have an opportunity to practice your coaching skill.

    B. *Orient Students*:

        1. *Lesson Tie-In*: The practical work which is conducted during this period of instruction, is the same type of exercise that will be the basis of the examination used in grading your ability as a rifle coach.

        2. *Motivation*: In order to receive the most benefit from this practical exercise, the coaching students will have to apply all the principles taught in the AA/RMCC to the coaching of your rifle team.

        3. *Scope*: This period of instruction will be practical exercises in which the student rifle coach will be able to apply all the material presented in the course that represents the proper techniques of coaching of the U.S. Army National Trophy Rifle Team.

V. BODY:

    A. *First Student Performance Objective*: Given the assignment as a National Trophy Rifle team coach, PERFORM correctly all the duties of the team coach, as outlined in Chapter 1, Section V, The USAMU Service Rifle Marksmanship Guide.

        1. The student body will be organized into 4-man groups, setting up a system of rotating functions so that each student performs the duties of the coach and the pit detail while two (2) members of the group are shooting.

        2. Course of fire will be one 10-round string at 200 yards, rapid fire, one 10-round string at 300 yards, and one 10-round string at 600 yards.

NOTE: MOVE STUDENTS TO FIRING LINE AND START THE PRACTICAL EXERCISE.

VI. CONCLUSION:

    A. *Retain Attention*: At this point, you have probably discovered that there is more to coaching than appears to the casual observer.

    B. *Summary*:

1. Critique class - point out errors in coaching duties and techniques.

2. Review Coaches Performance Check List and Examination as a reminder of the proper technique of coaching a rifle team.

C. <u>Application</u>: The practical exercises performed in this class will be repeated in a later class as a graded examination.

D. <u>Closing Statement</u>: Whether your team shoots and performs up to the level of their capability will depend on you - the leader of the team - the National Trophy rifle team coach.

L. NATIONAL TROPHY TEAM COACHING TECHNIQUES AA/RMCC 226
    (EXAMINATION) Four (4) Hours
    Apr 1975

I. LESSON OBJECTIVE: To enable Army Area Rifle Marksmanship Coaching Clinic students to demonstrate their ability to perform as a National Trophy Team Coach by a performance examination.

II. STUDENT PERFORMANCE OBJECTIVE: As a result of this instruction, the student must be able to accomplish the following student performance objectives:

   A. Assigned as a coach, DEMONSTRATE the proper techniques and sequence in the duties of a National Trophy Team Coach as outlined in Chapter I, Section 5, USAMU Service Rifle Marksmanship Guide.

III. ADVANCE ASSIGNMENT: Study and review Chapter I, Section 5, USAMU Service Rifle Marksmanship Guide.

IV. INTRODUCTION:

   A. <u>Gain Attention</u>: Your performance here today will determine how much you know about coaching a National Trophy Rifle Team through the high power rifle national match course.

   B. <u>Orient Students</u>:

      1. <u>Lesson Tie-In</u>: Throughout the Army, at the completion of any block of instruction, some method is used to test the students degree of learning. This period will be devoted to testing your high power rifle marksmanship degree of knowledge and ability to coach a rifle team in the National Trophy Rifle Match.

      2. <u>Motivation</u>: The score you receive on this performance examination, combined with your score on the written examination will determine whether or not you pass the course. The honor graduate will also be determined by the combined score from these examinations.

      3. <u>Scope</u>: As previously mentioned, this period will be concerned with a practical exercise performance type examination. You will be divided into groups of four students and one assistant instructor. The assistant instructor will grade the four students in his group.

V. BODY:

   A. <u>First Student Performance Objective</u>: Assigned as a coach, DEMONSTRATE the proper techniques and sequence in the duties of a National Trophy Team Coach as outlined in Chapter I, Section 5, USAMU Service Rifle Marksmanship Guide.

      1. Pass out examination sheets.

      2. This test has a total point value of 100.

      3. Break down students into 4-man team groups and start examination.

         a. Men are numbered 1-4. Number 1 man shoots, number 2 man coaches and 3 and 4 will conduct the pit operation.

         b. During slow fire, no. 1 and no. 2 will shoot, no. 3 will coach, and no. 4 in pits. When each slow fire stage is completed, team members will rotate until all team members have coached a stage of slow fire.

    c. Assistant instructors (graders) will be evaluating each of the four student coaches during the exercise.

 VI. CONCLUSION:

   A. <u>Retain Attention</u>: If you have a perfect score on this examination, you have the basic knowledge necessary to start coaching a rifle team.

   B. <u>Summary</u>:

    1. Review The Sequence of Coaching A National Trophy Rifle Team in Chapter I, Service Rifle Marksmanship Guide.

    2. Critique the entire match sequence for mistakes made and devise corrections in performance.

   C. <u>Application</u>: The coaching techniques which have been presented here should be used in training and coaching your rifle teams.

   D. <u>Closing Statement</u>: Your shooters are going to look to you for advice, training, and techniques for raising their scores. Apply what you have learned here and you will accomplish your coaching mission.

M. PANEL DISCUSSION   AA/RMCC 134
                                                        One Hr 30 Min
                                                        Apr 1975
                          LESSON OUTLINE

   I. LESSON OBJECTIVE: To enable Army Area Rifle Marksmanship Coaching Clinic students to ask questions pertaining to subjects taught during the Army Area Rifle Marksmanship Coaching Clinic.

   II. STUDENT PERFORMANCE OBJECTIVE: As a result of this instruction students must be able to accomplish the following student performance objective.

      A. As a panel moderator, INTRODUCE guest speaker(s), instructors and name the classes presented by them during the AA/RMCC as outlined in The USAMU AA/RMCC Program of Instruction.

      B. As panel moderator, CONDUCT a question and answer session between students and panel, as outlined in The USAMU AA/RMCC Program of Instruction.

   III. ADVANCE ASSIGNMENT: None

   IV. INTRODUCTION:

      A. <u>Gain Attention</u>: I am sure there are many of you that have questions to ask about certain points or phases of the instruction that you did not completely understand.

      B. <u>Orient Students</u>:

         1. <u>Lesson Tie-In</u>: In all previous periods of this clinic, we have been presenting instruction to you. During this period, we would like to hear from you, if you have questions or points to discuss pertaining to any subject taught during the clinic.

         2. <u>Motivation</u>: If there is doubt in your mind concerning anything taught during the clinic, now is the time for clarification. Vague comprehension on your part will be reflected in the teaching you conduct for your team.

         3. <u>Scope</u>: During this period, we hope to clear up any weak or vague points that you may have about rifle marksmanship principles taught during the clinic.

   V. BODY:

      A. <u>First Student Performance Objective</u>: As a panel moderator, INTRODUCE guest speaker(s), instructors, and name the classes presented by them during the AA/RMCC as outlined in The USAMU AA/RMCC Program of Instruction.

         1. Introduce guest speaker(s).

         2. Introduce instructors and name the classes presented by them.

         3. Questions asked should be directed to the principal instructor who presented the subject. You may ask for additional comments from other panel members.

         4. Please ask questions about subjects taught during the clinic only.

      B. <u>Second Student Performance Objective</u>: As a panel moderator, CONDUCT a question and answer session between students and panel, as outlined in The USAMU AA/RMCC Program of Instruction.

1. When you have a question, please stand and speak loud enough so that you may be heard by everyone in the class.

2. Conduct panel discussion.

VI. CONCLUSION:

A. <u>Retain Attention</u>: How will your team's performance in the next year's competition compare with the results attained this year?

B. <u>Summary</u>: It is our hope that the material we have presented to you will aid you in becoming a better rifle coach and the information you have learned will be used to train a better team that can win in competition.

C. <u>Closing Statement</u>: On behalf of the USAMU Service Rifle Instructor Group we hope that you have a successful shooting year. Thank you.

N. GRADUATION
AA/RMCC 135
1 Hour
Mar 1975

LESSON OUTLINE

I. LESSON OBJECTIVE: To enable Army Area Rifle Marksmanship Coaching Clinic students who attained a passing grade in the rifle coaching clinic to have their accomplishment recognized and to be awarded an appropriate certificate at a graduation ceremony.

II. STUDENT PERFORMANCE OBJECTIVE: As a result of this instruction, the students must be able to accomplish the following student performance objective:

A. As an instructor in the Army Area Rifle Marksmanship Coaching Clinic, RECOGNIZE the students' accomplishments by presenting the honor graduate certificates of completion to all students who attained a passing grade in the rifle coaching clinic.

III. ADVANCE ASSIGNMENT: None.

IV. INTRODUCTION:

A. Gain Attention: This period concludes the 40 hour clinic in rifle marksmanship coaching. It is our hope that you have increased your knowledge of rifle team coaching techniques.

B. Motivation: Unless you retain this knowledge and pass it on to your new shooters, this clinic will be of limited value to the US Army Marksmanship Program.

C. Scope: During this period, you will receive your certificates of accomplishment and be addressed by our guest speaker.

V. BODY:

A. First Student Performance Objective: As an instructor in the Army Area Rifle Marksmanship Coaching Clinic, RECOGNIZE the students' accomplishments by presenting the honor graduate certificate and certificates of completion to all students who attained a passing grade in the rifle coaching clinic.

1. Announce seating arrangement for the commander, his staff, marksmanship students and guest speaker.

2. Orient students of procedure to be followed in the graduation ceremony.

3. Coordinate with post commander or representative as to the starting time of the graduation ceremony and time needed for an opening or closing statement.

4. Have students seated five minutes prior to the graduation exercise.

NOTE: UPON ARRIVAL OF THE COMMANDING OFFICER, ANNOUNCE HIS ARRIVAL AND TAKE SEATS UPON HIS COMMAND.

5. Introduce guest speaker.

6. Present certificates as rehearsed, following the guest speaker's remarks.

NOTE: HONOR STUDENT IS INTRODUCED AT THIS TIME.

7. Allow guest speaker to present honor graduate certificate.

8. Rise as commanding officer and guest speaker leave the classroom.

9. Resume command of group. Ask for questions.

VI. CONCLUSION:

A. <u>Retain Attention</u>: If at any time you feel we can be of assistance in your program, don't hesitate to write us a letter.

<u>NOTE</u>: ADDRESS: Commander
United States Army Marksmanship Unit
Fort Benning, Georgia 31905

B. <u>Summary</u>: None.

C. <u>Application</u>: When you go back to your respective organizations, apply what you were taught during this clinic.

D. <u>Closing Statement</u>: If this clinic is to be of any value to the Army, you gentlemen will be the ones to make it so. By properly using what you were taught during this 40 hour block of instruction in your marksmanship program, you will be raising marksmanship standards throughout the Army.

CHAPTER XIX
LESSON OUTLINES FOR INTERNATIONAL RIFLE MARKSMANSHIP

A. INTERNATIONAL RIFLE MARKSMANSHIP IRMC 800
   (EQUIPMENT, PROCEDURES & TECHNIQUES OF TRAINING) 50 Min
   Apr 75

LESSON OUTLINE

I. LESSON OBJECTIVE: To enable the International Rifle Marksmanship student to acquire suitable shooting equipment and employ championship shooting techniques by learning about high quality, custom International Rifle equipment, procedures and techniques used in international rifle shooting.

II. STUDENT PERFORMANCE OBJECTIVES: As a result of this instruction students must be able to accomplish the following student performance objectives:

    A. GIVEN the criteria to be met in equipping an international rifle shooter and the reasons for use of each item, SELECT the proper highest quality, custom clothing, rifles and equipment to be used in international rifle shooting as contained in Chapters I and II, USAMU International Rifle Marksmanship Guide.

    B. FACED with the problem of breathing, aiming, and releasing the trigger properly, be able to APPLY the correct method of an integrated act of shooting as contained in Chapter III, USAMU International Rifle Marksmanship Guide.

    C. CONFRONTED with the problem of building good shooting positions, be able to APPLY the basic fundamentals that are common to all positions as contained in Chapters V, VI, VII and VIII, USAMU International Rifle Marksmanship Guide.

    D. FACED with the problem of training properly and maintaining a proper match attitude, DEVELOP a good training schedule, EXPLAIN why a shooter must retain his composure and maintain a good physical condition while engaged in competition as contained in Chapters IX, X, XI and XII, USAMU International Rifle Marksmanship Guide.

III. ADVANCED ASSIGNMENT: All previous references to USAMU International Rifle Marksmanship Guide.

NOTE: SHOW SLIDE #1 (USAMU CREST) WHILE STUDENTS ARE BEING SEATED.

IV. INTRODUCTION:

NOTE: SHOW SLIDE #2 (UNIT MARKER IN FRONT OF USAMU HEADQUARTERS).

    As you may know the US Army Marksmanship Unit, which includes the International Rifle, International Skeet and Trap, International Running Target, Service Rifle and Service Pistol are located at Fort Benning, Georgia. In addition, the US Army Marksmanship Unit has paved the way with research and development of current long range rifle equipment and advanced shooting techniques currently being employed by competitive shooters the world over. This expertise has been widely distributed due to our various marksmanship guides (hold one up for display) and the scores of shooting clinics conducted by various branches of the USAMU.

    A. <u>Gain Attention</u>: Welcome to students from USAMU International Rifle Team. Personnel from the USAMU International Rifle Team frequently are selected as members of the US International Rifle Team.

NOTE: SHOW SLIDE #3 (US INTERNATIONAL RIFLE TEAM).

    B. <u>Orient Students</u>:

1. **Lesson Tie-In:** Following this period of instruction, you will receive classes on the prone, kneeling and standing positions. A question and answer period will be held after the presentation on the standing position. Please hold all questions until that time.

2. **Motivation:** The techniques taught here are the ones that have been developed and used by the world champions and top shooters from all countries of the world.

NOTE: SHOW SLIDE #4 (USAMU INTERNATIONAL RIFLE MARKSMANSHIP GUIDE).

3. **Scope:** During this period of instruction, I will show you the clothing and equipment used in international rifle shooting. I will then explain the proper methods of breathing, aiming, trigger control and the basic fundamentals of position shooting. I will conclude with some ideas on training, match attitude and physical conditioning. All of this information is included in the USAMU International Rifle Marksmanship Guide.

V. BODY:

A. **First Student Performance Objective:** GIVEN the criteria to be met in equipping an international rifle shooter and the reasons for use of each item, SELECT the proper highest quality, custom clothing, rifles and equipment to be used in international rifle shooting as contained in Chapters I and II, USAMU International Rifle Marksmanship Guide.

NOTE: SHOW SLIDE #5 (SHOOTER'S CLOTHING).

1. Point out the items of special clothing required by the international rifle shooter.

NOTE: SHOW SLIDE #6 (INTERNATIONAL RIFLES).

2. Describe the rifles used by the international rifle team.

NOTE: SHOW SLIDE #7 (PALM RESTS).

3. Explain the use of the palm rest.

NOTE: SHOW SLIDE #8 (ADJUSTABLE HOOK).

4. Explain the use of the adjustable rail and hook.

NOTE: SHOW SLIDE #9 (SIGHTS).

5. Explain the features of adjustable sights.

NOTE: SHOW SLIDE #10 (EXPLODED VIEW OF GUN BOX EQUIPMENT).

6. Explain and show the shooting equipment needed by the international shooters.

NOTE: SHOW SLIDE #11 (GUN CASES).

7. Explain the need for both types of gun cases.

NOTE: SHOW SLIDE #12 (CARE & CLEANING).

8. Explain the need for care of equipment and proper cleaning of a rifle.

B. **Second Student Performance Objective**: FACED with the problem of breathing, aiming, and releasing the trigger properly, be able to APPLY the correct method of an integrated act of shooting as contained in Chapter III, USAMU International Rifle Marksmanship Guide.

NOTE: SHOW SLIDE #13 (INTEGRATED ACT OF SHOOTING).

    1. Explain how these acts are integrated into one sequence. For this discussion, each action will be taught separately.

        a. Breath control.

        b. Eye and sight alignment control.

        c. Trigger control.

NOTE: SHOW SLIDE #14 (BREATHING GRAPH).

    1. Breathing: The proper way to breathe is to hold the breath during the respiratory pause after exhaling, not after inhaling.

    2. Aiming.

NOTE: SHOW SLIDE #15 (EYE RELIEF).

        a. Explain the minimum of 2 inches eye relief.

NOTE: SHOW SLIDE #16 (EYE BLINDER).

        b. Explain use of blinders, eye patches, and importance of shooting with both eyes open.

NOTE: SHOW SLIDE #17 (GLASSES).

        c. Show proper use of eye glasses and explain that use of colored lens depend on individual shooter preferences.

NOTE: SHOW SLIDE #18 (APERTURE SIZES).

        d. Proper aperture sizes for prone (3.4-3.8) and other positions (3.6-4.4) depends on the shooter.

    3. Trigger Control:

NOTE: SHOW SLIDE #19 (SET TRIGGER).

        a. A set trigger requires the shooter to hold the rifle loosely. It is very delicate. The shooter must be consistent in the manner in which he touches the trigger in firing a shot.

        b. The two-stage trigger is used by some shooters in preference to the set trigger.

NOTE: SHOW SLIDE #20 (3 POUND TRIGGER).

        c. Three pound trigger is no longer required except for the International Dewar Match and by juniors. It is still used by some prone shooters. It is much too heavy for position shooting.

NOTE: SHOW SLIDE #21 (STANDARD RIFLE).

    d. One pound trigger is required in standard rifle competition.

NOTE: SHOW SLIDE #22 (FINGER PLACEMENT).

    e. Show proper placement of finger on trigger and explain how and when to release the shot.

  C. <u>Third Student Performance Objective</u>: CONFRONTED with the problem of building good shooting positions, be able to APPLY the basic fundamentals that are common to all positions as contained in Chapters V, VI, VII and VIII, USAMU International Marksmanship Guide.

NOTE: SHOW SLIDE #23 (POOR HOLD).

   1. "Hold" is the degree of steadiness with which a shooter is able to aim his rifle. A proper position should provide the shooter with a reasonably good hold.

NOTE: SHOW SLIDE #24 (GOOD HOLD).

   2. Any position used must be legal.

NOTE: SHOW SLIDE #24 (COMFORTABLE SHOOTER).

   3. A position must be reasonably comfortable.

NOTE: SHOW SLIDE #26 (SLING POSITIONS).

   4. The sling should be used correctly.

NOTE: SHOW SLIDE #27 (TELESCOPE PLACEMENT).

   5. The spotting scope should be positioned properly so that there is a minimum of head movement as the shooter scopes his target.

NOTE: SHOW SLIDE #28 (ERECT HEAD POSITION).

   6. An erect head position is important in all positions as the stability of the body is controlled by the balance mechanism in the inner ear.

   7. The center of gravity of the body and the rifle should be well distributed over the support area. Bone and ligament support should be utilized to the fullest extent. Muscles that support a rifle during competition are subject to frequent fatigue.

  D. <u>Fourth Student Performance Objective</u>: FACED with the problem of training properly and maintaining a proper match attitude, DEVELOP a good training schedule, EXPLAIN why a shooter must retain his composure and maintain a good physical condition while engaged in competition as contained in Chapters IX, X, XI and XII, USAMU International Rifle Marksmanship Guide.

   1. <u>Physical Training</u>:

NOTE: SHOW SLIDE #29 (ALCOHOL).

NOTE: SHOW SLIDE #30 (TOBACCO).

NOTE: SHOW SLIDE #31 (COFFEE).

NOTE: SHOW SLIDE #32 (DRUGS).

a. A proper diet that excludes harmful foods and beverages and personal habits that promote good health are of great benefit to championship competitive shooters.

NOTE: SHOW SLIDE #33 (DYNAMIC TENSION).

b. Dynamic tension exercises are good warm up exercises for the shooter to avoid high pulse rate prior to shooting.

NOTE: SHOW SLIDE #34 (RUNNING).

c. Running and swimming are the best. Quick reflex games and stretching exercises are also good.

NOTE: SHOW SLIDE #35 (SHOOTER).

d. A regular program of rifle practice is necessary.

2. <u>Mental Training</u>:

NOTE: SHOW SLIDE #36 (TARGET).

a. A shooter must learn to have confidence in his ability. He gains this through practice.

NOTE: SHOW SLIDE #37 (SNOWSTORM).

b. A competitor must be ready to cope with any and all unforeseen circumstances which might tend to disrupt his performance. He must retain his composure at all times. A shooter must be able to analyze his performance. He has to learn to <u>concentrate</u> on each shot and <u>never</u> quit!

NOTE: SHOW SLIDE #38 (SHOOTER CLEANING RIFLE).

3. <u>Maintenance</u>: A shooter must keep his equipment clean and functional at all times. All the training in the world will not be of any benefit if the sights or rifles are broken or hampered by dirt.

VI. CONCLUSION:

NOTE: SHOW SLIDE #39 (INT'L DISTINGUISHED MEDAL).

A. <u>Retain Attention</u>: In order to become a champion you must master all the fine techniques of competition with the free rifle.

B. <u>Summary</u>:

1. Know how to select proper clothing, rifles and equipment for use in competition.

2. Know how to breathe properly, aim correctly, and release the shot in exactly the same manner every time.

3. Know how to apply the basic characteristics of all positions to build a good position that will enable you to shoot winning scores.

4. Be able to set up a training program for mastering the mental and physical aspects of shooting.

C. <u>Application</u>: If you apply all the points taught here to the fundamental rules of each position you will be on the right tract to becoming a champion shooter.

D. <u>Closing Statement</u>: Many things contribute to the success of a champion. Contrary to what some people believe, there is no secret that will open the door to winning. The formula for winning in competition is using the proper techniques that have been developed and proven, in a training program of constant improvement through practice and frequent competition.

B. INTERNATIONAL RIFLE MARKSMANSHIP CLINIC   IRMC 801
    (FUNDAMENTALS OF THE PRONE, KNEELING & STANDING POSITIONS) 50 Minutes
                                                           Apr 1975
                          LESSON OUTLINE

I. LESSON OBJECTIVE: To enable international rifle marksmanship students to develop an ability to win in competition by understanding the fundamentals of the prone, kneeling, and standing positions.

II. STUDENT PERFORMANCE OBJECTIVES: As a result of this instruction the student must be able to accomplish the following student performance objectives:

    A. Given the characteristics of the prone position, USE the correct equipment and techniques and OBTAIN the best prone position for each student as contained in Chapter V, USAMU International Rifle Marksmanship Guide.

    B. Given the characteristics of the kneeling position, USE the correct equipment and techniques and OBTAIN the best kneeling position for each student as contained in Chapter VII, USAMU International Rifle Marksmanship Guide.

    C. Given the characteristics of the standing position, USE the correct equipment and techniques and OBTAIN the best standing position for each student as contained in Chapter VI, USAMU International Rifle Marksmanship Guide.

III. ADVANCED ASSIGNMENT: Chapters V, VI and VII, USAMU International Rifle Marksmanship Guide.

IV. INTRODUCTION:

    A. <u>Gain Attention</u>: In the past decade scores for the three position match have been constantly on the rise. A good score ten years ago would only be an average score today. It is true that ammunition and equipment have made advances and great improvement, but also the evolution of the position shooting techniques have been a major contribution toward better scores. In the future better scores will depend on the shooter's ability to refine and improve his positions and techniques.

    B. <u>Orient the Students</u>:

        1. <u>Lesson Tie-In</u>: The previous instruction was directed toward basic fundamentals common to all positions. The shooter must be able to apply these universal principles to a particular position taking into consideration the various personal physical characteristics of the average shooter.

        2. <u>Motivation</u>: The techniques that are going to be explained to you during this period of instruction will be those used by current and past champions and record holders at both National and World level.

        3. <u>Scope</u>: During this period of instruction, I will explain and demonstrate to you the United States Army Marksmanship Units method of employing the factors for the prone, kneeling and standing positions.

V. BODY:

    A. <u>First Student Performance Objective</u>: Given the characteristics of the prone position, USE the correct equipment and techniques and OBTAIN the best prone position for each student as contained in Chapter V, USAMU International Rifle Marksmanship Guide.

<u>NOTE</u>: SHOW SLIDE #1 (PRONE POSITION).

1. Orienting the position: The prone position is so steady that it may be said to have a single point of aim. The position should be oriented so the natural point of aim is the 10 ring. Fine changes can be made with slight movement of the leg or the breath control. Any major change should be made by reorienting the whole position You should strive for perfect scores.

NOTE: SHOW SLIDE #2 (PRONE SHOOTER VIEWED FROM THE REAR)

2. Assuming the prone position: To assume a good prone position the shooter lies between 5 and 30 degrees to the left of the line of fire. The body is not twisted but is stretched out and relaxed, the spine is straight.

NOTE: SHOW SLIDE #3 (LEFT LEG)

3. Position of the left leg: The left leg is roughly parallel to the spine with the toes pointed to the right or inward. The heel should not be forced down to the ground but relaxed. Pointing the toe outward places a strain on the muscles of the left leg and forces too much body weight onto the right side and elbow.

NOTE: SHOW SLIDE #4 (RIGHT LEG)

4. Position of the right leg: The right leg is drawn up with the knee bent and is roughly parallel to the left leg with the toe pointed outward. The leg is brought up to (1) level the shoulders (2) to free the right side of the diaphram from the ground to allow for easier breathing. However, drawing the right leg up too far adds too much pressure on the left elbow and can become painful quite rapidly.

NOTE: SHOW SLIDE #5 (LEFT ARM)

5. Position of the left arm: The left elbow is placed slightly to the left of the rifle. Placing the elbow under or to the right of the rifle strains the ligaments and muscle of the upper torso. The placement of the forend stop is determined by the length of the shooter's arm. A general guide is to start with the distance from the trigger to the forend stop. The left hand and fingers should be relaxed and the wrist straight. The left wrist is approximately six inches from the group to comply with international regulations.

NOTE: SHOW SLIDE #6 (SLING POSITION)

6. Adjusting the sling: The sling may be either high or low on the arm (See Chapter V) and should be adjusted so that it supports the weight of the rifle completely. The left hand should be snug against the forend stop. Adjust the sling on the arm the way it will produce the least amount of pulsebeat.

NOTE: SHOW SLIDE #7 (RIGHT ARM)

7. Positioning the right arm: The right elbow is placed a comfortable distance away from the body. By bringing the right elbow too close to the right side it will raise the shoulder to an uncomfortable position and unstable condition will result. However, placing the right arm too far out can result in an illegal position because the forearm cannot legally touch the ground. The right hand may grip the stock with any degree of pressure. The important factor is that it be the same for each shot. The thumb may be over the top or alongside the stock. Positioning the finger on trigger is also a shooter preference but it should be clear of the stock so when pressure is applied to the trigger it also is not applied to the stock.

NOTE: SHOW SLIDE #8 (HEAD & SHOULDER)

8. Positioning the rifle: The butt should fit snugly into the shoulder. Many shooters will place the rifle low in the shoulder and in order to achieve maximum rifle-shoulder contact, the butt plate will be raised in the stock. The point of greatest importance is that the rifle is placed in the same place each shot. Some shooters use the hook to accomplish this.

NOTE: SHOW SLIDE #9 (HEAD)

9. Positioning the head: The stock of the rifle should be constructed so when the shooter is in position and places his head on the stock he is looking through the sights. The head pressure on the stock should be constant. The head should be as erect as possible with good eye relief.

10. Refining the Position: A shooter should work constantly to refine his position and improve his hold. Very slight changes or adjustments can have a noticeable effect. A good way to check movement in your position is (usually from pulsebeat) with a telescopic sight. Learn to shoot without disturbing the position.

NOTE: HAVE DEMONSTRATOR REENACT EACH OF THE NINE POINTS OF PRONE POSITION AS DESCRIBED ON SLIDES. PRINCIPAL INSTRUCTOR WILL INDICATE IMPORTANT POINTS OF POSITION AND MAKE COMMENTS.

TRANSITION: Due to recent techniques developed by our own US shooters and some of the better kneeling shooters throughout the world, it is now possible to achieve a hold in the kneeling position for short periods of time that is as steady as the prone position.

B. Second Student Performance Objective: Given the characteristics of the kneeling, use the correct equipment and techniques and obtain the best kneeling position for each student as contained in Chapter VII, USAMU International Rifle Marksmanship Guide.

NOTE: SHOW SLIDE #10 (KNEELING POSITION)

1. As with all stable structures, the kneeling position should be built from the ground up.

NOTE: SHOW SLIDE #11 (KNEELING ROLL)

The kneeling roll should be made of firm comfortable material and conform to International rules of approximately 20 cms (7.88") in length and 18 cms (7.09") in diameter.

NOTE: SHOW SLIDE #12 (RIGHT FOOT AND KNEE)

2. The proper placement of the instep of the right foot on the kneeling roll is highly important for it carries the majority of weight of the entire position.

NOTE: SHOW SLIDE #13 (RIGHT FOOT AND KNEE)

The right knee should carry no more than its own weight and should be at an angle of between 30° and 45° to the line of fire.

NOTE: SHOW SLIDE #14 (TORSO)

NOTE: SHOW SLIDE #15 (HEAD)

3. The torso is erect and balanced over the kneeling roll sitting evenly on the buttocks. The spine should be slightly arched out over the kneeling roll to balance the forward hunch of the shoulders. The head should be erect and relaxed with from 3-4 inches of eye relief.

NOTE: SHOW SLIDE #16 LEFT ARM)

NOTE: SHOW SLIDE #17 (ELBOW)

NOTE: SHOW SLIDE #18 (KNEE)

    4. The left arm should be relaxed with the sling providing all of the support. The elbow is placed solidly on the knee. The elbow can be no more than 10 centimeters to the front or rear of the knee cap by International Rules. The left forearm and thigh should form a straight line when supporting the rifle.

NOTE: SHOW SLIDE #19 (LEFT LEG)

NOTE: SHOW SLIDE #20 (FOOT)

    5. The left leg should be vertical with the left foot $30°-45°$ angle from the line of fire.

NOTE: SHOW SLIDE #21 (RIGHT ARM AND HAND)

    6. The right arm should be relaxed with the hand in such a position as to give proper trigger control. The right hand grip may be firm or light whichever is preferred by the shooter.

NOTE: SHOW SLIDE #22 (SLING AND ACCESSORY ADJUSTMENTS)

    7. The sling should be adjusted to give proper support of rifle. However, it should not be so tight as to give pain or loss of blood circulation to the left hand or wrist. It may be placed high or low on the arm depending on which position gives the shooter a more stable hold. Adjustments to the accessories are mainly for producing a natural point of aim. This is made possible by an adjustable sling, sling swivel and butt plate assembly.

    8. Refining the positions: The entire process of developing a good kneeling position is starting from the basic or "classic" position and making your own individual refinements to suit your body characteristics and personal preferences.

NOTE: HAVE DEMONSTRATOR REENACT EACH OF THE ELEVEN POINTS OF KNEELING POSITION AS DESCRIBED ON SLIDES, PRINCIPAL INSTRUCTOR WILL INDICATE IMPORTANT POINTS OF THE POSITION AND MAKE COMMENTS.

TRANSITION: Standing is the last of the three positions to be taught because it is the most difficult to master. It is, however, by far the most rewarding of the three positions. The standing position can win or lose matches for you. It is here that more points are to be gained or lost to other shooters than in any of the other positions.

NOTE: SHOW SLIDE #23 (STANDING POSITION)

    C. **Third Student Performance Objective**: Given the characteristics of the standing position, use the correct equipment and techniques and obtain the best standing position for each student as contained in Chapter VI, USAMU International Rifle Marksmanship Guide.

NOTE: SHOW SLIDE #24 (POSITION OF FEET)

    1. As with all unstable structures, the standing position should be built on as large a foundation as possible. The feet should be placed comfortably apart, slightly less than shoulder width as a starting point. As your proficiency increases, you may experiment with different widths. The projected line of fire should cross each foot at the same point.

NOTE: SHOW SLIDE #25 (KNEE POSITION)

    2.   The knees are relaxed in a position similar to the one used when standing and talking to a friend.

NOTE: SHOW SLIDE #26 (HIP POSITION)

    3.   Next the hips must be kept level and 50% of the weight of the system should be kept on each foot.

NOTE: SHOW SLIDE #27 (BACK BEND)

    4.   In order to counterbalance the weight of the 17 lb rifle used in International Competition, we employ the back bend. The upper body is bent rearward at the waist so the center of gravity of the rifle - body - system will be located over the support area -- the feet!

NOTE: SHOW SLIDE #28 (BODY TWIST)

    5.   We immobilize the system at this point by rotating the upper body to the left at the waist. This is called body twist. What actually takes place is that the body is forced into using the bones and ligaments of the back to support the system instead of muscles which can become fatigued and cause movement.

NOTE: SHOW SLIDE #29 (LEFT ARM POSITION)

    6.   The left arm supports the main weight of the weapon. To ensure stability, the back of the left upper arm is rested on the left chest. For long limbed shooters, the left elbow may rest on the hip bone and is held directly under the rifle. The palm rest is on the heel of the hand, and the left wrist is comfortable. The wrist may or may not be bent back, but it must never be twisted.

NOTE: SHOW SLIDE #30 (RIGHT ARM POSITION)

    7.   The right arm is used solely to position the right hand for the act of undisturbed trigger manipulation. It should rest comfortably on the hook buttplace and the right hand should grip the pistol grip tight enough to enable the shooter to press the trigger and fire the shot without disturbing the weapon.

NOTE: SHOW SLIDE #31 (HEAD POSITION)

    8.   We now must position the head to the best advantage. In order not to disturb the balance apparatus of the inner ear, the head must be held erect. If the head is tilted, the inner ear reacts and signals the brain to have the body to make moves to reestablish equilibrium. The head should also be held erect so the eye looks straight ahead when aiming, not through the eyebrow or across the bridge of the nose. The most efficient area of the eye retina, the macula lutea, must register the sight impressions for greatest clarity. An upright head position, as you can see, is very important. It is necessary to adjust the rifle up to the head instead of the head down to the rifle. The rifle may be canted into the face to maintain an upright position provided the cant is the same each time.

    9.   Refining the position: The entire process of developing a standing position is a process of refinement of basic principles. It should be conducted with the care of a scientific experiment and by its very nature must be developed over an extended period of time.

NOTE: HAVE DEMONSTRATOR REENACT EACH OF THE EIGHT POINTS OF STANDING POSITION AS PRESCRIBED ON SLIDES. PRINCIPAL INSTRUCTOR WILL INDICATE IMPORTANT POINTS OF THE POSITION AND MAKE COMMENTS.

VI. CONCLUSION:

    A. <u>Retain Attention</u>: These three shooting positions just presented represent the type used by Olympic and World Champion shooters.

    B. <u>Summary</u>:

        1. The prone position is the steadiest of the three positions and the shooter should strive for a perfect score. There are four important points to remember in a good prone position - proper leg position, positioning arms and head and rifle/shoulder contact. Proper eye relief applies to all positions.

        2. Scores for the kneeling position approach those of the prone position There are four major points to remember in building a good kneeling position - the proper use of the kneeling roll, the body torso, the left leg and arm and the proper use of sling and accessories.

        3. There are three main points to remember in building a good standing position - the back bend, the body twist, and an upright head position.

    C. <u>Application</u>: If you learn to employ the principles and fundamentals disclosed to you in this class, you should be able to develop each of the three positions to give you a competitive advantage. The shooter should make his own individual refinements to suit his body characteristics and personal preferences.

    D. <u>Closing Statement</u>:

        1. The entire process of developing a position is a process of refinement of basic principles. It should be conducted with the care of a scientific experiment and by its very nature must be developed over an extended period of time. Your scores will improve to the point where you can compete with the best shooters anywhere. Gear your mind and training program to be a winner in competition.

C. INTERNATIONAL RIFLE MATCH PROGRAM IRMC 802
50 Minutes
Apr 75

LESSON OUTLINE

I. LESSON OBJECTIVE: To enable the International Rifle student to have a more comprehensive understanding of the US Army International Rifle Marksmanship Program and how it can provide opportunities for US Army personnel to be selected as members of US International Rifle Teams. Also to know the International Shooting Union rules for conducting international rifle matches and the modified NRA rules for conducting domestic international rifle matches.

II. STUDENT PERFORMANCE OBJECTIVES: As a result of this instruction students must be able to accomplish the following student performance objectives:

A. GIVEN the opportunity to participate in international matches, be able to apply the rules of International Rifle Shooting (ISU) and the NRA modified rules for domestic type international rifle matches as contained in The ISU Rule Book For International Rifle Competition and The NRA Rule Book For Modified International Rifle Competition.

B. FACED with the problem of sponsoring a domestic international rifle match, be able to describe how an international rifle match is conducted as contained in The NRA Rules For Modified International Rifle Competition.

C. CONFRONTED with the problem of promoting the sport of rifle shooting, be able to GIVE a 10 year summary of accomplishments of US International Rifle Teams, RELATE how goals and training methods of the US Army have placed members on US Rifle Teams, and REVEAL the opportunities for international competition through participation in the US Army International Rifle Program as contained in Annex 1, USAMU International Rifle Marksmanship Guide.

III. ADVANCE ASSIGNMENT: None

IV. INTRODUCTION:

A. **Gain Attention**: International rifle shooting has increased in popularity until there are 104 countries who are members of the International Shooting Union (ISU). The ISU type of match is the primary course of fire in the great majority of participating nations. It is also the standard type of match in all world-championship free rifle events. Headquarters for the International Shooting Union is located at Wiesbaden Klarenthal, West Germany.

NOTE: SHOW SLIDE #1 (ISU HEADQUARTERS, WEISBADEN, WEST GERMANY)

B. **Orient Students**:

1. **Lesson Tie-In**: The previous lesson discussed how the shooter can improve his skill and ability. Another important area of interest to the shooter is the match conditions and rules which he must abide by during an international (ISU) rifle match. ISU Matches are conducted in the United States under the auspices of the National Rifle Association. ISU rules are usually modified to allow the use of NRA approved equipment. Registration of ISU type matches at local, state, and regional levels can be made with the NRA in the regular manner.

2. **Motivation**: The NRA also conducts, under strict ISU rules, the United States National International Championships. The best international shooters in the US compete for awards in military, civilian, junior, and women's categories. This match is usually conducted simultaneously with the US tryouts for a World Championships Team, i.e.,

the Olympics, the World Championships, and the Pan American Games. The latter event is, as the name implies, a contest between nations of the two American Continents. The other two are open to all nations of the world. Each event is held once every four years, and they are so spaced that two do not fall on the same year.

3. <u>Scope</u>: This period will be used to explain the differences in international rifle shooting as compared to matches sponsored by the National Rifle Association. You will learn how international matches are conducted and how you can enjoy the opportunities of the US Army International Shooting Program. A more active international rifle program benefits the US International Rifle team selection by generating a broader base of choice in picking the best U.S. shooters.

<u>NOTE</u>: SHOW SLIDE #2 (NRA MODIFIED INTERNATIONAL RIFLE RULES)

V. BODY:

A. <u>First Student Performance Objective</u>: GIVEN the opportunity to participate in international matches, be able to apply the rules of International Rifle Shooting (ISU) and the NRA modified rules for domestic type international rifle matches as contained in the ISU Rule Book For International Rifle Competition and The NRA Rule Book For Modified International Rifle Competition.

1. <u>Time</u>: NRA matches require the competitor to shooter his positions in a limited amount of time. ISU matches allow a longer period of time for the shooter to work and concentrate on his performance. There is no need to hurry the shot.

2. <u>Targets</u>: The bull's-eye of the ISU target is larger than the NRA bull's-eye, and produces a sight picture that is much easier to see. However, the scoring rings on the ISU bull's-eye is easier to see, it is more demanding upon the shooters. It is a more discriminating measure of performance than the NRA bull's-eye.

3. <u>Firing Points</u>: In ISU matches, firing points are usually covered and enclosed on three sides. The purpose of this is to protect the shooters from the elements. Also, all shooters are equally protected, this is not always true in NRA matches when trees or buildings behind an open firing line protect some shooters from wind, while other shooters go unprotected.

4. <u>Equipment</u>: As pointed out in previous lessons there are several items of equipment which are approved by the NRA for domestic matches which cannot be used in ISU type competition. I will not mention each of these items again, but only remind you that in World and Olympic type competition the shooter must use ISU approved equipment.

B. <u>Second Student Performance Objective</u>: FACED with the problem of sponsoring a domestic international rifle match, be able to describe how an international rifle match is conducted as contained in The NRA Rules For Modified International Rifle Competition.

<u>NOTE</u>: SHOW SLIDE #3 (SCENE OF AN INTERNATIONAL SHOOTING MATCH WITH NATIONAL FLAGS)

1. <u>Description of how an international rifle match is conducted</u>.

a. <u>Squadding</u>: Squadding is accomplished by drawing lots. Normally, a team representative from each country draws a team squadding ticket which gives him one firing point for each of his team members. He then assigns each shooter to a specific firing point. Where there are no team matches, however, lots are drawn to assign a specific shooter directly to a specific firing point.

b. <u>Opening the Firing Line</u>: The firing line is opened to competitors for a specified and announced time before the match begins. When the firing line is officially opened shooters may go to the line and set up equipment.

2. <u>Competition</u>: In all matches, the positions must be shot in order of (1) prone, (2) standing and (3) kneeling. There are six types of ISU matches of interest to us. They are as follows:

   a. Smallbore 3-position (50 meters). The competitor fires a maximum of 10 sighting shots and 40 shots for match score in each position (prone, standing, kneeling). This makes a maximum total of 50 shots in each position. Time limits: Prone, 1 hr. 30 min; standing, 2 hrs; kneeling, 1 hr. 45 min.

   b. English Match (50 meters). The competitor fires a maximum of 15 sighting shots and 60 shots for match score from the prone position. Time limit is 2 hrs. 30 min.

   c. Standard Rifle (50 meters). The competitor fires a maximum of 6 sighting shots and 20 shots for match score in each position (prone, standing, kneeling) Total time 2 hrs. 30 min. A standard rifle must be used, i.e., no palm rest, no adjustable butt plate, no hook butt, no thumbhole stock, no set trigger.

NOTE: SHOW SLIDE #4 (A US ARMY INTERNATIONAL RIFLE SHOOTER IN COMPETITION)

   d. Free rifle (300 meters). The competitor fires a maximum of 10 sighting shots and 40 shots for match score in each position (prone, standing, kneeling). This makes a maximum total of 50 shots in each position. Time limits: prone, 1 hr. 30 min; standing, 2 hrs; kneeling, 1 hr. 45 min.

   e. Army Rifle (300 meters). The competitor fires a maximum of 6 sighting shots and 20 shots for match score in each position (prone, standing, kneeling). Total time: 2 hrs. 30 min. Each shooter fires the service rifle of his own country, or all shooters will fire the service rifle of the host country.

   f. Air Rifle (10) meters). The competitor fires a maximum of 10 sighting shots and 40 shots for match score from the standing position. Time: 2 hrs.

3. <u>Sighting shots</u> are made at specified sighting bull's-eye. They must be made before or between 10-shot strings for match score. Once a shooter begins shooting for match score, he must complete a 10-shot string in that position before he can return to a sighter bull's-eye in that position.

4. <u>Before the match begins, each competitor's targets are clean and marked by position and numbered</u>. It is the shooter's responsibility to see that he fires on the correct target.

5. In world championship smallbore competition, only one shot is fired at each target. Thus the changing of targets requires the firer to proceed at a much slower pace than is the custom in NRA matches. In 300 meter matches, 10 shots are usually fired at each target.

NOTE: SHOW SLIDE #5 (SCENE OF REGISTER KEEPER AT INTERNATIONAL RIFLE MATCH)

6. <u>Score Keeper</u>:

   a. Behind each competitor is a register keeper. He is responsible to:

   (1) Signal the pit detail to change targets.

   (2) Insure that the competitor does not fire more than the legal number of match or sighting shots.

(3) Record the value (as best he can) of each shot on a scoreboard for the benefit of spectators. His record of the <u>number</u> of shots fired is official. His records should be preserved, however, as it may be referred to by the Jury of Appeals in reviewing targets.

b. The position of Register Keeper is highly valued in European countries, where spectator interest in shooting events runs very high. Regulation ISU ranges have large areas reserved for spectators. During a match, the area behind a leading shooter is usually completely filled with spectators equipped with binoculars. Spectators are traditionally very courteous and sympathetic toward the shooters.

NOTE: SHOW SLIDE #6 (AWARDS CEREMONY)

7. Awards Ceremony: A traditional ceremony attends the completion of every ISU event. The first three place winners mount a 3-tier pedestal and are presented with gold, silver, and bronze medals, for first, second, and third place. Then the first place winner is honored by the playing of his national anthem and the raising of his national flag. The completion of this ceremony marks the official close of the match.

C. <u>Third Student Performance Objective</u>: CONFRONTED with the problem of promoting the sport of rifle shooting, be able to GIVE a 10 year summary of accomplishments of US International Rifle Teams, RELATE how goals and training methods of the US Army have placed members on US Rifle Teams, and REVEAL the opportunities for international competition through participation in the US Army International Rifle Program as contained in Annex 1, USAMU International Rifle Marksmanship Guide.

1. After World War II, the United States reentered international rifle competition for the first time at the 1948 London Olympics. Considering the difficulties that the United States Shooters were forced to overcome, they did a commendable job in representing their country.

2. The majority of American rifle matches have always been held according to American rules and using American targets. Such a great difference exists between our domestic competitions and true international matches, as conducted by other countries of the world, that the transition between the two is not performed without handicapping our shooters.

3. Until recent years, the United States has been a dominating power in prone shooting but has lacked the training and experience to excel in either kneeling or standing. Most of the shooters that comprised our teams were products of gallery training. The marksmen soon found that they must train on the 50 and 300 meters ranges with the ISU target if they were to be contenders with shooters of other countries. Also, they had to spend considerable practice time in the more difficult positions.

4. So completely were the world competitions dominated by Russia, Switzerland, Sweden, Finland, and the Germanies that the USA was not even considered a threat to win any medal except prone.

5. In 1956 the United States Army established a Marksmanship Unit at Fort Benning, Georgia. A special section was designated to develop a team to train for International Rifle Shooting. Through the years the personnel of this section have studied and worked on the techniques of ISU Shooting.

6. The match results given here will verify the success that the shooters have attained and the important role that the unit has played in establishing the United States as the dominating power in rifle marksmanship that it is today. Since its inception the unit has been responsible for winning 88.4% of all medals won in international rifle competition by the United States of America in the Olympics, World Championships and Pan American Games.

7. A nucleus of competitive talent has been formed and these people have been instrumental in successfully representing their country and in teaching their methods to newer shooters everywhere. It is hoped that the information passed out at clinics will benefit even a greater number of shooters and that a never ending flow of top notch marksmen will be produced to represent their country.

8. The USAMU sends a group of the best international rifle shooters on a training shooting trip each year to foreign countries, mainly in Europe and South America, to gain seasoning and maturity in facing up to the competition in world shooting events.

NOTE: SHOW SLIDE #7 (A GROUP OF US ARMY INTERNATIONAL RIFLE SHOOTERS ON A UNITED STATES RIFLE TEAM)

VI. CONCLUSION:

A. <u>Retain Attention</u>: You have been given a brief description of international shooting as it exists today. There are many fine points which cannot be covered due to the time frame we are limited to. Many potential international shooters here today will walk away saying that's nice, but I'll never be able to participate on a US International Rifle Team and with that type of an attitude you probably won't.

B. <u>Summary</u>: Let us recall for a minute what we have discussed. The ISU Rules for conducting an International Rifle competition are complicated and involved. Each shooter must understand these rules. Think of that new equipment which must be purchased to keep abreast of the innovation and changing rules. Contemplate the many hours of practice necessary to find better ways to shoot because we are not posting enough winning scores in world competition.

C. <u>Application</u>: We have come so far and worked so hard and some of you don't think you are capable of being an international shooter. I say any one of you can be as good a shot as the next man if you are willing to pick up the knowledge left behind by past champions and work hard and diligently toward your goal.

D. <u>Closing Statement</u>: The equipment, technology, and knowledge are ever present. The opportunity is here and the time to act is now. But you and only you can choose that opportunity and meet it with a burning desire to win. If you make the right choice, people will be reading about you as an Olympic Champion and your name will be in the record book.

CHAPTER XX

LESSON OUTLINES FOR INTERNATIONAL SKEET AND TRAP MARKSMANSHIP

A. INTRODUCTION TO INTERNATIONAL SKEET AND TRAP SHOOTING   USAISMC 900
                                                               50 minutes
                                                               Mar 1975

LESSON OUTLINE

I. LESSON OBJECTIVE: To enable the International Trap and Skeet marksmanship student to obtain a general knowledge of the US Army and the US International Trap and Skeet Programs.

II. STUDENT PERFORMANCE OBJECTIVES: As a result of this instruction the student must be able to accomplish the following student performance objectives:

   A. Scheduled to give instruction on the general aspects of International Skeet and Trap shooting, EXHIBIT a thorough knowledge of the International Skeet and Trap Program, as outlined in Chapters 1 and 6, USAMU International Skeet and Trap Guide.

   B. Scheduled to give instruction concerning US accomplishments in moving target competition, SUMMARIZE the accomplishments of the US International Skeet and Trap Teams in recent years as related in official records on file and published by the International Shooting Union.

   C. Scheduled to give instruction on the differences between domestic and international skeet and trap shooting, EXPLAIN the major differences between domestic skeet and trap and the international type match, as outlined in Chapters 4 and 5, USAMU Skeet and Trap Guide.

   D. Programmed to give a period of instruction on USAMU goals and training methods, RELATE the goals and training methods for placing US Army skeet and trap shooters on the US International Trap and Skeet Teams in the future as described in Chapters 4 and 5, USAMU Skeet and Trap Guide.

   E. Assigned the task of briefing prospective trainees for Skeet and Trap training program, DISCUSS the opportunities for participation in international competition through the US Army International Skeet and Trap Program.

III. INTRODUCTION:

   A. <u>Gain Attention</u>: Welcome to the International Skeet and Trap portion of the US Army Marksmanship Unit Clinic. International Clay Pigeon - often referred to as International Trap or Olympic Trap and International Skeet are relatively new to American Shotgun shooters. Neither game is well understood by the majority of the shooting public and clay pigeon is particularly foreign to US sportsmen. However, during the past ten years the interest in International shooting by American Shooters has steadily increased. International Clay Pigeon and International Skeet shooting can be challenging to the most skilled marksmen. Each, within its own field of shooting proficiency, is a much greater measurement of individual shooting ability. A score of one hundred straight in International Trap is an exception, not a rule. The same score in International Skeet is a greater rarity.

   B. <u>Orient Students</u>:

      1. <u>Lesson Tie-In</u>: Our purpose here today is to start you on your way to being a champion <u>International</u> Skeet and Trap Shooter by providing you with a working knowledge of the International Skeet and Trap Shooting Program supervised by the International Shooting Union (UIT).

2. _Motivation_: There are probably more clay targets thrown, more shotgun hulls dropped on shooting fields in this country in the course of a years time than in the rest of the world combined. There exists a boundless reserve of championship potential in our country, it is largely yet untapped. This talent has to be attracted into the International Skeet and Trap Program in order for the US Team to bring home more Gold Medals from World International competition.

3. _Scope_: This instruction will cover the United States International Skeet and Trap Shooting Program, the mission and goals of the US Army Marksmanship Unit in regard to international clay target shooting. We will discuss the performance of US Teams in recent years, briefly talk about the major differences between the American style of skeet and trap and the International style of shooting and reveal the opportunities for international competition by participating in the US Army International Skeet and Trap Shooting Program.

IV. BODY:

A. _First Student Performance Objective_: Scheduled to give instruction on the general aspects of International Skeet and Trap shooting, EXHIBIT a thorough knowledge of the International Skeet and Trap Program, as outlined in Chapters 1 and 6, USAMU International Skeet and Trap Guide.

1. International Skeet and Trap competitions are supervised by the International Shooting Union which was first established in 1907, and later dissolved in 1915. The Union was reestablished in 1921 under the name "Union International de Tir." The purpose of the Union is to promote and guide the sound development of the shooting sports and to strengthen the bonds of friendship between the shooting associations and federations of all nations, irrespective of political, racial and religious differences.

2. This is accomplished by:

a. Establishing permanent communications between the national shooting associations for the exchange of ideas on the development and perfection of the sport of shooting.

b. Organizing World Championships and encouraging and controlling the organization of continental or regional championships.

c. Supervising the shooting events of Olympic Championships and regional games organized under the auspices of the International Olympic Committee.

d. Issuing technical rules for the various shooting sports.

e. Awarding distinctions to shooters and to those who have worked for the development of shooting.

f. Publishing an official bulletin.

g. Encouraging in a general way all efforts to strengthen the comradeship between shooters of different nations, thereby creating international confidence and good will.

3. The organization that sponsors the International Skeet and Trap competition within the US is the International Committee of the National Rifle Association Shoots that are held in local gun clubs are either NRA sanctioned or NRA Registered. The NRA has the responsibility for conducting the US International Skeet and Trap Championships each year. From this competition the members of the US Team are selected and further trained to represent the United States. There are several major competitions

recognized by the ISU and the NRA in which the US Skeet and Trap Team competes. These championships are not the same each year but they are important to the development of a championships caliber US International Skeet and Trap Team.

NOTE: SHOW SLIDE #1 (MAJOR INTERNATIONAL COMPETITIONS).

  a. <u>Olympic</u> Games: Every four years.

  b. <u>World</u> Championships: Every four years alternating with the Olympic Games.

  c. <u>Pan American</u> Games: Every four years between the World Shooting Championships and Olympic Games.

  d. <u>World Moving</u> Target Championships: Each year except the year of the Olympic Games.

  These are the major competitions in which Gold Medals are won or lost for our country. To participate in one of these competitions is the highlight of an International Shooter's career. This should be your goal.

TRANSITION: Progress in international shooting is sometimes slow, depending on individual's desire to be a champion. The teams that have represented the United States in international competition during the past few years, have showed steady progress. However, the US Team still has some distance to go to achieve dominance of International Skeet and Trap competition.

NOTE: SHOW SLIDE #4 & 5 PERFORMANCE OF US TEAMS.

  B. <u>Second Student Performance Objective</u>: Scheduled to give instruction concerning US accomplishments in moving target competition, SUMMARIZE the accomplishments of the US International Skeet and Trap Teams in recent years as related in official record on file and published by the International Shooting Union.

  1. This slide shows the number of competitions that the US has participated in since 1912. As you can see the results are not too impressive in comparison to the number of events that we have participated in. Our progress has steadily increased in the team competitions but individual winners are few and far between. This is the area that we must improve in the future.

  2. The team results are a little better in comparison to the individual. This slide shows results since 1962, information prior to 1962 is not available, however, this will give you a good picture of the US Teams recent performance.

TRANSITION: To some shooters, both skeet and trap, the term "International" denotes competition beyond their level of skill. If anything, this type of competition is more challenging, and a greater measurement of an individual's ability.

  C. <u>Third Student Performance Objective</u>: Scheduled to give instruction on the differences between domestic and international skeet and trap shooting, EXPLAIN the major differences between domestic skeet and trap and the international type match, as outlined in Chapters 4 and 5, USAMU Skeet and Trap Guide.

  1. The international skeet field, rules, and shooting procedures approximate those in American skeet, however, there are some major differences that are significant.

NOTE: SHOW SLIDE #2 - DIFFERENCES BETWEEN INT'L AND AMERICAN SKEET.

    a. Flight distance of the targets.

    b. Gun position.

    c. Variable time release system.

    d. Specifications of the clay target.

    e. Shooting procedure.

  2. As in International Skeet, there are several major differences in the American Trap and International Trap that I would like to point out.

NOTE: SHOW SLIDE #3 - DIFFERENCE BETWEEN INT'L AND AMERICAN TRAP.

    a. Field layout.

    b. Distance of targets.

    c. Shooting procedure and rules.

  3. The prospective international shooter should be familiar with these major differences if he is going to progress in International Skeet and Trap competition. For this reason a more detailed explanation will be given in a later period of instruction.

TRANSITION: The United States Army Marksmanship Unit is constantly endeavoring to improve the skill of the Army shooters who tryout for the US Team to represent the United States of America in International Skeet and Trap competition.

  D. **Fourth Student Performance Objective:** Programmed to give a period of instruction on USAMU goals and training methods, RELATE the goals and training methods for placing US Army Team members on the US International Trap and Skeet Teams in the future as described in Chapters 4 and 5, USAMU Skeet and Trap Guide.

  1. The USAMU goal is to place as many members as we possibly can on the US International Skeet and Trap Teams each year. We don't always accomplish this for various reasons. All the training within the Shotgun Branch is oriented toward the international type match. Due to the lack of wide interest in International Trap and Skeet in the US, there are fewer matches to compete in, compared to the domestic type skeet and trap shooting.

  2. Our training methods are basically no different than those of an individual, or club team, training for an American match. There is one bit of difference, our training is more closely controlled and is designed to meet the individual shooters needs, rather than those of a group.

    a. In skeet, for example, the shooter is required to shoot a minimum number of targets for record each week. The number of targets required depends on the progress of the individual and the season of the year. Normally the number of targets required decreases as the shooting season progresses. In addition to record targets, the shooter is required to shoot station practice on the stations that he had difficulty with the previous week in regular practice and record practice. A sample training schedule for the skeet team would be as follows:

NOTE: SHOW SLIDE #6 - TRAINING SCHEDULE - SKEET TEAM.

    (1) Monday AM - Station Practice
       Monday PM - Regular Round Practice

    (2) Tuesday AM - Station Practice
       Tuesday PM - Regular Round Practice

    (3) Wednesday - Free (individual shoots as he desired)

    (4) Thursday - 100 tgts Record Practice

    (5) Friday - 100 tgts Record Practice

    b. In our International Trap training there is a slight problem in keeping our training directed toward the individual's needs. This is due primarily to the fifteen trap layout. In order to allow an individual to train on a particular angle target it required that the targets, or angles, be changed almost daily. We can still accomplish this by taking the time necessary to set the angles to a particular individual needs.

    As in the Skeet, Trap shooters are required to shoot a minimum of 200 targets per week for record, plus any additional targets outlined in our schedule. A sample weekly training schedule for a trap team would be as follows:

NOTE: SHOW SLIDE #7 - TRAINING SCHEDULE TRAP TEAM.

    (1) Monday AM - 1st Barrel Practice (load 1 shell)
       Monday PM - Regular Rd - with extra shots at tgts
         missed (score sheet will reflect which
         angle tgts missed if any).

    (2) Tuesday - Regular rd practice 50 tgts with shots at
        tgts missed. (Score sheet will reflect which
        angle tgts missed, if any.)

    (3) Wednesday - Free day (individual shoots as desired).

    (4) Thursday - 100 tgts Record Practice.

    (5) Friday - 100 tgts Record Practice. NOTE: Targets missed
        on record days are noted on score sheet as to
        which angle.

    The sample training schedules that you have just seen are our method of training. We have had a certain amount of success in the past and we hope to improve our training techniques. In order to gain the maximum performance from an individual, we must plan for a sufficient amount of training time to allow each shooter to improve and gain confidence.

TRANSITION: Training in international type shooting is important, in order for an individual to become a champion, and most of all training must be serious and controlled.

   E. __Fifth Student Performance Objective__: Assigned the task of briefing prospective trainees for skeet and trap training program, DISCUSS the opportunities for participation in international competition through the US Army International Skeet and Trap Program.

    1. Success in competition is the only true measure of an individual's training methods. Participation in as much competition as possible, is just as important in the development of a shooter, as is daily training. If he can meet certain qualifications, there exists within the Army Marksmanship Unit's International Program a great

opportunity for the young shooter who is contemplating entering the Army or sees in the Army a challenge. We don't say that everyone selected for a tryout with the team should be a champion. However, there are several areas that we look at very close prior to accepting an individual for a tryout:

NOTE: SHOW SLIDE #8 - AREAS CONSIDERED PRIOR TO REQUESTING TRYOUT

      a. International competition, if any.

      b. Competitive experience.

      c. Competitive averages (American and International).

      d. Willingness to stay with team for three shooting seasons if selected after tryout.

    2. If accepted for a tryout for the team, the prospect must have completed basic and AIT, prior to reporting to Fort Benning, Georgia. Upon arrival at Fort Benning, the prospect is assigned to the US Army Marksmanship Unit, in a tryout status for a period of 90 to 120 days. During the tryout status every effort is made to insure that the individual receives as much training as possible, to include competition away from Fort Benning. At the end of the tryout period a decision is made whether to assign the individual to the team, or request reassignment. This decision is made after a careful evaluation of the following:

NOTE: SHOW SLIDE #9 - AREAS CONSIDERED PRIOR TO ASSIGNMENT.

      a. Performance in training.

      b. Performance in competition.

      c. Progress during tryout period.

      d. Attitude.

      e. Future value to the team.

    3. Once the shooter is assigned to the team there is still no guarantee that he will attend every shoot that the team attends. Competition within the section is conducted to determine who will attend the shoots. It is not feasible for the team to attend every International Shoot scheduled in the United States. Only those shoots that will provide the most competition for the team are considered. This includes shoots out of country, such as Europe, where competitive experience can be gained against a majority of the better international Skeet and Trap shooters in the world. The opportunity to participate in a major competition each year is available depending on the individual performance in the annual US Team tryouts.

VI. CONCLUSION:

    A. <u>Retain Attention</u>: International type shotgun shooting is by far the most challenging shotgun sport available to the serious shooter today. It must be remembered that in order to be a champion, training must be serious, and strong competition must be available to the shooter.

    B. <u>Summary</u>:

    1. The International Shooting Union sponsors the major world International Skeet and Trap competitions in which the US Teams participate. In our country the International Program is sponsored and supervised by the National Rifle Association.

2. The major differences between the American style of shooting and International must be understood by the new International shooter if he is going to progress at a satisfactory pace.

3. The performance of US Teams in the past has not been too impressive. However, we are making progress in the team events, but individual medals are few and far between.

4. The goal of the United States Army Marksmanship Unit is to place as many U.S. Army members on the US Teams as possible. We accomplish this by supervised and controlled training, participating in major International competition in the US and abroad.

5. The opportunity to participate in International Skeet and Trap competition with the Army Team is available to the shooter about to enter military service If selected, after a tryout period, there are many opportunities for a shooter to compete in world competition as a member of the US Team.

C. <u>Application</u>: The knowledge gained during this period of instruction will give you a better understanding of the US Army International Skeet and Trap Program and the opportunities available.

D. <u>Closing Statement</u>: The United States Skeet and Trap Teams are competing in World International competition each year. The records indicate that we are far from being a leader in International Shotgun competition. If we expect to improve our standing it is imperative that we get more people to train and participate in the International Shotgun Program. By doing this, it will increase our chances of winning, and put our country on top which is where it should be.

THE UNITED STATES ARMY MARKSMANSHIP UNIT
Fort Benning, Georgia 31905

B. INTERNATIONAL CLAY PIGEON SHOOTING (TRAP)

USAISMC901
50 Minutes
Mar 1975

LESSON OUTLINE

I. LESSON OBJECTIVE: To enable the International Trap student to participate in International Clay Pigeon (Trap) Shooting competition.

II. STUDENT PERFORMANCE OBJECTIVES: As a result of this instruction the student will be able to accomplish the following student performance objectives:

A. Selected as a candidate for the US Army International Trap Team, EXHIBIT a general knowledge of the history of trap shooting as outlined by the principal instructor.

B. Selected as a candidate for the US Army International Trap Team, EXHIBIT a general knowledge of the International Clay Pigeon field layout and its equipment as described in Chapter V, USAMU International Skeet and Trap Guide.

C. Selected as a member of the US Army International Trap Team, DISPLAY a knowledge of shooting procedures and the rules of International Clay Pigeon shooting as described in Chapter V, USAMU International Skeet and Trap Guide.

D. As a member of the US Army International Trap Team, BE ABLE to select the best shooting equipment as described in Chapter V, USAMU International Skeet and Trap Guide.

E. As an International Trap competitor, DEVELOP a detailed understanding of and be able to DEMONSTRATE the proper techniques of shooting International Trap as described in Chapter V, USAMU International Skeet and Trap Guide.

III. ADVANCE ASSIGNMENT: None.

IV. INTRODUCTION:

NOTE: INSTRUCTOR THROWS A HANDFULL OF CLAY PIGEONS IN THE AIR, THEN "CRASH."

A. Gain Attention: The International Clay Pigeon. They are made hard but they break easy if you hit them. The terms "International Trap," "Olympic Trap" and "Clay pigeon" will be used synonymously in this class. The reason for this is that through common usage they have come to relate to the same thing. Officially though, the International Shooting Union identifies this phase of shotgun marksmanship as International Clay Pigeon. The most commonly used form, however, is International Trap.

B. Orient Students:

1. Lesson Tie-In: This lesson is the second part of a three hour course to acquaint you with the methods and techniques of shotgun shooting which the Army Team has found to be most effective.

2. Motivation: Your presence here this evening indicates that you have a distinct interest in international shooting, and if the chance arises many of you will be competing in International Trap Shooting Tournaments. I am sure that all of you want to be champions, but you must remember that to be a champion, you must have a thorough knowledge of the game you are shooting. You must be able to consistently establish the proper gun-target relationship that will result in breaking the target. You must have

the confidence and experience to be able to break targets under pressure. It is hoped that the following discussion will give those of you who are new to the game a foundation in shooting. For those of you who have had experience in shooting, maybe it will give you a surer approach to better scores that will help you become a World Champion or Olympic Gold Medal winner.

    3.   <u>Scope</u>: During this period I will describe the International Trap field layout, proper shooting procedures, several rules of International Clay Pigeon shooting, how to select the best types of shooting equipment, and trap shooting techniques.

V. BODY:

    A. <u>First Student Performance Objective</u>: Selected as a candidate for the US Army International Trap Team, EXHIBIT a general knowledge of the history of trap shooting as outlined by the principal instructor.

The term, Trap Shooting, dates back to England in the early 1700's. It comes from an early form of live bird shooting, in which the birds were imprisoned in a series of cages or traps. Cords were attached to levers on the traps so that the birds could be released on command. The shooters would stand behind the traps and signal when they were ready. The trap operator would then pull a cord releasing a bird for the shooter to kill.

By the 1750's many such trapshooting clubs had developed in England. However, in 1832 a particularly interesting English shooting club called the "High Hats" was founded. This club took its name from the trapshooting rules it enforced. Instead of releasing birds from traps, the shooter placed them under their hats and at a signal from the referee, they lifted their hats releasing the bird. But before they were allowed to shoot, they had to replace their hats on their heads. The club members felt that this gave the bird a fair chance to escape.

Trapshooting began in the United States about 1825. The first recorded trapshooting event is found in the Journal of the Sportsmen's Club of Cincinnati, Ohio, in 1831, and this date is often chosen as the beginning of organized Trapshooting in America.

Trapshooters in England and the United States soon found that they were depleting the supply of rare birds and they began to look for a substitute. In 1866, Charles Portlock of Boston, Massachusetts, introduced glass balls as targets. Most of the balls were about 2½ inches in diameter and hollow. The balls were placed one at a time in a cup containing a spring that threw them into the air. In order to create the impression of killing live birds, the balls were often covered or filled with feathers, so that when the shooter hit the ball the feathers would be scattered over the landscape. When feathers were not available, the balls were filled with a substance that gave off clouds of smoke when they broke.

The first clay targets were introduced about 1870. These were made of ordinary baked clay in the shape of a saucer. They proved unsuitable however, for they were too hard and brittle. About 1880, an Englishman named McCaskey made targets by mixing pitch with a binder such as river silt. These became known as "Blue Rocks" and are still on the market today.

With the advent of a suitable substitute for the live bird, trapshooting in the United States and on the European Continent took diverging paths. The American style of shooting has come to utilize a single auto-angling trap which throws the target at constantly changing angles but at a given height. In Europe, however, a fifteen trap layout was developed, over which targets were thrown from traps set at a different, specific height, and angle.

B. <u>Second Student Performance Objective</u>: Selected as a candidate for the US Army International Trap Team, EXHIBIT a general knowledge of the International Clay Pigeon field layout and its equipment as described in Chapter V, USAMU International Skeet and Trap Guide.

    1. The student must have a knowledge of the general layout of the International Clay Pigeon field and the legal settings of the traps. The International Traphouse is often referred to as a pit, trench, or bunker. The reason for this is obvious, for the roof of the traphouse has to be on the same level as the shooting station. This usually necessitates that the traphouse be an underground structure. In order for you to understand the trap field, let us look at the diagram.

<u>NOTE</u>: SHOW SLIDE #1 - STANDARD INT'L CLAY PIGEON FIELD PLAN.

    There is a total of 15 traps in parallel with the firing line. They are mounted in groups of three for each shooting. The center trap in each group is directly in line with its corresponding shooting station. Flank traps are mounted three and a half feet to either side of the center trap. This group of three traps is referred to as a "bank". The first bank corresponds to shooting station #1, the second bank to station #2, etc. Thus, the shooter may receive a target from any one of the three traps in the bank at his station. The distance between the center trap in each bank is normally 4 meters. Likewise, this is the corresponding distance between the center of the shooting stations along the firing line. Since this particular measurement is given considerable leeway, according to ISU rules, some bunkers tend to be longer or shorter than our example. The distance from the shooting stations to the traps is 15 meters. Each shooting station is one meter by one meter. Although the official referee and flank referee stands locations are quite arbitrary, this diagram shows where they are often positioned.

    The traps in each bank are adjusted so that the left trap in each group throws the target to the right and the right trap in each group throws the targets left. Such angles may vary from zero up to and including 45 degrees horizontally. To illustrate let us look at the next slide.

<u>NOTE</u>: SHOW SLIDE #2 - LEGAL ANGLE RIGHT TRAP.

This shows the legal limits of a left angle target (Explain limits again).

<u>NOTE</u>: SHOW SLIDE #4 - LEGAL ANGLE CENTER TRAP.

The center trap of each bank may not throw targets which vary more than 15° to either side of center.

    The height of the target's path above the trap shall be at least one meter and not more than four meters at a point ten meters beyond the traps.

<u>NOTE</u>: SHOW SLIDE #5 - COMPARATIVE EXAMPLE OF VERTICAL TARGET ANGLES.

    Please direct your attention to the lower half of the diagram, (Explain limits again). All targets on the field are set to travel 75 meters, $\pm$ 2 meters, on level ground at the best angle of elevation. Therefore all targets should be traveling at the same speed. However, depending on the height at which the trap is set, the distance will vary.

    In match competition, the angles and elevation of all trap settings is changed before each day's shooting. The new targets are selected from a chart established by the jury, thereby giving the competitors a completely new "spread" of targets for that day. Normally a valid spread will include a near equal number of similar targets, i.e. high, low and horizontal angles either left or right.

2. The student must be able to demonstrate a knowledge of the traps and releasing system.

It has already been established that a competitor may draw any one of three targets from his shooting station. How, then, is it possible to keep a shooter from knowing which target he will receive? This is accomplished by the use of an electrical selector computer.

NOTE: SHOW SLIDE #6 - SELECTOR COMPUTER.

The target selector is an electrical release system operated by the target puller. It consists of a console with five release buttons (one per station), a wheel used to "scramble" the order of target release, and a group of 15 buttons so that each trap may be released individually. The wheel, or "scrambler", is turned after each string of five targets. There are 30 combinations available on the control wheel, and a new shooter quickly learns that it is unwise to anticipate which target will appear next. When properly operated, the computer (Diviseur) will apportion every competitor an equitable share of similar targets over a 100 target event.

NOTE: SHOW SLIDE #7 - MICROPHONE, SHOOTER AND STATION.

In order to eliminate "slow" and "fast" pulls, most International ranges have installed a phono-pull system. It consists of a microphone at each shooting station that picks up the shooters call and converts it into an electrical impulse, releasing the trap.

NOTE: SHOW SLIDE #8 - PHONE-PULL ACCESSORIES.

This system is integrated directly into the "diviseur", thereby maintaining the normal target distribution. As you will notice, this particular model has a device for varying the sensitivity (amount of noise needed to set traps off) and the release delay time (time from call to target release). This group of 5 buttons is used to activate the microphone at the shooting station. The button is pressed down when the shooter raises his gun to his shoulder and released after he calls for the target.

3. The student must be able to differentiate between the International, Continental, and American field layouts.

Briefly let us clarify some of the major differences between International, American, and NRA Continental style trap.

NOTE: SHOW SLIDE #9 - COMPARATIVE EXAMPLE OF VERTICAL TARGET ANGLE.

Here we can see one of the major differences between American and International. The American targets remain on the same plane of flight and are much slower.

NOTE: READ MATERIAL ON THE SLIDE AND COMPARE THE DIFFERENCE.

NOTE: SHOW SLIDE #10 - NRA MODIFIED INT'L CLAY PIGEON FIELD PLAN.

As you can see, the NRA modified field utilizes the basic American layout, except that the trap not only oscillates horizontally but also vertically. The target must travel 65 yards rather than the normal 50 yards which the American target is thrown.

TRANSITION: You have now seen the major differences between the three sports.

C. Third Student Performance Objective: Selected as a member of the US Army International Trap Team, DISPLAY a knowledge of shooting procedures and the rules of International Clay Pigeon shooting as described in Chapter V, USAMU International Skeet and Trap Guide.

1. The student must know the function of the referee, the jury, and how they are chosen.

    a. Just like a game of football or basketball, the International Clay Pigeon Field also has a referee. It is his duty to make immediate decisions regarding hit or missed targets and decision whether a repeat target is to be thrown. Because of his responsibilities, the referee must have wide experience in clay pigeon shooting and a sound knowledge of shotguns.

    b. The referee shall be aided by two assistant referees. Usually these are appointed in rotation by the referee from among the competitors and preferably from those who have shot in the preceding squad. The referee shall always make decisions himself. If any of the assistant referees is in disagreement, it is his duty to raise his hand and advise the referee of this. The referee will then make his final decision.

    c. The referee's final decision, however, may be protested. Any protested decision goes before a jury appointed before the tournament. The jury consists of a representative from each country, with the organizing country's representative chairman. If more than five countries participate, the representative shall appoint a jury consisting of 5 members. The jury makes its decision by majority vote. Besides settling disputes that come up during the shoot, it is also the duty of the jury to verify that the ranges and targets conform to the regulations before shooting begins.

2. The student must know the shooting procedure, specifically what composes a squad and a round.

NOTE: SHOW SLIDE #11 - INTERNATIONAL CLAY PIGEON FIELD.

At the beginning of the competition, 6 shooters shall be ready to shoot, one at each shooting station, with the 6th shooter ready to take his place on No. 1 station. After the shooter at No. 1 station has fired at the target, he moves to station No. 2 as soon as the shooter at that station has fired, and so on. A sixth match is necessary to prevent delays caused by the walk to station 1 after shooting on station 5. Each round consists of 25 birds. Each squad member will shoot one round of 25 targets and retire until all remaining squads have also fired at 25 targets. This procedure is repeated until the daily event is finished. Normally 100 targets per shooter are shot daily, and 200 or 300 targets constitute a full match.

3. The student must have a knowledge of the rules including what "dead" and "lost" targets are.

The rules of International Clay Pigeon shooting are clear and are usually easily enforced. The following is a listing of the most important shooting rules. First of all, a shooter is allowed two shots at each target. Clearly this rule of shooting twice at each target distinguishes International Shooting from many of the other shotgun sports. The shooter must have both feet entirely within the boundaries of the shooting stand and must call "pull" or some other word of command when he is ready for the target to be released. The target is declared "dead" when it is thrown and shot at according to the rules and at least one visible piece is broken from the target. The target is declared "lost" when it is not hit during its flight, when it is only "dusted" (no visible piece falls off), when the shooter does not fire at a target for which he has called, and when the shooter is unable to fire because he has not released the safety catch or has forgotten to load or cock his gun.

Under particular situations a target is ruled a "No Bird" and another target will be allowed whether or not the competitor has fired. This happens if the target breaks when thrown, takes an irregular course on leaving the trap, is thrown from the wrong group, is a noticeably different color from the other targets, or if two or more targets are thrown simultaneously. These are some examples of "no birds" and "lost" targets. For a complete list, please consult the ISU rules or USAMU Skeet and Trap Guide.

4. The student must know the rules of conduct.

    a. By far the most important rule is the aspect of gun safety. All guns, even when empty, are to be handled with the greatest of care. Conventional double barrel guns are to be carried with the breech open. Magazine guns are to be carried with breech open with the muzzle pointed down. When a shooter puts his gun aside it must be placed vertically in a gunstand, muzzle up. It is also forbidden to touch another competitor's gun without the owner's consent. All guns must be carried in an open position between stations 1 and 5 and must be carried open and unloaded from station 5 to 1.

    b. Shooting and sighting may be practiced only from the shooting stations. Shots may be fired only when it is the shooter's turn and the target has been thrown. It is forbidden to sight at the other competitor's targets. It is also forbidden to sight at or shoot willfully, live birds or animals while on the shooting line.

    c. No shooter shall leave his stand before the shooter on the next stand has shot at a regular target. However, the shooter from stand 5 may move immediately to 1. After the shooters have fired their last shot in the round they are to remain standing on their stations until the last man in the squad has shot and the referee has announced "finished".

    d. The shooter is not allowed to put cartridges in the gun until he is at the shooting station facing the traps. He may not close his gun until it is the turn of the competitor to his left. These are a few of the major rules of conduct that are the most recognized and enforced.

TRANSITION: Let's look at some of penalties that may be incurred by infractions of the rules.

5. The student must be able to explain what penalties may be incurred by rule infractions.

The shooters are obligated to acquaint themselves with these rules and regulations. They bind themselves by their entry, in the match, to the penalties and disciplinary measures enforced upon competitors. However, the jury may fine the shooter one bird on repeated violations and in aggravating circumstances, may exclude the shooter from the round concerned or from the whole match. If a shooter leaves the squad for an unavoidable reason, he will be fined one target and with an opportunity of finishing his series later. Should the jury find that a shooter delays the shooting or conducts himself in an unsportsmanlike manner, it may give him a warning, fine him one bird or exclude him from the match. When the jury fines a shooter one bird and this decision does not refer to a special target, the first hit after the decision has been made known, is to be counted as a miss. If the shooter has completed the day's shooting, one bird shall be deducted from the score of the last series.

    D. Fourth Student Performance Objective: As a member of the US Army International Trap Team, BE ABLE to select the best shooting equipment available as described in Chapter V, USAMU International Skeet and Trap Guide.

1. The student must know the value of proper shooting clothing.

A good quality shooting coat, vest or shooting sweater is desirable. Two shots are allowed at international targets; therefore, the shooter will normally begin each round carrying thirty to forty shells. Hence, the garment selection should provide good pocket support for the additional weight of ammunition. Also, if preference is given to using different size loads for first and second shots, the two pocket arrangement for shells is essential in keeping loads separated during firing. Other considerations should necessarily be given to comfort, freedom of movement, and climatic conditions. If a shooting jacket of some type is not worn, a belt with pouch is recommended.

A cap or hat is also recommended for protection against the elements without interferring with clear vision. In this day of increasing hair length for men, a hat or some substitute might be an absolute necessity to keep the hair out of the shooter's eye. Of course we in the military don't have this problem.

NOTE: DEMONSTRATOR WILL MODEL ARMY TRAP SHOOTING UNIFORM AND REITERATE THE POINTS OF COMFORT AND FREEDOM OF MOVEMENT.

2. The student must be able to explain the need for ear protection and the type of protectors that accomplish this.

All shooters should be aware of the distinct damage to hearing that can be incurred by shooting without some form of ear protection. This hearing loss is irreplaceable. The loss occurs along specific frequency ranges and is often termed as Hunter's or Shooter's Syndrome. The only way to avoid this loss is to wear some type of ear protection. There are many ear protecting devices on the market today. Some offer only token protection while others are well designed and do an adequate job.

NOTE: SHOW SLIDE #12 - (OR ACTUAL EXAMPLES) - PROTECTION DEVICES USED BY SHOTGUN SHOOTERS IN THE USAMU.

NOTE: EXPLAIN TYPES SHOWN ON SLIDE.

There has been a number of studies conducted to determine what type of devices are best. However, little actual scientific proof is available. The best rule to use is that as most of the ear area is covered by the device, the more protection you receive. Hence the best type of protection would be the ear muff devices.

3. The student must be able to explain the value of shooting glasses.

It is obvious that some shooters need corrective glasses in order to see the target clearly. We are not concerned about these people because they wear their glasses faithfully. It is the person with good eyesight that needs to be encouraged to wear glasses for safety reasons. All of us have heard stories of guns blowing up and people being hurt or losing their sight. However, you must stop and think, granted these accidents only happen a very small percentage of the time but what if just this one unusual time it happened to you.

NOTE: SHOW SLIDE #13 (OR ACTUAL EXAMPLES) - SHOOTING GLASSES AVAILABLE.

As you can see the glasses have rather wide lenses. This keeps the frames or edges from obstructing your vision. The various colors are designed for specific lighting conditions (explain and demonstrate - gray and green for bright sunlight, etc), to prevent eyestrain from glare, thereby lessening the shooter tendency to squint. Remember, choose the proper glasses for the proper day.

NOTE: SHOW SLIDE #14 - GUNS ISSUED BY US ARMY MARKSMANSHIP UNIT FOR USE IN INTERNATIONAL TRAP COMPETITION (OR SHOW EXAMPLE).

4. The student must know the best type of shotgun to use and how the gun should be choked and stocked.

Selection of the proper gun for clay pigeon shooting should be made with three considerations in mind: reliability, durability, and speed of operation. Those three requirements are essential to successful international shooting. All shotguns including automatic models, 12 gauge or smaller, may be used. However, the necessity of firing a second shot at a missed target often results in a competitor firing 150 shells or more per 100 targets. The most practical weapon has been shown to be a double barrel shotgun. Consequently, there is a trend toward use of a breech loading superposed (over/under) single trigger weapon. Two barrel guns provide an opportunity to select a different choke for each shot. More experienced shooters find the most effective chokes to be a modified (60-65%) bottom barrel and a full choke (73-75%) top barrel. The percentages are pellet counts in a 30 inch circle at 35 yards.

As any experienced trap shooter realizes, correct fit in a trap gun is critical. The usual trap gun is stocked with a built in vertical lead to cover the rising target. However the International Gun may be stocked to point "dead on" so that when properly mounted, the eyes look straight down the sighting plane. This opinion may be subject to controversy by some trap shooters. However, because of the difficult one meter bird which is easily shot over, it is felt that a high pointing gun will increase this possibility. If this minimum low angle bird is missed on the first shot its velocity decreases and it begins to "fade" rapidly, thus compounding the error incurred with a high pointing gun.

Other factors in gun stocking such as pull, pitch, grip size and thickness at the comb should be fitted to the characteristics of the individual.

NOTE: SHOW SLIDE #15 - GUN WITH PITCH, COMB, ETC. ILLUSTRATED.

For those of you who are unfamiliar with these terms let us briefly review them.

5. The student must be able to select the proper ammunition in accordance with ISU rules.

QUESTION: What is the best ammunition for Trap?

ANSWER: Look at the International Clay Pigeon Rules. The International Shooting Union places no restriction on powder charge, however the shot is confined to a size no larger in diameter than $2\frac{1}{2}$mm (#7 European, about #$7\frac{1}{2}$ US) and total weight of 36 grams (1.270202). The shell may be no longer than 70mm (2 3/4 inches) in length. The Army Team has found that #$7\frac{1}{2}$ or #8 nickel plated shot has proven to be most effective. This hard shot produces excellent patterns and yields approximately 25 more pellets in a $1\frac{1}{2}$ oz load than an equivalent weight of ordinary lead shot. This in itself may be considered to be quite an advantage.

Most of the Army shooters prefer a #8 size load for the first barrel backed-up by size $7\frac{1}{2}$ for the second shot. However, most shotguns will pattern one particular load better than others. If tests prove that a gun possesses this characteristic to any significant degree between $7\frac{1}{2}$ and 8 loads, the most favorable size should be considered, for use in both barrels.

While no limitation is placed on the allowance volume of powder, $3\frac{1}{4}$ drams if believed to be optimum. It provides good uniformity of pattern and shot string, and produces sufficient load velocity to break the target while yielding a minimum of recoil.

NOTE: SHOW SLIDE #16 - INT'L TRAP SHELLS AVAILABLE ON THE MARKET (EXPLAIN EXAMPLE).

E. **Fifth Student Performance Objective**: As an International Trap competitor, DEVELOP a detailed understanding of and be able to DEMONSTRATE the proper techniques of shooting international trap as described in Chapter V, USAMU International Skeet and Trap Guide.

1. The student should be able to demonstrate proper gun-body relationship.

Obviously it is not possible in this or any other marksmanship training program to present iron clad guarantees or even claims that study and application will bring complete success. If this were possible, all scores would eventually be perfect and no challenge would remain. However, the search for improved clay target marksmanship basic fundamentals must be recognized and applied with varying techniques depending upon the individual. There is no secret formula. Fundamentals must be practiced if improvement is to be forthcoming.

The US Army Trap Team has developed a set of fundamentals which it uses to train new members of the team. These are broken down to fit into three main topics: BODY-GUN RELATIONSHIP, HITTING THE TARGET, AND GUN-TARGET RELATIONSHIP.

In order to attain the proper body-gun relationship the shooter must begin with his feet. Good body balance, ease of gun movement and smooth follow through begin here.

NOTE: DEMONSTRATOR WILL ACT OUT THE FOLLOWING UNTIL THE COMPLETE SHOOTING STANCE IS ATTAINED.

The feet should be a comfortable distance apart, approximately shoulder width. An imaginary line drawn from toe to toe should intersect the front boundary of the shooting station at an approximate 45° angle.

Insure that the body is relaxed, but remain mentally alert. The shooter may lean forward from the hips (into the gun), but not to the point of being tense or off balance. The knee of the forward leg should be bent slightly with a major portion of the body weight on this leg. This facilitates good balance when leaning into the gun.

The gun must be brought to the face and body in precisely the same position for each shot. It would be wise to try to bring the gun to your face rather than your face to the gun. But the emphasis should be placed on what you do, exactly the same, every time.

The hand on the forearm of the shotgun only supports the weapon. The grip must be sufficient to hold the gun but no tighter. The exact position of the forehand must be determined by the shooter, remembering two basic principles. First, the hand extended too far out on the forearm will slow down and restrict the swing. Second, the hand too far back on the forearm will reduce gun support and cause the shooter to "whip" the gun barrel and also restrict his vertical swing. We suggest that the hand be placed about half-way out on the forearm. We also suggest that the index finger of forehand be extended and be used to "naturally" point out the target. It is absolutely necessary to keep the head down on the gun stock throughout the shot and swing.

QUESTION: Are there any questions about the stance?

NOTE: DISMISS DEMONSTRATOR.

    2. The student must know the principles of hitting a moving target.

The entire content of this class on International Clay Target Marksmanship is presented for the one primary objective - HITTING A MOVING TARGET. There are three basic fundamentals involved in doing this.

NOTE: SHOW SLIDE #17 - FUNDAMENTALS.

These fundamentals are: SEE THE TARGET, MOVE WITH THE TARGET, FIRE AND FOLLOW THROUGH.

    a. The first of these, Seeing the Target, is the key to successful clay target shooting. Because of increased speed and variety of angles this point can not be over emphasized.

NOTE: SHOW SLIDE #18 - INTERNATIONAL TRAP HOUSE.

If you will direct your attention to the yellow stripe on the bunker roof, this indicates the position of the center trap. However, the clay target DOES NOT APPEAR AT ONE POINT IN RELATION TO THE TRAP HOUSE. With a bank of three machines in front of the shooter, the bird may appear at any point along a 7 foot horizontal span, depending upon which trap is released. The vision along the sighting plane must be EXPANDED to include the ENTIRE zone, from which a target may appear. The shooter must SEE the target at the earliest possible moment and determine it's exact direction of flight to obtain a smooth swing and a timely hit. The Army, also feels, that an integral part of OPENING the vision, is the relation of the gun muzzle to the trap house roof. We advise holding a lowgun, pointing directly at the area below the edge of the trap house roof.

Since it is not possible to present a step by step instruction on vision as in the mechanical technique of foot position, balance, swing, etc., this may be a list of errors which indicate that the shooter did not initially see the target correctly.

NOTE: SHOW SLIDE #19 - ERRORS & DESCRIPTIVE GUN BARREL MOVEMENTS (EXPLAIN SLIDE AND GUN BARREL MOVEMENTS).

  b. Move with the target. If vision has been expanded as precisely explained, the eyes will perceive the location of the target as the target appears. When the target does appear, the gun begins movement toward the target. The eyes fully focus and the brain interprets the correct horizontal and vertical flight path.

New shooters often jump or jerk their guns toward the target. This is a near panic reflex caused by their being unaccustomed to the increased speeds and angles of International targets. This characteristic will, or course, disappear with practice and a determined effort, towards a smooth, well controlled and coordinated gun movement.

  c. Since the rules of International Trap allow two shots to be fired at each target the accuracy of the second shot depends completely on follow through of the first. Follow through must be smooth and coordinated. At the instant of the trigger pull, the eyes should perceive if the gun-target relationship was correct or incorrect. The eyes should continue to follow the target path and if the first shot missed, compute corrections to the reflexes for an adjusted second-barrel shot. In the event both shots miss, the eyes should continue to follow the bird momentarily. Occasionally, this will give a mental picture of the error(s) for correction when the same target is drawn again.

When experience is gained and the new International shooter has developed a smooth, well coordinated gun swing, he should be breaking the target with the first shot at 30-32 meters from the gun barrel and with the second barrel load at 35 to 40 meters. The first round should be fired within seven-tenths (.70) of a second, with the second shot, if needed, following fifteen-hundredths (.15) of a second later. However, these figures are only relative and should be used only as a guide in developing good timing. Again, to reemphasize, the shooter must make an effort to develop a smooth, coordinated swing and follow through.

NOTE: SHOW SLIDE #20 - EXAMPLES OF GUN-TARGET RELATIONSHIP (FIG. 19 - USAMU SKEET AND TRAP GUIDE).

  3. The student must be able to understand Gun-Target Relationship.

To properly employ the swing through method at any target on the International shooting field, speed of swing (timing), is the critical factor. The gun must be moving at an apparent faster pace than the target and fired at the exact instant when gun muzzle passes through the target. Thus, at all times, the eyes are fixed upon a definite object and not measuring a projected horizontal and vertical lead at some relative distance ahead, or above the target. When movement of the gun is smooth, coordinated and correctly timed, good hits will follow. There is no rule for regulating the apparent speed of gun movement <u>with and through the target</u>. This factor can only be developed with each individual through practice.

NOTE: ANNOUNCE THE TIME THE CLASS WILL PROCEED AND ARRIVE AT THE SHOOTING FIELDS FOR PRACTICAL WORK.

VI. CONCLUSION:

    A. <u>Regain Attention</u>: QUESTION: How many of you feel that you have gained something from this block of instruction? QUESTION: Do any of you have suggestions on how we could improve this block of instruction or what material might be added or improved?

    B. <u>Application</u>: It is expected that you will use this information on International Clay Target shooting and pass it on to other shooters.

    C. <u>Summary</u>: We have discussed the trap field layout, proper shooting procedures, rules of International Clay Pigeon shooting, how to select the best types of shooting equipment and described trap shooting techniques.

NOTE: DISPLAY AN INTERNATIONAL TRAP GOLD MEDAL WON BY USAMU PERSONNEL.

    D. <u>Closing Statement</u>: "This is an Olympic (Pan American or Moving Target Championships) Gold (Silver or Bronze) medal won by _____ of the US Army Marksmanship Unit. Remember that your success in international trap will take diligent practice of proper shooting procedures and ACTUAL experience in competition. I hope you will continue to maintain your interest in International Shotgun Shooting. Your determination to train and participate to win in International Clay Target Shooting will result in a United States win of the Olympic Gold Medal in Trap Shooting.

C. INTERNATIONAL SKEET MARKSMANSHIP	USAISMC 902
	50 Minutes
	Mar 1975

LESSON OUTLINE

I. LESSON OBJECTIVE: To enable the International Skeet shooting student to become involved in the International Skeet Program.

II. STUDENT PERFORMANCE OBJECTIVES: As a result of this instruction the student should be able to accomplish the following student performance objectives:

  A. Given the requirement to compete in International Skeet competition EXHIBIT a general knowledge of the skeet shooting field and its equipment as described in Chapter IV, USAMU Skeet and Trap Guide.

  B. Faced with participation in an International Skeet competition, ACQUIRE a knowledge of shooting procedures and the rules of International Skeet Shooting as shown in Chapter IV, USAMU Skeet and Trap Guide.

  C. Faced with the task of selecting the best skeet shooting guns and equipment, EXERCISE discrimination in selecting equipment as prescribed in Chapter IV, USAMU Skeet and Trap Guide.

  D. Required to win in International Skeet competition, EXPLAIN and DEMONSTRATE the proper techniques of shooting International Skeet as outlined in Chapter IV, USAMU Skeet and Trap Guide.

III. ADVANCED ASSIGNMENTS: Chapter IV, USAMU International Skeet and Trap Guide.

NOTE: DEMONSTRATOR SAYS "PULL" AND FIRES A DOUBLE AS THE CLAY TARGETS APPEAR.

IV. INTRODUCTION:

  A. <u>Gain Attention</u>: In this group there may be a future International Skeet World Champion. Good afternoon. My name is_____. Welcome to The United States Army Marksmanship Unit International Skeet Shooting Clinic.

  B. <u>Orient Students</u>:

    1. <u>Lesson Tie-In</u>: Skeet shooting today is a favorite pastime of many people in this country and throughout the world. Unlike many competitive sports, skeet can be mastered by both sexes, young and old alike. International Skeet is a game that requires highly conditioned reflexes, intense concentration and more time spent in dedicated practice.

    2. <u>Motivation</u>: The term "International Style" denotes a greater challenge to sportsmen. This competition is a more exciting type of skeet and is a better measure of the individual shooters ability.

    For a U.S. shooter to stand on the top step at the award stand, hear his National Anthem being played, and be awarded a gold medal for winning in World International Skeet Competition, is the ultimate goal for every International Skeet Shooter.

    3. <u>Scope</u>: The purpose of this class is to assist the shooter in learning certain techniques used by The United States Army International Skeet Team in International competition. I will show you how to select proper guns and equipment, acquaint you with international skeet rules and show you the lay out of an International skeet field.

IV. BODY:

   A. <u>First Student Performance Objective</u>: Given the requirement to compete in international Skeet Competition, EXHIBIT a general knowledge of the history of skeet shooting, know the layout of the skeet shooting field and its equipment as described in Chapter IV, USAMU Skeet and Trap Guide.

   1. The sport of Skeet was started about half a century ago on the grounds of Glen Rock Kennels at Andover, Mass. The kennel was owned by the late Charles E. Davis, an avid upland game hunter and a crack shot. With the help of his son, Henry, and a close friend and hunting companion, Bill Foster, Mr. Davis designed the first skeet field. The design was laid out and consisted of a circle with a twenty-five yard radius that had twelve positions marked on the circumference like the numbers on the face of a clock. The trap was stacked down at 12 o'clock and positioned so it would throw clay targets toward 6 o'clock. This gave them 12 different angles to shoot at which helped them improve their field shooting. The idea soon gained popularity with many local hunters and the layout was revised to its present state because of safety precautions. An association was formed and skeet fields and members grew across the country.

   2. The International Skeet field layout is approximately the same as the American Skeet Field. There are only 3 major differences between them.

   a. The specifications of the clay target. The target has to be 11 cm in diameter, 25 to 28 mm in height, and must weigh between 105 ± 5 grams.

   b. The distance of the target in international is 77 yards as compared to 60 yards in American.

   c. The delay system used in International will delay a target up to 3 seconds before it emerges.

   B. <u>Second Student Performance Objective</u>: Faced with participation in an International Skeet competition, ACQUIRE a knowledge of shooting procedures and the rules of International Skeet Shooting as shown in Chapter IV, USAMU Skeet and Trap Guide.

   1. The procedures and rules of International Skeet differs slightly from that of American style. Although a squad is made up of 5 shooters and one round consists of 25 shots; 17 single and 4 pair of doubles; the squad does not shoot each round back to back as done in American Skeet. In International Skeet, before the 1st squad can start their second round, all squads must finish their 1st round, and so on. This factor alone makes the game much more difficult. At a large shoot, the shooter may have to wait up to 3 hours between rounds. It makes for a very long day and the concentration must be very intense.

   2. The shooting shall be conducted by a referee with wide experience in skeet shooting and a sound knowledge of shotguns and who should normally have a valid referee's license. His main function is to make immediate decisions regarding hit or missed targets, checking the shooter for illegal gun positions and determining "No Birds".

   a. The referees shall be aided by two assistant referees called Flanks. Usually these are to be appointed in rotation by the referee from among the competitors and preferably from among those who have shot in the preceding squad. All competitors are obligated, upon request to function as assistants. The main function of these flanks, is to give, immediately after a shot, a signal by raising his hand or flag if he considers a target lost. The referee shall make all decisions on the shooting field. If any of the assistants are in disagreement, it is his duty to advise the referee of this. The referee will then make his final decision.

3. Providing no other decision has been made, there shall be appointed a jury for international competitions consisting of a representative from each country, state or team represented. The organizing club will furnish the jury chairman and if more than 5 countries, states or clubs participate the chairman shall appoint a jury consisting of 5 members.

    a. It is the duty of the jury to verify before shooting begins that the range conforms with regulations and that the arrangements in general are suitable and correct; see during the shooting that the rules are adhered to and that guns and ammunitions and targets are examined by random tests; make decisions in connection with technical defects or other disturbances in the shooting; deal with protests; and make decisions regarding penalties if a shooter does not adhere to the rules deports himself in an unsportsmanlike manner.

    b. The jury decisions cannot be appealed against unless a special jury of appeal has been appointed for the competition.

4. In International shooting a dead target is determined by any visible piece broken off the target by the shooter and shall be ruled as such. There are several other factors in determining lost targets however. The four main reasons for losing a target other than missing it while shooting are:

    a. Upon the fourth and subsequent malfunction of the weapon in the same round.

    b. If the shooter, without legitimate reason does not fire at a regular double both targets are lost.

    c. If the shooter without legitimate reason does not fire at the second target of a regular double, the results of the first target shall be recorded and the target is lost.

    d. If in a regular double, the first is lost and the second cannot be fired upon because of a gun or shell malfunction. The 1st shall be ruled lost and the double will be repeated to determine the results of the second shot.

5. In International Skeet the rules of conduct of the shooter is to insure good competition, fair play, and most important, safety. Because a shotgun, at close range is very deadly, simple points of etiquette have been drawn up for the safety of all shooters.

    a. All guns, even when empty, shall be handled with the greatest of care. Conventional double barrel guns are to be carried with the breech open. Magazine guns are to be carried with action open and the muzzle pointed in a safe direction. Straps or slings on guns are prohibited. When a shooter puts his gun aside it must be placed in a gun stand or another place intended for this purpose. It is forbidden to touch or handle another competitor's gun without his specific permission.

    b. Shooting and sighting may only be practiced at the shooting station. Shots may be fired only when it is the shooters turn and the target has been thrown. It is also forbidden to willfully sight or shoot at live birds or animals.

    c. No member of a group shall advance to the shooting station until it is his turn to shoot and until the previous shooter has left the shooting station. No member of a group having shot from one station shall proceed toward the next station in such a way as to interfere with another shooter.

    d. It is prohibited to place a shell into any part of the gun before the shooter is standing on the station with the gun pointed toward the shooting field. During the shooting of singles only one shell may be loaded at a time.

6. Every competitor is obligated to acquaint himself with these rules insofar as they apply to all the shooters. By entering the competition he thereby agrees to submit to any penalty that may be incurred through failure to comply with the rules or with the referees' decisions.

International Skeet penalties are usually assessed by firing the shooter one or more targets depending on the seriousness of the offense. Illegal gun position after the first warning may cost a shooter one target. Failing to be present to shoot after the referee has called his name and competitor number 3 times shall cost the shooter 3 targets. If the penalty is of a serious nature, such as unsportsmanship or safety violation or willfully disobeying the rules, the shooter can be excluded from the match.

C. <u>Third Student Performance Objective</u>: Faced with the task of selecting the best skeet shooting guns and equipment, EXERCISE discrimination in selecting equipment as shown in Chapter IV, USAMU Skeet and Trap Guide.

Before the new shooter can begin to learn the techniques of International Shooting, he must not only know the rules and conduct of the game, but also the type of clothing and shooting equipment he should wear and use.

1. International rules require that the gun be brought to the shoulder from the hip. The jacket, sweater or vest selected should be of a design which permits a smooth upward movement of the gun butt, followed by a free swing and follow thru. A cap or hat which provides protection from the elements and other foreign particles without interfering with clear vision on all stations is recommended. Shoes and trousers should be comfortable and not bind the swing. A crepe or ripple sole shoe is recommended to prevent slipping and sliding on the station.

2. The protection of ones hearing is advisable in all forms of shooting. The type of protection device used is usually one of shooter preference. Proper fit and comfort are prime consideration in choosing either an ear plug, headphones or other anti-noise devices. The new shooter should experiment with several types of ear protectors and select the one best suited for him.

3. Protective shooting glasses are considered to be an essential safety measure in shooting also. Select your shooting glasses with care to assure proper fit and also be sure they are made of shatterproof safety glass. Besides protection from foreign particles, tinted shooting glasses may be used to help prevent eye strain and fatigue. On bright sunny days a dark green or gray tint is recommended. On overcast or dark rainy days, a yellow tint or clear glass may be used.

4. The shotgun is the most important piece of your shooting equipment; you should choose your gun accordingly. International shooters tend to favor the auto-loading and superposed guns as compared to the pump action shotgun.

a. When selecting a gun, choose one with balance at the forward end of the receiver. A gun which is muzzle heavy or extremely butt heavy is hard to handle and awkward to control. A heavy balanced gun is slow to swing but will move smoothly and follow through. Conversely, too light a gun can result in jerky swing, gun whip, and excessive recoil. When selecting a shotgun, take into consideration your physical strength. Whether it be a light or heavy gun, it must be balanced properly to insure good gun control.

b. The length and drop of the stock is pretty much up to the shooter depending on his physical size and shooting experience. The standard skeet length and drop are good dimensions to start out on and later after you get an idea of what you need it is advisable to change if warranted. Sharp combs, such as the English style combs, should be avoided, because any error in mounting the gun may cause undue punishment to the face during recoil.

    c. A rubber recoil pad can be disadvantageous to some shooters, particularly those whith short arms who may find the gun butt snagging in the clothing before it is properly seated onto the shoulder. If one is used, it should be painted with a lacquer or similar substance to provide a smooth surface and to minimize snagging.

    d. Because of the short range in which a skeet target is shot from, a skeet choke or spread choke is recommended. This will give a good open pattern at 25 yards, approximately the longest distance the skeet target will be shot from. The recommended barrel length is 26 to 28 inches. Some individuals have found that a 28 inch barrel offers a better sighting plane. This appears more noticeable in the superposed shotgun which lacks the additional sighting plane provided on the receiver type gun. However, many small-light statured shooters or older shooters who have slower reflexes prefer the 26 inch barrels because they are usually lighter and can be mounted quicker than the 28 inch barrelled gun.

   5. The rules on ammunition are very emphatic and are strictly enforced. I.S.U. rules state that the length of the shell shall not exceed 2 3/4 inches. The shot load is not to exceed 32 grams (1 1/8 oz). The pellets shall be only spherical in shape made of solid lead and 2 mm in diameter (American No 8 & 9). The standard American 3 dram 1 1/8 sheet loads are suitable for International shooting.

TRANSITION: To briefly summarize everything we have covered: A shooter must know all of the rules and etiquette of International Skeet and have a general idea how to select your shotgun, ammunition and other shooting equipment. Now is the time to see just how International Skeet is shot.

  D. <u>Fourth Student Performance Objective</u>: Required to win in International Skeet competition, EXPLAIN and DEMONSTRATE the proper techniques of shooting International Skeet as outlined in Chapter IV, USAMU Skeet and Trap Guide.

  The three techniques of shooting International Skeet that we will be concerned with are: THE BODY POSITION, MOUNTING THE GUN, AND THE THEORY OF HITTING MOVING TARGETS.

   1. Body position is perhaps the most important single factor in skeet shooting.

    a. Basically avoid any position which requires the use of muscles not normally used in a relaxed standing position while holding a shotgun. Extraneous movements or exaggerated shooting stances may impress the novice, but will contribute nothing to the scoreboard. Most champion skeet shooters are unimpressive to watch. They have eliminated every physical movement and mental process which does not contribute directly to the mechanics of breaking targets.

    b. A proficient skeet shooter must maintain a well balanced body while swinging, delivering the shot, and following through. To attain a well balanced body position, certain principles must be observed.

    c. Be sure that the body is relaxed but alert. Body tension is the main cause of a jerky swing. The feet should be placed a comfortable distance apart, about shoulder width is recommended. Placing the feet too close together will cause difficulty in maintaining proper balance. When placed too far apart they create unnecessary muscular tension about the body. Any stance which requires extra muscular control is incorrect. One foot should be placed slightly ahead of the other. For the right handed shooter it would be the left foot. The right foot for the left handed shooter. The knee of the forward leg should be bent slightly. The majority of the weight should be placed on this foot. The body may lean forward from the hips but do not over exaggerate to the point of being off balance or tense. Some shooters advocate that while waiting for the target to emerge, the majority of the weight should be on the back foot. Then as the target appears and the swing starts, the weight is shifted to the forward foot. This gives one a feeling of driving through the target and helps insure proper follow through. Although this technique calls for quite a bit more body motion, it can be advantageous for some to use this procedure. The basic idea to remember about body positions is the body must be relaxed, balanced, and comfortable.

    2. The second technique we will discuss is mounting the gun. The ready position in International Skeet is with the gun butt touching the hip bone. The movement from the starting position to the shoulder is the most critical movement in shooting skeet.

    a. It has been found that by placing the gun butt on the forward edge of the hipbone, it is easier to mount the gun with a flowing motion. Having the gun butt too far alongside the body, or directly on the hip, requires both a forward and upward motion in order to shoulder the gun. This creates two separate movements causing the swing to be late and erratic. Mounting the gun must be coordinated with the swing as one single movement. The gun is always brought up to the face and not the face to the gun. Before calling for the target, the shooter must be prepared for the target to appear instantly of any time within 3 seconds. Be alert and concentrate upon the appearance of the target. If the shooter receives a long (3 second) pull, he must continue intense concentration until the target emerges or the referee rules "no bird". Never permit the thought that a pull is too long to interrupt concentration. Also be ready for an instant target before calling for the bird. It is imperative that the shooter employ the same motion of bringing the gun to his shoulder on each shot, regardless of when the target appears.

    b. If the shooter attempts to bring the gun to his shoulder before moving with the bird he will find that considerably more time is required to catch up with the target. Again it is essential to develop a single, smooth, flowing motion.

    c. International Skeet does not lend itself to the technique of picking a point to break the target, better known as "spot shooting". Regardless of practice or skill in gun handling, the gun cannot be mounted from the international position with the same precision on every shot. The good international shooter is aware of this fact and is mentally prepared to adjust his shot accordingly. The target breaking point may vary with each shot and the shooter who tries to "groove" himself will not improve his score.

    3. The 3rd and last technique I will discuss with you, the theory of hitting a moving target. Whether it's skeet or wing shooting, there are four fundamentals necessary for optimum results.

    a. Seeing the target as quickly as possible. In International Skeet this means looking back at the opening from which the target will emerge. Shooters who are accustomed to looking for the target on a line of sight somewhere between the house and target crossing point must develop the technique of looking toward the general area of, but not into the trap opening. The eyes must pick up and begin tracking the target at the earliest possible instant of its flight, avoid being cued by the sound of the trap release. Sooner or later the listening habit will cause a shooter to jump at the wrong sound, resulting in a false start. Look for the target - don't listen to the trap.

    b. Move with the target. As soon as a bird emerges from the target house opening, start moving the gun with the target as the gun is brought up to the shoulder. Movement of the gun with the bird must be coordinated with the movement of the gun to the face and shoulder in one smooth flowing motion.

c. See the head and pull the trigger when the eyes, the gun, and the target are brought together and the correct lead is established, pull the trigger. It makes no difference whether you pass shoot, come-from-behind, or shoot a substained lead, always stay-in-front. Each shooter should know the method best suited for him and continue to use that system. Both work satisfactorily; however, it is advisable to use only one method. Use it on every shot, and do not switch at different stations.

d. Keep the gun moving. It is useless to shoot at a moving target and stop the movement of the gun. Remember the target is moving and the gun must move with it. Most targets are missed by shooting behind because the shooter tends to slow his swing just as the trigger is pulled. Follow through is ever more essential in International Skeet because of the greater speed of the target. If the shooter will remember these four basic points of hitting a moving target, he will find that his number of hits will increase steadily.

V. CONCLUSION:

A. <u>Retain Attention</u>: USAMU International Skeet Branch needs five more members who can hit 25 out of 25 consistently. If you can fire 199 out of 200 in a U.S. match, we can use you on the U.S. International Team. I could stand here and talk about shotgun competition for hours, but the best way to learn how to shoot international shotgun competitively is by going out and trying it yourself. See just how challenging the game really is.

B. <u>Summary</u>:

Regardless of your present shotgun proficiency, do not discard any theory or suggestion contained here as not for you. Each factor has been proven over a period of time and can assist even the more proficient shooters.

The three main points in this class to remember are:

1. Shooting equipment:

   a. Clothing.

   b. Ear-protectors.

   c. Shooting glasses.

   d. Proper shotguns.

   e. I.S.U. Ammunition.

2. Shooting Techniques:

   a. Body position.

   b. Mounting the gun.

   c. Hitting a moving target.

3. Differences in field lay-out and target throwing equipment.

For those of you that would desire further information on International Skeet, The United States Army Marksmanship Unit has published a complete detailed guide on International Skeet and Trap Shooting. This is a step by step, fully illustrated guide which covers all aspects of the game.

NOTE: DISPLAY A COPY OF THE USAMU SKEET AND TRAP GUIDE.

      C.   <u>APPLICATION</u>: The good U.S. shotgun shooters here have a chance to become international champions if they start participating in International Skeet with the interest they now show in American Skeet.

      D.   <u>Closing Statement</u>: Do not interpret a position or technique as "the only way." Each theory is flexible. Because of different body conformations, an individual may find it more advantageous to modify certain suggestions to fit his particular need. When your technique is perfected and you can consistently follow it, put in your formal request to the International Olympic Committee for our National Anthem to be played for you in the next Olympic Skeet Shoot.

CHAPTER XXI
LESSON OUTLINE FOR INTERNATIONAL RUNNING TARGET MARKSMANSHIP
RTCC 903
1 Hour
Mar 75

A.   RUNNING TARGET MARKSMANSHIP PROGRAM

LESSON OUTLINE

I.   LESSON OBJECTIVE: To enable the marksmanship student to be familiar with the US Army Running Target Marksmanship Program.

II. STUDENT PERFORMANCE OBJECTIVE: As a result of this instruction the students will be able to accomplish the following student performance objectives:

   A. Aspiring to become a Running Target shooter, be familiar with the origin of this sport, accomplishments of Army members of the US Running Target Teams and opportunities for international competition through participation in the US International Running Target Program as described in the USAMU International Running Target competitive records.

   B. Aspiring to become a Running Target International Competitor, be familiar with the Running Target Range, target mechanism, range operating procedures, course of fire and regulations as described in Chapters V and VI, USAMU International Running Target Guide.

   C. Aspiring to become a Running Target shooter, be familiar with the specialized equipment used by Running Target shooters as described in Chapter V, USAMU International Running Target Guide.

   D. Aspiring to become a Running Target shooter, be familiar with the shooting techniques used to produce a winning score as described in Chapter V, USAMU International Running Target Guide.

III. ADVANCE ASSIGNMENT: Chapters V and VI, USAMU Running Target Guide.

IV. INTRODUCTION:

   A. Gain Attention: Do you think you can hit a 2 inch circle going across a 10 meter opening in 2½ seconds from a distance of 50 meters? If you can you are a prospective Running Boar shooter.

   B. Orient Students:

      1. Lesson Tie-In: In previous marksmanship instruction you have been taught to hit a stationary target with a single projectile or a moving target with multiple projectiles.

      2. Motivation: Would you like to learn to hit a moving target with a single projectile? By following the instructions given today you will be able to accomplish this goal.

      3. Scope: During this period of instruction you will become familiar with the origin of Running Target shooting, accomplishments of USAMU Running Target personnel, opportunities for international competition, the Running Target Range, target mechanism, range operating procedure, specialized equipment and techniques used by a Running Target shooter to produce winning scores.

V.   BODY:

A. **First Student Performance Objective**: Aspiring to become a Running Target shooter, be familiar with the origin of this sport, accomplishments of the Army members of the US Running Target Teams and opportunities for international competition through participation in the US International Running Target Program as described in the USAMU Running Target Competitive Records.

TRANSITION: A person would be hard pressed to find any written record on the origin of the Running Target sport. Knowledge of the competitive accomplishments by U.S. Army Running Target shooters is, however, a matter of record.

    1. Originated in Northern Europe as a simulated hunting sport.

    2. 1958 World Championships - Moscow, USSR - First US participation (Running Deer), the US had its first world champion in the doubles with world record score of 223X250.

        a. 1961 World Championships - Oslo Norway - (Running Deer). US shooters were World Champions in the singles and World Champions in the doubles.

        b. 1962 World Championships - Cairo Egypt - (Running Deer). US shooter won a Silver Medal in singles and a Bronze Medal in the doubles.

        c. 1966 World Championships - Weisbaden, W. Germany, US shooter won Bronze Medal.

        d. US Army shooters won National Running Boar Champion 1966, 1967, 1968 and 1969.

        e. Nine Army shooters have been members of medal winning United States Running Target Teams in World Championship competition.

    3. The event is shot in the full World Shooting Championships every four years and in the World Moving Target Championships held in the years between Olympics and the World Championships. A European Training Trip is made yearly for seasoning of US Army shooters who will compete in tryouts for US Running Target Team and as members of the US International Running Target Team.

TRANSITION: Before we reach this apex of shooting competition we have many things to learn.

B. **Second Student Performance Objective**: Aspiring to become a Running Target International competitor, be familiar with the Running Target <u>Range</u>, target <u>mechanism</u>, range operating <u>procedures</u>, course of fire and regulations as described in Chapters V and VI, USAMU International Running Target Guide.

    1. The ISU Regulations for running target competition are in Chapter VI, USAMU Running Target Guide.

    2. Describe Running Target Range.

NOTE: SHOW SLIDE #1 - The Running Target Range

    3. Describe Target Mechanism.

NOTE: SHOW SLIDE #2 - The Target Mechanism

    4. Describe Range Operation and Procedure.

    5. Describe Course of Fire.

NOTE: SHOW SLIDE #3 Course of Fire

TRANSITION: Now that we know where and what we are going to shoot, let us discuss the equipment used to accomplish this.

    C. <u>Third Student Performance Objective</u>: Aspiring to become a Running Target shooter, be familiar with the specialized <u>equipment</u> used by Running Target shooters as described in Chapter V, USAMU International Running Target Guide.

        1. Describe Weapons Authorized.

NOTE: SHOW SLIDE #4 - Weapons Authorized

            a. Center Fire

            b. Rimfire

            c. Importance of Stock Fit

        2. Describe sights used

            a. Telescope Sights

            b. Iron Sights

        3. Reticules used in Telescopes

NOTE: SHOW SLIDE #5 - Common Reticules Used

            a. Crosshair

            b. Crosshair with single dot

            c. Crosshair with diamond

            d. Crosshair with three dots

        4. Describe clothing used

            a. Sweaters

            b. Shooting vests

            c. Shooting coat

TRANSITION: All of this sophisticated equipment is useless unless it is in the hands of a person who knows how to deliver a shot to the right place on the target.

    D. <u>Fourth Student Performance Objective</u>: Aspiring to become a Running Target shooter, be familiar with the shooting techniques used to produce a winning score as described in Chapter V, USAMU International Running Target Guide.

        1. Describe preparation

            a. Sufficient Ammo

            b. Telescope on correct setting

            c. Position

            d. Mental preparation

2. Describe ready position

NOTE: SHOW SLIDE #6 - Ready Position

    a. Position of Weapon

    b. Position of Body

3. Describe Shooting Position

NOTE: SHOW SLIDE #7 - Shooting Position

    a. Position of Weapon

    b. Position of Body

4. Describe aiming points and what they must be

    a. Consideration in choosing an aiming point

    b. Examples of some aiming points that are used.

NOTE: SHOW SLIDE #8 - Crosshair, Slow Run; #9 Crosshair, Fast Run; #10 - 3 dot, Slow Run; #11 - 3 dot, Fast Run

5. Discuss Concentration

    a. On Aiming Point

    b. On smooth delivery of shot

6. Describe Trigger Control

    a. Shooting the shot when you *want* to shoot it.

QUESTION: Why must you shoot the shot when you want to?

ANSWER: Because no one can track on the aiming point for the complete run across the opening.

7. Describe Follow Through

QUESTION: Why is follow through so important?

ANSWER: Because the aiming point is calculated from where the bullet will strike the target when you are tracking at the exact speed and direction of the target. Any movement deviation from the exact line and speed of the target will effect the strike of the bullet

VI. CONCLUSION

NOTE: SHOW SLIDE #12 - USAMU Shooting Team Crest

    A. <u>Retain Attention</u>: I want to fire on the United States Team in the next International Running Target Competition. The shooting world respects the wearers of this crest because the U.S. International Shooting Teams are made up by a majority of shooters from USAMU.

B. <u>Summary</u>:

    1. Review U.S. Army Running Target shooters accomplishments.

    2. Review description of running target range, target mechanism, range procedures, courses of fire and ISU regulations.

    3. Review Telescope and Reticule selection.

    4. Review weapon selection.

    5. Review shooting techniques.

        a. Stress trigger control.

        b. Stress follow through.

C. <u>Application</u>: This instruction will give the beginner a start in Running Target competition. It will give the expert shooter some of the finer points that can improve his competitive performance. The end result will be the strengthening of the United States Running Target Team in world competition.

D. <u>Closing Statement</u>: Remember that this information on becoming a winner is worth very little unless it is applied diligently and practiced at every opportunity.

SECTION FOUR.   TEAM ADMINISTRATION

CHAPTER XXII

TEAM ORGANIZATION AND ADMINISTRATION

All administrative burdens must be handled by administrative and supervisory personnel and not by the shooter if best results are to be achieved. The most routine and simple administrative concern is a measurable distraction to the shooter. The organization and administration of a pistol team requires extensive planning if highly qualified personnel and optimum results are to be obtained. As a minimum requirement the pistol division of a marksmanship unit should include the following:

A. ORGANIZATION

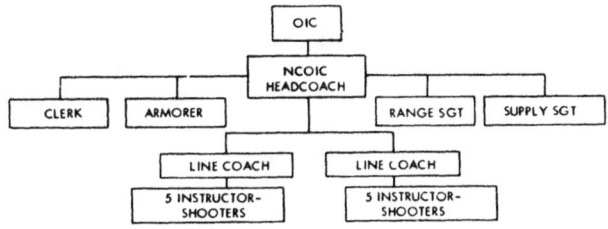

B. ADMINISTRATIVE RESPONSIBILITIES

  1. <u>Officer-In-Charge.</u>

      a. Plans, directs and supervises the performance of the Pistol Branch.

      b. Recommends matches for participation in and by the Pistol Branch.

      c. Makes recommendations to the Commanding Officer concerning training policy and keeps him informed of the activities of the Pistol Branch.

  2. <u>Head Coach, Pistol Branch.</u>

      a. Assists in the planning of training.

      b. Schedules training.

      c. Directs the training by personal supervision.

      d. Initiates shooting equipment checks that serve to pinpoint faulty mechanical functioning, cleanliness, proper lubrication, security and inherent accuracy. Accountability is maintained through periodic check of equipment serial numbers.

      e. Constantly maintain a current evaluation of each shooter regularly assigned to USAMU Pistol Branch. In addition, an evaluation and an estimate of potential should be exercised concerning outstanding shooters in major command level competition.

      f. Conduct a continuous analysis of the rate of team progress, note individual morale and attitudes and the degree of team effort exercised, all of which should be current and decisive.

      g. Propagate doctrine and performance standards by constant review of training, manuals, training materials and methods so as to reflect new ideas and methods proven to be sound and reliable and weed out unsound techniques.

h. Improve the team potential by conducting periodic organized, group instruction and tests. Improve the individual potential by personal and private interview and conducting individual coaching sessions.

i. Supervises the team preparation for match participation.

j. Supervises the coaching technique of individual line coaches.

k. Assists in preparation of instructor courses.

l. Assists in rehearsal of instructors.

m. Participates in all registered competition.

n. Exercises a profound influence on the morale, attitudes and promote an enthusiastic desire to the shooter to win by exhibiting individual consideration, stimulation of confidence and creation of an atmosphere of inevitable success.

3. <u>Non-Commissioned Officer-In-Charge.</u>

a. In charge of team administration.

b. Accountable to OIC of the duty status of each individual in Pistol Branch.

c. Responsible for team preparation for match competition and daily practice.

d. Supervises and participates in organized group instruction. Assists in rehearsal of instructors.

e. Personally supervises security, maintenance and accountability of all weapons and equipment.

f. Conducts physical conditioning program.

g. Conducts briefings, notifies Pistol Branch personnel of appointments, duties and administrative requirements.

h. Attends to administrative requests of Pistol Branch personnel.

i. Participates in registered competition and practical training activity.

4. <u>Line Coach, Pistol Branch.</u>

a. Supervises the physical and mental preparation of members of the team. This function includes assignment of target, relay and additional duties in support of team effort.

b. Guides and influences the performance of the shooter on the firing line.

(1) Stimulates morale, attitudes, enthusiasm and confidence of team members.

(2) Develops and improves coaching technique and recommends new training ideas to Head Coach.

(3) Evaluates current performance and potential of each team member so as to be able to advise Head Coach on any necessary changes in composition of team.

c. Responsible for posting team scores on AMU Master Score board during practice firing and during registered matches.

d. To be able to remain with and maintain control of his team on the firing line during matches, the coach is required to thoroughly brief the designated scorer that he furnishes to score for the adjacent team. This may well be one of his Off-relay team members if there are insufficient support personnel available. The team captain is responsible for supervising the scoring of his team which is generally accomplished by a Representative from an adjacent team.

e. Assists in preparation of instructor courses and is a principal instructor in these courses.

f. Participates in individual portion of registered competition.

g. Is available for advice during periods between individual matches.

h. Continuous checking of equipment and weapons for proper functioning, cleanliness, accuracy, security and safety in handling.

i. Supervise by direct observation during individual matches, the scoring responsibility of individual shooters. Remind shooters to post scores on AMU Master Score board.

j. Observe and intercede if an argument, protest or an infraction of the rules concerns an Army team member.

5. <u>Instructor Shooter, Pistol Branch.</u>

a. The shooter represents the US Army in all competition.

b. Instruct and coach shooters of less experience and ability during instructor training courses.

c. Continuously check weapons and equipment for proper functioning, cleanliness, accuracy, security and safe handling. NCOIC must be notified of all exchanges drawing or turn-in or any issue weapons.

d. Responsible for scoring adjacent team member or competitor during practice and matches.

e. Actively coach a team member on firing line in the absence of the regular coach.

6. <u>Armorer or Gunsmith, Pistol Branch.</u>

a. Responsible for the proper mechanical functioning, operative safety devices, inherent accuracy, repair and certain preventative maintenance checks of weapons.

b. Research and development of accuracy and dependability of weapons and ammunition so as to remain abreast of improved techniques in the use of weapons.

7. <u>Range And Supply Sergeant, Pistol Branch.</u>

a. Operates Pistol Range and controls all range firing.

        b.   Issues all ammunition for training and competition.

        c.   Controls issue of all supplies during trips to matches.

        d.   NCOIC of transportation during trips to matches.

        e.   Assists in conduct of periodic serial number checks of all issue weapons.

    8.   **Clerk Pistol Branch.**

        a.   Assistant to NCOIC in administrative functions.

        b.   Maintains file and does all typing.

        c.   Assistant to Range Sergeant in issue of supplies and ammunition on trips.

        d.   Assistant driver on trips to matches.

C.   ADMINISTRATIVE REQUIREMENTS

    1.   **At Home Station.**

        a.   All TDY personnel must have personnel, pay and medical records.

        b.   Regular pay and travel pay must be brought up-to-date.

        c.   No E.T.S. during training period.

        d.   NRA membership for all pistol team personnel.

        e.   All personnel must have proper travel orders to cover travel. (Turn-in copy of orders on arrival, to Sergeant Major.)

        f.   Eye examinations.

        g.   If shooter's family is present, quarters address.

        h.   Check for adequate billets (Senior resident NCO is charge).

        i.   Check for adequate mess facilities.

        j.   Register personal weapons with Provost Marshal.

        k.   Register automobile with Provost Marshal.

        l.   Orient shooters on security of weapons. Use of unit storage facilities required.

        m.   Sick call.

        n.   Mailing address - locally and on trips.

        o.   Mail call.

        p.   Unit Bulletin Board.

        q.   PIO forms and pictures.

r. Traffic Law Enforcement.

s. Punctuality at formations.

t. Leaves and passes.

2. <u>Away From Home Station.</u>

    a. Prior to Departure.

        (1) Training schedule - Head Coach.

        (2) Strip Map - NCOIC.

        (3) Match Programs - NCOIC.

        (4) Advance Travel Pay - HQ USAMU.

        (5) Travel Orders - HQ USAMU.

        (6) Announce Sign Out and Sign In times - NCOIC.

        (7) Truck Loading List - NCOIC.

        (8) Shooter's Equipment List - NCOIC.

        (9) Personnel Notice of Emergencies - NCOIC.

        (10) Have triggers weighed on all weapons, periodically and immediately prior to any match - NCOIC.

        (11) Shooters are notified they are personally responsible for transportation, off-post billets and finances while on trips. NCOIC.

        (12) Prompt departure and arrival time will conform to designated Sign Out and Sign In time. NCOIC.

        (13) Shooter's name and all passenger's names be filled in on car card which requires description and tag number of automobiles. Turn-in to NCOIC.

        (14) All personnel will follow route designated on strip map unless otherwise directed.

        (15) Reservations for family billets at any match locality including National Matches at Camp Perry is the responsibility of the individual concerned.

    b. At the Match Site.

        (1) Armorer will be present on range during all match firing and immediately available on firing line during team matches - NCOIC.

        (2) Register Billets - NCOIC.

        (3) Issue Ammunition - Range/Supply Sergeant.

        (4) After Action Report Information - NCOIC.

        (5) Weekly Report to S3 - OIC.

(6) Squadding Tickets - NCOIC.

(7) Pick-up Team Awards after completion of match - Range/Supply Sergeant.

(8) Invoice to Match officials - NCOIC or Headcoach.

(9) Lunches for Enlisted Personnel - NCOIC.

(10) Proper police of Government billets - NCOIC.

(11) Make Team Entries - Head Coach.

(12) Arrange for government transportation to and from range if match is held at a military installation - NCOIC.

(13) Use same lot number of ammunition in match competition as used during practice firing - Range/Supply Sergeant.

(14) Competitor numbers recorded on master score board before match shooting starts - Head Coach.

(15) Each team's assigned target numbers for team matches will be posted on team score sheets located on master score board before team match starts - Head Coach.

(16) Each firing member of each team will have his relay assignment on team score sheet before team match starts - Head Coach.

(17) All personnel will furnish NCOIC with address of billet and telephone number if match location requires off-post billets.

(18) Personnel desiring leaves or delay en route at completion of TDY must have consent from parent unit.

(19) A report of match attendance by category and total (NCOIC).

    (a) Army personnel.

    (b) Air Force, Navy, Marines, Coast Guard etc.

    (c) Reserve components.

    (d) National Guard.

    (e) Police.

    (f) Civilian

(20) A report of match results of each competition participated in to include team aggregate standings. Head Coach.

<u>SHOOTING EQUIPMENT</u>

1. Gun box, with cover
2. Guns - 22 Cal., 38 Cal., 45 Cal. W.C. & H.B.
3. Ammunition
11. Carbide light & carbide
12. Rosin powder
13. Screwdriver
14. Chair or stool

4.	Shooting glasses	15.	Extra magazines - All calibers
5.	Scorebook & pencil	16.	Binoculars (coaches only)
6.	Cleaning rod and brush	17.	Sight adjustment card
7.	Cleaning patches	18.	Ear plugs or ear protectors
8.	Watch (stop)	19.	Oil, lubricating
9.	Spotting scope	20.	Stapler with staples
10.	Barrel bushing wrench		

ACCOMPLISH PRIOR TO DEPARTURE

1. Copies of Special Orders on your person and in baggage
2. Haircut
3. Safety check of automobile (lights, tires, brakes, wipers, etc.)
4. Consult strip map for route to destination.
5. **SIGN OUT**

REMEMBER AT ALL TIMES
CONTINUOUS MAXIMUM SECURITY OF WEAPONS

Keep your car locked and weapons in locked truck during daylight hours when weapons are not being used.

Keep weapons in secure guarded facilities overnight. Do not leave weapons in unattended vehicle. (AR 190-11)

DRIVE SAFELY AND COURTEOUSLY

We can't win team matches with you in the hospital.

DON'T LET THIS HAPPEN TO YOU!

The following item from the National Safety Council Newsletter concerning what happens when an automobile collides with an immovable object, is recommended reading.

When a car hits a large tree or other immovable objects at 55 m.p.h. here is the fatal course of events that take place in seven-tenths of a second:

"First tenth of a second: The front bumper and radiator grill collapse, steel slivers penetrate radiator to a depth of 1 1/2 inches."

"Second tenth of a second: The hood crumples and smashes against the windshield. Fenders make contact, forcing the rear parts over the front doors. The structural members of the car begin to act as a brake on the forward momentum, but the driver's body is still moving at 55 m.p.h. Legs, straight as arrows begin to buckle at the knee joints. The noise of rending and crumpling metal is deafening."

"Third tenth of a second: Driver's body rises off the seat, broken knees impact against the instrument panel. Steering wheel frame begins to bend under his grip. His head is near sun visor, chest over steering column."

"Fourth tenth of a second: First 24 inches of a car is demolished. The rearend still is traveling at 35 m.p.h., the driver's body continues to catapult along at 55 m.p.h. Motor block makes contact; rear-end of car begins to lift from ground."

"Fifth tenth of a second: Force of forward inertia starts to impale driver on steering wheel shaft. The shaft punctures the lungs, heart and arteries; blood enters lungs. The driver's body speed begins to decelerate at this point. Now only 35 m.p.h.

"Sixth tenth of a second: Driver's feet are bursting from laced shoes; brake pedal shears off at floor boards; chassis bends in middle; driver's head smashes into windshield His chest cavity now contains the first eighteen inches of the steering column, enough to penetrate through the spinal column and the muscle tissue of the back."

"Seven tenth of a second: Door latches and hinges tear loose; doors fly open. Front seat, ripped from floor, moves forward pinning impaled driver onto steering wheel column. Blood spurts from mouth; bladder bursts, shock freezes heart. The driver is dead."

In the subsequent minutes of comparative quiet, the first passersby stop and rush to the wrecked vehicle to render any aid possible and recoil in horror at the gory sight, noticing that the bloody, new statistic is sitting on his safety belt.

ANNEX I TO TEAM ORGANIZATION AND ADMINISTRATION

SUBJECT: Administrative Check List

NAME OF MATCH

	EXPECTED DATE	ACTUAL DATE	INITIALS
1. Program and entry cards requested on			
2. Program and entry cards received on			
3. Program and entry cards dispatched			
4. Hotel or Government Reservations			
5. Commercial Travel Reservations Confirmed			
6. Special Orders Requested			
7. Score sheets prepared			
8. Equipment inspection prior to departure			
9. Personnel briefed on safe driving			
10. Personnel briefed on conduct and appearance			
11. Sign out time at USAMU			
12. Sign in time at match			
13. Sign out time at match			
14. Match officials furnished sample of invoice for entry fees with instructions for preparation			
15. Arrangements made to pick up team awards			
16. Sign in time at USAMU on return			
17. Preliminary Report of Match made to COMMANDER, USAMU			
18. Equipment inspection on return			
19. Equipment lost/damaged reported to S4			
20. Attendance report of match by category			
21. After Action Report			
22. Bulletin of match results			

ANNEX II TO TEAM ORGANIZATION AND ADMINISTRATION

SUBJECT: Pistol Team Vehicles, Equipment and Supplies for Pistol Match Participation.

ITEM	NR.	ITEM	NR.
Ammunition		Tacks, thumb	
Box, Gun w/spare match weapons		Tape, masking	
Binoculars, 20X		Tape, scotch	
Boxes, Tool, Pistol Armorers		Truck, Carryall	
Cans, Trash (50 gal)		Truck, 1/2-ton pickup	
Cans, Trash, Blue (5 gal)		Truck, 1 1/2-ton van	
Cans, Water, (5 gal)		Truck, 2 1/2-ton cargo	
Carbide, 1b		Truck, 10-ton semi trlr van	
Cards, POL Credit		Typewriter	
Chairs, folding, aluminum		Umbrella, beach	
Cleaner, bore		Oil, lubricating	
Clipboards		Paper, bond, ream	
Clips, paper		Paper, carbon, package	
Cups, paper, hot and cold		Pads, Writing, ruler	
Dispenser, drinking, 10-gal		Pencils, colored	
Erasers, rubber		Pencils, grease	
Guns, staple w/extra staples		Pencils, lead	
Guidons with staff		Patches, cleaning	
Ice (Purchase and obtain receipt)		Pills, Vitamin w/dispenser)	
Kit, first aid		Public address set	
Hammer, claw		Stapler, machine, desk w/staples	
Mallet, wood			
Markers, magic		Rod, cleaning, pistol	
Awning, attach to van		Rags, cleaning	
Scissors		Rule books, NRA Pistol	
Stakes, steel		Rope, 100 ft	
Streamers, guidon, team colors		Stencils	
Supply records, TDY personnel		Scoresheets, individual	

ITEM	NR.	ITEM	NR.
Table, folding, steel		Scoresheets, team	
Table, folding, wood		Tablet, Salt (w/dispenser)	
Telescope, M49		Tissue, lens, cleaning	
Target Centers, Repair 50 & 25 yds		Tissue, toilet	
Manuals, Pistol		Scoreboard w/easel	

ANNEX III TO TEAM ORGANIZATION AND ADMINISTRATION

THE UNITED STATES ARMY MARKSMANSHIP UNIT
Fort Benning, Georgia 31905

SUBJECT: Information Questionnaire for NTG or Pistol Marksmanship Instructors' and Coaches' Clinic Students.

NAME_____RANK_____SN_____

Parent Organization_____Location_____

New Man National Trophy Team Match (Yes) (No)_____

Medal Winner (Yes) (No)_____

Last year you fired in National Trophy Team Match_____

Distinguished (Yes) (No)_____

Number of points toward distinguished award_____

Last overseas assignment_____

ETS (Date of expiration of present term of service)_____

On orders for overseas assignment (Yes) (No)_____

Can you stay through the complete training period?_____

Are you a member of the National Rifle Association?_____

What is your NRA Pistol Classification?_____

How many years of experience in competitive pistol shooting?_____

How many years of experience in coaching pistol marksmanship?_____

What is your present connection with marksmanship in your parent unit and/or home station?_____

Do you presently hold or have previously held a NRA National Pistol record?

Individual_____

Team_____

What is the highest score you have fired in registered NRA Match Competition:

.22 Cal Aggregate_____Center Fire Aggregate_____

.45 Cal W. C. Aggregate_____.45 Cal Service Ammunition Agg_____

Three (3) Gun 2700 Aggregate_____

How many Army Area Pistol Marksmanship Instructors' and Coaches' Clinics have you attended in the last three (3) years?_____

Please indicate the years and give location: 19---at _____

19---at _____ 19---at _____

Have you given pistol marksmanship instruction?_____

Individual_____Team or Group_____

Do you have a NRA Instructor's Certificate?_____

Have you fired the International Pistol Courses? (Yes) (No)_____

If yes, circle the type of course and give location. Int'l Slow Fire_____

Int'l Center Fire_____Int'l Rapid Fire_____

Int'l Standard Pistol_____Int'l Air Pistol_____

ANNEX IV TO TEAM ORGANIZATION AND ADMINISTRATION

SUBJECT: Example of a weekly Training Schedule

HEADQUARTERS
THE UNITED STATES ARMY MARKSMANSHIP UNIT
Fort Benning, Georgia 31905

TRAINING SCHEDULE
US ARMY PISTOL BRANCH
From <u>12 Feb</u> through <u>16 Feb</u>   DATE

DATE & TIME		AREA	INSTRUCTOR	UNIFORM
Mon 12 Feb				
0800-1200	900 Aggregate, Cal .22	Pistol Range	Head Coach	Fatigues
1300-1430	NMC Cal .22 Team Match	Pistol Range	Head Coach	Fatigues
1430-1600	NMC Cal .45 H.B.	Pistol Range	Head Coach	Fatigues
1600-1700	Physical Training	Pistol Range	Line Coach	Fatigues
Tue 13 Feb				
0800-1200	900 Aggregate, Cal .38	Pistol Range	Head Coach	Fatigues
1300-1430	NMC, Cal .38 Team Match	Pistol Range	Head Coach	Fatigues
1430-1600	NMC, Cal .45 H.B.	Pistol Range	Head Coach	Fatigues
1600-1700	Physical Training	Pistol Range	Line Coach	Fatigues
Wed 14 Feb				
0800-1200	Mandatory Training, (Map Reading)	Unit Classroom	Training Officer	Fatigues
1300-1700	Organized Athletics	Unit Area	NCO In Charge	Fatigues
Thu 15 Feb				
0800-1200	900 Aggregate, Cal .45 W.C.	Pistol Range	Head Coach	Fatigues
1300-1430	NMC, Cal .45 W.C. Team Match	Pistol Range	Head Coach	Fatigues
1430-1600	NMC, Cal. 45 H.B.	Pistol Range	Head Coach	Fatigues
1600-1700	Physical Training	Pistol Range	Line Coach	Fatigues
Fri 16 Feb				
0800-0930	NMC, .22 Cal Team Match	Pistol Range	Line Coaches	Fatigues
0930-1100	NMC, Cal.38 Team Match	Pistol Range	Line Coaches	Fatigues
1100-1200	NMC, .45 Cal W.C. Team Match	Pistol Range	Line Coaches	Fatigues
1300-1430	NMC, .45 Cal HB Team Match	Pistol Range	Line Coaches	Fatigues
1430-1530	Maintenance of Weapons	Pistol Range	Armorer	Fatigues
1530-1600	Inspection of Weapons and Equipment	Pistol Range	Shop Officer	Fatigues
1600-1700	Physical Training	Pistol Range	Line Coach	Fatigues

DISTRIBUTION "D"

HINDS
Colonel

OFFICIAL:
S3

ANNEX V TO TEAM ORGANIZATION AND ADMINISTRATION

SUBJECT: LIST OF FACILITIES AND EQUIPMENT TO BE MADE AVAILABLE AT THE SITE OF PISTOL MARKSMANSHIP COACH/INSTRUCTOR CLINIC

Vu-graph or 35mm slide projector with extra 750 watt bulb

Screen for projection

3 blackboards with chalk and eraser

3 easels ("A" frame type for hanging charts)

Classroom facilities sufficient to seat entire class

Indoor area to be used for dry fire and practical exercises during inclement weather

Range facilities, 1 firing point per 2 students

Podium

1 ream bond paper

1 front sight "Wobble stick" or "Sling shot device" if available

1 clip board per 4 students

1 stop watch per 4 students

1 pair binoculars (10 x 50) for each coach

1 straight edge (18" or 24" ruler)

1 pair scissors

1 hammer

Magic markers, assorted colors

Pencils grease, assorted colors

22 cal, 38 cal and 45 cal ammunition as per individual student requirement on enclosed training schedule

Bleachers, 100 man set on range

Assistant instructors, 10 qualified pistol competitors and/or coaches

Targets, 50 yd and 25 yd, and repair centers, standard American

Targets, 50 meter int'l slow fire

Targets, 25 meter silhouette, int'l rapid fire

Public address system for range firing

1 microphone, chest type for bleacher instruction with 100' extension

1 pistol cal .22, per student

1 pistol cal .45, service, per student

ANNEX V TO TEAM ADMINISTRATION (CONTINUED)

1 case, pistol and accessories per student (cleaning rod, oil, score book, spotting scope, screwdriver, cleaning patches, etc.)

1 stapler per 4 students

1 roll 2" masking tape

1 trophy, appropriate for award to honor student, undistinguished (chosen on basis of high 45 HB scores and examination grade)

2 pointers, 36" long

3 tables, folding

50 copies NRA Rule book Pistol

LIST OF EQUIPMENT TO BE TAKEN BY PISTOL INSTRUCTORS TO CONDUCT PISTOL MARKSMANSHIP COACHES' AND INSTRUCTORS' CLINIC

Appropriate training aids and charts

Pertinent lesson plans and notes

Uniform, instructor-shooters

Podium cards for each subject in course

Podium cards of each instructor's name and rank

LIST OF EQUIPMENT AND TRAINING MATERIAL TO BE SHIPPED TO HOST HEADQUARTERS FOR CONDUCTING PISTOL MARKSMANSHIP COACH/INSTR CLINICS

50 Manuals, Pistol Marksmanship

50 Manuals, Marksmanship Coaches' and Instructors'

100 Certificates of attendance including honor graduate certificate

100 Questionnaires for students

100 Comment sheets for students

100 Mental discipline handout

4 Scoresheets, individual match

25 Scoresheets, team match

200 Scorecards, pistol NMC

50 Scorecards, Pistol NMC Team Match

4 Visitors' folders

50 copies NRA Pistol Rule Book

100 Worksheet booklet, pistol shooter

100 Worksheets, pistol coach

100 Questionnaire, NRA Pistol Match Rules

100 Format for lesson plan

100 Examination, pistol marksmanship

100 Schedules of Marksmanship Coach/Instructor Clinic

1 Letter of instruction to Host headquarters stating time of arrival, number of instructors w/rank, duration of course, request for mess and billets, and reserve a day of preparation for clinic.

ANNEX VI TO TEAM ORGANIZATION AND ADMINISTRATION

THE UNITED STATES ARMY MARKSMANSHIP UNIT
Fort Benning, Georgia 31905

SUBJECT: PISTOL MARKSMANSHIP STUDENT COMMENTS:

1. Did you receive a proper orientation and instruction concerning this course prior to attending this class? If so, by whom?

2. Which period of instruction in this course gave you the most benefit? Explain.

3. Which period(s) of instruction in the course do you feel can be improved? Tell how.

4. Do you have any suggestion(s) in training better coaches and instructor-shooters which this course failed to bring out? If so, explain briefly.

5. Do you feel that you have an improved understanding of pistol instructor/coaching and shooting from this course? How?

6. What are your comments about the course in general? (Continue on the reverse side if this space is insufficient.)

| (RANK) | NAME | (SERIAL NUMBER) | (ORG) |

(PLEASE PRINT)

NOTE: PLEASE TURN THIS COMMENT SHEET INTO ONE OF THE COACH-SHOOTER-INSTRUCTORS PRESENT OR MAIL TO COMMANDER, U.S. ARMY MARKSMANSHIP UNIT
FORT BENNING, GEORGIA 31905

CHAPTER XXIII

PROCUREMENT, MAINTENANCE AND SECURITY OF MATCH WEAPONS, EQUIPMENT AND AMMUNITION

The Department of the Army has made provision for organic support of a comprehensive marksmanship training program for all types of small arms used in the combat branches.

A. SECURITY

Organizational responsibility for security of weapons, equipment and ammunition is guided by directives tailored to the security requirements of local posts, camps and stations. These directives cover the responsibility of the individual to safeguard US Government property and registration of personal weapons.

A. PROCUREMENT

TA 60-18, is the reference source of weapons and equipment allowances. This TA lists types and number of weapons authorized for various command level teams. Items of equipment other than weapons are also listed.

1. <u>Travel, Per Diem and Entry Fees</u> are authorized out of appropriated funds while the awards and trophies are purchased with nonappropriated funds.

2. <u>Expendables Items</u> are purchased at the self-service store. Each unit is authorized a certain amount of money for this purpose.

3. <u>Ear Plugs and Prescription Glasses</u> are obtained through medical channels. Medical authorities encourage wearing of ear plugs or other protective devices for the conservation of hearing.

4. <u>Ammunition</u> allowances for various command level teams are listed in TA 23-100. In this TA you will find that you can usually draw all the .22 caliber and hardball .45 caliber ammunition that your team needs to implement a marksmanship training program. Commercial .38 and .45 caliber ammunition is also authorized in limited quantities depending upon the command level of the team concerned. Pay particular attention to the authorization for off-duty marksmanship practice for .22 caliber and .45 caliber service ammunition.

5. <u>Match Grade Weapons</u> should be procured in all three calibers.

    a. Any match grade .22 caliber automatic target pistol is good. There are several good models and it is up to the individual to choose which one he prefers. A unit purchasing these weapons will usually get more than one type so as to give the shooter a choice.

    b. Center fire weapons may be any center fire pistol or revolver .32 caliber or larger. A revolver is more difficult to master than either a .45 caliber automatic or a .38 caliber super automatic. As a result, the majority of center fire match shooters on the firing lines today shoot a .38 caliber super automatic. A .38 super automatic may be accurized by the unit gunsmith or they may be purchased commercially from any number of competent gunsmiths.

    c. Government issue match grade .45 caliber automatics are available in marksmanship units and most are very accurate.

    d. Commercial .45 caliber automatics may be purchased and accurized by various civilian gunsmiths.

6. <u>Miscellaneous Equipment.</u> Any number of gun boxes are available and one of good solid construction should be chosen. Spotting scopes M49 are available for issue and are of excellent quality. Stop watches are available for issue and other miscellaneous items such as scoring plugs, carbide lamps, etc., are avaialble either from TA 60-18 or local purchase.

C. CAUSES OF WEAPONS MALFUNCTIONS AND CARE AND MAINTENANCE

Weapons used in competition are by necessity closely fitted. Occasionally a weapon will malfunction until it has been fired a few hundred rounds. Some magazines may not work in a weapon due to the lips being either too wide or too close together. The followers may be at an improper angle, causing the rounds to either ride too high and hit the upper edge of the chamber or not high enough in which case it may not properly slide up the feed ramp and into the chamber. Magazines should be handled carefully so that they do not become bent and they should be taken apart and cleaned occasionally. All wearing surfaces of the weapon and the magazines should be well oiled at all times with oil preservative, light, as issued. Lack of lubrication cause excess wear, malfunctions and materially reduces the accuracy life, making it necessary to reaccurize the weapon. Cleaning materials as issued are adequate and should be used daily after firing. A dirty weapon also causes stoppages and excessive wear. When transporting weapons, they should be fastened securely within the box so as not to damage the sights. It is mandatory that weapons be locked up at all times when not in use and should never be left unattended.

D. THE MARKSMANSHIP UNIT SHOP IS PART OF THE TEAM EFFORT

Each Marksmanship Unit should have a shop where the unit gunsmith has facilities to:

1. <u>Accurize Weapons.</u>

2. <u>Test Weapons for Inherent Accuracy.</u>

3. <u>Make Major Repairs.</u>

4. <u>Do Preventive Maintenance.</u>

5. <u>Make Special Adjustments on Weapons to Meet the Desires of Individual Shooters.</u>

A machine rest or static testing device is essential for testing of newly accurized weapons and for checking weapons to determine if they need reaccurization after extensive use. The shop is a very important part of the team effort. The quality of work turned out is directly reflected in the scores fired by the team. Target pistols accurized especially for competitive shooting undergo careful checking and adjustment of parts and mechanisms under precise conditions. However, they may require special adjustment, to a greater or lesser extent. The tooling to fit and the adjustment of the pistol, which is intended to improve the interaction of the parts and mechanisms, as carried out by armorers, fails in many instances to completely meet the individual shooters' requirements. For example, in the adjustment of the trigger, the altering of the sighting devices to suit the shooter's sharpness of vision, etc. It is necessary for the shooter to inform the armorer just how he wants the weapon adjusted. This responsibility is the shooters' and cannot be delegated to anyone else. In addition, when a pistol is sent to the shop for reaccurizing, the pistol should be reissued to the same shooter because he is more than likely delicately attuned and coordinated to the trigger action and the operating rate of that particular weapon.

6. <u>All Match Grade Ammunition Used in Competition Must be Tested for Uniform Accuracy by Lot Number.</u> Match weapons should be fired using the ammunition lot that gives the greatest accuracy to the particular weapon.

7. **Shot Group Areas of Dispersion.**

If it were possible to fire a series of rounds under identical conditions, the bullets would describe the same trajectory in the air and strike the same point.

However, in practice it is almost impossible to have absolute uniformity of all firing conditions, since there are always small, practically indiscernible variations in size of powder grains, weight of charge and bullet, shape of bullet; different igniting ability of primer; various conditions of bullet motion in the barrel and outside it, such as constant fouling and heating of the barrel, wind gusts, changing air temperature, errors committed by the shooter in aiming, assuming the firing position, etc. As a result, even under the most favorable firing conditions, each bullet describes a trajectory which differs from the trajectories of other bullets. This phenomenon is called dispersion.

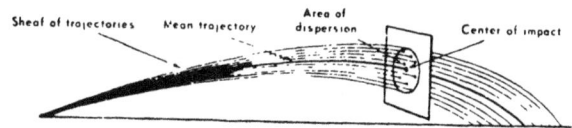

Figure 22-2. Sheaf of Trajectories, Mean Trajectory, Area of Dispersion.

If a sufficiently large number of rounds are fired, the trajectories form a sheaf of trajectories, which produces a number of shot holes separated by various distances in the target. The area the holes occupy is called the <u>Area of Dispersion.</u>

Without going into detail to explain the laws of dispersion and the concepts of the probability of hitting, let us say that in shooting any type of firearm, the character of the shot pattern in the area of dispersion always remains the same, even though the size of the areas of dispersion can differ to a great extent, depending upon the model of weapon and ammunition used.

It follows from the above that dispersion is an objective process occurring independently of the will and desires of the shooter. It is senseless to require all bullets to strike the same point.

However, dispersion is not an unavoidable and invariable amount which is definitely fixed for a certain weapon under a definite firing condition. The art of accurate shooting and gunsmithing success lies in recognizing the causes of dispersion and diminishing their effect.

One of the actions that can be taken by the individual shooter to reduce dispersion, in addition to the normal mechanical fitting of the pistol components by the gunsmith, is the maintenance of a clean, unleaded pistol bore. Lead begins to build up in .38 cal., and .45 cal. Pistol barrels after twenty (20) to thirty (30) rounds fired. It is recommended that a brass wire brush be swabbed through the bore after each ten (10) rounds fired so as to remove the small initial lead accumulations before a larger build-up effects the accuracy of the pistol. A suitable solvent should be used to thoroughly clean and remove all lead fouling after the completion of the days firing.

Marksmanship makes very high demands upon a gun's accuracy of fire and upon the ammunition. No matter how well trained and capable the shooter may be technically, his competitive results will depend to a large extent upon the quality of the gun and the ammunition.

SECTION FIVE. COMPETITIVE REGULATIONS

CHAPTER XXIV
COMPETITIVE REGULATIONS II
NRA PISTOL RANGE PROCEDURE AND SAFETY RULES

It is the duty of each competitor to sincerely cooperate with tournament officials in the effort to conduct a safe, efficient tournament. Competitors are expected to promptly call the attention of proper officials to any infraction of rules of safety or good sportsmanship.

The safety of competitors, range personnel and spectators requires a high degree of self-discipline, constant attention to range control commands, the careful handling of firearms, exercising caution in moving about the range area, acting and speaking with consideration toward other competitors.

A. RANGE CONTROL (Rule 10) (NRA Pistol Rule Book.)

1. (10.1) <u>Discipline</u>. The safety of competitors, range personnel and spectators requires continuous attention by all to the careful handling of firearms and caution in moving about the range. Self-discipline is necessary on the part of all. Where such self-discipline is lacking is the duty of range personnel to enforce discipline and the duty of competitors to assist in such enforcement.

2. (10.2) <u>Loud Language</u>. Loud or abusive language will not be permitted. Competitors, scorers, and range officers will limit their conversation directly behind the firing line to official business.

3. (10.3) <u>Delaying a Match</u>. No competitor may delay the start of a match through tardiness in reporting or undue delay in preparing to fire.

4. (10.3.1) <u>Preparation Period</u>. In all cases competitors will be allowed three minutes to take their places at their firing point and prepare to fire after the firing point has been cleared by the preceeding competitor.

5. (10.4) <u>Policing Range</u>. It is the duty of competitors to police the firing points after the completion of each string. The range officer will supervise such policing and see that the firing points are kept clean.

6. (10.5) <u>Competitors Will Score</u>. Competitors will act as scorers when requested to do so by the Executive Officer or Chief Range Officer, except that no competitor will score his own target.

7. (10.6) <u>Repeating Commands</u>. A range officer will repeat the Chief Range Officer's commands only when those commands cannot be clearly heard by competitors under his supervision.

8. (10.7) <u>Firing Line Commands</u>. When ready to start firing a match the range officer (usually the Chief Range Officer) commands "RELAY NO. 1 MATCH NO. _____ (or naming the match) ON THE FIRING LINE." Each competitor in that relay then immediately takes his assigned place at his firing point and prepares to fire but does not load. If this is a new relay, the range officer will announce "THE PREPARATION PERIOD STARTS NOW." At the end of three (3) minutes, the range officer states "THE PREPARATION PERIOD HAS ENDED," then proceeds with the match.

a. The range officer having made sure that the range is clear (in timed and rapid fire, targets must also be turned out of firing position) then commands "WITH 5 ROUNDS, LOAD."

b. The range officer then asks, "IS THE LINE READY?" Any competitor who is not ready or whose target is not in order will immediately raise his arm and call, "Not ready on target....." The range officer will immediately state, "The line is not ready," and the range officer will immediately investigate the difficulty and assist in correcting it. When the difficulty has been corrected, the range officer calls, "THE LINE IS READY."

c. When the range officer asks, IS THE LINE READY? and the line is ready, he then calls, "THE LINE IS READY."

d. The range officer then commands, "READY ON THE RIGHT. READY ON THE LEFT." Competitors may point their guns toward the target after the command, "Ready on the right." The range officer will then command, "READY ON THE FIRING LINE." The targets will be exposed or the signal to commence firing will be given in approximately three seconds.

e. The range officer then commands "COMMENCE FIRING" which means to start firing without delay as timing of the string is started with this command. "Commence Firing" may be signalled verbally, by a short blast on a whistle or by moving the targets into view.

f. "CEASE FIRING" is the command given by the Range Officer at the end of time limit for each string or at any other time he wishes all firing to cease. Firing must cease immediately. Even if a competitor is about to let off a carefully aimed shot he must hold his fire and open the action of his gun. Failure to immediately obey this command is one of the worst infractions of range discipline as it may result in the wounding or death of some man, woman or child who has wandered into the line of fire somewhere on the range or behind the targets. On this command cylinders shall be opened or slides locked back and all guns placed on the shooting stand and not handled until the next command of the Range Officer. "Cease firing" may be signalled verbally, by a short sharp blast on a whistle or by moving the targets out of view.

g. After the command to "cease firing" is given at the end of a string, a subsequent command is given, "CEASE FIRING - UNLOAD - SLIDES BACK - GUNS ON THE TABLE. On this command all assistant range officers and scorers check their competitors to make sure each one obeys the command before signalling the Range Officer that their portion of the firing line is clear.

h. When all assistant range officers and scorers have given the "clear" signal the range officer commands "SCORE TARGETS AND PASTE" (or "CHANGE").

i. Other commands used less frequently are:

(1) "POLICE FIRING POINTS" means pick up fired cartridge cases, empty cartridge cartons and "tidy-up" the firing line.

(2) "AS YOU WERE" means disregard the command just given. For example, if the commands were given "Ready on the right" followed by "As you were" it would mean someone was not ready.

(3) "CARRY-ON" means proceed with whatever was being done before some interruption occurred.

B. PISTOL RANGE SAFETY RULES

The coach and the shooter must exercise utmost care in all phases of handling weapons and ammunition. Cleaning weapon, dry fire practice, transporting guns and ammunition to and from quarters or on trips to matches seem to be more dangerous than when the equipment is being used in a match or organized practice. The average person is less apt to commit a dangerous act when in a crowd where many weapons are present. This, no doubt, is why you seldom hear of anyone being wounded or killed at pistol or rifle matches. The constant supervision and caution in handling firearms is well served A coach or range official would not hesitate to admonish when observing even the slightest infraction.

The following items are listed to enable the shooter to avoid jeopardizing the safety of others or himself:

1. <u>Clear the Pistol Every Time it is Picked Up for Any Purpose.</u> Never trust your memory. Consider every pistol as loaded until you have proved it otherwise.

2. <u>Always Unload and Clear the Pistol if it is to be Left Where Someone Else May Handle It.</u> A cleared pistol has the slide back, magazine out and the chamber is inspected for unextracted round.

3. <u>Always Point the Pistol up When Snapping</u> it after examination. Keep the hammer fully down when the pistol is not loaded.

4. <u>Never Place the Finger Within the Trigger Guard Until You Intend to Fire.</u>

5. <u>Never Point the Pistol at Anyone You do not Intend to Shoot</u>, nor in a direction where an accidental discharge may do harm. On the range, do not "dry fire or snap" for practice while standing back of the firing line. "Dry fire or snap shoot" only with a weapon that has been cleared and point weapon only at a backstop that has been approved by range officials.

6. <u>Before Loading the Pistol</u>, draw back the slide and look through the bore to see that it is free from obstruction. When loading pistol, insert loaded magazine until fully seated, place left thumb on hammer, release slide stop with left index finger while gripping stock firmly with right hand.

7. <u>On the Range, Do Not Insert a Loaded Magazine Until the Time for Firing.</u>

8. <u>Never Turn Around at the Firing Point While You Hold a Loaded Pistol</u> in your hand, because by so doing you may point it at the man firing alongside you.

9. <u>On the Range, Do Not Load the Pistol</u> (with a cartridge in the chamber) <u>Until Immediate Use is Anticipated.</u> If there is any delay, lock the pistol and only unlock it while extending the arm to fire. Do not lower the hammer on a loaded cartridge; the pistol is much safer cocked and locked.

10. <u>In Reducing a Jam First Remove the Magazine.</u>

11. <u>To Remove a Cartridge Not Fired</u>, first remove the magazine, and then extract the cartridge from the chamber by drawing back the slide.

12. <u>Safety Devices Should be Frequently Tested.</u> A safety device is a dangerous device if it does not work when expected.

13. <u>Don't Mix Alcohol With Gun Powder.</u>

14. <u>Make Sure Range is Clear Before Practice Firing</u> if you are not participating in organized firing or when firing alone. Fire only on command of range officer if organized practice is being conducted.

15. <u>Do Not Handle Weapon.When Any Person is Forward of Firing Line.</u>

16. <u>No Government Issue Ammunition to Leave the Firing Range.</u> Return all excess ammunition to supply personnel.

17. <u>Do Not Fire Over and Above Barrier or Backstop</u>.

18. <u>Protect Weapons From Theft</u>, thereby preventing lost weapons from falling into the hands of irresponsible persons. Do not leave weapons unattended either on the range or in a car. Lock all shooting equipment in trunk of car when transporting. Equipment exposed in an unlocked car is a temptation needlessly advertised.

19. <u>Register Personal Weapons</u> on post, camp, or station and with your local police department.

CHAPTER XXV
COMPETITIVE REGULATIONS III
NATIONAL TROPHY PISTOL MATCH RULES

In order to compete in the National Trophy Pistol Match, it is necessary that the service pistol competitor have a working knowledge of the provisions of AR 920-30 as applicable to national trophy individual and team pistol competition. By successfully competing in the National Trophy Pistol Match, points can be earned toward the award of the Distinguished Pistol Shot Badge.

1. <u>National Trophy Matches</u>.

    a. The National Trophy Matches are conducted by the National Match Director for the National Board for the Promotion of Rifle Practice in accordance with the provisions of AR 920-30.

    b. These matches consist of the following events:

        (1) The National Trophy Individual Match, both rifle and pistol.

        (2) The National Trophy Team Match, both rifle and pistol.

2. <u>Mandatory Participation</u>: (See Para 12, AR 920-30) Military competitors whose transportation or other travel expenses are paid, either partially or wholly from Government funds (appropriated) or quasipublic funds (exchange, recreation, welfare and morale) and/or who uses Government issued equipment will be required to:

    a. Attend the appropriate Small Arms Firing School.

    b. Enter and complete firing in the appropriate National Trophy Individual Match, unless eliminated under the provisions of paragraph 20.

    c. Enter and complete firing in the appropriate National Trophy Team Matches as eligible under this regulation.

    d. Fire a service rifle authorized by paragraph 25 at ranges of 200 to 600 yards inclusive in the NRA as well as National Trophy Rifle Matches, except as may be specifically exempted in the current National Match program. Military personnel will fire the type service rifle as directed by their parent Service. The firing of other weapons at greater ranges or in international-type matches is permissible.

3. <u>Rules and Regulations</u>: (See Para 18, AR 920-30)

    a. Except as specifically set forth herein the National Trophy Matches will be conducted under the rules and regulations as prescribed in the latest edition of Official Rules of the NRA for the conduct of high power rifle and pistol competitions.

    b. The Executive Officer, National Matches, may alter the rules or conditions of the National Matches only if necessary for the successful completion of the matches. In the event such alteration becomes necessary, the Executive Officer, National Matches, will immediately submit a written report of details to the President of the National Board for the Promotion of Rifle Practice.

4. <u>Competitor's Responsibilities</u>: (See Para 21, AR 920-30)

    a. It is the competitor's responsibility to know and abide by the provisions of:

        (1) AR 920-30.

(2) The Official National Match Program.

(3) NRA rules as applicable.

(4) Appropriate National Match Executive Officer's Bulletins.

b. In team matches, the team captain is charged with this responsibility for his team.

5. <u>Entries</u>: (See para 19, AR 920-30)

a. Entries of individuals and teams in the National Trophy Matches will be made for each match as provided hereinafter. No entries will be accepted after the closing time for the submission of entries for each match.

b. Competitors are urged to make advance entry in the Small Arms Firing School and individual matches by mail to the Statistics Director, National Matches, Camp Perry, OH 43452. The following entry blanks for individual matches may be obtained from the Statistics Director:

(1) DA Form 1342 (Entry and Score Card for NBPRP Individual Service Pistol Match).

(2) DA Form 1344 (Entry and Score Card for NBPRP Individual Service Rifle Match).

c. Mail entries will not be accepted for the team matches. Entry in team matches must be made at the National Match site by the team captain on official entry blanks provided by the Statistics Director as follows:

(1) DA Form 1343 (Entry and Score Card for Pistol Team Match) for National Trophy Pistol Team Match.

(2) DA Form 1345 (Entry and Score Card for Rifle Team Match) for National Trophy Rifle Team Match.

(3) DA Form 1690 (Entry and Score Card for Infantry Team Match) and DA Form 1690-1 (Infantry Team Match Scorer's Card) for the National Trophy Infantry Team Match.

d. Entries for individual matches will close not later than twelve noon 10 days prior to the day on which the match is scheduled to be fired. (Consult current National Match Program for exact date.)

e. Entries for team matches will close not later than twelve noon 2 days prior to the day on which the match is scheduled to be fired.

f. Entries must be complete and accurate. Incomplete or inaccurate entries may be refused or cause disqualification.

g. Entries will be accepted to range capacity on a first-come, first-served basis.

6. <u>Entry Fee</u>: (See para 6, AR 920-30) When approved by the President, NBPRP, an entry fee will be charged for any National Trophy Match. No entry fee for the Small Arms Firing Schools will be charged except to defray necessary expenses for ammunition, if required.

7. <u>Individual Eligibility</u>: (See Paras 10 and 11, AR 920-30)

a. The National Trophy Match competitions are open to all United States citizens, male and female, who are at least 16 years of age, and to all members of the United States Armed Forces, both active duty and Reserve components.

b. Certain competitive events and certain awards are restricted to special categories of shooters as a matter of grouping contestants in the best interest of fair competition.

c. The Small Arms Firing School is open to individuals, both military and civilian, 16 years of age or older, attending but not necessarily competing in the National Matches.

8. **Competitor category:** (See para 13, AR 920-30)

a. In National Trophy Matches, competitors will register and participate in only one of the following categories: Regular Service, National Guard, Reserve, ROTC, civilian, police (police category is for pistol matches only).

b. The following eligibility requirements will govern in determining proper categorical match participation:

(1) Regular Service Category. Regular Service, National Guard, and Reserve personnel on extended active duty (90 days or more) with the regular Service will participate in this category only.

(2) Reserve category. Members of any Reserve branch of the U.S. Armed Forces (Naval and Marine Corps Reserve includes Fleet and Fleet Marine Corps Reserve), except Army and Air National Guard will participate in this category unless qualified as regular Service under (1) above, or, if a bona fide member, an individual may select and participate only in police or ROTC category. If qualified, individual may elect to fire as a civilian within the criteria specified in (6) (c) below.

(3) National Guard category. Members of the Army or Air National Guard will participate in this category unless qualified as regular Service under (1) above, or, if a bona fide member, an individual may select and participate only in police or ROTC category. If qualified, may elect to fire as a civilian within the criteria specified in (6) (c) below.

(4) Police category (Pistol Trophy Matches only). Full-time, paid members of any police organization will participate in the police category or, if a bona fide member, an individual may select and participate only in National Guard or Reserve category. A policeman if qualified as a civilian within the criteria specified in (6) (c) below, may also participate in the civilian category if he attends as one of two NBPRP-authorized police members of a State Civilian Pistol Team provided he is not a member of a police organization that enters a team in the National Trophy Pistol Team Match. In the National Trophy Individual Pistol Match all policemen, except those who enter as National Guard or Reserve, will participate in the police category.

(5) ROTC category. Members of any high school or college ROTC unit (Army, Navy, or Air Force) and service academies of the Armed Forces will participate in this category or, if a bona fide member, an individual may select and participate in the National Guard, Reserve, police, or civilian category.

(6) Civilian category.

(a) Individuals who are not in any of the above-listed categories will fire as civilians.

(b) Retired service personnel will fire as civilians or, if a bona fide member, an individual may select and participate only in police category.

(c) Individuals of any Reserve or National Guard component who, during the present calendar year, have not competed as Reservists, National Guardsmen, or members of a Service component team, and have not been provided Service support for competition, in the form of weapons, ammunition, payment of travel or other expenses, wholly or in part, may fire as civilians. The provision of weapons and ammunition for a specific competition, i.e., National Matches or NBPRP Regional Leg Matches, when such is available to both military and civilian competitors, is not considered Service support or a bar for subsequent competition in the civilian category.

(d) Reserve and National Guard personnel who elect to compete as "civilians" may not revert to their Service category to compete in open competition within any one calendar year, unless called to active Federal service.

9. <u>Ammunition</u>: (See para 27, AR 920-30)

a. In the National Trophy Matches service ammunition will be issued by range personnel at the firing line. Competitors are required to fire this ammunition and none other. A competitor will be disqualified if any other ammunition of the same caliber as issued is found about the person while in position at the firing line. Exception: Until match grade 5.56 ammunition is available to match sponsor, competitors using M16 rifle or equivalent, shall use safe ammunition provided by themselves.

b. Under certain circumstances, when approved by the President, NBPRP, it may be necessary to charge competitors for the ammunition expended.

10. <u>Weapons</u>: (See para 25, AR 920-30)

a. U.S. Rifle, Caliber .30 M1, as issued by the U.S. Army, or modified with a 7.62 barrel, having not less than a 4½ pound trigger pull, with standard-type stock and standard-type leather or web sling. Sling cuffs and sling pads are not permitted. External alterations to the stock will not be allowed. The application of synthetic coatings, which includes those containing powdered metal, to the interior of the stock to improve bedding is authorized provided the coatings does not interfere with function or operation of safety features. The front and rear sights must be of U.S. Army design but may vary in dimensions of rear sight aperture and front sight blade. The internal parts of the rifle may be specially fitted and include alterations which will improve the functioning and accuracy of the arm, provided such alterations in no way interfere with the proper functioning of the safety devices as manufactured. Personnel concerned are responsible that weapons are inspected to insure adequate safety, and otherwise meet the requirements of this paragraph.

b. U.S. Rifle, Caliber 7.62mm, M14 as issued by the U.S. Army, having not less than 4½ pound trigger pull, with standard-type stock and standard-type leather or web sling. Sling cuffs and sling pads are not permitted. The rifle must be so adjusted as to be incapable of automatic fire without removing the stock and changing parts. In all courses and in all positions the 20-round box magazine will be attached. The hinged butt plate will be used only in the folded position. The gas system must be fully operational. External alterations to the stock will not be allowed. The application of synthetic coatings which includes those containing powdered metal to the interior of the stock to improve bedding is authorized provided the coating does not interfere with the function or operation of safety features. The front and rear sights must be of U.S. Army design, but may vary in dimensions of rear sight aperture and front sight blade. The internal parts of the rifle may be especially fitted and include alterations which will improve the functioning and accuracy of the arm, provided such alterations in no way interfere with the proper functioning of the safety devices as manufactured. Personnel concerned are responsible that weapons are inspected to insure adequate safety, and otherwise meet the requirements of this paragraph.

c. U.S. Pistol Caliber .45 M1911 or M1911A1 or the same type and caliber of commercially manufactured pistol. The pistol must be equipped with issue or similar factory-type standard stocks, i.e., without thumb rests. Trigger pull must be not less than 4 pounds. This pistol must be equipped with open sights. The front sight must be nonadjustable. The pistol may be equipped with an adjustable rear sight with open U or rectangular notch, the distance between sights measuring not more than 7 inches from the apex of the front sight to the rear face of the rear sight. The forestrap of the receiver grip may be checkered. The mainspring housing may be either the flat or arched type, checkered or uncheckered. Trigger shoes may be used. Trigger stops internal or external, are acceptable. Otherwise, external alterations or additions to the arm will not be allowed. The internal parts of the pistol may be specially fitted and include alterations which will improve the functioning and accuracy of the arm, provided such alterations in no way interfere with the proper functioning of the safety devices as manufactured. All standard safety features of the weapon must operate properly. Personnel concerned are responsible that weapons are inspected to insure adequate safety, and otherwise meet the requirements of this paragraph.

d. U.S. Rifle, Caliber 5.56mm M16 Series, as issued by the US Armed Forces, or commercial equivalent, without bipod, grenade launcher, or other attachments, having not less than 4½ pound trigger pull, with standard-type stock, pistol grip, handguard, and leather or web sling. Sling cuffs and sling pads are not permitted. The rifle must be so modified as to be incapable of automatic fire without replacing or altering parts. In all courses of fire and in all positions the 20-round or 30-round box magazine will be attached. The gas system must be fully operational. External alterations to the stock, pistol grip, or handguard will not be allowed, except that extension of the stock maintaining contour similar to the standard stock and butt plate is permitted. The application of synthetic coatings to the interior of the stock, pistol grip, and handguard to improve bedding is authorized provided the coating does not interfere with the function or operation of safety features. The front sight shall be of open metallic post or blade type mounted on the standard front sight assembly. The rear sight shall be metallic, not containing a lens or system of lenses, mounted on the carrying handle, so mounted that the carrying handle may be used as intended. Distance between front and rear sights shall be not less than 19 nor more than 22 inches, measured from the apex of the front sight to the rear face of the rear sight. The internal parts of the rifle may be especially fitted and include alterations which will improve the functioning and accuracy of the arm, provided such alterations in no way interfere with the proper functioning of the safety devices as manufactured.

11. <u>Use of weapon by more than one competitor</u>: (See para 28, AR 920-30) Two or more competitors may fire the same rifle or pistol in any NBPRP Excellence-In-Competition Match if tournament squadding permits.

12. <u>Firing positions authorized</u>: (See para 35, AR 920-30) In the National Trophy Matches no special consideration will be given to physically handicapped shooter. NRA rules will apply except as shown below.

a. Standing. Standing erect on both feet, no other portion of the body touching the ground or any supporting surface. The sling must be attached to the rifle but may not be used for support. The rifle will be supported by both hands and one shoulder only. The elbow of the forward arm may be placed against the body or rested on the hip.

b. Sitting. The sitting position will be any of those authorized under NRA rules.

c. Kneeling. The kneeling position will be any of those authorized under NRA rules; however, use of the kneeling roll or pad is not authorized.

d. Ground cloths or ground pads.

(1) Ground cloths or ground pads normally will not be authorized for use during the National Trophy Matches.

(2) Exceptions to this rule may be by the Executive Officer National Matches if, in his opinion, the ground conditions resulting from inclement weather result in undue disadvantage to the competitor.

(3) Ground cloths or ground pads, when authorized for use, must not be constructed or used in a manner that provides artificial support.

e. With the M14 or M16 series, the magazine may touch the person of the shooter, but may not touch the ground when in firing position.

13. <u>Loading and reloading</u>: (See para 40, AR 920-30)

a. Rifle caliber .30 M1. In sustained fire stages of National Trophy Rifle Matches, individual or team, the rifle will be loaded initially with a clip containing two rounds. After the two rounds are fired, the rifle is reloaded with a clip containing eight rounds from the belt or the ground at the option of the firer.

b. Rifle, 7.62 mm, M14 and rifle 5.56 mm, M16 series. In sustained fire stages of National Trophy Matches, individual or team, the rifle will be loaded initially with a magazine containing two rounds. After the two rounds are fired, the rifle is reloaded with a magazine containing eight rounds from the pouch or the ground at the option of the firer.

14. <u>Field glasses and telescopes</u>: (See para 41, AR 920-30) Competitors may use field glasses or telescopes on the firing points unless otherwise prescribed in the conditions of the match.

15. <u>Competitor must verify score</u>: (See para 42, AR 920-30) The competitor (team captain in team matches) must verify the correctness of the score card, including the name on the card, the value of each individual shot and all other data on the card. The competitor acknowledges the correctness of all data on the card when he signs it. Should a competitor or team captain sign an incorrect card or leave the firing line without signing the score card, no challenge or protest will be allowed. If the competitor or team captain desires to protest, he will write the word "Protested" on the score card above his signature.

16. <u>Competitors present punctually</u>: (See para 34, AR 920-30) Competitors will be present at the firing points punctually at the time stated on their squadding tickets. It is the competitors responsibility to appear at the firing point to which he has been assigned, prepared to fire, when his relay in any match is called to the firing line. Any competitor who fails to report on the proper relay or firing point will forfeit the right to fire in that match unless the competitor presents satisfactory evidence to the National Match Executive Officer that he is late through fault of the National Match staff. In team matches, the first pair only need be present at the time set for firing to begin.

17. <u>Station of competitors</u>: (See para 32, AR 920-30) Each competitor will remain on or in rear of the assembly line in rear of his firing point until called by the range officer to take his position at the ready line or firing point. No one except the officials of the matches, members of the National Board for the Promotion of Rifle Practice, team officials, the competitors on the ready line and on the firing points, scorers and others on duty will be permitted in front of the assembly line without special permission of the Range Director.

18. <u>Elimination of teams or individuals</u>: (See para 20, AR 920-30) The Executive Officer, National Matches may, at his and by such standards as he may prescribe eliminate teams and/or individuals of the lowest standing at any time after the first stage of a team or individual trophy match is completed.

19. __Teams authorized__: (See para 15, AR 920-30) The following teams are authorized in the National Trophy Team events:

    a. Active Service Teams -- One each, representing the major Services, each U.S. Army Region, each Naval District, each Marine Corps competition area, each Air Force area, each Coast Guard area and the Defense Atomic support area.

    b. National Guard Teams -- One each representing each state, the Commonwealth of Puerto Rico and the District of Colombia. Such teams may be composed of mixed Army National Guard, Air National Guard personnel.

    c. Reserve Teams -- Not to exceed two each representing the Army Reserve the Naval Reserve, the Air Force Reserve, the Marine Corps Reserve, the Coast Guard Reserve, and one each representing the Army Reserve within any one major oversea command; the Naval Reserve within any one Naval District; the Air Force Reserve within any one Continental Air Command Reserve Region; the Marine Corps Reserve within any one Marine Corps Reserve District; and the Coast Guard Reserve within any one Coast Guard area.

    d. ROTC Teams -- One each representing the Reserve Officer's Training Corps (Army, Navy or Air Force) from each of the respective Army Area commands. Naval Districts and Air Force ROTC Liaison Areas within the Continental United States. ROTC students from units in Alaska and Hawaii are authorized to compete as members of the Sixth US Army Area ROTC Rifle Team. ROTC students from Puerto Rico are authorized to compete as members of the First US Army Area ROTC Rifle Team.

    e. Service Academy Teams.

    f. Civilian Teams:

    (1) One each representing each of the States, the Commonwealth of Puerto Rico and the District of Colombia.

    (2) Civilian shooting clubs organized under the rules of the NBPRP and in good standing on the rolls of the DCM as of 30 June 1968, or enrolled with the DCM subsequent to that date, may enter one or more teams. However, entry of more than one team will be subject to available range facilities as determined by the Match Executive Officer.

    g. Police Teams in Pistol Team Match -- One or more teams representing each regularly organized Federal, State, County or Municipal law enforcement agency in the United States. Subject to available range facilities.

20. __Team Membership Requirements__: (See para 16, AR 920-30)

    a. At least one of the firing members of each team will be an individual who has never before fired as a member of any team which has competed in that particular event. Firing members who meet the requirements of this paragraph may be counted against the requirements of paragraphs c and d below.

    b. No one who has fired on any team which has placed in the top 15% of competing teams in the same event in any two out of three National Trophy Team Matches immediately proceeding may be a firing member or alternate.

    c. At least three of the firing members of rifle teams and two of the firing members of pistol teams selected to represent the Armed Forces (Army, Navy, Air Force, Marine Corps and Coast Guard), active, Reserve or National Guard, will be individuals who have never before fired on any team which has placed in the top 15% of teams competing in the same event in past respective rifle or pistol National Trophy Team Matches.

    d. Teams representing ROTC, Service Academies, civilian or police organizations, will be required to have at least two firing members on rifle teams and one firing member on pistol teams, who have never before fired on any team which has placed in the top 15% of teams competing in the same event in past respective rifle or pistol National Trophy Team Matches.

    e. Participation as a shooting member of a Service Academy team or ROTC team will not be considered as previous participation within the above eligibility requirements provided the team did not place in the top 15% of winning teams.

    f. No individual may be a member of any of the authorized teams unless he has been a bona-fide member of that particular group or organization from which the team is selected for at least 30 days prior to the opening date of the National Matches except that: Members of State civilian teams will be confined to bona-fide residents of the state which the team represents, who have lived in the state for at least 30 days prior to the opening date of the National Matches and are otherwise qualified.

21. <u>Alternates</u>: (See para, 48 AR 920-30) Not later than the time designated for the closing of entries, each team captain will submit to the Statistics Director at his office, on score cards furnished by the Statistics Director for the purpose, a legible list of members and alternates. A team captain and/or team coach may also serve as firing member or alternate, provided he is so listed on the entry form and is otherwise eligible. Substitution for a firing member may be made at any time before he fires his first shot in a match, provided the individual has been listed on the score card as an alternate. Once a team member has fired a shot, substitution may be made for him only in case of incapacitating injury and upon approval by the National Match Executive Officer

22. <u>Team Certification</u>: (See para 17, AR 920-30) Team captains will present to the Statistics Director lists of their respective team members to include designation of team officials and other members of the team, the correct first name, middle initial, last name, service number and grade of members as applicable, their individual home addresses, official military mailing addresses if applicable, and proper certification as to the eligibility of the members of their team under these rules and regulations at the time entry in the match is made. Civilian teams will be recognized as representing a particular state only if attested to by the State Adjutant General or President, Vice President or Secretary of the respective State Rifle or Pistol Association, as accredited by the National Rifle Association or under conditions established by the DCM. Individual club presidents will attest to the teams representing their organizations. Teams representing law enforcement agencies and service teams of the Armed Forces of the United States will be certified as provided under appropriate regulations governing each such agency or service. ROTC and Service Academy teams will be certified by appropriate military commanders.

23. <u>Coaching</u>: (See para 49, AR 920-30)

    a. Individual Matches: No coaching is permitted.

    b. Team Matches: Coaching is permitted in all team matches within the team only; any member of a team (captain, coach, firing member or alternate) may function as coach provided such member fulfills the eligibility requirements set forth below:

        (1) A team captain or coach must be at least 16 years of age on his last birthday prior to the beginning date of the National Matches and must otherwise meet the eligibility rules of the National Matches;

        (2) Officers, warrant officer, or enlisted personnel of the active military services may act as coaches for teams of their respective services and for civilian, ROTC and Service Academy teams;

(3) Teams representing the National Guard and teams representing other components of the Armed Forces Reserves may be coached only by officers, warrant officer, or enlisted personnel of the particular reserve component concerned;

(4) ROTC and service academy teams will be coached only by members of the Armed Forces;

(5) A civilian rifle or pistol team may request the National Match Executive Officer to assign a coach, and upon application, the National Match Executive officer will assign a coach from the officers and enlisted men or resident civilians;

(6) A coach once assigned will not be changed except in an emergency or other valid reason and upon approval of the National Match Executive Officer.

24. <u>Station of Team Officials</u>: (See paras 50 and 51, AR 920-30)

a. In team matches a team coach may take position on the firing line between the competitors of the pair firing in slow fire stages and immediately in rear of the firing member in rapid fire stages.

b. In pistol team matches one coach will be allowed for each target assigned to the team.

c. The coach cannot shift position nor shift the position of the competitors of the pair firing for the purpose of forming a windshield for the firer.

d. The coach must confine himself to the normal position of a coach and his activities to those normally expected of a coach.

e. The coach may assist team members by calling shots, checking time, checking scoring, ordering sight changes, etc., but must control his or her voice and actions so as not to disturb other competitors. Natural voice commands will be unaided by electrical, mechanical, or other means.

f. The coach will not physically assist in loading or making sight corrections.

g. In team matches the team captain and one assistant may be seated in front of the assembly line, but not in advance of a line established three paces in rear of the line of scorers except as provided herein. A team captain may coach but only if he actually occupies the coaching position on the line.

25. <u>Safety Precautions</u>: (See para 22, AR 920-30)

a. No arms will be loaded except at the firing point and under direction and command of a range officer.

b. During and after loading, the pistol will be kept at raised (or ready) pistol or on the bench, muzzle down range until unloaded, except when aimed at the target for authorized practice or firing.

c. Pistols will have their magazines withdrawn and the slides pulled back except when the competitor is at the firing point in the act of loading, firing, or ready to fire.

d. During and after loading, the muzzle of the rifle will be kept pointed in the direction of the target until unloaded.

e. Rifles will be carried at all times with actions open except when the competitor is at the firing point in the act of firing or ready to fire.

26. <u>Challenges</u>: (See para 54, AR 920-30) The procedure outlined in NRA rules for scoring challenges will be modified to the extent indicated below.

    a. No change for rifle matches.

    b. In pistol matches the competitor, when dissatisfied with the score announced by the scorer and the line officer, may upon payment of the challenge fee have the disputed score settled by the scoring referee whose decision is final.

    c. The official referee is appointed by the National Match Director.

    d. The challenge fee in these matches will be $1. All challenge money will be delivered to Executive Officer, National Matches, for payment to the National Match Fund.

27. <u>Protest</u>: (See para 55, AR 920-30)

    a. A competitor may formally protest:

        (1) Any injustice it is felt has been exercised. (This excludes evaluation of a target which may be challenged as outlined elsewhere.)

        (2) The conditions under which another competitor has been permitted to fire.

        (3) The equipment which another competitor has been permitted to use.

    b. A protest must be initiated immediately unpon the occurence of the protested incident. Failure to comply with the following procedure will automatically void the protest:

        (1) State the complaint verbally to the Chief Range Officer or the Statistics Director, as appropriate. If not satisfied with that decision then--------

        (2) State the complaint orally to the official referee. If not satisfied with his decision, then within 2 hours------------.

        (3) File a written protest with the official referee stating all the facts in the case. If not satisfied with that decision, then within 5 hours-----

        (4) File with the NBPRP Protest Committee a protest in writing stating all facts in the case. The decision of the Protest Committee will be final.

    c. The NBPRP Protest Committee will consist of not less than three members. The members of the committee shall be appointed by the President, NBPRP.

    d. Challenges and protests in team matches must be made by the team captain.

28. <u>Penalties</u>: (See para 57, AR 920-30)

    a. Any person interfering with a competitor on the firing line or annoying the competitor in any way will be warned to desist. If upon said warning, the offense is repeated, such person will be ordered off the range at once.

    b. Any competitor or team will be disqualified from competing further in the matches and may be denied any prize won during the current matches when found guilty by the National Match Executive Officer of any of these offenses listed below:

(1) Firing under a name other than that under which entered;

(2) Firing twice for the same prize;

(3) Falsifying scores or being an accessory thereto;

(4) Offering a bribe of any kind to any official or other person;

(5) Evading any of the conditions prescribed for the conduct of any match;

(6) Refusing to obey any instructions of the National Match Director, Range Director or Range Officer;

(7) Being guilty of disorderly conduct;

(8) Violating range safety regulations;

(9) Being guilty of any conduct considered by the National Match Executive Officer to be discreditable.

CHAPTER XXVI
COMPETITIVE REGULATIONS IV
REQUIREMENTS FOR EARNING A DISTINGUISHED PISTOL, RIFLE OR INTERNATIONAL SHOOTER BADGE

QUALIFICATION IN ARMS
SECTION I MILITARY PARTICIPATION
(Extracts from AR 672-5-1, Condensed)
3 June 1974

*   *   *   *   *   *   *   *   *   *

5-31. United States Distinguished International Shooter Badge.

   a. Purpose. The United States Distinguished International Shooter Badge is awarded to military and civilian personnel in recognition of an outstanding degree of achievement in international shooting. Winners of this badge will not part with them without authority of the President of the National Board for the Promotion of Rifle Practice and will hold them subject to inspection at any time.

   b. Eligibility requirements.

   (1) Members of US International Teams who have won a 1st, 2d, or 3d place medal in individual events or as a firing member of a 1st, 2d, or 3d place Medal Winning Team in any International Shooting Union World Championship, Olympic, or Pan American Games Shooting event will be awarded the United States Distinguished International Shooter Badge.

   (2) Retroactive award may be made to qualified individuals upon their personal request and provision of evidence of eligibility for the badge.

   (3) Retroactive award, in certificate form in lieu of a badge, for qualified deceased individuals may be made to next of kin upon request and provision of evidence of eligibility for the award.

   c. Who may award. The President of the National Board for the Promotion of Rifle Practice.

   d. Engraving. The name of the recipient and year of attainment will be engraved on the reverse of the pendant.

5-32. Distinguished designation badges.

a. Purpose. A Distinguished Rifleman Badge or a Distinguished Pistol Shot Badge is awarded to a member of the Army or to a civilian in recognition of a preeminent degree of achievement in target practice firing with the military service rifle or pistol. Winners of Distinguished designation badges will not part with them without authority of the Secretary of the Army and will hold them subject to inspection at any time.

b. Eligibility requirements.

(1) A member of the Army will be designated as a Distinguished Rifleman or Distinguished Pistol Shot when he has earned 30 credits toward the Distinguished designation. See AR 350-6.

(2) A civilian will be designated by the Army as a Distinguished Rifleman or Distinguished Pistol Shot when he has earned 30 credit points toward the Distinguished designation provided that at least one credit leg was won in the National Matches, or, in lieu thereof, the civilian competitor must earn an 8- or a 10-leg in any other authorized match (major command or NRA Regional). (See AR 350-6.) Badges awarded prior to 1948 will be considered toward achievement of the Distinguished designation under the rules of the match in which won. A credit granted by the National Board for the Promotion of Rifle Practice under rules in effect for matches prior to 1948 will be considered toward the award of this badge the same as though an Excellence in Competition Badge had been awarded.

(3) The year in which a person first became eligible for designation by the Army as a Distinguished Rifleman or Distinguished Pistol Shot is the year in which he is regarded as having attained the Distinguished designation and for which he will be so designated.

c. Who may award.

(1) To Army personnel. Commanding Generals of TRADOC and FORSCOM. Copies of Special Orders authorizing awards will be forwarded to HQDA (DAPC-PAR), 200 Stovall Street, Alexandria, VA 22332, for record purposes.

(2) To all others. The Director of Civilian Marksmanship, Department of the Army, Washington, DC 20314.

d. Engraving. The name of the recipient and the year of attainment will be engraved on the reverse of the metal pendant.

5-33. Excellence in Competition Badges.

a. Purpose. Excellence in Competition Badges are awarded to individuals in recognition of an eminent degree of achievement in target practice firing with the standard military service rifle or pistol.

b. Types of badges. Two types of Excellence in Competition Badges will be awarded to denote the outstanding performance in target practice and the progress toward distinguished designation. A bronze Excellence in Competition Badge will be awarded to Army competitors who earn their first credit points regardless of credit value. A silver Excellence in Competition Badge will be awarded an individual when 20 credit points have been earned. All badges consist of a bar, clasp, and pendant and are identical in design except for clasp which is crossed pistols for pistol matches and crossed muskets for rifle matches.

c. Eligibility requirements. The number of badges which will be awarded in recognition of achievement in the National Matches, the US Army Championships, Major Command Championships, Interservice Championships, USAR (ARCOM/GOCOM)-US Army Installation/Division Matches, State National Guard or All-National Guard Championships or in National Rifle Association Regional Championships will depend primarily upon the number of "nondistinguished" participants in the match (see AR 350-6). In all competitions the badges will be awarded only for excellence in individual competition. Comparable badges will be awarded to civilians by the Director of Civilian Marksmanship in accordance with regulations prescribed by the National Board for the Promotion of Rifle Practice. A badge for excellence in competition in a match conducted subsequent to 1947 will be awarded only to a person whose score in authorized competition constitutes a credit toward a Distinguished Designation badge. The determination as to whether a badge for excellence in competition which has been awarded for achievement in a match conducted prior to 1948 may be considered toward the award of a Distinguished Designation Badge will be in accordance with Army regulations in effect at the time such match was conducted.

    d.   Limit on award.

        (1)   In no case will an individual be awarded more than one badge of each type. Credits will be given in lieu of additional awards of the same badge.

        (2)   Individuals who have either qualified for or attained the Distinguished designation are ineligible for further awards of this badge. Any such individual who fraudulently accepts an additional award of the Excellence in Competition Badge when he is aware of his eligibility for Distinguished designation, or has been designated as a Distinguished Rifleman or Distinguished Pistol Shot will be subject to revocation of the award.

    e.   Who may award.

        (1)   To Army personnel. Commanding Generals of TRADOC and FORSCOM. Copies of special orders authorizing awards will be awarded for posting to the official military personnel file (OMPF).

        (2)   To all others. The National Board for the Promotion of Rifle Practice.

5-34. President's Hundred Tab. A President's Hundred Tab is awarded to each person who qualified among the top 100 successful contestants in the President's Match held annually at the National Rifle Matches.

* * * * * * * * * * * * * * * * * *

(Extracts from AR 350-6, Condensed)

8. Army participation in competitive marksmanship activities.

    a. Army participation in the National Matches will be governed by the Rules and Regulations for National Matches (AR 920-30).

    b. Army participation in civilian and interservice competition at Government expenses should be supervised by major Army commanders. Major Army command participation in special international level competitions (e.g., Prix Leclerc type matches) is authorized and encouraged.

    c. Participation in the annual US Army Rifle and Pistol Championships will be regulated and supervised by CGFORSCOM. US Army, Europe; US Army, Pacific; FORSCOM; and any other major Army command conducting a competitive marksmanship program for which funds are programmed and budgeted will each provide a minimum of one service rifle team and one service pistol team for participation. Participation by other major Army commands is optional and must be coordinated with CGFORSCOM.

    d. Competitors from other services and civilian citizens of the United States may be invited to participate in all Army sponsored competitions, insofar as range facilities permit, except the US Army Championships.

9. Award of Army Excellence-in-Competition and Distinguished Badges.

    a. Award of Excellence-in-Competition badges and credit toward Distinguished badges will be made on the basis of individual unassisted performance in recognized individual matches.

    b. Award of the appropriate Distinguished badge will be made when an individual has earned a minimum of 30 credit points for Excellence-in-Competition in recognized matches.

    c. Credit points will be awarded to the highest scoring 10 percent of all nondistinguished participants firing in the match ranked in order of merit. Fractions of .5 and over will be resolved to the next higher whole number. Smaller fractions will not be considered. (Total credit points for any one match constitute a "leg".)

    d. Credit points will be awarded to winning personnel as determined in c above, on the following basis:

        To the highest scoring 1/6 ....10 points.
        To the next highest scoring 1/3............................. 8 points.
        To the remaining personnel authorized credit points.......... 6 points.

Exception: The top 10 percent of authorized competitors in US Army installation/division, USAR (ARCOM and GOCOM), and NG State Championship Excellence-in-Competition matches may be awarded a maximum of four credit points (table 1).

    e. Each individual is authorized to fire for credit points in not more than four recognized matches with each weapon (service rifle and service pistol) during each calendar year. An individual who fires any portion of a match will be considered to have fired that match. Participation eligibility in National Board for the Promotion of Rifle Practice (NBPRP) "Leg" (credit point) matches, conducted in accordance with AR 920-30 as a part of official competitive events, is shown in tab

f. A bronze Excellence-in-Competition badge will be awarded to Army competitors who earn their first credit points regardless of credit point value. A silver Excellence-in-Competition badge will be awarded individuals when 20 credit points have been earned.

g. Nondistinguished personnel who hold "leg" credits for Distinguished award, as of 1 January 1963, will be credited with 10 points for each such "leg" not to exceed a total of 20 points for each weapon (rifle or pistol). Award of silver badge is not retroactive.

h. Badges will be provided eligible Army personnel by CG FORSCOM in accordance with AR 672-5-1.

i. Members of the other Services and civilian competitors who qualify for credit points toward Distinguished award in major command matches will be reported by CG FORSCOM as follows:

(1) Members of other Services direct to the Service concerned.

(2) Civilians, to include ROTC students and retired military personnel, to the Director of Civilian Marksmanship, Department of the Army, Room 1E053 Forrestal Building, Washington, DC 20314.

j. Entry of nondistinguished personnel in Excellence-in-Competition matches on a noncredit basis for practice is not authorized.

k. Nondistinguished military competitors entering NBPRP Regional and State Championship Excellence-in-Competition matches must sign a statement for the match sponsor which confirms that they have not earned the maximum credit points allowed in that level of competitions, e.g., USN 20 credit points; USMCR and all components of the Army 10 credit points (table 1).

TABLE 1. PARTICIPATION ELIGIBILITY IN NBPRP "LEG" MATCHES
(Conducted in accordance with the Rules and Regulations for the National Matches, AR 920-30/
OPNAVINST 3590.7B/AFR 50-17/MCO P3590.13)

Competitive Event (of which "Leg" match is a part)	Army	USAR ARNG	USN USNR	USMC USMCR	USAF USAFR ANG	Svc Academy/ ROTC Cadets	US Coast Guard	Civ
				Applicable Footnote				
US Army Installation/Division, USAR (ARCOM and GOCOM) and NG State Championships	(1)	(1)	(2)	(2)	(1)	(1)	(2)	(2)
US Major Army Command/CONUS Army Championships	(3)	(3)	(4, 5)	(2)	(5, 6)	(7)	(3)	(3)
US Army Championships	(3)	(3)	(2)	(2)	(2)	(8)	(3)	(2)
Armed Forces (Interservice) Championships	(3)	(3)	(3)	(3)	(3)	(3)	(3)	(2)
NBPRP Approved Matches (NRA Regional and State Championships)	(9)	(9)	(10)	(9)	(9)	(7, 10, 11)	(9)	(12)
Major Command Championships of Other Services	(9)	(9)	(4, 5, 13)	(2)	(9)	(7)	(13)	(13)
National Trophy Individual Championships	(3)	(3)	(3)	(3)	(3)	(3)	(3)	(3)
National Guard Championships	(14)	(14)	(2)	(2)	(14)	(14)	(2)	(14)

1. Limited to Army, USAR, ARNG, ANG, USMA and Army ROTC personnel who have not earned a Bronze Excellence-in-Competition badge (para 9d).
2. Entry not authorized.
3. One per calendar year.
4. USN personnel require written authority from Chief of Naval Technical Training (NAVPERS 93683).
5. Entry authorized only when circumstances preclude participation in parent Service major command championship or equivalent.
6. ANG participation authorized.
7. Parent service restrictions apply.
8. USMA and Army ROTC cadets only as range capacity permits.

9. Entry limited to one NBPRP approved match or one major command championship of other service per calendar year (footnote 5 applies). Personnel of the USMC, USMCR and all components of the Army can earn a maximum of 10 credit points in NBPRP approved matches (para 9k). USAF personnel require written authority of HQ USAF (AFR 50-20). US Coast Guard personnel require written authority from commanding officer. USAR and ARNG entry in lieu of major Army command/CONUS. Army or US Army or Armed Forces championships only when circumstances preclude participation in those four parent service events.

10. USN personnel may compete in not more than one NBPRP approved rifle and/or pistol match in NRA Regional Championship per calendar year if 20 or more credit points have not been earned in either NRA Regional or other service command championships (OPNAV P34-03).

11. ROTC Cadets compete in civilian category.

12. Two per calendar year.

13. Entry limited to one service sponsored major command championship per calendar year.

14. Participation in the annual Winston P. Wilson (National Guard) Championships will be allowed for rifle an pistol "leg" matches only.

All footnotes apply separately to pistol and rifle "leg" matches. No competitor is eligible to compete in more than four "leg" matches for pistol and four "leg" matches for rifle, annually.

www.ingramcontent.com/pod-product-compliance
Lightning Source LLC
Chambersburg PA
CBHW080720230426
43665CB00020B/2562